Modern Methods in Phytomedicine

The Editor

Dr. Parimelazhagan Thangaraj is a Professor in the Department of Botany, Bharathiar University, Coimbatore. He completed his graduation from St. Joseph's College Trichy and Ph.D in Botany from Bharathiar University, Coimbatore, Tamil Nadu, India. His areas of specialization include Bioprospecting of medicinal plants. He is currently engaged in compilation and documentation on available knowledge on valuable plant resources to prove their utility scientifically through detailed phytochemical, biological and pharmacological investigations, isolation and characterisation of bioactive compounds from potential medicinal plants. Parimelazhagan has worked as Scientist in DRDO, Ministry of Defence, Govt. of India for four years. He has also worked as Scientist in Rubber Research Institute of India, Ministry of Commerce, Govt. of India, Tura, Meghalaya for four years. He has published 5 edited books and has filed four patents and has more than 50 research publications in international and national peer reviewed journals. He serves as an editorial board member of Progressive Horticulture Journal. He is also the reviewer of many peer reviewed journals. He has bagged National Science Day Medal 2005 from DRDO HQ, New Delhi and Laboratory Scientist of the Year award 2005 from DRDO. Recently, he was awarded Raman Post-Doctoral Fellowship for the year 2014-2015 by UGC, New Delhi under INDO-USA collaboration - to visit University of Hawaii, Hilo, USA. Under his supervision, five students awarded Ph.D degree.

Modern Methods in Phytomedicine

Editor

Dr. Parimelazhagan Thangaraj

Bioprospecting Laboratory,
Department of Botany,
Bharathiar University,
Coimbatore – 641 046,
Tamil Nadu

2015

Daya Publishing House®

A Division of

Astral International Pvt. Ltd.

New Delhi – 110 002

Cataloging in Publication Data--DK
Courtesy: D.K. Agencies (P) Ltd. <docinfo@dkagencies.com>

Modern methods in phytomedicine / editor, Dr. T. Parimelazhagan.
pages cm
Includes bibliographical references and index.
ISBN 978-93-5130-684-9 (International edition)

1. Materia medica, Vegetable. 2. Herbs--therapeutic use. I. Parimelazhagan, T., editor.

DDC 615.321 23

Published by : **Daya Publishing House®**
A Division of
Astral International Pvt. Ltd.
– ISO 9001:2008 Certified Company –
4760-61/23, Ansari Road, Darya Ganj
New Delhi-110 002
Ph. 011-43549197, 23278134
E-mail: info@astralint.com
Website: www.astralint.com

Laser Typesetting : **Classic Computer Services**, Delhi - 110 035

Printed at : **Thomson Press India Limited**

PRINTED IN INDIA

Preface

Phytomedicine or the use of herbal medicine with therapeutic properties has played a significant role throughout history. Although its usage greatly diminished during the dawn of the scientific era, there is a revival of interest in its potential by late 20th century, especially in the development of new drugs. Traditional medical practice remains the largest healthcare system in the world, also in recent years, the search for drugs and dietary supplements derived from plants has picked up pace in drug discovery. Herbal medicine can be categorized into three general groups, namely phytotherapy, over-the-counter herbal medicine and traditional herbalism. Among these, phytotherapy is the one that adheres to scientific methodology and generates reasonably sound data. "**Modern Methods in Phytomedicine**" aims at compilation of recent trends in herbal medicine and explores the hidden medicinal potentials of significant plants. This book covers 26 research and review articles encompassing the valuable works of eminent academicians and research scientists on different aspects of the subject. Researches pertaining to the field of plant tissue culture, plant diversity, DNA barcoding, microbiology, cancer biology and investigations of plants possessing medicinal properties like antioxidant, antimicrobial activities are included in this book in a simple and clear way so as to help in easy understanding by students, researchers, scientists and industrialists. The review processes of the articles have been carried out by experts from Universities and Research Institutes. The editor conveys his heartfelt thanks to Prof. Dr. G. James Pitchai, Vice-Chancellor, Bharathiar University for his unbound guidance and encouragement. The support rendered by Dr. V. Narmatha Bai, Professor & Head, Department of Botany, Bharathiar University is also acknowledged. The editor also thanks the Bioprospecting research team, Mr. M. Iniyavan, Mr. Rahul Chandran, Ms. P. Revathi, Mr. Sajeesh Thankarajan, Mr. K. Murugan, Mr. M. Kasipandi, Mr. S Saikumar, Ms. S. Harini and Ms. S. Divya for their contribution in the compilation of the manuscripts in a well-designed manner. We would like to express our special appreciation for the publishers and their team for the sincere efforts in bringing out the book in time.

Parimelazhagan Thangaraj
Editor

M.S. Swaminathan Research Foundation

M.S. Swaminathan
Founder Chairman
Ex-Member of Parliament
(Rajya Sabha)

Foreword

There is increasing realization of the value of traditional wisdom and ecological prudence. Also tribal and rural families, particularly women, are great conservers of biodiversity. They give special emphasis to cultural, culinary and curative diversity. As a result we have a wide range of medicinal plants available to us. There is need for strengthening our conservation traditions, both insitu and exsitu. Dr Parimelazhagan has done great service by gathering valuable information on medicinal plants and biodiversity. He has thus helped to promote phytomedicine. The articles in this book further reinforce the urgent need for the conservation and enrichment of biodiversity and for serious attention to traditional knowledge and wisdom. We owe a deep debt of gratitude to Dr Parimelazhagan for his labour of love in the area of strengthe~ing community conservation and traditional wisdom.

M S Swaminathan

3rd Cross Road, Taramani InstitutionalArea, Chennai (Madras) - 600113, India
Phone: +91-44-22542790,22541698 Fax: +91-44-2254 1319
E-mail: founder@mssrf.res.in, swami@mssrf.res.in

Contents

List of Contributors

Anandhan, S.
Directorate of Onion and Garlic, ICAR, Rajgurunagar – 410 505, Pune

Arumugam, Rajendran
Department of Botany, School of Life Sciences, Bharathiar University, Coimbatore – 641 046, Tamil Nadu

Ayyanar, Muniappan
Department of Botany, Pachaiyappa's College, Chennai – 600 030, Tamil Nadu

Balaguru, B.
Department of Botany, Jamal Mohamad College, Tiruchirappalli – 620 020, Tamil Nadu

Balakumaran, Manickam Dakshinamoorthy
Centre for Advanced Studies in Botany, School of Life Sciences, University of Madras, Guindy Campus, Chennai – 600 025, Tamil Nadu

Baskaran, X.
Department of Botany, St. Joseph's College, Tiruchirapalli – 620 002, Tamil Nadu

Bhoyar, Manish S.
Defence Institute of High Altitude Research, Defence R&D Organisation, C/o 56 APO, Leh, Ladakh – 901 205, J&K

Bhuvaneawari, T.

Plant Genetic Engineering and Molecular Biology Lab, Department of Biotechnology, Periyar University, Periyar Palkalai Nagar, Salem – 636 011, Tamil Nadu

Binu, T.V.

Department of Botany, Avinashilingam Institute for Home Science and Higher Education for Women, Coimbatore – 641 043, Tamil Nadu

Chandran, Rahul

Bioprospecting Laboratory, Department of Botany, Bharathiar University, Coimbatore – 641 046, Tamil Nadu

Chitraselvi, R. Pemila Edith

Department of Biotechnology, School of Biotechnology and Health Sciences, Karunya University, Coimbatore – 641 114, Tamil Nadu

Das, M. Sudalai

Department of Yoga, Bannari Amman Institute of Technology, Sathyamangalam – 638 401, Tamil Nadu

Dhivya, Sivaraj

Bioprospecting Laboratory, Department of Botany, Bharathiar University, Coimbatore – 641 046, Tamil Nadu

Geetha, N.

Department of Biotechnology, Mother Teresa Women's University, Kodaikanal – 624 101, Dindugal, Tamil Nadu

George, Blassan P.

Bioprospecting Laboratory, Department of Botany, Bharathiar University, Coimbatore – 641 046, Tamil Nadu

Gomathi, R.

Department of Botany, PSG College of Arts and Science (Autonomous), Peelamedu, Coimbatore – 641 004, Tamil Nadu

Harini, S.

Bioprospecting Laboratory, Department of Botany, Bharathiar University, Coimbatore – 641 046, Tamil Nadu

Iniyavan, Murugaiyan

Bioprospecting Laboratory, Department of Botany, Bharathiar University, Coimbatore – 641 046, Tamil Nadu

Jeyachandran, R.

Department of Botany, St. Joseph's College, Tiruchirapalli – 620 002, Tamil Nadu

Johnson, M.

Department of Botany, St. Xavier's College (Autonomous), Palayamkottai – 627 002, Tamil Nadu

Kalaichelvan, Puthupalayam Thangavel
Centre for Advanced Studies in Botany, School of Life Sciences, University of Madras, Guindy Campus, Chennai – 600 025, Tamil Nadu

Kasipandi, Muniyandi
Bioprospecting Laboratory, Department of Botany, Bharathiar University, Coimbatore – 641 046, Tamil Nadu

Kothandapani, Chanthru
Department of Biotechnology, Bharathiar University, Coimbatore – 641 046, Tamil Nadu

Kumar, Abhay
Directorate of Groundnut Research, PB No 05, Ivnagar Road, Junagadh – 362 001, Gujarat

Manian, S.
Department of Botany, Bharathiar University, Coimbatore – 641 046, Tamil Nadu

Margret, Kerna Book Leena
Department of Botany, Avinashilingam Institute for Home Science and Higher Education for Women, Coimbatore – 641 043, Tamil Nadu

Mishra, Gyan P.
Defence Institute of High Altitude Research, Defence R&D Organisation, C/o 56 APO, Leh, Ladakh – 901 205, J&K

Mishra, Gyan P.
Directorate of Groundnut Research, PB No 05, Ivnagar Road, Junagadh – 362 001, Gujarat

Mohanasundaram, Saravanan
Plant Genetic Engineering Laboratory, Department of Biotechnology, Bharathiar University, Coimbatore – 641 046, Tamil Nadu

Murkute, Ashutosh A.
Directorate of Onion and Garlic, ICAR, Rajgurunagar – 410 505, Pune

Murkute, Ashutosh
Defence Institute of High Altitude Research, Defence R&D Organisation, C/o 56 APO, Leh, Ladakh – 901 205, J&K

Murugan, Rajan
Bioprospecting Laboratory, Department of Botany, Bharathiar University, Coimbatore – 641 046, Tamil Nadu

Murugesan, G.S.
Department of Biotechnology, Bannari Amman Institute of Technology, Sathyamangalam – 638 401, Tamil Nadu

Muthukumar, B.
P.G. and Research Department of Botany, National College, Tiruchirappalli – 622 001, Tamil Nadu

Nagamurugan, N.
Department of Biotechnology, Kurinji College of Arts and Science, Tiruchirappalli – 622 002, Tamil Nadu

Natarajan, E.
P.G. and Research Department of Botany, National College, Tiruchirappalli – 622 001, Tamil Nadu

Parimelazhagan, T.
Bioprospecting Laboratory, Department of Botany, Bharathiar University, Coimbatore – 641 046, Tamil Nadu

Parthipan, M.
Floristics Laboratory, Department of Botany, School of Life Sciences, Bharathiar University, Coimbatore – 641 046, Tamil Nadu

Paulsamy, P.
Department of Botany, Kongunadu Arts and Science College, Coimbatore – 641 029, Tamil Nadu

Prakash, Baby John
Bioprospecting Laboratory, Department of Botany, Bharathiar University, Coimbatore – 641 046, Tamil Nadu

Preethi, K.
Department of Microbial Biotechnology, Bharathiar University, Coimbatore – 641 046, Tamil Nadu

Radha, S.R.
Department of Botany, Avinashilingam University, Coimbatore – 641 043, Tamil Nadu

Radhakrishnan, T.
Directorate of Groundnut Research, PB No. 05, Ivnagar Road, Junagadh – 362 001, Gujarat

Radhakrishnan, T.,
Directorate of Groundnut Research, PB No. 05, Ivnagar Road, Junagadh – 362 001, Gujarat

Raja, D. Patric
Department of Botany, St. Xavier's College (Autonomous), Palayamkottai – 627 002, Tamil Nadu

Rajavel, L.
P.G. and Research Department of Botany, National College, Tiruchirappalli – 622 001, Tamil Nadu

Ramachandran, Rajan
Centre for Advanced Studies in Botany, School of Life Sciences, University of Madras, Guindy Campus, Chennai – 600 025, Tamil Nadu

Ramalingam, Sathishkumar
Plant Genetic Engineering Laboratory, Department of Biotechnology, Bharathiar University, Coimbatore – 641 046, Tamil Nadu

Revathi, P.
Department of Botany, Bharathiar University, Coimbatore – 641 046, Tamil Nadu

Sajeesh, T.
Bioprospecting Laboratory, Department of Botany, Bharathiar University, Coimbatore – 641 046, Tamil Nadu

Saravanan, Shanmugam
Bioprospecting Laboratory, Department of Botany, Bharathiar University, Coimbatore – 641 046, Tamil Nadu

Sarma, Rajeevkumar
Plant Genetic Engineering Laboratory, Department of Biotechnology, Bharathiar University, Coimbatore – 641 046, Tamil Nadu

Selvaraj, Dhivya
Plant Genetic Engineering Laboratory, Department of Biotechnology, Bharathiar University, Coimbatore – 641 046, Tamil Nadu

Shanmughanandhan, Dhivya
Plant Genetic Engineering Laboratory, Department of Biotechnology, Bharathiar University, Coimbatore – 641 046, Tamil Nadu

Singh, Raghwendra
Defence Institute of High Altitude Research, Defence R&D Organisation, C/o 56 APO, Leh, Ladakh – 901 205, J&K

Soosairaj, S.
Department of Botany, St. Joseph's College, Tiruchirappalli – 620 002, Tamil Nadu

Srivastava, R.B.
Defence Institute of High Altitude Research, Defence R&D Organisation, C/o 56 APO, Leh, Ladakh – 901 205, J&K

Subramaniam, Kalidass
Department of Biotechnology, School of Biotechnology and Health Sciences, Karunya University, Coimbatore – 641 114, Tamil Nadu

Sudha, P. Prema
Department of Nanoscience and Technology, Bharathiar University, Coimbatore – 641 046, Tamil Nadu

Suriyamoorthy, Sembian

Department of Biotechnology, School of Biotechnology and Health Sciences, Karunya University, Coimbatore – 641 114, Tamil Nadu

Thomas, Binu

Department of Botany, School of Life Sciences, Bharathiar University, Coimbatore – 641 046, Tamil Nadu

Vanitha, R.

Department of Botany, Avinashilingam University, Coimbatore – 641 043, Tamil Nadu

Venkatachalam, P.

Plant Genetic Engineering and Molecular Biology Lab, Department of Biotechnology, Periyar University, Periyar Palkalai Nagar, Salem – 636 011, Tamil Nadu

Vigila, A. Geo

Department of Zoology, S.T. Hindu College, Nagercoil – 629 002, Tamil Nadu

Vijayakumari, B.

Department of Botany, Avinashilingam Institute for Home Science and Higher Education for Women, Coimbatore – 641 043, Tamil Nadu

Vijayakumari, B.

Department of Botany, Avinashilingam University, Coimbatore – 641 043, Tamil Nadu

Viji, Zereena

Department of Botany, NSS College, Nemmara, Palakkad – 678 508, Kerala

Xavier, G. Sahaya Anthony

Department of Botany, St. Xavier's College (Autonomous), Palayamkottai – 627 002, Tamil Nadu

2015, Modern Methods in Phytomedicine *Pages* ***1–14***
Editor: **T. Parimelazhagan**
Published by: **DAYA PUBLISHING HOUSE, NEW DELHI**

1

Antifreeze Proteins: Cold Tolerance and other Applications

Gyan P. Mishra[1]*, Ashutosh A. Murkute*[2]*, S. Anandhan*[2]*, Abhay Kumar*[1] *and T. Radhakrishnan*[1]

[1]*ICAR-Directorate of Groundnut Research, P.B. No 05, Ivnagar Road, Junagadh – 362 001, Gujarat*
[2]*ICAR-Directorate of Onion and Garlic, Rajgurunagar – 410 505, Pune*

1.0 INTRODUCTION

Cold temperature is one of the major factors affecting crop yield in temperate climates, with the farming industry loosing billions of dollars each year to freezing temperatures. Much research has focused on ways to improve crops' tolerance to cold and/or freezing temperatures, with the aim to both increase productivity and broaden geographical range. Low temperature is one of the major limiting environmental factors which constitute the growth, development, productivity and distribution of plants. Over the past several years, the proteins and genes associated with freezing resistance of plants have been widely studied (Yuan-zhen *et al.,* 2005). The damaging effects of low temperatures and freezing conditions on plant material consist of mechanical injury (cell and tissue disruption), which results from ice formation, and dehydration injury caused by water loss associated with ice formation. Under deep-freezing conditions, intracellular bulk water and water oriented on the surface of macromolecules and on the polar heads of lipids in cellular membranes are effectively removed, causing severe dehydration and structural and functional damage to plasma membranes (Gattiker *et al.,* 2002).

Antifreeze proteins (AFPs) or ice structuring proteins (ISPs) refer to a class of polypeptides produced by certain vertebrates, plants, fungi and bacteria that permit their survival in subzero environments. AFPs bind to small ice crystals to inhibit growth and recrystallization of ice that would otherwise be fatal. AFPs and Antifreeze glycoproteins (AFGPs) are produced as a specialised adaptation by certain fish, insects, plants and bacteria. AFPs and AFGPs are capable of lowering the freezing point of a solution. Antifreeze proteins are found in a wide range of overwintering plants where they inhibit the growth and recrystallization of ice that forms in intercellular spaces. Unlike antifreeze proteins found in fish and insects, plant antifreeze proteins have multiple, hydrophilic ice-binding domains. Surprisingly, antifreeze proteins from plants are homologous to pathogenesis-related (PR) proteins and also provide protection against psychrophilic pathogens (Griffith and Yaish, 2004).

In 1992, Griffith *et al.,* documented their discovery of AFP in winter rye leaves. Around the same time, Urrutia *et al.* (1992) documented thermal hysteresis protein in angiosperms. In 1993, Duman and Olsen noted that AFPs had also been discovered in over 23 species of angiosperms, including ones we eat. As well, they reported their presence in fungi and bacteria. Recent attempts have been made to re-label antifreeze proteins as *'ice structuring proteins'* in order to more accurately represent their function and to rid of any assumed negative relation between AFPs and automotive antifreeze, ethylene glycol. These two things are completely separate entities bearing loose similarity only in their function.

1.1 Functions

Plant AFPs may have four functions in the antifreeze process of plant:

1. Lowering the freezing point;
2. Inhibiting ice-recrystallization;
3. Modifying ice morphology;
4. Regulating the supercooling state of protoplasm. And it is the last one that may be the key role of AFPs to beneficiate the plant undergoing an antifreeze physiological process (Yong *et al.,* 1987).

2.0 SOURCES OF AFPS

2.1 Plant AFPs

Overwintering plants produce AFPs having the ability to adsorb onto the surface of ice crystals and modify their growth. Recently, several AFPs have been isolated and characterized and five full-length AFP cDNAs have been cloned and characterized in higher plants. The derived amino acid sequences have shown low homology for identical residues. Theoretical and experimental models for structure of *Lolium perenne* AFP have been proposed. In addition, it was found that the hormone ethylene is involved in regulating antifreeze activity in response to cold. It is seen that the physiological and biochemical roles of AFPs may be important to protect the plant tissues from mechanical stress caused by ice formation (Atýcý and Nalbanto, 2003).

The classification of AFPs became more complicated when antifreeze proteins from plant were discovered (Griffith *et al.,* 1992). Plant AFPs are rather different from the other AFPs in the following aspects: They have much weaker thermal hysteresis activity when compared to other AFPs. Their physiological function is likely in inhibiting the recrystallization of ice rather than in the preventing ice formation. Most of them are evolved pathogenesis-related proteins, sometimes retaining antifungal activities (Griffith and Yaish, 2004). Recently, five AFPs in plants are purified: Sd67 (in *Solanum dulcamara*), three antiftmgal proteins (in *Secale cereale*) and AFP (in *Ammonpiptanthus mongolicus*). Their THA (thermal hysteresis activity) is lower than that of fish and insect AFPs.

2.2 Winter Rye (*Secale cereale*)

In winter rye, AFPs accumulate in response to cold, short daylength, dehydration and ethylene, but not pathogens. Transferring single genes encoding antifreeze proteins to freezing-sensitive plants lowered their freezing temperatures by approximately 1°C. Genes encoding dual-function plant antifreeze proteins are excellent models for use in evolutionary studies to determine how genes acquire new expression patterns and how proteins acquire new activities (Griffith and Yaish, 2004).

During cold acclimation, AFPs that are similar to pathogenesis-related proteins accumulate in the apoplast of winter rye (*Secale cereale* L. cv Musketeer) leaves. AFPs have the ability to modify the growth of ice. The proteins were assayed and immunolocalized in winter rye leaves, crowns and roots. Each of the total soluble protein extracts from acclimated plants exhibited antifreeze activity, whereas no antifreeze activity was observed in extracts from non-acclimated rye plants. Antibodies raised against three apoplastic rye AFPs, corresponding to a glucanase-like protein (GLP, 32 kD), a chitinase-like protein (CLP, 35 kD), and a thaumatin-like protein (TLP, 25 kD), were used in tissue printing to show that the AFPs are localized in the epidermis and in cells surrounding intercellular spaces in cold-acclimated plants. Although GLPs, CLPs, and TLPs were present in nonacclimated plants, they were found in different locations and did not exhibit antifreeze activity, which suggests that different isoforms of pathogenesis-related proteins are produced at low temperature. The location of rye AFPs may prevent secondary nucleation of cells by epiphytic ice or by ice propagating through the xylem. The distributions of pathogenesis-induced and cold-accumulated GLPs, CLPs, and TLPs are similar and may reflect the common pathways by which both pathogens and ice enter and propagate through plant tissues (Antikainen, *et al.,* 1996).

2.3 Carrot

A protein purified from carrot shares many functional features with antifreeze proteins of fish. Expression of the carrot complementary DNA in tobacco resulted in the accumulation of antifreeze activity in the apoplast of plants grown at greenhouse temperatures. The sequence of carrot antifreeze protein is similar to that of polygalacturonase inhibitor proteins and contains leucine-rich repeats (Worrall *et al.*, 1998).

3.0 MECHANISMS OF ACTION

Unlike the widely used automotive antifreeze, ethylene glycol, AFPs do not lower freezing point in proportion to concentration. Rather, they work in a non colligative manner. This allows them to act as antifreeze at concentrations 300-500 times lower than other dissolved solutes. This minimizes their effect on osmotic pressure (Fletcher *et al.*, 2001). AFPs are thought to inhibit growth via an adsorption–inhibition. They adsorb to non-basal planes of ice, inhibiting thermodynamically favored ice growth (Raymond *et al.*, 1989) (Figure 1.1). The presence of a flat, rigid surface in AFPs seems to facilitate its interaction with ice via van der waals force surface complementarity (Yang *et al.*, 1998).

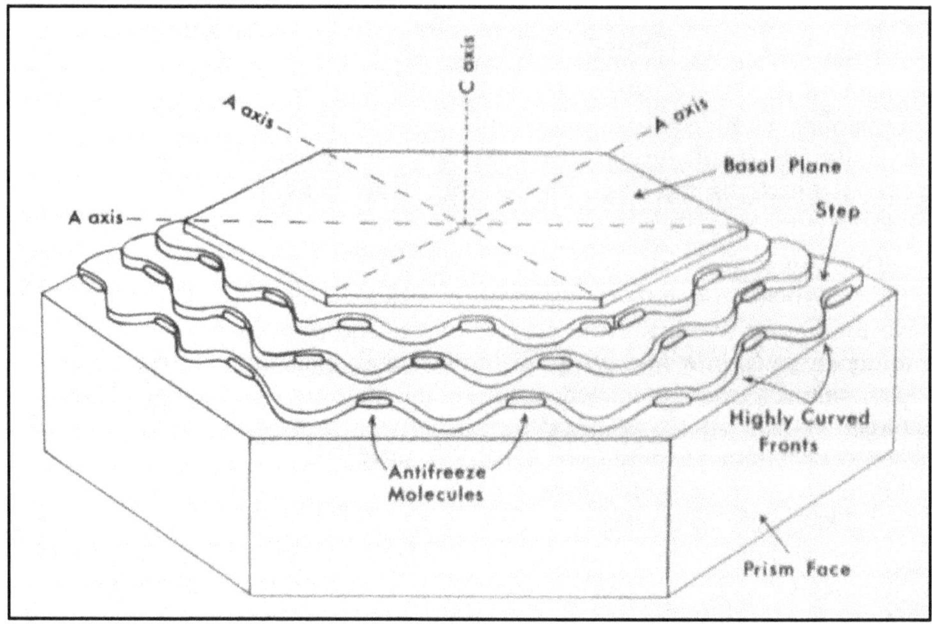

Figure 1.1: Adsorption-Inhibition of Ice (Raymond *et al.*, 1989).

3.1 Binding to Ice

The unusual capabilities of AFPs are attributed to their binding ability at specific ice crystal surfaces (Jorov *et al.*, 2004). AFPs depress the freezing temperature of a solution in a non-colligative manner, by arresting the growth of ice crystals. Ice crystal grown in water normally grows along the a-axis while in presence of AFPs growth along the c-axis is favored. This is due to the thermodynamic effect which ultimately results in antifreeze property (Figure 1.2). Normally ice crystals grown in solution only exhibit the basal (0001) and prism faces (1010) and appears as round and flat

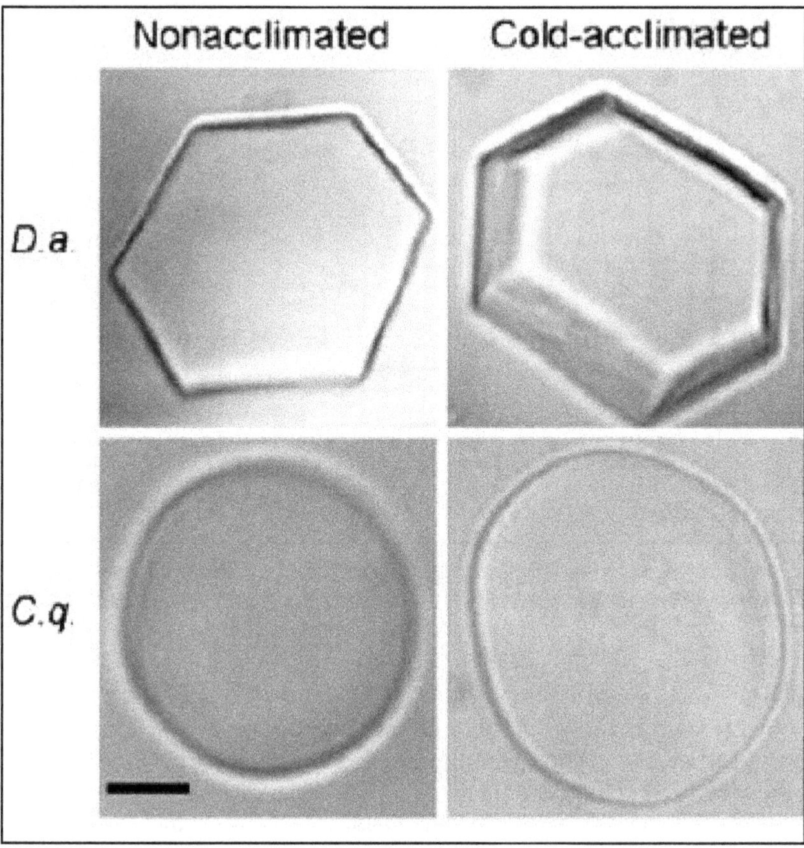

Figure 1.2: Mechanism of Ice Crystal Modification (Jorov *et al.,* 2004).

discs (Jorov *et al.,* 2004). However, it appears the presence of AFPs exposes other faces. It now appears that the ice surface is the preferred binding surface, at least for AFP type I (Knight *et al.,* 1991). Through studies on type I AFP, it was initially thought that ice and AFP interacted through hydrogen bonding (Raymond and DeVries, 1977). However, when parts of the protein that were thought to facilitate this hydrogen bonding were mutated, the hypothesized decrease in antifreeze activity was not observed. Recent data suggests that hydrophobic interactions could be the main contributor (Haymet *et al.,* 1998).

It is difficult to discern the exact mechanism of binding because of the complex water-ice interface. Currently, attempts to uncover the precise mechanism are being made through use of molecular modelling programs (molecular dynamics or Monte Carlo method) (Jorov *et al.,* 2004). A proposed mechanism of antifreeze binding to ice surfaces is given which requires: first, that the dipole moment from the helical structure dictates the preferential alignment of the peptide to the c-axis of ice nuclei; second, amphiphilicity of the helix; and third, torsional freedom of the side chains to facilitate hydrogen bonding to ice surfaces (Yang *et al.,* 1988). The antifreeze activity in

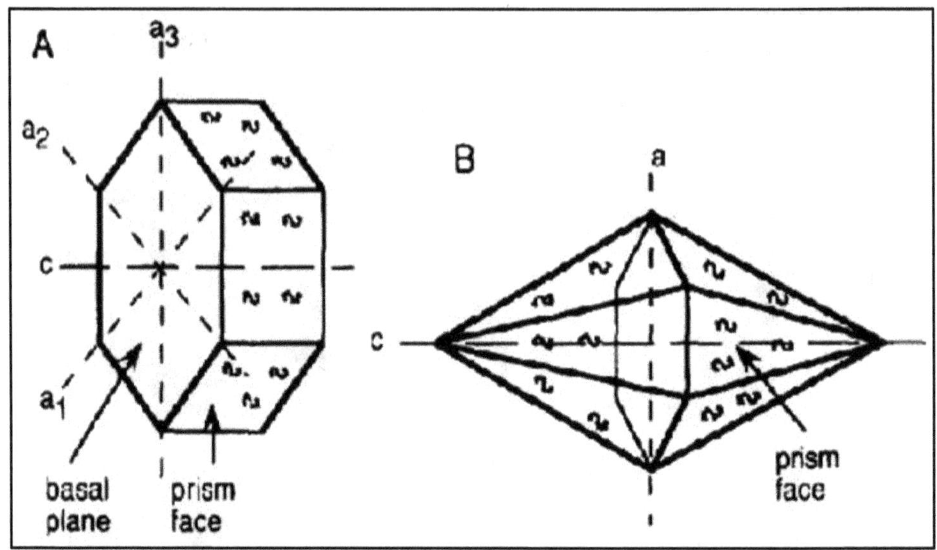

Figure 1.3: Antifreeze Activity in Apoplastic Extract of Nonacclimated and Cold-acclimated Antarctic Vascular Plants *D. antarctica* (*D.a.*) and *C. quitensis* (*C.q.*) (Leon and Griffith, 2005).

apoplastic extract of nonacclimated and cold-acclimated antarctic vascular plants *D. antarctica* (*D.a.*) and *C. quitensis* (*C.q.*) were also studied (Figure 1.3). The kinetics of this effect, studied using a new technique called temperature gradient thermometry is consistent with an adsorption-mediated inhibitory mechanism. The results obtained by this approach provide a new experimental basis for understanding AFP interaction with ice (Chapsky and Rubinsky, 1997).

4.0 GENE(S) AND PROTEINS FOR AFP

4.1 Chitinase Genes Encoding Antifreeze Proteins

Antifreeze proteins similar to two different chitinases accumulate during cold acclimation in winter rye. Chitinase-antifreeze proteins purified from the plant were similar in mass to the predicted mature products of *CHT9* and *CHT46*, thus indicating that there was little chemical modification of the amino acid sequences in plants. Two novel cold-responsive genes encoding chitinases with ice-binding activity may have arisen in winter rye and other cereals through gene duplication (Yeh *et al.*, 2000).

4.2 Carrot

The gene encoding a 1,099-bp carrot AFP was amplified by the polymerase chain reaction (PCR) using genomic DNA from carrot seedlings as the template. Sequencing indicated a difference of three bases between this cloned gene and that published in GenBank. The cloned antifreeze protein gene was expressed in *Escherichia coli*, and a fusion protein of about 60 kDa was detected after isopropyl thiogalactoside induction. This AFP gene was also cloned into binary vector pCAMBIA2300 with the CaMV 35S promoter and used to transform tobacco NC82. PCR and Southern blot results verified integration of this gene into the genome of tobacco and reverse transcription-PCR verified that this gene had been expressed in transgenic tobacco. Experiments confirmed that transgenic tobacco plants displayed greater stamina than wild-type ones when subjected to cold treatment. The genetic stability of the transgenic lines was analyzed, and findings confirmed that transgenic expression of the carrot AFP gene could enhance the tolerance of plants to cold or frigid conditions (Fan *et al.*, 2002).

4.3 *Arabidopsis thaliana*

Kurkela and Franck (1990) identified by differential screening a novel *Arabidopsis thaliana* gene, called kin1, which is induced at +4°C. The nucleotide sequences of both the genomic clone and the corresponding cDNA were determined. The deduced 6.5 kDa polypeptide has an unusual amino acid composition being rich in alanine, glycine and lysine. The gene belongs to a family of at least two genes. Northern blot analysis revealed that the level of kin1 mRNA is increased 20-fold in cold-treated plants. In addition to being expressed in cold, kin1 is also induced by water stress and the plant hormone abscisic acid (ABA) which has been suggested to be a common mediator for osmotic stress responses and cold acclimation in plants. Sequence comparisons showed that the kin1 gene product has similarities to fish antifreeze proteins (AFPs).

4.4 AFP Accumulation in Freezing-Tolerant Cereals

To determine whether the accumulation of antifreeze proteins is common among herbaceous plants, Antikainen and Griffith (1997) assayed antifreeze activity and total protein content in leaf apoplastic extracts from a number of species grown at low temperature, including both monocotyledons (winter and spring rye, winter and spring wheat, winter barley, spring oats, maize) and dicotyledons (spinach, winter and spring oilseed rape [canola], kale, tobacco). Apoplastic polypeptides were also separated by SDS-PAGE and immunoblotted to determine whether plants generally respond to low temperature by accumulating pathogenesis-related proteins. Their results showed that significant levels of antifreeze activity were present only in the apoplast of freezing-tolerant monocotyledons after cold acclimation at 5/2°C. Moreover, only a closely related group of plants, rye, wheat and barley, accumulated antifreeze proteins similar to pathogenesis-related proteins during cold acclimation. The results indicate that the accumulation of antifreeze proteins is a specific response that may be important in the freezing tolerance of some plants, rather than a general response of all plants to low temperature stress (Antikainen and Griffith, 1997). Accumulation of a high molecular weight protein in the range of 200 kDa in cold-tolerant cultivars (*Triticum aestivum* L. cv. Frederick and cv. Norstar) compared to the cold sensitive spring wheat was observed in its electrophoretic pattern (*T. aestivum* L. cv. Glenlea (Sidebottom *et al.*, 1998).

4.5 Calcium Signaling during Cold-Induction of the Kin Gene

The involvement of calcium signaling during cold-induction of the kin genes of *Arabidopsis thaliana* (L.) Heynh. was examined. Treatments with chemicals which either chelate extracellular calcium (EGTA) or block the plasma-membrane calcium channels (La^{3+}, Gd^{3+}) inhibited cold acclimation as well as kin gene expression. Ruthenium red, an inhibitor of calcium release from intracellular stores partially inhibited kin gene expression and development of freezing tolerance. An inhibitor of calcium-dependent protein kinases (CDPKs) and calmodulin prevented cold acclimation as well as the cold induction of kin genes. Using restriction fragment length polymorphism-coupled domain-directed differential display, five CDPK clones were identified which showed differential regulation by cold. The amplified fragments showed homology to known plant CDPKs. The involvement of calcium and calcium-binding proteins in cold acclimation of *A. thaliana* is discussed (Tähtiharju *et al.*, 1997).

5.0 ANTIFREEZE PROTEINS AND THEIR GENES: FROM BASIC RESEARCH TO BUSINESS OPPORTUNITY

AFPs and their genes can be used in fish and plants to enhance resistance to freezing. AFPs can be used in medicine to improve the cold protection of blood platelets (to extend their shelf life prior to transfusion); paradoxically, when used in conjunction with cryosurgery, they can help destroy malignant tumors. Their ability to inhibit recrystallization can improve the quality of frozen foods. In addition, antifreeze gene promoters are uniquely suited to drive the expression of functional genes, such as growth hormone that results in enhanced growth rates of salmonids (*e.g.*, salmon, trout) and other fish species valuable to aquaculture (Fletcher *et al.*, 1999).

Because commercial applications of these proteins are still in the R&D phase, the economic viability of such ventures must be left to future evaluation. Two excellent reviews provide additional details about the potential uses of AFPs in food products (de Maagd, 1989). On the basis of our current knowledge about AFPs and AFP genes, commercial applications have been identified in the following areas (DeVries, 1986):

- ✰ Cold protection of mammalian cells, tissues, and organs;
- ✰ Enhanced tumor cell destruction during cryosurgery;
- ✰ Protection of fish and plants against cold and freezing temperatures;
- ✰ Longer shelf life for and better quality of frozen foods and
- ✰ Improved growth characteristics in transgenic fish by using afp gene promoters.

5.1 Commercial Use

Commercially, there appears to be infinite uses for antifreeze proteins. Numerous fields would be able to benefit from the protection of tissue damage by freezing. Businesses are currently investigating the use of these proteins in:

- ✰ Increasing freeze tolerance of crop plants and extending the harvest season in cooler climates
- ✰ Improving farm fish production in cooler climates
- ✰ Lengthening shelf life of frozen foods

- ☆ Improving cryosurgery
- ☆ Enhancing preservation of tissues for transplant or transfusion in medicine
- ☆ A therapy for hypothermia

Currently two companies market AFPs: A/F Protein Inc and Ice Biotech Inc.

6.0 CROP PROTECTION WITH AFPs

Although many plants are killed by low and freezing temperatures, some species and varieties have developed physiological processes to help them survive the winter. With the onset of cold conditions, some plants produce colligative cryoprotectants such as sucrose and proline; in others, changes in membrane lipids and proteins that render membranes more cold stable have been reported (Gattiker, 2002). A few plants are able to produce cold-regulated cryoprotective proteins or AFPs, which protect them from cold damage under freezing conditions (Gilbert *et al.*, 2004).

AFPs similar to those found within the animal kingdom have been identified in plants and appear to behave similarly at freezing temperatures (Green *et al.*, 1988). In addition to their protective action at subzero temperatures, fish AFPs protect cold-sensitive cells at hypothermic temperatures, possibly by interacting with the cell membrane (Griffith *et al.*, 1992). A similar effect may be exerted by plant cold protective or antifreeze proteins in that they may provide a protective environment in the proximity of the membranes that stabilizes the membrane or inhibits ice propagation through cell walls (Gurian-Sherman *et al.*, 1993).

At least 30 species of angiosperms are capable of antifreeze activity after they become acclimated to low temperature, and antifreeze activity has been found in commercially grown species *e.g.* winter rye, spring and winter wheat, winter barley, spring oat, winter canola, potato, carrot, cabbage, kale, and brussels sprouts (Dayhoff *et al.*, 1978).

6.1 Producing Cold-Hardy Varieties

The ability of plants to withstand low temperatures appears to be a function of three separate heritable traits: freezing tolerance, freezing avoidance, and speed of acclimation (Laemmli, 1970). Although some improvements in cold hardiness have been achieved (Lifshitz *et al.*, 1986), understanding the control of these traits and their manipulation at the genome level is a vast undertaking and is still in the early stages for most crop species (Lloret *et al.*, 1998). Whether the lack of speed is due to the difficulty in identifying the appropriate genetic markers (Laemmli, 1970), the limitation of naturally occurring genetic variability in cold-hardy traits (Lindow *et al.*, 1978), or other factors such as the vagaries of the environment during selection trials, the desired level of cold tolerance needs to be determined for commercially important crops, increased cold tolerance is still required in commercially important crops.

In recent years, with the advent of transgenic technology, effort has been directed toward enhancing the gene complement of crop species with novel genetic material. AFP genes are now being introduced into plant and animal species to extend their geographic range, increase their chances of overwintering survival, and improve the texture of products that might be stored frozen after harvest. In one experiment, leaves of potato, canola, and *Arabidopis thaliana* plants were vacuum infiltrated with AFP Type I from winter flounder (Lopez *et al.,* 2003). The experimental plants and the control (water-infiltrated) plants of the same three species then were exposed to freezing conditions. The AFP-infiltrated plants were found to be more cold-hardy than the controls. The experimenters found that this method of AFP introduction depressed the freezing point to a level that would significantly improve crop survival under agricultural conditions.

The use of transgenic technology to produce plants capable of synthesizing their own antifreeze has yielded transgenic tobacco, tomatoes, and potatoes. In these transgenic plants, researchers have observed increased freeze resistance and inhibition of electrolyte leakage from cold-stressed cells. Depending on the role the AFPs are expected to play in transgenic plants, the levels of antifreeze expression required will differ. For example, higher levels (~1000-fold) are needed to inhibit ice crystal formation and propagation than are needed to protect frozen products from ice recrystallization during storage (Nelson *et al.,* 2002). Consideration must be given to the type of antifreeze molecule best suited to the recipient organism. In plant systems, plant antifreeze or synthetic molecules may be more acceptable to plants, whereas fish antifreeze molecules would seem to be more acceptable in fish species (Dayhoff *et al.,* 1978).

6.2 Low-Temperature Preservation of Cells, Tissues and Organs

Until 1990, it was generally believed that the sole function of AFPs was to protect fish from freezing; however, in a series of experiments published after 1990, Rubinsky and colleagues discovered that all of the antifreeze types known at that time could improve the cold tolerance of cold-sensitive mammalian cells (Nielsen *et al.,* 1997). In initial experiments, the researchers incubated bovine oocytes at 4°C for 24 h in the presence and absence of AFPs, then rewarmed both the test and control cells. Oocytes stored cold in the presence of AFPs were as capable of normal maturation, fertilization, and embryonic development as were freshly collected oocytes; however, oocytes stored cold without AFPs lost their membrane integrity and died (Reese, 2001).

6.3 More Effective Cryosurgery

Cryosurgery is a procedure in which probes cooled to very low temperatures are used to freeze and destroy solid tumors, such as those of the prostate and liver. One of the problems with cryosurgery as it is currently practiced is the variation in its effectiveness in destroying tumor cells. The number of target cells that survive cryosurgery is highly variable and depends on the thermal parameters used during surgery (*e.g.,* rate of cooling, number of freezing cycles, and final freezing temperature before rewarming) (Wolber *et al.,* 1986). Because of the nature of the cryosurgical procedure, it may be difficult, if not impossible, to control the freeze-thaw thermal parameters accurately.

One potential solution to the problems associated with the efficacy of current cryosurgical practice lies with the manner in which AFPs interact with ice crystals and modify their morphology and growth. In a dilute solution, AFPs adsorb onto the prism faces and prevent or limit growth along the preferred a-axis. When the temperature of the solution is lowered appropriately, the crystal grows rapidly along the c-axis, resulting in a bipyramidal crystal. Crystals grown in high concentrations (5-10 mg/mL) of antifreeze are long and needlelike.

6.4 Improved Storage of Frozen Foods

It is well known that some foods (*e.g.*, strawberries, raspberries, and tomatoes) cannot be frozen without loss of quality caused by cellular destruction; even food products that freeze well deteriorate to varying degrees over time. One of the causes of deterioration during frozen storage is the growth of large ice crystals within the product when refrigeration temperatures are not optimal. This process, known as ice recrystallization, occurs when the products are stored at subzero temperatures or are subject to fluctuating subzero temperatures such as occur during freezer defrost cycles. Food quality and texture would be improved if a means could be found to reduce or eliminate this phenomenon.

6.5 Antifreeze Protein Sources

At present, relatively small amounts of AFP are being obtained for R&D, primarily from their most easily accessed natural source: the plasmas of cold ocean teleost fish, in which concentrations of 1-30 g/L have been reported (de Maagd *et al.,* 1989). However, as commercial applications are developed, the demand for greater amounts of AFP will increase. Clearly, the amount of AFP required will depend on the application. How much protein can reasonably be obtained from a natural fish source? Right now, the answer to this question can be no more than a rough estimate. Factors such as fish species, time of year, plasma antifreeze concentration, and AFP losses that occur during purification affect the yield considerably.

Before we arrive at a mismatch between AFP supply and demand, alternative sources of antifreeze production must be investigated and developed. Recombinant AFP production that uses fermentation technologies, mammalian cell lines, transgenic plants, and transgenic dairy animals are currently being explored as means of meeting this future demand.

7.0 RECENT DEVELOPMENTS

One recent, successful business endeavor has been the introduction of AFPs into ice cream and yogurt products. This ingredient, labelled ice-structuring protein, has been approved by the Food and Drug Administration. The proteins are isolated from fish and replicated, on a larger scale, in yeast. The ISPs have been approved for human consumption following diligent tests. Intake of AFPs in diet is likely substantial in most northerly and temperate regions already. Given the known historic consumption of AFPs, it is safe to conclude that their functional properties do not impart any toxicologic or allergenic effects in humans (Crevel *et al.,* 2002). As well, the transgenic process of ISP production is widely used in society already. This is how mass amounts of insulin are made to treat people with type I diabetes each year. The process does not impact the product; it merely makes production more efficient and prevents the death of many fish that would, otherwise, be killed for the extraction of such protein. Currently Unilever incorporates AFPs into some of its products including some popsicles and a new line of Breyers Light Double Churned ice cream bars. In ice cream, AFPs allow the production of very creamy, dense, reduced fat ice cream with fewer additives. They control ice crystal growth brought on by thawing on the loading dock or kitchen table which drastically reduces texture quality (Regand *et al.*, 2006).

REFERENCES

Antikainen, M., Griffith, M., 1997. Antifreeze protein accumulation in freezing-tolerant cereals. *Physiol Plant* 99, 423-432.

Antikainen, M., Griffith, M., Zhang, J., Hon, W.C., Yang, D.S.C., Pihakaski-Maunsbach, K., 1996. Immunolocalization of Antifreeze Proteins in Winter Rye Leaves, Crowns, and Roots by Tissue Printing. *Plant Physiol* 110(3), 845–857.

Atýcý, O., Nalbanto, B., 2003. Antifreeze proteins in higher plants. *Phytochemistry* 64 (7), 1187-1196.

Chapsky, L., Rubinsky, B., 1997. Kinetics of antifreeze protein-induced ice growth inhibition. *FEBS Letters* 412 (21), 241-244.

Crevel, R.W.R., Fedyk J.K., Spurgeon, M.J., 2002. Antifreeze proteins: characteristics, occurrence and human exposure (Review). *Food and Chemical Toxicology* 20, 899-903.

Dayhoff, M.O., Schwartz, R.M., Orcutt, B.C., 1978. A model of evolutionary change in proteins. Matrices for detecting distant relationships, p. 345-358. In M. O. Dayhoff (ed.), Atlas of protein sequence and structure, vol. 5:. National Biomedical Research Foundation, Washington, D.C.

de Maagd, R.A., Wijfjes, A.H.M., Spaink, H.P., Ruiz-Sainz, J.E., Wijffelman, C.A., Okker, R.J.H., Lugtenberg, B.J.J., 1989. *nodO*, a new nod gene of the *Rhizobium leguminosarum* biovar viciae sym plasmid pRL1JI, encodes a secreted protein. *J. Bacteriol* 171, 6764-6770.

DeVries, A.L., 1986. Antifreeze glycopeptides and peptides: interactions with ice and water. *Methods Enzymol* 127, 293-303.

Duman, J.G., Olsen, T.M., 1993. Thermal hysteresis protein activity in bacteria, fungi, and phylogenetically diverse plants. *Cryobiology* 30, 322–328.

Fan, B.L., Wang, H., Wang, S., Wang, J., 2002. Cloning of an antifreeze protein gene from carrot and its influence on cold tolerance in transgenic tobacco plants. *Plant Cell Reports* 21, (4), 296-301.

Fletcher, G.L., Goddard, S.V., Wu, Y., 1999. Antifreeze proteins and their genes: From basic research to business opportunity. *Chemtech* 30 (6), 17-28.

Fletcher, G.L., Hew, C.L., Davies, P.L., 2001. Antifreeze Proteins of Teleost Fishes. *Annu. Rev. Physiol* 63, 359–90.

Gattiker, A., Gasteiger, E., Bairoch, A., 2002. ScanProsite: a reference implementation of a PROSITE scanning tool. *Appl. Bioinformatics* 1, 107-108.

Gilbert, J.A., Hill, P.J., Dodd, C.E.R., Laybourn-Parry, J., 2004. Demonstration of antifreeze protein activity in Antarctic lake bacteria. *Microbiology* 150, 171-180.

Green, R.L., Corotto, L.V., Warren, G.J., 1988. Deletion mutagenesis of the ice nucleation gene from *Pseudomonas syringae* S203. *Mol. Gen. Genet* 215, 165-172.

Griffith, M., Ala, P., Yang, D.S., Hon, W.C., Moffatt, B.A., 1992. Antifreeze protein produced endogenously in winter rye leaves. *Plant Physiol* 100, 593–596.

Griffith, M., Yaish, M.W., 2004. Antifreeze proteins in overwintering plants: a tale of two activities. *Trends Plant Sci* 9(8), 399-405.

Gurian-Sherman, D., Lindow, S.E., Panopoulos, N.J., 1993. Isolation and characterization of hydroxylamine-induced mutations in the *Erwinia herbicola* ice nucleation gene that selectively reduce warm temperature ice nucleation activity. *Mol. Microbiol* 9, 383-391.

Haymet, A.D.J., Ward, L.G., Harding, M.M., Knight, C.A., 1998. Valine substituted winter flounder 2antifreeze': preservation of ice growth hysteresis. *FEBS Lett* 430, 301-306.

Jorov, A., Zhorov, B.S., Yang, D.S., 2004. Theoretical study of interaction of winter flounder antifreeze protein with ice. Protein. *Science* 13, 1524-1537.

Knight, C.A., Cheng, C.C., DeVries, A.L., 1991. Adsorption of alpha-helical antifreeze peptides on specific ice crystal surface planes. *Biophys J* 59(2), 409-18.

Kurkela, S., Franck, M., 1990. Cloning and characterization of a cold- and ABA-inducible *Arabidopsis* gene. *Plant Mol Biol* 15(1), 137-44.

Laemmli, U.K., 1970. Cleavage of structural proteins during the assembly of the head of bacteriophage T4. *Nature* 227, 680-685.

Leon, A.B., Griffith, M., 2005. Characterization of antifreeze activity in Antarctic plants. *J. Exp. Bot* 56, 1189-1196.

Lifshitz, R., Kloepper, J.W., Scher, F.M., Tipping, E.M., Laliberte, M., 1986. Nitrogen-fixing pseudomonads isolated from roots of plants grown in the Canadian High Arctic. *Appl. Environ. Microbiol* 51, 251-255.

Lindow, S.E., Arny, D.C., Upper, C.D., Barchet, W.R., 1978. The role of bacterial ice nuclei in frost injury to sensitive plants, p. 249-263. In P. Li (ed.), Plant cold hardiness and freezing stress. Academic Press, New York, N.Y.

Lloret, J., Wulff, B.B.H., Rubio, J.M., Downie, J.A., Bonilla, I., Rivilla, R., 1998. Exopolysaccharide II production is regulated by salt in the halotolerant strain *Rhizobium meliloti* EFB1. *Appl. Environ. Microbiol* 64, 1024-1028.

Lopez, R., Silventoinen, V., Robinson, S., Kibria, A., Gish, W., 2003. WU-Blast2 server at the European Bioinformatics Institute. *Nucleic Acids Res* 31, 3795-3798.

Nelson, K., Paulsen, I., Weinel, C. *et al.,* 2002. Complete genome sequence and comparative analysis of the metabolically versatile *Pseudomonas putida* KT2440. *Environ. Microbiol* 4, 799-808.

Nielsen, H., Engelbrecht, J., Brunak, S., von Heijne, G., 1997. Identification of prokaryotic and eukaryotic signal peptides and prediction of their cleavage sites. *Protein Eng* 10, 1-6.

Raymond J. *et al.,* 1989. Inhibition of growth of nonbasal planes in ice by fish antifreezes. *Proc. Natl. Acad. Sci* 86, 881-885.

Raymond, J., DeVries, A.L., 1977. Adsorption inhibition as a mechanism of freezing resistance in polar fishes. *Proc. Nati. Acad. Sci* 74(6), 2589-2593.

Reese, M.G., 2001. Application of a time-delay neural network to promoter annotation in the *Drosophila melanogaster* genome. *Comput. Chem* 26, 51-56.

Regand, A., Goff, H.D., *et al.,* 2006. Ice recrystallization inhibition in ice cream as affected by ice structuring proteins from winter wheat grass. *J Dairy Sci.* 89(1), 49-57.

Sidebottom, C., Lillford, P., Telford, J., Holt. C., 1998. Accumulation of a High Molecular Weight Protein during Cold Hardening of Wheat (*Triticum aestivum* L.)

Tähtiharju, S., Sangwan, V., Monroy, A.F., Dhindsa, R.S., Borg, M., 1997. The induction of kin genes in cold-acclimating *Arabidopsis thaliana.* Evidence of a role for calcium. *Planta* 203(4),442-7.

Urrutia, M., Duman, J.G., Knight, C.A., 1992. Plant thermal hysteresis proteins. *Biochem Biophys Acta* 1121, 199–206.

Wolber, P. K., Deininger, C.A., Southworth, M.W., Vandekerckhove, J., Montagu, M.V., Warren, G.J., 1986. Identification and purification of a bacterial ice-nucleation protein. *Proc. Natl. Acad. Sci.* USA 83, 7256-7260.

Worrall, D., Elias, E., Ashford, D., Smallwood, M., Sidebottom, C., Lillford, P., Telford, J., Holt, C., Bowles, D., 1998. A carrot leucine-rich-repeat protein that inhibits ice recrystallization. *Science* 282, 115–117.

Yang, D.S., Sax, M., Chakrabartty, A., Hew, C.L., 1988. Crystal structure of an antifreeze polypeptide and its mechanistic implications. *Nature* 333(6170),232-7.

Yeh, S., Barbara A. Moffatt, Marilyn Griffith, Fei Xiong, Daniel S.C. Yang, Steven B. Wiseman, Fathey Sarhan, Jean Danyluk, Yi Qi Xue, Choy L. Hew, Amanda Doherty-Kirby, Gilles Lajoie. 2000. *Plant Physiol* 124(3), 1251–1264.

Yong, J., Jia Shi-Rong, Fei Yun-Biao Tan Ke-Hui, 1987. Antifreeze Proteins and Their Role in Plant Antifreeze Physiology. *Plant and Cell Physiology* 28(7), 1173-1179.

Yuan-Zhen, L., Shan-Zhi, L., Zhi-Yi, Z., Wei, Z., Wen-Feng, L., 2005. Plant antifreeze proteins and their expression regulatory mechanism. *Forestry Studies in China* 7(1), 46-52.

2015, Modern Methods in Phytomedicine *Pages* ***15–32***
Editor: **T. Parimelazhagan**
Published by: **DAYA PUBLISHING HOUSE, NEW DELHI**

2

Antimicrobial Activities of *Pteris tripartita* Sw.: A Critically Endangered Fern

X. Baskaran[1]*, *R. Jeyachandran*[1] *and A. Geo Vigila*[2]

[1]*Department of Botany, St. Joseph's College, Tiruchirapalli, Tamil Nadu– 620 002*
[2]*Department of Zoology, S.T. Hindu College, Nagercoil, Tamil Nadu – 629 002*

1.0 INTRODUCTION

In general, the widespread use of antibiotics has led to the decimation of sensitive organisms from the population with the consequent increase in the number of resistant microorganisms. This situation has forced scientists to search for new antimicrobial substances in various sources like medicinal plants (Kumar *et al.*, 2006). Plant extracts have been widely used for centuries as a popular method for treating several health disorders. Numerous studies have been carried out on various natural products screening their antimicrobial activity (Bhattacharjee *et al.*, 2006; Parekh and Chanda, 2006, 2007; Vaghasiya *et al.*, 2008).

The medicinal values of the pteridophytes have been known to human for more than 2,000 years. The antibiotic activities of pteridophytes are still in their early years and found very little application in modern chemotherapy and researches. The Greek botanist Theophrastus (ca. 372– 287 B.C.) had referred the medical values of ferns.

* *Corresponding Author.* E-mail: fernsbaskar@gmail.com

Dioscorides (ca. 50 A.D.) also referred in his De Materia Medica to a number of ferns including *Pteridium aquilinum* and *Dryopteris filixmas* having medicinal values. In ancient Indian medicine systems, several ferns were used to cure number of human ailments. Sushruta (ca. 100 A.D.) and Charaka (ca. 100 A.D.) recommended the medicinal use of some ferns in their Samhitas and several ferns have been used by Unani physicians in India and Western Asia. Literally, several studies have been carried out to explore antimicrobial activities of ferns such as *Adiantum caudatum*, *Ampelopteris prolifera* (Banerjee and Sen, 1980), *Pteridium aquilinum* (Francisco and Driver, 1984), *Nephrolepis* sp. (Basile *et al.*, 1997), *Adiantum lunulatum* (Niranjan Reddy et al., 2001), *Equisetum arvense* (Joksic *et al.*, 2003; Radulovic *et al.*, 2006), *Adiantum capillus-veneris* (Guha *et al.*, 2004; 2005; Besharat *et al.*, 2008), *Athyrium pectinatum* (Parihar *et al.*, 2006), *Adiantum* sp. (Singh *et al.*, 2008 a), *Pteris vittata* (Singh *et al.*, 2008 b), *Pteris multifida* (Hao-bin *et al.*, 2008), *Drynaria quercifolia* (Kandhasamy *et al.*, 2008), *Mecodium exsertum* (Maridass *et al.*, 2009), *Selaginella involvens*, *Selaginella inaequalifolia* (Haripriya *et al.*, 2010), *Asplenium scolopendrium*, *Cystopteris fragilis*, *Polypodium vulgare* (Soare *et al.*, 2012), *Adiantum caudatum*, *Angiopteris evecta*, *Pteris confusa*, *Pteris argyraea*, *Lygodium microphyllum* (Herin Sheeba Gracelin *et al.*, 2012), *Pteris biaurita* (Dalli *et al.*, 2007; John De Britto *et al.*, 2012) and some more Pteridophytes (Maruzzella, 1961; McCutcheon, *et al.*, 1995). Thus, the objective of this present investigation was to screen the antimicrobial activity of *P. tripartita* extracts against pathogenic strains.

2.0 MATERIALS AND METHODS

2.1 Plant Powder and Plant Extracts Preparations

The fresh and healthy fronds of *P. tripartita* (free from insect damaged and fungus infected) collected from Alagar hills, Madurai district and the fresh material was washed in running tap water and shade dried in the laboratory at room temperature for 5–8 days. Once completely dried, plant parts were ground to a fine powder using an electronic blender. Plants were stored in a closed container at room temperature until required.

10 grams of the dried and powdered plant materials were soaked (1:6 w/v) separately with 60 mL of each solvents *viz.* hexane, chloroform, acetone, ethyl acetate, ethanol, methanol and water in a Soxhlet apparatus for 48 hours (according to the boiling point of each solvents used) until complete extraction. At the end of 48 hours, each extract was filtered through Whatman No.1 filter paper and filtrates were concentrated at room temperature in order to reduce the volume. All the extracts were stored at 4°C in air tight bottles for further studies.

2.2 Qualitative Phytochemical Analysis

The presence of various phytoconstituents such as, carbohydrates, terpenoids, saponins, phenolic compounds, tannins, phlobatannins, flavonoids and alkaloids were analyzed qualitatively using standard protocols (Mace, 1963; Harborne, 1973; Evans, 1997; Kokate, 1999).

2.3 Test Bacteria

The bacteria used in this study were collected from Microbial Type Culture Collection (MTCC), Institute of Microbial Technology, Chandigarh, Punjab, India.

The bacteria include *viz.* gram positive *Bacillus subtilis* Ehrenberg (MTCC 441), *Bacillus cereus* Bizio (MTCC 1272), *Bacillus megaterium* A. deBary (MTCC 2444) and gram negative *Escherichia coli* Escherich (MTCC 1195), *Proteus vulgaris* Hauser (MTCC 1771), *Serratia marcescens* Bizio (MTCC 8780), *Salmonella typhi* Eberth (MTCC 733), *Klebsiella pneumoniae* Friedlander (MTCC 2405), *Vibrio cholerae* Pacini (MTCC 3904), *Shigella sonnei* Carl Olaf Sonne (MTCC 2957), *Enterobacter aerogenes* Bizio (MTCC 2823) and *Pseudomonas aeruginosa* Schroeter (MTCC 2642).

2.4 Test Fungi

The fungal moulds consist of *Aspergillus niger* van Tieghem (MTCC 2425), *Aspergillus flavus* Link ex Gray (MTCC 3396), *Fusarium oxysporum* Schlecht (*MTCC* 2480), *Penicillium chrysogenum* Thom. (MTCC 2725) and *Rhizopus oryzae* Went and Prins. Geerl. (MTCC 262) were also collected from Microbial Type Culture Collection, Institute of Microbial Technology, Chandigarh, Punjab, India.

2.5 Methods of Antibacterial Screening

The antimicrobial screening of the crude extracts of *P. tripartita* were investigated through different methods. The disc diffusion assay was carried out by standard method (Iennette, 1985 as described by Rosoanaivo and Ratsmanaga-Urverg, 1993 and Rabe and Van Staden, 1997). Sterile liquid nutrient agar medium (pH 7.4 ± 0.2) was poured (20 mL) into each sterile petriplates. After solidification, 100 µl of suspension containing 10^5 CFU/mL of each test bacteria were spread over nutrient agar plates. The sterile filter paper discs (6 mm in diameter) were impregnated with 25 µg/mL and 50 µg/mL of extracts were placed on the inoculated agar plates. All the extracts were dissolved in DMSO (2 per cent). Chloramphenicol (10 µg/disc) was used as positive reference control to determine the sensitivity of plant extract on each bacterial species. The inoculated plates were incubated at 37°C for 24 hours. Antibacterial activity was evaluated by measuring the zones of inhibition against the test organisms. Each assay was conducted in triplicate.

2.6 Antifungal Assay

The antifungal activities of frond crude extracts of *P. tripartita* were tested by disc diffusion method (Taylor *et al.*, 1995). The potato dextrose agar plates were inoculated with each fungal culture (10 days old) by point inoculation. The filter paper discs (Whatman No. 1, 6 mm diameter) impregnated with 25 and 50 µg/mL concentrations of the extracts were placed on test organism-seeded plates. DMSO (2 per cent) was used to dissolve the extracts before application on test organism-seeded plates. Nystatin (10 µg/disc) was used as positive control and the activity was determined after 72 hours of incubation at 28°C. The diameters of the inhibition zones were measured with a caliper and the mean values are presented.

2.7 Time Course Growth Assay

The time course growth assay was used to evaluate the antibacterial sensitivity of plant extract at different time intervals on pathogenic bacterial strains. All the extracts were dissolved in 2 per cent DMSO for time course growth assay. The hundred microlitres of overnight bacterial culture in Nutrient Broth (NA) were added to 1 mL of NA containing 100 µl of plant extracts (25 µg/mL and 50 µg/mL) in sterile test tubes. The tubes were incubated at 37°C with shaking and the optical density at 550 nm was measured in Biotek microquant spectrophotometer (USA) after 0, 1, 2, 3 and 5 hours. Control tubes without plant extract were incubated under same conditions. All the assays were carried out in triplicate. A decrease in optical density was observed in the incubated cultures suggesting a bacteriostatic effect (Palombo and Semple, 2001).

2.8 Minimum Inhibitory Concentration (MIC)

In MIC assay, different concentrations of samples were taken in separate test tubes and labeled accordingly. Each of the test tubes were added with 3 mL of the nutrient broth (NA), 0.1 mL of the bacterial suspension and different concentrations of plant extracts (25 µg/mL and 50 µg/mL) dissolved in 2 per cent of DMSO. A positive control tube containing the growth medium and the bacterial suspension without the plant extracts was also prepared. The tubes were incubated at 37°C for 24 hours. After 24 hours, the turbidity in each of the test tube was measured in Biotek microquant spectrophotometer (USA) at 420 nm. The turbidity was taken as an indicator of the bacterial density. The rate of inhibition was directly related to the turbidity of the medium. The lowest concentration which did not permit any visible microbial growth when compared with that of the control was recorded as the MIC value (Barry and Thornsberry, 1991).

2.9 Statistical Analysis

All experiments were carried out in triplicates and their results are expressed as mean±SE (n=3). The statistical analysis was performed by one way ANOVA with SPSS 17.0 software (Chicago, USA) and relationships were considered to be statistically significant when $P<0.05$.

3.0 RESULTS AND DISCUSSION

3.1 Preliminary Phytochemical Investigation

Table 2.1 revealed that the primary metabolite carbohydrate is almost present in all the crude extracts of *P. tripartita*. Except hexane crude extract, all the extract contains phenolic compounds. Likewise, tannins are present in all the extracts except hexane extract. Secondary metabolites both flavonoids and terpenoids are present in all the extracts. The phlobatannins is present in ethyl acetate, ethanol, methanol and water extracts, respectively. None of the extracts showed the presence of both alkaloids and saponins. However, the extracts of *P. tripartita* showed the presence of metabolites like carbohydrate, phenolic compounds, tannins, flavonoids, terpenoids and phlobatannins. According to John de Britto *et al.* (2012) *P. biaurita* crude extracts contains steroids, triterpenoids, reducing sugars, alkaloids, phenolics, flavonoids, saponins and tannins. The hydro alcoholic extract of a fern, *Equisetum arvense* showed the presence of tannins, saponins, sterols and flavonoids (Santos *et al.*, 2005 a, b).

3.2 Inhibitory Activity of *P. tripartita* Extracts

The *in vitro* susceptibility of micro-organisms by various extracts (hexane, acetone, chloroform, ethyl acetate, ethanol, methanol and water) and positive control

Table 2.1: Preliminary Phytochemical Analysis of Various Extracts of *Pteris tripartita* Sw.

Plant Extracts	*Hexane*	*Chloroform*	*Acetone*	*Ethyl Acetate*	*Ethanol*	*Methanol*	*Water*
Carbohydrates	+	+	+	+	+	+	+
Alkaloids	–	–	–	–	–	–	–
Saponins	–	–	–	–	–	–	–
Phenolic compounds	–	+	+	+	+	+	+
Tannins	–	+	+	+	+	+	+
Flavonoids	+	+	+	+	+	+	+
Terpenoids	+	+	+	+	+	+	+
Phlobatannins	–	–	–	+	+	+	+

Note: +: Presence, –: Absence.

chloramphenicol were evaluated by various assays such as disc diffusion assay, time course growth assay and MIC methods against different bacteria and fungi. The bacteria include both gram-positive and gram-negative. Plants produce a diverse range of bioactive molecules making them a rich source of different types of medicines. Attention has been focused on phytochemicals as potential sources of functional substances such as antimicrobial substances (Rojas *et al.*, 2003; Duraipandiyan *et al.*, 2006). Moreover, plants can resist parasitic attacks using several defense mechanisms. Plant defense substances belong to a wide range of different chemical classes including flavonoids, terpenoids, alkaloids, steroidal saponins, tannins, phenolic acids, lactones, quinines essential oil and polyphenols (Cowan, 1999).

Significant inhibitions of both gram positive and negative strains were noticed and the diameter of inhibitory zones ranged from 7 mm– 22.33 mm by *P. tripartita* extracts. The positive control showed inhibitory zones of 24 mm to 33.33 mm. Of the seven different extracts of *P. tripartita*, 50 µg/mL of ethanol extract showed highest antibacterial activity against both *S. typhi* (22.33 mm) and *V. cholerae* (21.66 mm), respectively. Banerjee and Sen (1980) also proved *V. cholerae* was significantly inhibited by the extracts of four ferns species namely, *Dryopteris cochleata*, *D. odontoloma*, *Pteris longipes* and *Tectaria macrodonta*. Furthermore, *Adiantum caudatum* and *Ampelopteris prolifera* extracts showed best antibacterial activity against *Staphylococcus aureus*, *Salmonella typhi* and *Bacillus subtilis*. The ethyl acetate crude extract of *P. tripartita* showed inhibition zone against *Shigella sonnei* (7.66 mm, 13.66 mm) and *B. cereus* (7 mm). Ripa *et al.* (2009) observed that the ethyl acetate extract of *Marsilea quadrifolia* showed inhibition zone against both *S. sonnei* and *B. cereus*. The present results directly coincide with earlier study that chloroform and ethyl acetate extracts of a fern, *Blechnum orientale* showed antibacterial activity against gram-positive bacterial strains namely, *Bacillus cereus*, *Micrococcus luteus*, *Staphylococcus aureus* and *Stapylococcus epidermidis* due to the presence of total phenolic contents, flavonoids, terpenoids and tannins (Lai *et al.*, 2010).

Table 2.2 revealed that maximum inhibition zones (16.96 mm and 20 mm) were observed against *P. aeruginosa* in both 25 µg/mL and 50 µg/mL of chloroform crude extract, while chloroform extract showed inhibition zone against *B. subtilis* (8.66 mm), *S. typhi* (7.33 mm, 9.66 mm) and *S. sonnei* (7 mm). Likewise, the chloroform extract of *M. quadrifolia* inhibits the growth of *B. subtilis, S. typhi* and *S. sonnei* (Ripa *et al.*, 2009). On the other hand, 50 µg/mL of ethanol extract showed best inhibitory zone against *K. pneumoniae* (18.33 mm). Both 25 µg/mL of ethanol and methanol extracts demonstrated similar inhibitory zone against *S. typhi* (15.66 mm). The acetone extract inhibited only *V. cholerae* (12.33 mm, 14.33 mm), *E. coli* (7 mm) and *B. megaterium* (7 mm).

Table 2.2: Antibacterial Activity of *P. tripartita* (P.T) Extracts against Various Bacterial Strains

Test Samples	*Zone of Inhibitions (including disc diameter 6 mm)*					
	B. cereus	*B. subtilis*	*B.megaterium*	*S. soneii*	*V. cholerae*	*P. aeruginosa*
P.T extracts (µg/mL)						
Hexane 25	–	–	–	–	9.66±0.33	–
Hexane 50	7.00±0.00	–	–	–	9.33±1.85	–
Acetone 25	–	–	–	–	12.33±1.76	–
Acetone 50	–	–	7.00±0.00	–	14.33±2.72	–
E. acetate 25	7.00±0.00	–	–	7.66±0.66	7.66±0.33	–
E. acetate 50	–	–	–	13.66±0.88	14.00±3.6	7.66±0.66
Chloroform 25	–	–	–	–	–	16.96±0.38
Chloroform 50	–	8.66±1.20	–	7.00±0.00	7.33±0.33	20.00±1.00
Ethanol 25	–	–	–	–	16.33±1.85	–
Ethanol 50	–	7.33±0.33	–	17.33±2.84	21.66±2.90	13.33±1.20
Methanol 25	–	–	–	–	–	–
Methanol 50	–	–	–	11.33±0.88	9.66±2.66	–
Water 25	–	–	–	–	–	–
Water 50	–	–	–	7.00±0.00	–	10.00±0.57
Positive control (10µg/mL)					–	–
Chloramphenicol	33.66±2.18	33.33±0.66	24.66±2.02	28.00±0.00	25.66±0.66	28.66±0.33

The positive control chloramphenicol inhibits both positive and negative bacterial strains. In comparison with positive control, totally ten inhibitory zones by *P. tripartita* extract against *V. cholerae* and nine zones against *K. pneumoniae* were noticed. Of them, highest inhibition zone (18.33 mm) was observed against *K. pneumoniae* with 50 µg of ethanol extract, followed by 50 µg of methanol extract (13.33 mm), 50 µg of ethyl acetate, methanol extracts (10.66 mm, 10 mm), 50 µg of chloroform extract (9.33 mm), 25 µg of chloroform extract (8 mm) and very least inhibition zone was observed with 50 µg of water extract (7.33 mm). Gracelin *et al.* (2012) proved that the methanol extract of *Adiantum caudatum, Angiopteris evecta, Pteris confusa, Pteris argyraea* and

Lygodium microphyllum showed significant inhibitory effects against bacterial strains. Singh *et al.* (2008b) proved that 70 per cent of aqueous methanol extract of *Pteris vittata* showed potent activity against *P. aeruginosa, E. coli, B. cereus* and *Klebsiella pneumoniae,* in addition to that the antimicrobial rutin was also present. Lai *et al.* (2010) proved that ethyl acetate fraction of *Blechnum orientale* inhibited the growth of *B. cereus* with 11 mm zone.

The positive control Chloramphenicol inhibited *S. typhi* (29.33 mm) and 50 µg of ethanol extract inhibited at 22.33 mm of inhibition zone. The minimum inhibition zones were observed in acetone, ethyl acetate, ethanol and methanol extracts against *V. cholerae* (16.33 mm, 14.33 mm, 14 mm, 12.33 mm), *S. sonnei* (17.33 mm, 13.66 mm, 11.33 mm), *P. vulgaris* (13 mm, 10.66 mm), *S. typhi* (11.66 mm, 15.66 mm), *P. aeruginosa* (13.33 mm, 10 mm), *K. pneumoniae* (13.33 mm, 12.33 mm, 11.66 mm, 10.66 mm, 10 mm) and *E. coli* (10.33 mm), respectively (Table 2.3). The *P. aeruginosa, E. coli* and *S. typhi* were susceptible to ethanol and chloroform extracts of *Drynaria quercifolia.* Moreover, gram-negative bacteria were more susceptible to the crude extracts compared to gram-positive bacteria (Kandhasamy *et al.,* 2008). Milovanovic *et al.* (2007) examined that five Serbian *Equisetum* species showed zone of inhibition ranging from 12.10 mm to 17.9 mm against *E. coli* followed by, *K. pneumoniae* (10.6 mm to 18.7 mm) and *S. aureus* (7.8 mm to 15.2 mm) which may be due to the presence of flavonoids and caffeic acid derivatives namely, apigenin, quercetin, kaempferol, caffeoyl shikimic acid,

Table 2.3: Antibacterial Activity of *P. tripartita* Extracts against Various Bacterial Strains

Test Samples	*Zone of Inhibitions (including disc diameter 6 mm)*					
	K. pneumoniae	*S. morcescense*	*E. coli*	*E. aerogens*	*S. typhi*	*P. vulgaris*
P.T extracts (µg/mL)						
Hexane 25	–	–	–	–	–	
Hexane 50	–	–	–	11.66±0.33	8.33±0.88	
Acetone 25	–	–	7.00±0.00	–	–	–
Acetone 50	–	–	7.00±0.00	–	–	–
E. acetate 25	1 0.66±1.45	–	–	–	–	–
E. acetate 50	12.33±2.40					
Chloroform 25	8.00±0.57	–	–	–	7.33±0.33	–
Chloroform 50	9.33±0.33	–	–	–	9.66±1.20	–
Ethanol 25	11.66±2.60	–	–	–	15.66±2.72	10.66±0.88
Ethanol 50	18.33±l.66	–	10.33±2.84	–	22.33± 1.66	13.00±3.21
Methanol 25	10.00±0.57	–	–	–	–	–
Methanol 50	13.33±2.33	–	–	–	15.66±2.33	10.66±0.88
Water 25	–	–	–	–	–	–
Water 50	7.33±0.33	–	–	–	–	–
Positive control (10 µg/mL)						
Chloramphenicol	24.00±0.57	31.00±2.08	26.33±0.88	26.00± 1.00	29.33±1.76	30.00±0.00

monocaffeoyl meso-tartaric acid and dicaffeoyl meso-tartaric acid. The chloroform and ethanol extracts of *P. tripartita* showed inhibition zone against *B. subtilis* with least activity.

In previous study, *Pteris multifida* was proved to be having antimicrobial activity by micro dilution method which showed moderate activity against both gram-positive and gram-negative bacteria. Bioactive compounds such as *sesquiterpenoid, ludongnin V* and *isoneorautenol* were isolated from *P. multifida* root which showed activity against *S. aureus* and *P. aeruginosa* (Hu *et al.,* 2009). The presence of sesquiterpene in *P. tripartita* frond acetone extract using GC-MS was also reported by Baskaran and Jeyachandran (2010). Moreover, hexane and acetone extracts exhibited only least inhibition zones (7 mm) against *E. coli, B. cereus* and *B. megaterium.* Solvents like ethanol, hexane, acetone, chloroform and methanol are generally used to extract bioactive principles from medicinal plants and most of them are able to exhibit inhibitory effects on both gram positive and gram negative bacterial strains (Shyamala Gowri and Vasantha, 2010; Vimala *et al.,* 2012). The secondary metabolites including glycosides, saponins, tannins, flavonoids, terpenoids and alkaloids have been previously reported to have antimicrobial activity (Ebi and Ofoefule, 1997; Okeke *et al.,* 2001; Singh and Singh, 2003; Kaur and Arora, 2009). Moreover, lipophilic flavonoids may also disrupt microbial membranes (Tsuchiya *et al.,* 1996). The chloroform, ethyl acetate, ethanol and methanol crude extracts of *P. tripartita* showed the presence of antimicrobial potent bio-molecule compounds.

3.3 Disc Diffusion Antifungal Assay

Generally, aromatic and medicinal plants are well known to produce certain bioactive compounds which react with other organisms in the environment and inhibiting bacterial or fungal growth by antimicrobial activity (Sengul *et al.,* 2009). The antifungal activity of *P. tripartita* range was measured based on their diameter of inhibition zones (Table 2.4). The results of fungal tests showed that hexane, acetone, ethyl acetate and chloroform crude extracts exhibited weak inhibitory activities (7.33 mm to 7.66 mm) against *P. chrysogenum,* while 50 µg/mL of ethyl acetate exhibited least zone against *A. flavus* (7.33 mm). The ethyl acetate crude extract (50 µg/mL) showed moderate antifungal activity against *P. chrysogenum* (10.33 mm). The methanol extract did not inhibit any of the fungal species which correlate with previous reports (Moulin-Traffort *et al.,* 1990; Garcia *et al.,* 2003; Phongpaichit *et al.,* 2004; Duraipandiyan and Ignacimuthu, 2011). There was no antifungal activity of *P. tripartita* crude extracts against *A. niger, F. oxysporum* and *R. oryzae.*

3.4 Time Course Growth Assay

The hexane, chloroform, acetone, ethyl acetate, ethanol, methanol and water extracts of *P. tripartita* exhibited significant optical density values in time course growth assay (Tables 2.5–2.7). The hexane, acetone and ethyl acetate extracts inhibited the growth of *V. cholerae, S. typhi, P. vulgaris, B. cereus, E. coli, B. megaterium, S. sonnei, P. aeruginosa* and *K. pneumoniae.* The chloroform extract showed optical density (OD) between the ranges of 0.16 to 0.80 against seven different bacterial strains. All the extracts except water extract inhibited *V. cholerae* where the OD ranges between from 0.15 to 0.61. The chloroform extract strongly inhibited the growth of *P. aeruginosa* and

Table 2.4: Antifungal Activities of *P. tripartita* extracts

Test Samples (µg/mL)	*Zone of Inhibitions (including disc diameter 6 mm)*				
	A. niger	*A. flavus*	*F. oxysporum*	*P. chrysogenum*	*R. oryzae*
Hexane 25	–	–	–	–	–
Hexane 50	–	–	–	7.33±0.33	–
Acetone 25	–	–	–	–	–
Acetone 50	–	–	–	7.66±0.33	–
E. acetate 25	–	–	–	–	–
E. acetate 50	–	7.33±0.33	–	10.33±0.33	–
Choloroform 25	–	–	–	7.33±0.33	–
Chloroform 50	–	–	–	7.66±0.33	–
Ethanol 25	–	–	–	–	–
Ethanol 50	–	–	–	–	–
Methanol 25	–	–	–	–	–
Methanol 50	–	–	–	–	–
Water 25	–	–	–	–	–
Water 50	–	–	–	–	–
Positive control (10µg/mL)					
Nystatin	16.36±0.37	14.70±0.51	14.23±0.34	18.80±0.20	13.90±0.40

S. typhi. The ethanol frond extract of *P. tripartita* showed strong inhibition capacity against bacterial strains with the optical density ranging from 0.15 to 0.44. In addition, methanol extract inhibited *S. typhi*, *P. vulgaris*, *S. sonnei* and *K. pneumoniae* between 0.17 to 0.52 OD ranges. The water extract (50 µg/mL) inhibited only three bacterial strains namely, *K. pneumoniae*, *S. sonnei* and *P. aeruginosa* with their optical density between 0.22 to 0.43.

3.5 Minimum Inhibitory Concentration

Generally, MIC values are interpreted as the lowest concentration that inhibits visible microbial growths and expressed in terms of mg/mL (Roy *et al.*, 2010). The assays involving MIC methodology are widely used and an accepted criterion for measuring the susceptibility of organisms to inhibitors (Lambert and Pearson, 2000). The crude extracts of *P. tripartita* showed inhibitory activities against number of microorganisms. Among them, *P. aeruginosa*, *S. typhi*, *P. vulgaris*, *V. cholerae*, *E. coli*, *S. sonnei* and *K. pneumoniae* were found to be more susceptible to *P. tripartita* crude extracts (Tables 2.5–2.7). MIC values of *P. tripartita* chloroform crude extract showed significant inhibitory effect (0.19 µg/mL) against bacterial strain *P. aeruginosa* as compared with the rest (Table 2.6). The MICs of chloroform extract at 25 µg/mL and 50 µg/mL exhibited bactericidal activity ranging from 0.19 µg/mL to 2.45 µg/mL in which, 50 µg/mL of extract inhibited *P. aeruginosa* (0.19 µg/mL) and *S. typhi* (0.48 µg/mL). Basile *et al.* (1997) reported that the absence of bacterial growth inhibition at

Table 2.5: Antibacterial Efficacy of *P. tripartita* Sw. (P.T) Extracts against different Bacterial Strains Using Time Course Growth Assay and Minimum Inhibitory Concentration (MIC)

Test Samples and Bacterial Strains	*Time Course Growth Assay (OD 550 nm) Inhibition of Bacterial Growth at Time Intervals*					*MIC (OD 420 nm) 24 hours*
	0 Hour	*1 Hour*	*2 Hours*	*3 Hours*	*5 Hours*	
P.T extracts (µg/mL)						
Hexane 25 *V. cholera*	0.15±0.00	0.15±0.00	0.15±0.00	0.19±0.00	0.31±0.01	2.21±0.04
Hexane 50 *S. typhi*	0.16±0.01	0.15±0.00	0.15±0.00	0.21±0.00	0.22±0.00	1.65±0.03
Hexane 50 *P. vulgaris*	0.17±0.03	0.15±0.01	0.13±0.00	0.13±0.00	0.17±0.04	1.45±0.02
Hexane 50 *V. cholerae*	0.16±0.00	0.16±0.00	0.17±0.00	0.20±0.00	0.22±0.00	1.96±0.02
Hexane 50 *B. cereus*	0.18±0.04	0.12±0.00	0.12±0.00	0.13±0.00	0.14±0.01	1.53±0.07
Acetone 25 *E. coli*	0.22±0.00	0.24±0.00	0.34±0.01	0.37±0.01	0.50±0.02	1.60±0.09
Acetone 25 *V. cholerae*	0.19±0.00	0.20±0.00	0.21±0.00	0.29±0.00	0.50±0.01	2.06±0.08
Acetone 50 *V. cholerae*	0.25±0.00	0.26±0.00	0.28±0.00	0.35±0.00	0.48±0.03	2.42±0.11
Acetone 50 *E. coli*	0.28±0.02	0.29±0.00	0.39±0.00	0.43±0.00	0.49±0.01	2.41±0.21
Acetone 50 *B. megaterium*	0.30±0.01	0.29±0.01	0.30±0.01	0.35±0.01	0.52±0.03	2.07±0.06
E. acetate 25 *S. soneii*	0.22±0.00	0.25±0.00	0.30±0.00	0.43±0.00	0.64±0.01	1.74±0.01
E. acetate 25 *V. cholerae*	0.24±0.00	0.27±0.00	0.29±0.00	0.37±0.01	0.54±0.03	2.30±0.10
E. acetate 25 *P. aeroginosa*	0.25±0.01	0.23±0.01	0.23±0.01	0.23±0.01	0.23±0.02	1.89±0.10
E. acetate 25 *K. pneumoniae*	0.33±0.02	0.35±0.01	0.42±0.01	0.55±0.00	0.64±0.04	1.91±0.02
E. acetate 25 *B. cereus*	0.22±0.00	0.23±0.00	0.24±0.00	0.35±0.01	0.52±0.02	1.75±0.08
E. acetate 50 *V. cholerae*	0.49±0.00	0.41±0.00	0.37±0.00	0.40±0.00	0.50±0.04	2.61±0.01
E. acetate 50 *K. pneumonia*	0.43±0.02	0.45±0.00	0.53±0.01	0.63±0.02	0.60±0.02	2.35±0.06
E. acetate 50 *S. soneii*	0.30±0.00	0.33±0.01	0.35±0.00	0.49±0.02	0.64±0.05	1.72±0.02

Table 2.6: Antibacterial Efficacy of *P. tripartita* Sw. (P.T) Extracts against different Bacterial Strains Using Time Course Growth Assay and Minimum Inhibitory Concentration (MIC)

Test Samples and Bacterial Strains	*Time Course Growth Assay (OD 550 nm) Inhibition of Bacterial Growth at Time Intervals*					*MIC (OD 420 nm)*
	0 Hour	*1 Hour*	*2 Hours*	*3 Hours*	*5 Hours*	*24 hours*
P.T extracts (µg/mL)						
Chloroform 25 *S. typhi*	0.26±0.00	0.23±0.00	0.25±0.00	0.31±0.00	0.53±0.01	2.31±0.03
Chloroform 25 *P. aeroginosa*	0.16±0.00	0.18±0.00	0.21±0.00	0.22±0.00	0.27±0.00	1.10±0.04
Chloroform 25 *K. pneumoniae*	0.25±0.00	0.34±0.00	0.41±0.01	0.49±0.01	0.62±0.02	2.00±0.06
Chloroform 50 *P. aeroginosa*	0.19±0.00	0.20±0.00	0.28±0.03	0.20±0.00	0.49±0.01	0.19±0.00
Chloroform 50 *S. typhi*	0.26±0.00	0.27±0.00	0.27±0.00	0.33±0.01	0.46±0.01	0.48±0.01
Chloroform 50 *B. subtilis*	0.38±0.00	0.45±0.00	0.48±0.00	0.60±0.01	0.73±0.03	2.11±0.04
Chloroform 50 *S. soneii*	0.40±0.00	0.41±0.01	0.48±0.02	0.40±0.00	0.52±0.01	2.10±0.10
Chloroform 50 *V. cholerae*	0.31±0.01	0.32±0.01	0.35±0.01	0.44±0.02	0.61±0.06	2.45±0.07
Chloroform 50 *K. pneumoniae*	0.53±0.02	0.59±0.02	0.69±0.01	0.78±0.02	0.80±0.06	2.21±0.11
Ethanol 25 *S. typhi*	0.15±0.01	0.16±0.01	0.18±0.01	0.21±0.01	0.27±0.02	1.72±0.03
Ethanol 25 *B. subtilis*	0.26±0.00	0.26±0.00	0.27±0.01	0.29±0.01	0.35±0.02	1.38±0.02
Ethanol 25 *P. vulgaris*	0.23±0.01	0.22±0.01	0.22±0.01	0.22±0.01	0.23±0.01	1.36±0.21
Ethanol 25 *V. cholerae*	0.16±0.00	0.15±0.00	0.15±0.00	0.15±0.00	0.15±0.00	1.02±0.04
Ethanol 25 *K. pneumoniae*	0.29±0.03	0.32±0.03	0.43±0.04	0.44±0.02	0.44±0.01	1.19±0.06
Ethanol 50 *S. typhi*	0.25±0.00	0.23±0.01	0.23±0.01	0.24±0.01	0.24±0.00	1.37±0.12
Ethanol 50 *P. vulgaris*	0.20±0.02	0.20±0.02	0.21±0.02	0.21±0.02	0.22±0.02	0.36±0.02
Ethanol 50 *S. soneii*	0.21±0.00	0.21±0.00	0.23±0.00	0.24±0.01	0.27±0.03	1.10±0.23
Ethanol 50 *V. cholerae*	0.17±0.01	0.16±0.01	0.16±0.01	0.16±0.01	0.17±0.01	0.72±0.04
Ethanol 50 *P. aeroginosa*	0.15±0.00	0.15±0.00	0.16±0.00	0.17±0.00	0.20±0.00	1.13±0.00
Ethanol 50 *K. pneumoniae*	0.21±0.01	0.22±0.01	0.23±0.01	0.24±0.01	0.26±0.01	0.37±0.01

Table 2.7: Antibacterial Efficacy of *P. tripartita* Sw. (P.T) Extracts against different Bacterial Strains Using Time Course Growth Assay and Minimum Inhibitory Concentration (MIC)

Test Samples and Bacterial Strains	*Time Course Growth Assay (OD 550 nm) Inhibition of Bacterial Growth at Time Intervals*					*MIC (OD 420 nm)*
	0 Hour	*1 Hour*	*2 Hours*	*3 Hours*	*5 Hours*	*24 hours*
P.T extracts (µg/mL)						
Methanol 50 *S. typhi*	0.23±0.00	0.24±0.00	0.25±0.00	0.25±0.01	0.24±0.01	0.50±0.01
Methanol 50 *P. vulgaris*	0.17±0.01	0.18±0.01	0.22±0.01	0.26±0.01	0.30±0.02	0.89±0.02
Methanol 50 *V. cholerae*	0.18±0.00	0.20±0.00	0.27±0.00	0.38±0.00	0.45±0.01	1.07±0.04
Methanol 25 *K. pneumoniae*	0.18±0.01	0.26±0.00	0.37±0.00	0.43±0.00	0.52±0.00	1.48±0.00
Methanol 50 *S. soneii*	0.23±0.00	0.24±0.00	0.25±0.00	0.25±0.01	0.24±0.01	0.50±0.01
Methanol 50 *K. pneumoniae*	0.23±0.00	0.25±0.01	0.25±0.00	0.27±0.00	0.26±0.00	0.56±0.01
Water 50 *K. pneumoniae*	0.22±0.01	0.23±0.01	0.23±0.01	0.24±0.01	0.24±0.00	1.37±0.12
Water 50 *S. soneii*	0.24±0.00	0.25±0.00	0.29±0.00	0.40±0.00	0.43±0.01	2.40±0.01
Water 50 *P. aeroginosa*	0.26±0.00	0.26±0.00	0.27±0.01	0.29±0.01	0.35±0.02	1.38±0.02
Positive control (10 µg/mL) (Chloramphenicol)						
S. typhi	0.16±0.00	0.12±0.00	0.49±0.01	0.11±0.00	0.11±0.00	0.34±0.03
B. subtilis	0.16±0.01	0.14±0.00	0.38±0.01	0.13±0.00	0.13±0.00	0.25±0.03
P. vulgaris	0.14±0.00	0.14±0.00	0.13±0.00	0.34±0.00	0.45±0.01	0.89±0.01
S. soneii	0.18±0.00	0.15±0.02	0.44±0.00	0.15±0.02	0.15±0.02	0.30±0.01
V. cholerae	0.16±0.00	0.12±0.00	0.34±0.07	0.12±0.00	0.12±0.00	0.29±0.01
P. aeroginosa	0.16±0.01	0.12±0.00	0.37±0.01	0.13±0.00	0.12±0.00	0.26±0.01
K. pneumoniae	0.17±0.00	0.13±0.00	0.49±0.01	0.13±0.00	0.13±0.00	0.37±0.01
S. marcescens	0.16±0.00	0.12±0.00	0.55±0.02	0.13±0.00	0.13±0.00	0.46±0.03
E. coli	0.18±0.00	0.12±0.00	0.41±0.01	0.12±0.00	0.12±0.00	0.35±0.01
E. aerogense	0.16±0.01	0.12±0.00	0.42±0.02	0.12±0.00	0.12±0.00	0.37±0.00
B. cereus	0.15±0.00	0.12±0.00	0.45±0.00	0.12±0.00	0.12±0.00	0.32±0.02
B. megaterium	0.17±0.00	0.13±0.00	0.49±0.01	0.14±0.00	0.14±0.00	0.44±0.03

highest concentration (1000 µg/mL) of acetone crude extract of *Nephrolepis* species using MIC method against *K. pneumoniae, E. aerogenes* and *P. vulgaris*, but showed antibiotic activity with *E. coli, S. typhi*, and *S. aureus*. In the present study also, acetone extract of *P. tripartita* showed less inhibition activity against the growth of *B. megaterium* and *E. coli* while moderate activity was found against *V. cholerae*. The lowest inhibition effects were observed in hexane, acetone and ethyl acetate crude extracts ranged from 1.45 to 2.61 µg/mL. The ethyl acetate fraction of *B. orientale* showed minimum inhibitory concentration at 250 mg/mL against *B. cereus* (Lai *et al.*, 2010).

The ethanol extract of *P. tripartita* (50 µg/mL) showed best bacterial activities against *P. vulgaris* (0.36 µg/mL), *V. cholerae* (0.72 µg/mL) and *K. pneumoniae* (0.37 µg/mL). The MICs values of both 25 µg/mL and 50 µg/mL concentrations of ethanol extracts exhibited bioactivity ranged from 0.36 to 1.72 µg/mL (Table 2.6). The broad spectrum of antibacterial activity of these extracts was due to the presence of active principle present in the extracts. Among 16 extracts of medicinal plants tested for the antibacterial activity, 10 extracts were not showed antibacterial activity against *B. subtilis* (Janovska *et al.*, 2003). Similarly, very fewer extracts of *P. tripartita* showed inhibitory activity against *B. subtilis*. The methanol crude extract was found to be more susceptible at MIC values ranged from 0.50 to 1.48 µg/mL (Table 2.7). Both methanol extract and flavonoid rutin of *P. vittata* showed less MIC values in *P. aeruginosa, E. coli, B. cereus* and *K. pneumoniae* (Singh *et al.*, 2008b). The presence of flavonoid rutin in *Pteris tripartita* using HPLC was also being reported (Baskaran, 2013). The methanol extracts of *A. philoxeroides, P. obtusa, P. cerasoides* and *I. acuminata* showed best MIC values ranged from 35.25 to 97.25 µg/mL (Rawani *et al.*, 2011). The active principles of crude extracts lose their activity when used in isolation and the bioactivity of crude extract is always assumed to be synergistic effects of mixture of compounds (Tirupathi Rao *et al.*, 2011). The least inhibitory activity was observed in water extract against *K. pneumoniae* (1.37 µg/mL), *S. sonnei* (2.40 µg/mL) and *P. aeruginosa* (1.38 µg/mL). Furthermore, numerous works have been reported on antibacterial effects of secondary metabolites against wide range of bacteria (Ahmad and Beg, 2001; Rodriguez *et al.*, 2009). In particular, hydrolysable tannins have potent antibacterial effects on various bacteria including *B. subtilis* and *S. aureus* (Taguri *et al.*, 2006; Rodriguez *et al.*, 2009; Boulekbache-Makhlouf *et al.*, 2010; 2013). Usually, the bioactive secondary metabolite phenolic compounds can act at two different levels in cell membrane and cell wall of the microorganisms (Taguri *et al.*, 2006). The phenolic compounds can interact with the membrane proteins of bacteria by means of hydrogen bonding through their hydroxyl groups which can result in changes in membrane permeability and cause cell destruction and also penetrate into bacterial cells and coagulate cell content (Tian *et al.*, 2009). Phenolic compounds are known to be synthesized by plants in response to microbial infection (Doughari *et al.*, 2008; Sengul *et al.*, 2009). The results indicated that bioactivity of *P. tripartita* extracts could be due to its richness in phenolic compounds, flavonoids and tannins.

4.0 CONCLUSION

The results of investigation concluded that seven extracts of *P. tripartita* exhibited good antimicrobial activity due to the presence of phenolic compounds, flavonoids and tannins, substantially. However, the results are essential to account the selected ethnomedicinal fern in human health care system and also confirmed the use in traditional folk medicine being safe and helpful to cure numerous disease caused by microbes.

REFERENCES

Ahmad, I., Beg, A. Z., 2001. Antimicrobial and phytochemical studies on 45 Indian medicinal plants against multi-drug resistant human pathogens. *Journal of Ethnopharmacology* 74, 113– 123.

Banerjee, R. D., Sen, S. P., 1980. Antibiotic Activity of Pteridophytes. *Economic Botany* 34(3), 284– 298.

Barry, A. L., Thornsberry, C., 1991. Susceptibility test: diffusion test procedures, In: Balows A, Hazsler WJ, Herrmann KL, Isenberg HD, Shadomy HJ (Eds): Manual of Clinical Microbiology, 5th ed., Am Soc Microbiol, Washington, pp. 463– 474.

Basile, A., Spagnuolo, V., Giordano, S., Sorrentino, C., Lavitola, A., Castaldo-cobianchi, R., 1997. Induction of antibacterial activity by α-D-oligogalacturonides in *Nephrolepis* sp. (Pteridophyta). *International Journal of Antimicrobial Agents* 8(2), 131– 134.

Baskaran, 2013. *DNA sequencing, micropropagation and pharmacological evaluation of Pteris tripartita Sw. – a critically endangered fern*. PhD dissertation, Bharathidasan University, Tiruchirapalli, India.

Baskaran, X., Jeyachandran, R., 2010. Evaluation of Antioxidant and Phytochemical Analysis of *Pteris tripartita* Sw.- a critically endangered fern from South India. *Journal of Fairylake Botanical Garden* 9(3), 28-34.

Besharat, M., Rahimian, M., Besharat, S., Ghaemi, E., 2008. Antibacterial effects of *Adiantum capillus-veneris* ethanolic extract on three pathogenic bacteria *in vitro*. *J. Clin. Diagn. Res.* 2, 1242– 1243.

Bhattacharjee, I., Chetterjee, S. K., Chetterjee, S. N., 2006. Antibacterial potentiality of *Argemone mexicana* solvent extracts against some pathogenic bacteria. *Mem. Ins. Oswaldo. Cruz.* 101, 645– 648.

Boulekbache-Makhlouf, L., Meudec, E., Chibane, M., Mazauric, J. P., Cheynier, V., Slimani, S., Henry, M., Madani, K., 2010. Analysis of phenolic compounds in fruit of *Eucalyptus globulus* cultivated in Algeria by high-performance liquid chromatography diode array detection mass spectrometry. *Journal of Agricultural and Food Chemistry* 58, 12615–12624.

Boulekbache-Makhlouf, L., Slimani, S., Madani, K., 2013. Total phenolic content, antioxidant and antibacterial activities of fruits of *Eucalyptus globulus* cultivated in Algeria. *Industrial Crops and Products* 41, 85– 89.

Cowan, M. M., 1999. Plant products as antimicrobial agents. *Clinical Microbiology Reviews* 12, 564– 582.

Dalli, A. K., Saha, G., Chakraborty, U., 2007. Chracterization of Antimicrobial compounds from a common fern *Pteris biaurita. Ind J Expt. Biol.* 45, 285–290.

Doughari, J. H., El-mahmood, A. M., Tyoyina, I., 2008. Antimicrobial activity of leaf extracts of *Senna obtusifolia* (L.). *African Journal of Pharmacy and Pharmacology* 2, 7–13.

Duraipandiyan, V., Ayyanar, M., Ignacimuthu, M., 2006. Antimicrobial activity of some ethnomedical plants used by Paliyar tribe from Tamil Nadu, India. *BMC complement. Altern. Med.* 6, 35–39.

Duraipandiyan, V., Ignacimuthu, S., 2011. Antifungal activity of traditional medicinal plants from Tamil Nadu, India. *Asian Pacific Journal of Tropical Biomedicine* S204–S215.

Ebi, G. C., Ofoefule, S. I., 1997. Investigating into folkloric antimicrobial activities of *Landolphia owerrience. Phytotherapy Research* 11, 149–151.

Francisco, M. S., Driver, G. C., 1984. Anti-Microbial Activity of Phenolic Acids in *Pteridium aquilinum. American Fern Journal* 74(3), 87–96.

Garcia, V. M. N., Gonzalez, A., Fuentes, M., Aviles, M., Rios, M. Y., Zepeda, G., Rojas, M. G., 2003. Antifungal activities of nine traditional Mexican medicinal plants. *J Ethnopharmacol.* 87, 85–88.

Gracelin, D. H. S., John De Britto, A., Benjamin, P., Kumar, J. R., 2012. Antibacterial screening of a few medicinal ferns against antibiotic resistant phyto pathogen. *International Journal of Pharmaceutical Sciences and Research* 3(3), 868–873.

Guha, P., Mukhopadhyay, R., Gupta, K., 2005. Antifungal activity of the crude extracts and extracted phenols from gametophytes and sporophytes of two species of *Adiantum. Taiwania* 50, 272–283.

Guha, P., Mukhopadhyay, R., Pal, P. K., Gupta, K., 2004. Antimicrobial activity of crude extracts and extracted phenols from gametophyte and sporophytic plant part of *Adiantum capillus-veneris* Linn. *Allopathy J.* 1, 57–66.

Hao-bin, H., Hong, C., Yu-feng, J., Xu-dong, Z., Jian-xin, L., 2008. Chemical constituents and antimicrobial activities of extracts from *Pteris multifida. Chemistry of Natural Compounds* 44(1), 106–108.

Haripriya, D., Selvan, N., Jeyakumar, N., Periasamy, R., Johnson, M., Irudayaraj, V., 2010. The effect of extracts of *Selaginella involvens* and *Selaginella inaequalifolia* leaves on poultry pathogens. *Asian Pacific Journal of Tropical Medicine* 3(9), 678–681.

Herin Sheeba Gracelin, D., John De Britto, A., Benjamin Jeya Rathna Kumar, P., 2012. Antibacterial screening of a few medicinal ferns against antibiotic resistant phyto pathogen. *International Journal of Pharmaceutical Sciences and Research* 3(3), 868–873.

Hu, H., Zheng, X., Hu, H., Cao, H., 2009. Three antibacterial compounds from the roots of *Pteris multifida. Chemistry of Natural Compounds* 45(1), 45–48.

Iennette, E. H., 1985. *Manual of clinical microbiology*. 4th (ed). American Association for Microbiology, Washington. pp. 978– 987.

Janovska, D., Kubikova, K., Kokoska, L., 2003. Screening for antibacterial activity of some medicinal plants species of traditional Chinese medicine. *Czech J. Food Sci.* 21(3), 107–110.

John De Britto, A., Herin Sheeba Gracelin, D., Benjamin Jeya Rathna Kumar, P., 2012. *Pteris biaurita* L.: A potential antibacterial fern against *Xanthomonas* and *Aeromonas* bacteria. *Journal of Pharmacy Research* 5(1), 678– 680.

Joksic, G., Stankovic, M., Novak, A., 2003. Antibacterial medicinal plants *Equiseti herba* and *Ononidis radix* modulate micronucleus formation in human lymphocytes *in vitro. J Environ Pathol Toxicol Oncol.* 22, 41– 48.

Kandhasamy, M., Arunachalam, K. D., Thatheyus, A. J., 2008. *Drynaria quercifolia* (L.) J.Sm: A potential resource for antibacterial activity. *African Journal of Microbiology Research* 2, 202– 205.

Kaur, G. J., Arora, D. S., 2009. Antibacterial and phytochemical screening of *Anethum graveolens, Foeniculum vulgare* and *Trachyspermum ammi. BMC Compl. Altern. Med.* 9, 30.

Kumar, R. S., Sivakumar, T., Sundaram, R. S., Sivakumar, P., Nethaji, R., Gupta, M., Mazumdar, U. K., 2006. Antimicrobial and Antioxidant Activities of *Careya arborea* Roxb. Stem Bark. *Iranian Journal of Pharmacology and Therapeutics* 5, 35– 41.

Lai, H. Y., Lim, Y. Y., Kim, K. H., 2010. *Blechnum orientale* Linn - a fern with potential as antioxidant, anticancer and antibacterial agent. *BMC Complementary and Alternative Medicine* 10(15), 1– 8.

Lambert, R. J. W., Pearson, W., 2000. Susceptibility testing: accurate and reproducible minimum inhibitory concentration (MIC) and non-inhibitory concentration (NIC) values. *Journal of Applied Microbiology* 88, 784– 790.

Maridass, M., 2009. Antibacterial Activity of *Mecodium exsertum* (Wall.ex Hook) Copel –A Rare Fern. *Pharmacology online* 1, 1– 7.

Maruzzella, J. C., 1961. Antimicrobial substances from ferns. *Nature* 191, 518.

McCutcheon, A. R., Roberts, T. E., Gibbons, E., Ellis, S. M., Babiuk, L. A., Hancock, R. E. W., Towers, G. H. N., 1995. Antiviral screening of British Columbian medicinal plants. *J. Ethnopharmacol.* 49, 101– 110.

Milovanovic, V., Radulovic, N., Todorovic, Z., Stankovic, M., Stojanovic, G., 2007. Antioxidant, Antimicrobial and Genotoxicity Screening of Hydro-alcoholic Extracts of Five Serbian *Equisetum* Species. *Plant Foods Hum Nutr.* 62, 113– 119.

Moulin-Traffort, J., Giordani, R., Regli, P., 1990. Antifungal action of latex saps from *Lactuca sativa* and *Asclepias curassavica. Mycoses* 33, 383– 392.

Niranjan Reddy, V. L., Ravikanth, V., Prabhakar Rao, T., Diwan, P. V., Venkateswarlu, Y., 2001. A new triterpenoid from the fern *Adiantum lunulatum* and evaluation of antibacterial activity. *Phytochemistry* 56(2), 173– 175.

Okeke, M. I., Iroegbu, C. U., Eze, E. N., Okoli, A. S., Esimone, C. O., 2001. Evaluation of extracts of the root of *Landolphia owerrience* for antibacterial activity. *Journal of Ethnopharmacology* 78, 119– 127.

Palombo, E. A., Semple, S. J., 2001. Antibacterial activity of traditional Australian medicinal plants. *Journal of Ethnopharmacology* 77, 151– 157.

Parekh, J., Chanda, S., 2006. Screening of some Indian medicinal plants for antibacterial activity. *Indian J Pharm Sci* 68, 835– 838.

Parekh, J., Chanda, S., 2007. Antibacterial and phytochemical studies on twelve species of Indian medicinal plants. *Afr J Biomed Res.* 10, 175– 181.

Parihar, P., Parihar, L., Bohra, A., 2006. Antibacterial activity of *Athyrium pectinatum* (Wall.) Presl. *Natural Product Radiance* 5(4), 262– 265.

Phongpaichit, S., Pujenjob, N., Rukachaisirkul, V., Ongsakul, M., 2004. Antifungal activity from leaf extracts of *Cassia alata* L., *Cassia fistula* L. and *Cassia tora* L. *Songklankarin J Sci Tech.* 26, 741– 748.

Rabe, T., Staden, V., 1997. Isolation of an antibacterial sesquiterpenoid from *Warbugia salutaris. Journal of Ethnopharmacology* 73, 171– 174.

Radulovic, N., Stojanovic, G., Palic, R., 2006. Composition and antimicrobial activity of *Equisetum arvense* L. essential oil. *Phytother Res.* 20, 85– 88.

Rawani, A., Pal, S., Chandra, G., 2011. Evaluation of antimicrobial properties of four plant extracts against human pathogens. *Asian Pacific Journal of Tropical Biomedicine* S71–S75.

Ripa, F. A., Nahar, L., Haque, M., Islam, M. M., 2009. Antibacterial, Cytotoxic and Antioxidant Activity of Crude Extract of *Marsilea quadrifolia. European Journal of Scientific Research* 33(1), 123– 129.

Rodriguez, H., Curiel, J. A., Landete, J. M., De las Rivas, B., De Felipe, F. L., Gomez-Cordoves, C., Mancheno, J.M., Munoz, R., 2009. Food phenolics and lactic acid bacteria. *International Journal of Food Microbiology* 132, 79– 90.

Rojas, R., Bustamante, B., Bauer, J., Fernandez, I., Albn, J., Lock, O., 2003. Antimicrobial activity of selected Peruvian medicinal Plants. *J. Ethnopharmocol.* 88, 199– 204.

Rosoanaivo, Ratsimanaga- Urverg, 1993. Biological evaluation of plants with reference to the Malagasy flora. Monograph for the IFs. NAPRECA Workshop on Bioassays. Antananavivo, Madagascar, 72– 79.

Roy, S., Rao, K., Bhuvaneswari, C., Giri, A., Mangamoori, L. N., 2010. Phytochemical analysis of *Andrographis paniculata* extract and its antimicrobial activity. *World J Microbiol Biotechnol.* 26, 85– 91.

Santos, J. G. J., Blanco, M. M., Monte, F. H. M., Russi, M., Lanziotti, V. M. N. B., Leal, L. K. A. M., Cunha, G. M., 2005a. Sedative and anticonvulsant effects of hydroalcoholic extract of *Equisetum arvense. Fitoterapia* 79(6), 508– 513.

Santos, J. G. J., Monte, F. H. M., Blanco, M. M., Lanziotti, V. M. N. B., Maia, F. D., Leal, L. K. A. M., 2005b. Cognitive enhancement in aged rats after chronic administration of *Equisetum arvense* L. with demonstrated antioxidant properties *in vitro*. *Pharmacology, Biochemistry and Behaviour* 81, 593– 600.

Sengul, M., Yildiz, H., Gungor, N., Cetin, B., Eser, Z., Ercili, S., 2009. Total phenolic content, antioxidant and antimicrobial activities of some medicinal plants. *Pakistan Journal of Pharmaceutical Sciences* 22, 102– 106.

Shyamala gowri, S., Vasantha, K., 2010. Free radical scavenging and antioxidant activity of leaves from Agathi (*Sesbania grandiflora*) (L.) Pers. *Am-Euras. J. Sci. Res.* 5, 114– 119.

Singh, B., Singh, S., 2003. Antimicrobial activity of terpenoids from *Trichodesma amplexicaule* Roth. *Phyto. Res.* 17(7), 814– 816.

Singh, M., Govindarajan, R., Rawat, A. K. S., Khare, P. B., 2008b. Antimicrobial Flavonoid Rutin from *Pteris vittata* L. against Pathogenic Gastrointestinal Microflora. *American Fern Journal* 98(2), 98– 103.

Singh, M., Singh, N. and Khare, P. B., Rawat, A. K. S., 2008a. Antimicrobial activity of some important *Adiantum* species used traditionally in indigenous systems of medicine. *J. Ethnopharmacol.* 115(2), 327– 329.

Soare, L. C., Ferdes, M., Deliu, I., Gibea, A., 2012. Studies regarding the antibacterial activity of some extracts of native Pteridophytes. *U.P.B. Sci. Bull., Series B.* 74(1), 21–26.

Taguri, T., Tanaka, T., Kouno, I., 2006. Antibacterial spectrum of plant polyphenols and extracts depending upon hydroxyphenyl structure. *Biological and Pharmaceutical Bulletin* 29, 2226– 2235.

Taylor, R. S. L., Manandhar, N. P., Hudson, J. B., Towers, G. H. N., 1995. Screening of selected medicinal plants of Nepal for antimicrobial activities. *J. Ethnopharmacol.* 546, 153– 159.

Tian, F., Li, B., Ji, B., Zhang, G., Luo, Y. 2009. Identification and structure–activity relationship of gallotannins separated from *Galla chinensis*. *LWT–Food Science and Technology* 42, 1289– 1295.

Tirupathi Rao, G., Suresh Babu, K., Ujwal Kumar, J., Sujana, P., Veerabhadr Rao, A., Sreedhar, A. S., 2011. Anti-microbial principles of selected remedial plants from Southern India. *Asian Pacific Journal of Tropical Biomedicine* 298– 305.

Tsuchiya, H., Sato, M., Miyazaki, T., Fujiwara, S., Tanigaki, S., Ohyama, M., Tanaka, T., Iinuma, M., 1996. Comparative study on the antibacterial activity of phytochemical flavanones against methicillin-resistant *Staphylococcus aureus*. *J. Ethnopharmacol.* 50(1), 27– 34.

Vaghasiya, Y., Nair, R., Baluja, S., Chanda, S., 2008. Antibacterial and preliminary phytochemical analysis of *Eucalyptus citriodora* Hk. Leaf. *Nat Prod Res* 22, 754– 762.

Vimala, T. A., Johnson, M., Solomon, J., 2012. Anti-bacterial studies on *Hemigraphis colorata* (Blume) H.G. Hallier and *Elephantopus scaber* L. *Asian Pacific Journal of Tropical Medicine* 5(1), 52– 57.

2015, Modern Methods in Phytomedicine *Pages* ***33–57***
Editor: **T. Parimelazhagan**
Published by: **DAYA PUBLISHING HOUSE, NEW DELHI**

3

Antioxidant and Antipyretic Activities of *Rubus fairholmianus* Gard.

Blassan P. George, S. Harini and T. Parimelazhagan

Bioprospecting Laboratory, Department of Botany, Bharathiar University, Coimbatore – 641 046, Tamil Nadu

1.0 INTRODUCTION

Antioxidants or "free radical scavengers" are nutrients as well as enzymes that are believed to play a vital role in preventing the development of chronic diseases such as cancer, heart disease, Alzheimer's, diabetics etc. by blocking or slow down the oxidation process by neutralizing free radicals. The antioxidant agents are found in foods such as vegetables and fruits. Well known antioxidants, such as Vitamin E (α-tocopherol), Vitamin C and polyphenols/flavonoids, have been investigated for their possible use to prevent the diseases (Nunez-Selles, 2005). The link between free radicals and disease processes led to considerable research to develop nontoxic drugs that can scavenge the free radicals. Several plant extracts and products have been shown to possess significant antioxidant potential (Sabu and Kuttan, 2003; Halliwell, 1995). An antioxidant compound inhibits or delays the oxidation of substrates even if the compound is present in a significantly lower concentration than the oxidized substrate. The scavenging of reactive oxygen species (ROS) is one of possible mechanism of action. Others include the prevention of ROS formation by metal binding or enzyme inhibition. The antioxidant compounds can be recycled in the cell or are irreversibly damaged, but their oxidation products are less harmful or can be further converted to harmless substances (Halliwell, 1995; Halliwell and Gutteridge, 2007).

Natural products are believed to be important sources of new chemical substances which have potential therapeutic effects. Medicinal plants are one of the important sources are extensively investigated both *in vitro* and *in vivo* to examine for their potential activities. Most people living in developing countries are almost completely dependent on traditional medicinal practices for their primary health care needs and higher plants are known to be the main source of drug therapy in traditional medicine (Calixo, 2005). The available analgesic drugs exert a wide range of side effects and either too potent or too weak, however the search for new analgesic compound has been a priority of pharmacologists and pharmaceutical industries (Mattison *et al.*, 1998). Drugs of natural origin continue to be important for the treatment of many diseases worldwide and are believed to be an important source of new compounds. Many medicinal plants have been investigated for their antinociceptive, antipyretic and anti-inflammatory activities.

The genus *Rubus* is very diverse, includes over 750 species in 12 subgenera, and is found on all continents except Antarctica (Finn, 2008). Due to useful Ethnomedicinal and pharmacological properties; *Rubus* species has been used in folk medicine (Patel *et al.*, 2004). *R. fairholmianus* (*R. moluccanus* L.) leaf extract is taken early in the morning to reduce headache by Koch- Rajbongshi and Rangia tribes (Das *et al.*, 2006). The leaves possess insecticidal properties, fruits are edible and stimulant (Barukial, 2011). Use of medicinal plants as a source of relief and cure from various illness is as old as humankind itself. Even today, medicinal plants provide a cheap source of drugs for majority of world's population. Plants have provided and will continue to provide not only directly usable drugs, but also a great variety of chemical compounds that can be used as a starting points for the synthesis of new drug with improved pharmacological properties (Ballabh *et al.*, 2008).

Even though this plant has immense ethnomedicinal value; a survey of literature revealed that the antioxidant and analgesic properties of this plant using animal models have not yet been evaluated fully. Keeping this in view, the main objectives of the present study was to carry out the *in vitro* antioxidant assays and *in vivo* analgesic activity using Eddy's hot plate mediated pain reaction in rats to put forward a scope to develop an effective drug from *R. fairholmianus*.

2.0 EXPERIMENTAL METHODS

2.1 Plant Collection and Extraction

The fresh plant parts of *R. fairholmianus* were collected from Marayoor Shola forest, Kerala, India, during the month of September 2010. The collected plant material was identified and authenticated by (Voucher specimen No. BSI/SRC/5/23/2010-11/Tech. 1657) Botanical Survey of India, Southern circle, Coimbatore, Tamil Nadu. The powdered leaf, stem and root were extracted successively using Soxhlet apparatus. The extracts were concentrated to dryness under reduced pressure in a rotary evaporator to yield dried petroleum ether, chloroform, acetone and methanol.

2.2 Chemicals

6-hydroxy-2,5,7,8-tetramethylchromane-2-carboxylic acid (Trolox), 2,2'-azinobis(3-ethyl-benzothiozoline-6-sulfonic acid disodium salt (ABTS) from Sigma Chemical Co., Ethylene diamine tetra-acetic acid (EDTA) disodium salt, potassium persulfate, Butylated hydroxy toluene (BHT), Butylated hydroxy anisole (BHA) and α-tocopherol were obtained from Merck India Ltd. All the other chemicals and solvents used were of analytical grade.

2.3 *In vitro* Antioxidant Activities

2.3.1 Trolox Equivalent Antioxidant Capacity (TEAC) Assay

The total antioxidant activity of the samples was measured by ABTS radical cation decolourization assay according to Re *et al.* (1999).

ABTS was dissolved in water to a 7 mM concentration. ABTS radical cation ($ABTS^{\bullet+}$) was produced by reacting ABTS stock solution with 2.45 mM potassium persulfate (final concentration) and allowing the mixture to stand in the dark at room temperature for 12–16 h before use. Because ABTS and potassium persulfate react stoichiometrically at a ratio of 1:0.5, this will result in incomplete oxidation of the ABTS. Oxidation of the ABTS commenced immediately, but the absorbance was not maximal and stable until more than 6 h had elapsed. The radical was stable in this form for more than two days when stored in the dark at room temperature. For the study of phenolic compounds and food extracts, the $ABTS^{\bullet+}$ solution was diluted with ethanol and for plasma antioxidants with PBS, pH 7.4, to an absorbance of 0.70 (60.02) at 734 nm and equilibrated at 30°C.

Triplicate determinations were made at each dilution of the standard, and the percentage inhibition was calculated against the blank (ethanol) with an absorbance at 734 nm and then was plotted as a function of Trolox concentration. The unit of total antioxidant activity (TAA) is defined as the concentration of Trolox having equivalent antioxidant activity expressed as μM/g sample extracts.

2.3.2 Metal Chelating Activity

The chelation of ferrous ions by various samples was estimated by Dinis *et al.* (1994). Briefly the extracts (100 μl) were added to a solution of 2 mmol/l $FeCl_2$ (0.05 mL). The reaction was initiated by the addition of 5 mmol/l ferrozine (0.2 mL) and the mixture was shaken vigorously and left standing at room temperature for 10 min. Absorbance of the solution was then measured spectrophotometrically at 562 nm. The chelating activity of the extracts was evaluated using EDTA as standard. The results were expressed as mg EDTA equivalent/g extract. Ferrozine can quantitatively form complex with Fe^{2+}, in the presence of other chelating agents, the complex formation is disrupted which results in the decreased intensity of the red colour of the complex. The metal chelating capacities of the extracts were evaluated using the following equation:

Metal chelating capacity (per cent) = (Control OD – Sample OD)/Control OD × 100

2.4 Animals and Acute Toxicity Study

Healthy Swiss albino mice (25-30g) of either sex and of approximately the same age were used for the study. They were fed with standard chow diet and water *ad libitum* and were housed in polypropylene cages in a well maintained and clean environment. The experimental protocol was subjected to scrutiny of institutional animal ethical committee for experimental clearance (KMCRET/Ph.D/03/2011).

The acute toxicity was performed as per Organization for Economic Co-operation and Development guidelines (OECD guidelines 423, 2001). Swiss albino mice were used to assess the toxicity level. The root acetone extract at dose of 100, 500, 1000 and 2000 mg/kg was administered to 3 male and 3 female mice in a single dose orally. The mice were fasted 3 h prior to the dosage. Animals are observed individually after drug administration at least once during the first 30 minutes, periodically during the first 24 h, with special attention given during the first 4 h, and daily thereafter, for a total of 14 days.

2.5 Analgesic Activity

2.5.1 Eddy's Hot Plate Mediated Pain Reaction

The hot-plate test was performed to measure response latencies according to the method described by Eddy and Leimback (Eddy and Leimback, 1953). Male Swiss albino mice were divided into four groups of six animals each. Group 1 served as control; group 2 served as standard which received morphine (10 mg/kg); group 3 and 4 served as plant extract at a dose of 200 mg/kg and 400 mg/kg respectively. The animals were placed on the hot plate, maintained at (55±1) °C. The pain threshold is considered to be reached when the animals lift and lick their paws or attempt to jump out of the hot plate. The time taken for the mice to react in this fashion was obtained using a stopwatch and noted as basal reaction time (0 min). A latency period of 15 seconds (cut-off) was defined as complete analgesia and the measurement was terminated if it exceeded the latency period in order to avoid injury (Awaad *et al.*, 2011). The reaction time was reinvestigated at 30, 60 and 120 min after the treatment and changes in the reaction time were noted.

3.0 RESULTS AND DISCUSSION

3.1 Antioxidant Activities

3.1.1 Trolox Equivalent Antioxidant Capacity (TEAC) Assay

The results of $ABTS^{\cdot+}$ cation radical scavenging activities are presented in Table 3.1. The acetone extracts of root showed significant ($p<0.001$) radical scavenging activity (14431.42 µM TE/g) followed by stem methanol and leaf acetone (9146.20 and 8687.20 µM TE/g extract). ABTS radical is used to study the radical scavenging effects of different organic extracts because TEAC is operationally simple. Since, the total phenolic content in different extracts of *R. fairholmianus* seems appreciable, the total antioxidant activity (TAA) of such samples are sufficient enough for functioning as potential neutraceuticals when they are ingested along with nutrients.

Similar activities were observed in *R. sanctus* plant extracts (Motamed and Naghibi, 2010). Cai *et al.*, 2004 reported the total antioxidant capacity TEAC values of the methanolic and aqueous extracts of *R. chingii* fruits, which were found to be 946.1 and 817.0 (µmol Trolox/100g DW). The DPPH and ABTS radical scavenging activities of *R. ulmifolius* (TEAC 3.8 ± 0.3mM Trolox and DPPH 5.10 ± 0.5 µg/mL) were reported by Dall'Acqua *et al.* (2008). Raspberry (*R. idaeus*) leaves, collected in different locations of Lithuania were extracted with ethanol and were tested for their antioxidant activity by using ABTS and DPPH scavenging methods. All extracts were active, with radical scavenging capacity at the used concentrations from 20.5 to 82.5 per cent in DPPH reaction system and from 8.0 to 42.7 per cent in ABTS reaction (Venskutonis *et al.*, 2007).

Table 3.1: ABTS Radical Scavenging and Metal Chelating Activities of *R. fairholmianus*

Plant/Part	*Extracts*	*TEAC (µM Trolox Equivalents/g extract)*	*Metal Chelating (µg EDTA Equivalents/mg extract)*
Rubus fairholmianus/	Pet. Ether	971.99±21.46[c]	1.67±0.52[d]
Leaf	Chloroform	749.25±11.50[c]	4.47±0.37[c]
	Acetone	8687.20±9.24[a]	8.61±0.37[b]
	Methanol	6722.96±8.04[b]	8.76±0.32[b]
Rubus fairholmianus/	Pet. ether	270.00±5.09[c]	0.74±0.27[e]
Stem	Chloroform	587.25±8.27[c]	3.51±0.14[d]
	Acetone	3766.48±4.10[b]	9.41±0.37[b]
	Methanol	9146.20±9.63[a]	7.14±1.02[c]
Rubus fairholmianus/	Pet. ether	783.00±54.81[c]	1.88±0.27[c]
Root	Chloroform	810.00±7.15[c]	3.00±0.49[c]
	Acetone	14431.42±6.76[a]	10.40±0.25[b]
	Methanol	8140.45±4.31[b]	9.36±0.44[b]

Values are mean of triplicate determination (n=3) ± standard deviation.

EDTA: Ethylene Diamine Tetra Acetic Acid. TEAC: Trolox Equivalent Antioxidant Capacity p<0.001, b- p<0.01, c- p<0.05, d- not significant compared with standard.

3.1.2 Metal Chelating Activity

The metal chelating activities of *R. fairholmianus* were given in Table 3.1. Methanol, hot water and acetone extracts showed better scavenging ability in all the parts. The metal chelating capacity of acetone extracts of root, stem and methanol extract of root were found to be 10.40, 9.41and 9.36µg EDTA Equivalents/mg extract respectively. Metal chelating capacity was significant (p<0.05 to p<0.001) as they reduced the concentration of the catalyzing transition metal in lipid peroxidation. It was reported that the chelating agents which form α-bonds with a metal are effective as secondary antioxidants because they reduce the redox potential, thereby stabilizing the oxidized form of the metal ion (Duh *et al.*, 1999). Antioxidants inhibit interaction between metal and lipid through formation of insoluble metal complexes with ferrous ion.

Hence, the results obtained for *R. fairholmianus* reveals that some of the extracts demonstrate an effective capacity for iron binding, suggesting that its action as antioxidant may be related to its iron binding capacity.

3.2 Acute Toxicity

In the acute toxicity studies; four groups of mice were administered with root acetone extract in graded doses of 100, 500, 1000 and 2000 mg/kg p.o., respectively. The animals were kept under observation for the change in behavior or death up to 14 days following the drug administration. The extract administration neither caused any significant change in the behaviors nor the death of animals in all the test groups. This indicates that the root acetone extract of *R. fairholmianus* was safe up to a single dose of 2000 mg/kg body weight. Hence 200 and 400 mg/kg oral doses were selected for analgesic study.

3.3 Analgesic Activity

3.3.1 Eddy's Hot Plate Mediated Pain Reaction

As shown in Table 3.2, the root acetone extract produced significant analgesic activity in a dose-dependent manner. In this model, the higher dose (400 mg/kg) prolonged significantly the reaction time of animal with relatively extended duration of stimulation. At the higher dose level; the animals could withstand on the hot plate for 11.4, 12.2 and 12.5 seconds at 30, 60 and 120 min reaction time which was the highest and comparable with that of the reference drug morphine10mg/kg (7.8, 9.6 and 12.4 sec.). The basal reaction time of the high dose and standard drug were 6.8 and 5.8 seconds. The anti-inflammatory, analgesic and antipyretic activities of the related species of *Rubus* (*R. ellipticus* and *R. niveus*) have been reported by George *et al.* (2013a, 2013b).

Table 3.2: Effect of *R. fairholmianus* Root Acetone Extract on Hot Plate Mediated Pain Reaction

Groups	*Basal Reaction Time 15 min cut off (After drug administration)*	*Reaction Time 15 min cut off (After drug administration)*		
		30 min	*60 min*	*120 min*
Control	7.2±0.23	9.4±0.43	8.8±0.53	9.8±0.6
Morphine (10mg/kg)	5.8±0.87	7.8±0.34***	9.6±0.8***	12.4±0.1***
RFRA (200mg/kg)	6.4±0.7	6.9±0.78	7.2±0.55*	8.0±0.65**
RFRA (400mg/kg)	6.8±0.5	11.4±1.9**	12.2±1.5**	12.5±1.2**

Values are expressed as mean ± SEM. (n=6), significantly different at * $p<0.05$, ** $p<0.01$, *** $p<0.001$ when compared to control. RFRA - *R. fairholmianus* Root Acetone.

The classic hot plate model was followed to evaluate the analgesic activity of *R. fairholmianus* root acetone extract. The hot plate model has been found suitable to investigate central antinociceptive activity because of several advantages, particularly the sensitivity to antinociceptives and limited tissue damage (Kou *et al.,* 2005). Proinflammatory mediators like prostaglandins and bradykinins were suggested to play an important role in analgesia (Vinegar *et al.,* 1969). Pain is an unpleasant sensory and emotional experience associated with actual or potential tissue damage, or described in terms of such damage (Nicholas and Moore, 2009). Pain may felt because of inflammation, infection, tissue necrosis, chemicals or burn. In the stomach and intestines, pain may result from inflammation of mucosa of from distension or muscle spam. Depending on the cause, pain may be sudden or short term marked primary by reflex withdrawal (Gould, 2002). The pain induced by thermal stimuli included hot plate and tail flick tests is known to be selective to centrally but not peripherally acting analgesics (Chau, 1989). Hot plate and tail flick tests are considered to be the specific tests of the evaluation of the central pain (Marchiaro *et al.,* 2005) at a supraspinal and spinal levels (Wong *et al.,* 1994), respectively. RFRA extract increased the latency of nociceptive responses in hot plate test. Lower dose of RFRA was less potent in analgesic activity than higher dose as indicated by a significant delay response against pain at lower dose in mice while higher dose exhibited more potent effects and had a rapid analgesic action.

The hot plate method is one of the most common tests used for evaluating the analgesic efficacy of drugs in rodents. The drug that reduces the nociceptive response indicated by cutaneous thermal stimuli in the hot plate test might exhibit central analgesic properties or supraspinal analgesia (Matheus *et al.,* 2005). Thermic painful stimuli are known to be selective centrally, but not peripherally acting analgesic drugs (Chau, 1989). Therefore the hot plate test which is usually used to determine the involvement of central nociceptive mechanism has, at least, conformed the ability of FFRA to influence the central mechanism as seen in other analgesic models such as formalin or acetic acid induced writhing test. The antinociceptive action of RFRA was partly mediated by opioid mechanism. This could be due to direct agonist activity of opioididomimetric constituents in RFRA and due to the increase release of endogenous opioid peptides (Deraniyagala *et al.,* 2003).

The obtained results confirmed that root acetone extract at the dose 200 and 400mg/kg has a central analgesic effect, which was compared with reference drug (Aspirin 100mg/kg). The analgesic effect of *R. fairholmianus* might be attributed to the inhibition of the synthesis of some pro-inflammatory mediators, such as prostaglandins and cytokines.

4.0 CONCLUSION

The present study confirms the promising antioxidant activities of *R. fairholmianus.* Five different extracts from the leaf, stem and root of this plant were screened for *in vitro* antioxidant potentials and one which gave maximum activity was selected for testing the *in vivo* analgesic activity. Among the 15 extracts tested for *in vitro* antioxidant assays, RFRA (*R. fairholmianus* root acetone) extracts showed superior activities in ABTS and metal chelating activities. Therefore RFRA was taken for acute oral toxicity and *in vivo* analgesic screening using Eddy's hot plate mediated pain reaction in rats at 200 and 400 mg/kg b. wt concentrations. RFRA extracts higher dose showed superior activity compared to lower dose. This antioxidant and analgesic effect could be due to the phenolic/flavonoid constituents. The increased exposure to free radicals or the impaired efficiency of the protective enzymes and molecules may lead to many free radicals generated diseases including cancer. Further studies are warranted in order to characterize the exact compound responsible for the antioxidant activity. In summary *R. fairholmianus* can act as a valuable natural antioxidant and has immense scope as an effective source to fight against the free radical generated diseases. It was inferred from the results RFRA can act as a natural drug which reduces pain reactions in rat.

REFERENCES

Awaad, A.S., El-meligy, R.M., Qenawy, S.A., Atta, A.H., Soliman, G.A., 2011. Anti-inflammatory, antinociceptive and antipyretic effects of some desert plants. *JSCS* 15, 367–373.

Ballabh, B., Chaurasia, O.P., Amed, Z., Singh, S.B., 2008. Traditional medicinal plants of cold desert Ladakh-Used against kidney and urinary disorders. *Food Chem.* 118, 331-339.

Barukial, J., Sarmah, J.N., 2011. Ethnomedicinal plants used by the people of Golaghat district, Assam, India. *Int J* Med *Arom Plants* 1, 203-211.

Cai, Y., Luo, Q., Sun, M., Corke, H., 2004. Antioxidant activity and phenolic compounds of 112 traditional Chinese medicinal plants associated with anticancer. *Life Sci.* 74, 2157–2184.

Calixo, J.B., 2005. Twenty five years of research on medicinal plants in latin America. A personal view. *J Ethnopharmacol.* 100, 131-134.

Chau, T., 1989. Pharmacology methods in the control of inflammation. In modern method in pharmacology, Vol V, New York, USA. Alan R. liss, pp.195-212.

Dall' Acqua, S., Cervellati, R., Loi, M.C., Innocenti, G., 2008. Evaluation of *in vitro* antioxidant properties of some traditional Sardinian medicinal plants: Investigation of the high antioxidant capacity of *Rubus ulmifolius. Food Chem.* 106, 745–749.

Das, N.J., Saikia, S.P., Sarkar, S., Devi, K., 2006. Medicinal plants of north kamrup district of Assam used in primary health care system. *IJTK* 5, 489-493.

Deraniyagala, S.A., Ratnasoorya, W.D., Goonasekara, C.L., 2003. Antinociceptive effect and toxicological study of the aqueous bark extract of *Barrangtinia recemosa* on rats. *J Ethnopharmacol* 86, 21- 26.

Dinis, T.C.P., Madeira, V.M.C., Almeida, L.M., 1994. Action of phenolic derivatives (acetoaminophen, salycilate and 5-aminosalyciliate) as inhibitors of membrane lipid peroxidation and as peroxyl radical scavengers. *Arch Biochem Biophys* 315, 161-169.

Duh, P.D., Tu, Y.Y., Yen, G.C., 1999. Antioxidant activity of water extract of Harng Jyur (*Chrysanthemum morifolium* Ramat). *Food sci technol* 32, 269-277.

Eddy, N.B., Leimback, D., 1953. Synthetic analgesics: II. Dithyienylbutenylamines and dithyienylbutylamines. *J Pharmacol Exp Ther* 3, 544–547.

Finn, C.E., 2008. *Rubus* spp., blackberry. In: Janick, J., Paull, R.E. (Eds.), The Encyclopedia of Fruits and Nuts. (pp. 348–351). CABI, Cambridge, MA.

George, B.P., Parimelazhagan, T., Saravanan, S., 2013 a. Anti-Inflammatory, Analgesic and Antipyretic Activities of *Rubus ellipticus* Smith. Leaf Methanol Extract. *IJPPS* 5, 220-224.

George, B.P., Parimelazhagan, T., Saravanan, S., Chandran, R., 2013 a. Anti-inflammatory, Analgesic and Antipyretic Properties of *Rubus niveus* Thunb. Root Acetone Extract. *Pharmacologia* 4, 228-235.

Gould, B.E., 2002. Inflammation and healing. Gould BB (ed) pathophysiology of the health professions (2nd ed), Philadelphia USA, WB Saunder pp. 192-199.

Halliwell, B., Gutteridge, J.M.C., 2007. Free radicals in biology and medicine. 4th edition. Oxford: Oxford University Press.

Halliwell, B., 1995. Antioxidant characterization. Methodology and mechanism. *Biochem Pharmacol.* 49, 1341–8.

Kou, J., Ni, Y., Li, N., Wang, J., Liu, L., Jiang, Z.H., 2005. Analgesic and anti-inflammatory activities of total extract and individual fractions of Chinese medicinal plant *Polyrhachis lamellidens. Biol Pharm Bull.* 28, 176–180.

Marchiaro, M., Blank, M.F.A., Maurao, R.H.V., Antoniolli, A.R., 2005. Antinociceptive activity of the aqueous extract of erthenia velutina leaves. *Fitoterapia* 76, 637-642.

Matheus, M.E., Berrondo, L.F., Vietas, E.C., Menzes, F.S., Fernandus, P.D., 2005. Evaluation of antinociceptive properties from *Brillantaisia palisotti* Lindau stems extract. *J ethnopahrmacol* 102, 377-381.

Mattison, N., Trimple, A.G., Lasgana, I, 1998. A new drug development in United States 1963 through 1983. *Clinical pharmacology and therapeutics* 43, 290-301.

Motamed, S.M., Naghibi, F., 2010. Antioxidant activity of some edible plants of the Turkmen Sahra region in northern Iran. *Food Chem.* 119, 1637–1642.

Nicholas, Moore, N.D., 2009. In search of an ideal analgesic for common acute pain. *Acute pain* 11, 129-137.

Nunez-Selles, A.J., 2005. Antioxidants therapy: myth or reality. *J Brazil Chem Soc* 16, 699- 710.

Organization for Economic Co-operation and Development, revised draft guidelines 423, 2000. "OECD Guidelines for the testing of chemicals" Revised document-October.

Patel, A.V., Rojas-Vera, J., Dacke, C.G., 2004. Therapeutic constituents and actions of *Rubus* species. *Curr Med Chem* 11, 1501-1512.

Re, R., Pellegrini, N., Proteggente, A., Pannala, A., Yang, M., Rice-Evans, C., 1999. Antioxidant activity applying an improved ABTS radical cation decolorization assay. *Free Radic Biol Med.* 26, 1231–7.

Sabu, M.C., Kuttan, R., 2003. Antioxidant activity of Indian herbal drugs in rats with alloxan induced diabetes. *Pharm Biol* 41, 500–505.

Venskutonis, P.R., Dvaranauskaite, A., Labokas, J., 2007. Radical scavenging activity and composition of raspberry (*Rubus idaeus*) leaves from different locations in Lithuania. *Fitoterapia* 78, 162–165.

Vinegar, R., Schreiber, W., Hugo, R., 1969. Biphasic development of carrageenan edema in rats. *J Pharmacol Exp Ther* 166, 96–103.

Wong, C.H., Day, P., Yarmush, J., Wu, W., Zbuzek, U.K., 1994. Nifedipine- induced analgesic after eoidural injections in rat. *Anesthesia and analgesic* 79, 303-306.

2015, Modern Methods in Phytomedicine *Pages 43–42*
Editor: **T. Parimelazhagan**
Published by: **DAYA PUBLISHING HOUSE, NEW DELHI**

Capparis spinosa: Potential Medicinal and Non-traditional Vegetable Plant from Trans-Himalayan Region of India

Gyan P. Mishra[1,2]*, *Raghwendra Singh*[1,3], *Ashutosh Murkute*[1,4], *Manish S. Bhoyar*[1], *T. Radhakrishnan*[2] *and R.B. Srivastava*[1]

[1]*Defence Institute of High Altitude Research, Defence R and D Organisation, C/o 56 APO, Leh, Ladakh – 901 205, J&K*
[2]*Present address: Directorate of Groundnut Research, PB No 05, Ivnagar Road, Gujarat*
[3]*Directorate of Weed Science Research (DWSR), Maharajpur, Jabalpur – 482 004, M.P.*
[4]*Directorate of Onion and Garlic Research, Rajgurunagar, Pune – 410 505, M.S.*

1.0 INTRODUCTION

For centuries, traditional societies have exploited edible wild plant resource to obtain their nutritional requirements in the period of food crisis (Chweya, 1985). There has been renewed or increasing interest in consuming wild food plants (Nebel *et al.*, 2006; Johns *et al.*, 2006). Various reports from different parts of the world also noted that many wild edibles are nutritionally rich (Ogle and Grivetti, 1985; Maundu

* *Corresponding Author.* E-mail: gyan.gene@gmail.com

et al., 1999) and can supplement nutritional requirements, especially vitamins and micronutrients. In addition, wild-plants are valuable source of a wide range of secondary metabolites which are used as pharmaceuticals, agrochemicals, flavor, fragrance, color, biopesticides, food additives and many more. Domestication and cultivation of wild edible plants is, therefore, essential in broadening the food base in developing countries. This will lead to diversification, which will ensure a dietary balance and the intake of micronutrients.

Ladakh is a part of Indian Himalaya at an altitude of 8,787-20,000 ft. above mean sea level, is characterized by diverse and complex land formations. It has many unconquered peaks of impregnable heights, uncharted glaciers and vallies. It is located at the latitude of 31° 44′ 57" – 32° 59′ 57" N and longitude of 76° 46′ 29"– 78° 41′ 34" E. The temperature ranges between 40° C in summer and -40° C during winter and annual precipitation is around 20 – 30 mm along with low relative humidity between 20-40 per cent. These climatic features make this region a typical cold arid dessert. Under these unique geographical position and adverse climatic conditions many plant species were able to establish themselves and majority of such plants were nutritionally as well as medicinally potential and suitable to use as non conventional vegetables. These plants were identified by local people through over the years of experience perhaps via trial and error method which led to the selection of plants which are edible. These selected plants were gradually added to the repertoires of the edible plants (Mishra *et al.,* 2009, Bhoyar *et al.,* 2010; Singh *et al.,* 2009; Chourasia *et al.,* 2007).

The caper bush (*Capparis spinosa*) is one such wild plant, which grows as an underutilized wild edible plant in Ladakh region, but is a semi-cultivated popular leafy vegetable in many parts of European countries (Bhoyar *et al.,* 2010; 2011). *Capparis* is evergreen perennial bush, grows in roadside, on the slopes, dry, rocky and stony soils, can withstand extreme temperature (-30 °C to +30 °C) of Ladakh and is highly drought tolerant. This plant has multiple uses in cuisine as salad, leafy vegetable, pickle and condiments. Besides these qualities it brings many environmental benefits, including soil and water conservation, desertification control and land reclamation in fragile cold ecosystem of Ladakh. *Capparis* has all the potential to meet the calorie requirement of the army deployed in the 'Ladakh' sector during road close period (November to April) and can play a significant role both in the national and international spice trade in the future (Mishra *et al.,* 2009; Bhoyar *et al.,* 2010).

2.0 BOTANY AND TAXONOMY

Caper bush (2n=38) belongs to the botanical family Capparaceae (formerly Capparidaceae) which has a very close phylogenetic relationship with Brassicaceae. *Capparis spinosa* is a winter-deciduous, diffuse, prostrate, glabrescent, perennial shrub or climber armed with divaricated light yellow thorn. Leaves are alternate, variable in texture, round to ovate, orbicular to elliptic, base rounded with mucronate apex. Leaf stipule may be formed in to spine, this is the reason it is called spinosa. Flower white to purple, solitary axillary and very pleasing in appearance. Reputed to be quite fragrant, the flowers open at dawn and close by late afternoon, during which time they are magnet for pollinators. Pollination is by insect. Fruits are fleshy, oblong ellipsoid with red flesh and many brown seeds. Seeds are kidney shaped and brown to black in colour which is 3-4 mm in diameter (Chaurasia *et al.,* 2007).

3.0 PLANT PROFILE

Capparis spinosa also called 'Caper' and locally known as '*Kabra*' is one of the oldest known medicinal plant of Amchi system (Local doctors) which is occasionally used by local people of Ladakh as a leafy vegetable and forage. It requires a hot, well drained dry position in full sun. The species is not hardy in the colder areas of the country, it tolerates temperature down to between -5 and -10pc (Huxley, 1992; Phillips and Rix, 1998).

3.1 Phytochemical and Nutritional Properties

Capparis spinosa is highly nutritious plant having immense medicinal properties. A sample of 100 g of prepared Capers contains energy (23 kcal), carbohydrates (5 g), sugars (0.41 g), dietary fiber (3.2 g), fat (0.9 g), protein (2.36 g), vitamin C (4.3 mg), iron (1.7 mg) and sodium (2964 mg) (USDA, 2008). Previous chemical studies on caper have shown the presence of alkaloids, lipids, polyphenols, flavonoids, indole and aliphatic glucosinolates (Bhoyar *et al.,* 2011; Rodrigo *et al.,* 1992; Sharaf *et al.,* 2000). Hydroxy cinnamic acids like caffeic acid, ferulic acid, p-coumaric acid and cinnamic acid were reported. p-methoxybenzoic acid was isolated from aqueous extract of *C. spinosa* aerial parts (Gadgoli and Mishra, 1999). Calis *et al.* (1999, 2002) have isolated from mature fruits of *C. spinosa* two glucose-containing 1H-indole-3-acetonitrile compounds, capparilosides A and B and two (6S)-hydroxy-3-oxo-aionol glucosides, corchoionoside C and a phenyl glucoside. A qualitative and quantitative analysis of rutin from leaves, fruits and flowers of *Capparis* growing wild was achieved (Ramezani *et al.,* 2008).

Immature flower buds are pickled in vinegar, sauces, or preserved in salt have very much demand in European countries (Bown, 1995). Previous chemical studies have reported the presence of alkaloids, lipids, flavenoides and glucosinolates, which are known as flavour compound, cancer preventing agent and biopesticide (Bhoyar *et al.,* 2010; Mikkelsen *et al.,* 2000; Germano *et al.,* 2002). *Capparis* flower buds contain 100.51 mg of rutin equivalent/gm methanolic extract which exhibit antioxidant activity (Germano *et al.,* 2002). The main glucosinolate, glucocapparin amounted to 90 per cent of total glucosinolates (Schraudolf, 1989) in caper buds. If taken before meal it will increase the appetite (Genders, 1994). Unopened flower buds are laxative.

Flavanoid glycosides like, rutin, quercetin, quercetin-3- rutinoside were isolated from the floral buds (Giuffrida *et al.,* 2002). Glucocapperin has been shown to be the main glucosinolate in the floral buds (Matthäus and Özcan, 2002). Presence of both flavonoids and hydroxycinnamic acids has also been demonstrated in capers (Bonina *et al.,* 2002). New compounds like sitosteryl glucoside-6′-octadecan-o-ate and 3-methyl-2-butenyl- glucoside were isolated (Khanfar *et al.,* 2003).

The seeds oil yield ranged from 27.3–37.6 g/100g and it contains high Vit-E (134 mg/100g), tocopherol (4961.8–10009.1 mg/kg) which act as natural antioxidant. Sterol an important constituent of oil, is capable of lowering plasma cholesterol (Matthaus and Ozcan, 2005). The leaf oil was composed of isothiocyanates, n-alkanes, terpenoids, a phenyl propanoid, an aldehyde and a fatty acid. The main components of this oil were thymol (26.4 per cent), isopropyl isothiocyanate (11 per cent), 2-hexenal (10.2 per cent) and butyl isothiocyanate (6.3 per cent). The volatile oils of the ripe fruit and the root were composed mainly of the methyl, isopropyl and sec-butyl isothiocyanates. A protein exhibiting an N-terminal amino acid sequence with some similarity to imidazole glycerol phosphate synthase was purified from fresh *Capparis* seeds (Lam *et al.,* 2009). A dimeric 62-kDa lectin exhibiting a novel N-terminal amino acid sequence was purified from caper seeds (Lam *et al.,* 2009).

In Ladakh region it grows as wild especially around 10,000 ft above mean sea level. It can be consumed as a cooked vegetable, salad, pickle, and condiments. This leafy vegetable is a reservoir of vitamins, minerals, dietary fibers and also as the major source of bioactive compounds having health promoting properties as listed in Table 4.1.

Table 4.1: Nutritional Properties of *Capparis spinosa* (Leaves)

Nutrient	*Mean Value*
Crude protein (per cent)	17.9
Fats (per cent)	04.9
Crude fiber (per cent)	06.8
Total carbohydrates (per cent)	47.7
Calcium (mg/100g)	550.0
Phosphorous (mg/100g)	135.0
Iron (mg/100g)	72.0
Sodium (mg/100g)	85.0

3.2 Medicinal Properties

Many workers have worked on this plant for both nutrients and various bioactive compounds from different parts of this plant. In Ladakh, various medicinal preparations from *Capparis* are used by the Amchis (Local traditional doctors) for treatment of various ailments since ages. *C. spinosa* extract was able to counteract the inflammatory process induced *in vitro* by IL-1β in human chondrocyte cultures (Panico *et al.,* 2005). Capers are a hepatic stimulant that has been used for improving the functional efficiency of the liver. The recent experimental work also confirms its protective action on the histological architecture of the liver and its positive effect on liver glycogen and serum proteins (Subhose *et al.,* 2005). Other glucosinolates like sinigrin, glucoiberin and glucocleomin were isolated from the seeds and leaves of *C. spinosa* (Romeo *et al.,* 2007). Three new alkaloids capparispine, capparispine 26-O-d-glucoside and cadabicine 26-O-d-glucoside hydrochloride were isolated from the roots of *C. spinosa* (Fu *et al.,* 2007, 2008).

In Ayurveda given in splenic, renal and hepatic complaints and root bark is analgesic, anthelmintic, deobstuent, diuretic, expectorant, and vaso-constructive (Chiej, 1984). Internally it is used in the treatment of gastrointestinal infection, diarrhoea, and rheumatism (Bown, 1995; Chopra *et al.,* 1986). From the roots, indole glucosinolates like Glucobrassicin, Neoglucobrassicin and 4- methoxyglucobrassicin were isolated (Ahmed *et al.,* 1972). The homologous polyphenols Cappaprenol-12, Cappaprenol-13 and Cappaprenol-14 with 12, 13 and 14 isoprene units respectively were also isolated (Dhurandhar, 1973; Germano *et al.,* 2002). It is used as drug for acute viral hepatitis and is a major constituent of the herbal formulation Liv 52, useful in liver disorders (Sama *et al.,* 1976; Mathur *et al.,* 1986).

The root-bark harbours analgesic, anthelmintic, anti heamorhidal, aperients, deobstruent, depurative, diuretic, emmenagouge, expectorant, tonic and vasoconstrictive properties (Chiej, 1984). The bark is used internally in the treatment of gastro intestinal infection, diarrhea, and rheumatism. Externally, it is used to treat skin condition, capillary weakness and easy bruising (Chopra *et al.,* 1986; Genders, 1994; Bown, 1995). Alkaloids to the tune of 0.91 per cent from root-bark and 0.86 per cent from seeds are isolated and maximum alkaloid content was found in the roots of which stachydrine constituted 87.43 per cent of the total alkaloid present (Sadykov *et al.,* 1981). In Unani medicine, the decoction of root bark is prescribed as deobstruent to liver and spleen, as anthelmintic and anti-inflammatory agent (Chopra *et al.,* 1999). Four bacterial strains (*viz. Pseudomonas stutzeri* var. mendocina, *Comamonas* sp., *Agrobacterium tumefaciens* bivar. 2 and *Sphignobacterium* sp.) isolated from the rhizosphere of caper were found to be able to fix N_2 (Andrade *et al.,* 1997).

From ancient times, the floral buds were employed as flavouring in cooking and are also used in traditional medicine for their diuretic, antihypertensive, poultice and tonic properties (Baytop, 1984; Çalis *et al.,* 1999). From the floral buds 3-O-Rhamnorutinosin Kaempferol (Tomas and Ferreres, 1978) and Kaempferol-3-rutinoside, quercetin-7-Oglucorhamnoside were isolated (Artemeva *et al.,* 1981). The buds are a rich source of compound known as aldose-reductase inhibitors it has been shown that these compounds are effective in preventing the formation of cataracts. Also the flower buds and roots are used as renal disinfectants, coughs, diuretics, and tonic and for arteriosclerosis and as compresses for the eyes (Genders, 1994; Bown, 1995: Batanouny, 1999).

The leaves are used for the treatment of gout, coughs, ear ache, expelling stomach worm, rheumatism, paralysis, toothache, as diuretic and for diabetes control (Chopra *et al.,* 1986; Andrade *et al.,* 1997; Sharma, 2003; Gunther, 1959; Shahina, 1994). The important medicinal properties are listed in Table 4.2.

Table 4.2: Medicinal Properties of *Capparis spinosa*

Plant Parts	*Bioactive Compound*
Flower buds	Quercetin-3-rutinocide,Rutic acid, Pectic acid, Kaemferol-3-rutinoside, Isothiocynate glycoside, Kaemferol-3-rhamnorutinoside, Saponin, Pentosan.
Leaves	Alkaloids, Reducing sugar, Fats, Resin, Ascorbic acid, Tannin
Fruits	Alkaloids, Glycosides, Fat, Reducing sugar, Resin, Ascorbic acid, titrable acid
Root/Bark	Rutic acid, Stachydrin, Glucosinolate, Glucopagolin
Seeds	Palmitic acid and Stearic acid, Oelic acid, Vit-E, Glucocapparin, Glucocleomin

4.0 ECOLOGY AND DISTRIBUTION OF *CAPPARIS SPINOSA*

Capparis spinosa is said to be native to the Mediterranean basin, but its range stretches from the Atlantic coast of the Canory Island and Morocco to the Black Sea of the Crimea and Armenia, and Eastward to the Caspian Sea in to Iran. *Capparis* probably originated from dry region in West or Central Asia (Chourasia *et al.,* 2007). In India it is found in inner vallies of Himalaya between 3020-3890 m AMSL which include Indus, Nubra, Suru and Zanskar vallies of Ladakh region and Spiti valley of Himachal Pradesh (Mishra *et al.,* 2009; Bhoyar *et al.,* 2010).

In Ladakh region it grows as wild especially around 11,000 ft above mean sea level. The plant can withstand extreme temperature from -30pc to +30p and is drought tolerant. Dry heat and intense sunlight make the preferred environment for caper plants. *Capparis* plant is hardy and grows in roadside, on the slopes, dry, rocky and stony soil. These characteristics can be explored as a source of important genes for developing abiotic stress tolerant plants (Bhoyar *et al.,* 2010, 2011, 2012). Caper plants grow well in nutrient poor sharply-drained gravelly soils. Mature plants develop large extensive root systems that penetrate deeply into the earth.

5.0 PHARMACOLOGICAL ACTIVITIES

5.1 Antioxidant Activity

The methanolic extract of flower buds showed *in vitro* antioxidant, antiviral, immunomodulatory activity and *in vivo* photoprotective and antiallergic activity (Germano *et al.,* 2002). Inhibition of lipid oxidation has been demonstrated *in vitro*; the mechanism is attributed to a cooperative interaction between the tocopherol, flavonoid, and isothiocyanate chemical constituents (Tesoriere *et al.,* 2007).

5.2 Hepatoprotective, Anti-hepatotoxic and Anti-inflammatory Activity

Capparis spinosa has been reported to possess hepatoprotective activity (Romeo *et al.,* 2007; Gadgoli and Mishra, 1995) and p-methoxy benzoic acid was found to be responsible for this activity (Gadgoli and Mishra, 1995). Approximately 600 mg of dried whole plant extract per day has been used in a mixed preparation in experiments investigating hepatoprotective effects. Similarly, a clinical trial investigating the efficacy of a mixed preparation containing caper extract combined with other extracts found an improvement in liver function laboratory values (Huseini *et al.,* 2005).

The aqueous extract of the aerial parts were reported to possess anti-inflammatory activity and Cappaprenol –13 isolated as an anti-inflammatory principle from caper was found to inhibit carrageenan induced paw oedema in rats by 44 per cent compared to 67 per cent by standard oxyphenbutazone (Al-Said, 1988; Ageel, 1985).

5.3 Hypolipidemic, Antiallergic and Antihistaminic Activity

In normal and diabetic (induced) rats fed aqueous extracts of the powdered caper fruits for a 2-week period, a reduction in plasma cholesterol and triglycerides was demonstrated (Eddouks, 2005). A methanol extract of *C. spinosa* buds, rich in flavonoids, including several quercetin and kaempferol glycosides, was demonstrated to possess strong antioxidant/free radical scavenging effectiveness in different *in vitro* tests; *in vivo* this extract showed a noteworthy antiallergic effectiveness against bronchospasm in guinea-pigs. A 2 per cent aqueous gel has been used for antihistaminic effects (Trombetta *et al.*, 2005). The ethanolic extract of *C. spinosa* root bark reported anthelmintic activity. It acts as anti allergic and antihistaminic agent (Zhan, 1978).

5.4 Cardiovascular, Antihypertensive, Antimicrobial and Photoprotective Activity

Aqueous extract of *Capparis spinosa* shows the hypotensive activity in spontaneously hypertensive rat (Ali *et al.*, 2007). The alcoholic extract of *C.spinosa* reported anti microbial activity (Mahasneh, 1996). Juice of leaves and fruits are used as anticystic, fungicide and bactericide. A methanol extract of *C. spinosa* buds afforded significant *in vivo* protection against UVB light-induced skin erythema in healthy human volunteers (Bonina *et al.*, 2002).

5.5 Miscellaneous Activity

It is used as a metabolic corrective in new-borne (Dhurandhar, 1973). Rutin or flavanoids from Caper, prevented increase in serum levels of alanine amino transferase, aspartase amino-transferase and aldolase caused due to J-irradiation, thereby reducing radiation sickness. It also acted as a coagulation enhancer and diuretic (Altymyshev, 1981).

6.0 PROPAGATION

Caper plants can be propagated through seeds or stem cuttings, however, both methods present serious problems and restrictions to the commercial expansion of this crop. Well matured 8-10 cm long shoot having 1 cm diameter with 5-8 buds are best suited. The cuttings are dipped in IBA solution of 0.5 to 3.0 ppm (15 sec) for better rooting. Capers are widely grown on dry and poor soil, where environmental conditions are not easy for the cultivation of other crops. It is difficult to propagate through seedlings because caper seed have both physical (seed coat) and physiological dormancy and is difficult to germinate (Bhoyar *et al.*, 2010).

6.1 Prospects of Propagation through Seeds

Seed germination is poor and hence requires extra measures to germinate. Dried seeds are immersed in warm water (40°C) and then soaked for 1 day. Seeds wrapped in moist cloth is placed in sealed jar and kept in refrigerator for 2-3 months. After refrigeration the seed is again soaked in warm water overnight, and then plant the seed about 1 cm deep in loose soil. Fresh caper seeds germinate readily, but in low percentages (1–2 per cent), whereas drying of seeds induces severe dormancy, which is difficult to overcome naturally (Olmez *et al.*, 2004). As the dormancy in this crop is due to the hard seed coat, external treatments are necessary to overcome the prevailing dormancy. The structure of the seed and the mucilage which develops when the seed is placed in contact with water could impose an effective barrier against the diffusion of oxygen to the embryo (Soyler and Khawar, 2007).

Figure 4.1: The Plant *Capparis spinosa* and its different Parts.

A: Habitat; B: Flowering plant at natural habitat; C: Flower buds; D-E: Flowers; F: Fruits; G: Seeds; H: Root system.

Pre–chilling, scarification, and treatments with GA_3 or KNO_3 are the standard procedures used to enhance seed germination of dormant seeds. To obtain higher germination (per cent) in *Capparis* various treatments were reported *viz.* gibberellic acid+KNO_3 (Fernandez *et al.,* 2002; Pupalla and Fowler, 2002), pretreatment with H_2SO_4 (Kara *et al.,* 1996), H_2SO_4+GA_3 (Sozzi and Chiesa, 1995) and warm water+chilling (Kontaxis 1997). The treatment of the seeds with plant growth regulators may influence root formation and rapid germination. The physiological explanation for this phenomenon may be the breakdown of germination inhibitors.

Fruit weight, position on the mother plant and maturation stage also affects caper seed germination besides, an efficient method has been standardized for ensuring satisfactory seed germination by breaking the physical and/or physiological dormancy (Pascual *et al.,* 2004). Pascual *et al.* (2006) reported that freshly harvested caper seeds showed highest germination rate and the shortest time to reach 50 per cent of the final germination percentage. The effect of seed soaking treatments and soaking time individually or in combination was studied with the addition of GA to maximize the seed germination percentage (Pascual *et al.,* 2009). The effects of temperature, light, pre-soaking treatment and removal of seed coat have been reported to effect germination of various crops (Travlos *et al.,* 2007; Shaik *et al.,* 2008). Research studies on the seed germination of caper are scanty and insufficient across the world.

6.2 Standardization of Caper Seed Germination at Ladakh

Commercial propagation of capers is complicated by limited and variable seed germination under natural conditions of Ladakh. One of the main problems that prevent sustainable use of the plant under Ladakh is that both under laboratory or natural conditions the plant shows highly variable germination, due to deep seed dormancy. High germination percentage is a must to ensure high plantation and viability.

Considering the potential of caper for rural livelihood, as erosion control agent, as new crop and new income sources, Bhoyar *et al.* (2010) investigated the importance of various dormancy breaking treatments that might affect germination of wild caper naturally occurred in Ladakh. Bhoyar *et al.* (2010) examine the role of various dormancy breaking treatments, *viz* hot water treatment, scarification, stratification, concentrated acids (H_2SO_4, HNO_3 and HCl), gibberellic acid (GA_3), potassium nitrate (KNO_3), alcohol, acetone and gamma-rays irradiation on the germination of caper (*Capparis spinosa* L.) seeds, which were collected from wild plants occurred in Ladakh. Dried seeds were placed in the germination chambers for 20–28 days at constant temperature of 25±2°C under continuous light (20 hr) photoperiod after its treatments. Highest germination of 62 per cent was obtained when seeds were pretreated with H_2SO_4 (40 min.), followed by 400 ppm of GA_3 soaking (2 hr). The results revealed that the seeds were epitomized by both physical and physiological type of dormancy, which should be overcome to have maximum germination percentage.

7.0 GENETIC DIVERSITY

Without a continuous source of variability, the ability of plant breeders to improve agronomic performance that is based on complex genetic combination could decline

(Smith and Smith, 1992). Hence, for efficient conservation and successful breeding programmes, it will be prudent to study the populations of *C. spinosa* at genetic and molecular levels. Molecular genetic markers would aid the long-term objective of identifying diverse parental lines to generate segregating populations for tagging important traits, such as gene(s) for high content of rutins, tocopherol etc from these *Capparis spinosa* genotypes.

There is very scanty information available on the genetic diversity of caper germplasm from different parts of the world. Most capers are gathered from natural habitat with little attention to the preservation of germplasm resources and genetic diversity. Preliminary studies under Ladakh conditions were carried out by Bhoyar *et al.* (2012) to assess the extent of genetic variation between and within natural population of caper using markers like RAPD and ISSR. Investigation and characterization of capers genetic diversity has been done by DNA marker, including molecular characterization of Italian and Tunisian population of caper (Khouildi *et al.,* 1999).

A genetic fingerprinting technique (AFLP) was studied by Inocencio *et al.,* 2005 to determine the relationship among *Capparis* sp. Genetic distances, based on AFLP data were estimated for 45 accessions of *Capparis* species, from Spain, Morocco and Syriya. The result of this analysis support the differentiation of four of the five taxa involved. The group of plants recognized as *C. spinosa* on the basis of morphological characters, includes several cultivars and appears in an intermidiat position between *C. orientalis* and *C. sicula* and overlaps with *C. orientalis.* The other two species *C. aegyptia* and *C. ovata* are separate from the rest. *Capparis spinosa* had a low number of unique bands in comparison with the other species. Although these results cannot confirm the hybridorigion of *C. spinosa,* the distribution of the bands supports this hypothesis, the most likely parental species being *C. orientalis* and *C. sicula* (Inocencio, 2005).

8.0 CONCLUSION

Capparis spinosa has multiple uses in cuisine as a salad, leafy vegetable, pickles and condiments. The wild resources of nutritive and medicinal plant have been depleted from their natural homes. The balance between demand and supply may require mass cultivation. Since the plant grows in the barren land of Ladakh region, cultivation practices needs to be standardized for commercial cultivation. The hardy nature of the plant along with its nutritional and medicinal properties makes it an ideal plant for commercial exploitation in the cold arid region of Ladakh.

Even today very little is known about the indigenous cultivation techniques, knowledge and utilization, the extent and structure of genetic variation, and the potential crop improvement through domestication, selection or breeding of *Capparis spinosa* in Ladakh region. The urgent need is standardization of seed germination as well as agronomical practices, genetic diversity study for future breeding programme and proper phytochemical investigation of selected plant species so as to have its nutritional profile. This plant should be included in the list of our agricultural/ horticultural crops as a new potential source of food as well as medicinal plants after proper analysis.

Due to lack of proper records and over-exploitation of wild edible plants like caper by local people; the natural resources along with related indigenous knowledge are depleting day by day (Mishra *et al.*, 2009; Jain, 1996; Roy, 2003). There is need to create social awareness among local tribal communities about ecological importance, cultivation and sustainable harvesting of non-conventional wild high altitude plants like caper which are used for culinary as well as medicinal purposes.

REFERENCES

Ageel, A.M., Parmar, N.S., Mossa, J.S., Al-Yahya, M.A., Al- Said, M.S., Tariq, M., 1985. Anti-inflammatory activation (IUCN): Switzerland: Academy of Scientific Research and Technology; *Wild Medicinal Plants in Egypt*, pp. 130 -131.

Bhoyar, M., Mishra, G.P., Singh, R., Singh, S.B., 2010. Effects of various dormancy breaking treatments on the germination of wild caper (*Capparis spinosa* L.) seeds from the cold arid desert of trans–Himalayas. *Ind. J. Agril. Sci* 80, 620-4.

Bhoyar, M.S., Mishra, G.P., Naik, P.K., Srivastava, R.B., 2011. Estimation of antioxidant activity and total phenolics among natural populations of *Capparis spinosa* leaves collected from cold arid desert of trans-Himalayas. *Aust. J. Crop Sci* 5, 912-919.

Bhoyar, M.S., Mishra, G.P., Naik, P.K., Murkute, A.A., Srivastava, R.B., 2012. Genetic variability studies among natural populations of *Capparis spinosa* from cold arid desert of trans-Himalayas using DNA markers. *Nat. Acad. Sci. Lett* 35, 505-515.

Bonina, F., Puglia, C., Ventura, D., Aquino, R., Tortora, S., Sacchi, A., Saija, A., Tomaino, A., Pellegrino, M.L., De Caprariis, P., 2002. *In vitro* and *in vivo* photo protective effects of lyophilized extract of *Capparis spinosa*. L. buds. *J. Cosmetic Sci* 53, 321-335.

Bown, D., 1995. *Encyclopaedia of Herbs and their Uses.* Dorling Kindersley, London, ISBN 0-7513-020-31.

Çalis, I., Kuruüzüm, A., Rüedi, P., 1999. 1H-Indole-3 acetonitrile glycosides from *Capparis spinosa* fruits. *Phytochem* 50, 1205-1208.

Çalis, I., Kuruüzüm, A., Lorenzetto, P.A., Rüedi, P., 2002. (6S)- Hydroxy-3-oxo-ionol glucosides from *Capparis spinosa* fruits. *Phytochem* 59, 451-457.

Chaurasia, O.P., Ahmed, Z., Ballabh, B., 2007. Ethno botany and Plants of Trans-Himalaya. pp. 181.

Chiej, R., 1984. *Encyclopaedia of Medicinal Plants,* MacDonald, ISBN 0-356-10541-5.

Chopra, R.N., Nayar, S.L., Chopra, I.C., 1999. *Glossary of Indian Medicinal Plants,* CSIR, New Delhi, India.

Chopra, R.N., Nayar, S.L., Chopra, I.C., 1986. *Glossary of Indian Medicinal Plants,* CSIR, New Delhi, India.

Chweya, J.A., 1985. Identification and nutritional importance of indigenous green leafy vegetables in Kenya. *Acta Hort* 153, 99-108.

Dhurandhar, J., 1973. Bonnisan - A metabolic corrective in gastrointestinal disorders of the newborn (a study of 100 cases). *Probe* 12, 73-78.

Eddouks, M., Lemhadri, A., Michel, J.B., 2005. Hypolipidemic activity of aqueous extract of *C. spinosa* L. in normal and diabetic rats. *J. Ethnopharmacol* 98, 345-350.

Fernandez, H., Perez, C., Revilla, M.A., Perez–Gar-cia, F., 2002. The levels of GA_3 and GA_{20} may be associated with dormancy release in *Onopordum nervosum* seeds. *J. Plant Growth Regula-tion* 38, 141–3.

Fu, X.P., Aisa, H.A., Abdurahim, M., Yili, A., Aripova, S.F., Tashkhodzhaev, B., 2007. Chemical composition of *Capparis spinosa* fruit. *Chem. Natl Comp* 43, 181-185.

Fu, X.P., Wu, T., Abdurahim, M., Su, Z., Hou, X.L., Aisa, H.A., Wu, H., 2008.New spermidine alkaloids from *Capparis spinosa* roots. *Phytochem Lett* 1, 59-62.

Gadgoli, C., Mishra, S.H., 1995. Preliminary screening of *Achillea millefolium, Cichorium intybus* and *Capparis spinosa* for antihepatotoxic activity. *Fitoterapia* 66, 319-323.

Gadgoli, C., Mishra, S.H., 1999. Anti hepatotoxic activity of p-methoxy benzoic acid from *Capparis spinosa. J. Ethnopharmacol* 66, 187-192.

Genders, R., 1994. *Scented Flora of the World.* Robert Hale. London.

Germano, M.P., Pasquale, R.D., D'Angelo, V., Catania, S., Silvari, V., Costa, C., 2002. Evaluation of extracts and isolated fraction from *Capparis spinosa* L. buds as an antioxidant source. *J. Agric. Food Chem* 50, 1168-1171.

Giuffrida, D., Salvo, F., Ziino, M., Toscano, G., Dugo, G., 2002. Initial investigation on some chemical constituents of Capers from the island of Salina. *Ital. J. Food Sci* 14, 25-33.

Gunther, R., 1959. *The Greek Herbal of Dioscorides.* New York: Hafner Publishing Co. 215.

Huseini, H.F., Alavian, S.M., Heshmat, R., Heydari, M.R., Abolmaali, K., 2005. The efficacy of Liv-52 on liver cirrhotic patients: a randomized, double-blind, placebo-controlled first approach. *Phytomedicine* 12, 619-624.

Huxley, A., 1992. The New RHS Dictionary of Gardening. MacMillan Press.

Inocencio, C., Cowan, R.S., Alcaraz, F., Rivera, D., Fay, M.F., 2005. AFLP fingerprinting in *Capparis* subgenus *Capparis* related to the commercial sources of capers. *Genet. Resour. Crop Evol* 52, 137-144.

Jain, S.K., 1996. *Ethnobotany in Human Welfare.* Deep Publications, New Delhi.

Johns, T., Eyzaguirre, P.B., 2006. Linking biodiversity, diet and health in policy and practice. *Proceedings of the Nutrition Society* 65, 182-189.

Kara, Z., Ecevit, F., Karakaplan, S., 1996. Toprak koruma elemani ve yeni bir tarismal urur olarak ka-pari (*Capparis spinosa* spp.).Tarim–Cevre Iliskileri Sempozyyumu.Mersin: 919–29 (in Turkish).

Khanfar, M.A., Sabri, S.S., Zarga, M.H.A., Zeller, K.P., 2003.The chemical constituents of *Capparis spinosa* of Jordanian origin. *Nat. Prod. Res* 17, 9-14.

Khouildi, S., Pagnotta, M.A., Tanzarella, O.A., Porceddu, E., Ghorbel, A., 1999. Assessement of the genetic variation in natural populations of *Capparis spinosa* L. using RAPD analysis. *CWANA Newsletter,* pp. 8.

Kontaxis, D.G., 1997. Caper: Specialty and Minor Crops Handbook. Small Farm Center, University of California, Davis, pp. 4.

Lam, S.K., Han, Q.F., Ng T.B., 2009. Isolation and characterization of a lectin with potentially exploitable activities from caper (*Capparis spinosa*) seeds. *Biosci Rep* **29,** 293-299.

Mahasneh, A.M., Abbas, J.A., El-Oqlah, A.A., 1996. Antimicrobial activity of extracts of herbal plants used in the traditional medicine of Bahrain. *Phytotherapy Res* 10, 251-253.

Mathur, S., Prakash, A.O., Mathur, R., 1986. Protective effect of Liv-52 against beryllium toxicity in rats. *Curr. Sci* 55, 899- 901.

Matthäus, B., Özcan, M., 2002. Glucosinolate composition of young shoots and flower buds of capers (*Capparis* species) growing wild in Turkey. *J. Agric. Food. Chem* 50, 7323–7325.

Matthaus, B., Ozcan, M., 2005. Glucosinolates and fatty acid, sterol, and tocopherol composition of seed oils from *Capparis spinosa* Var. *spinosa* and *Capparis ovata* Desf. Var. *canescens* (Coss.) Heywood. *J. Agric. Food Chem* 53, 7136-7141.

Maundu, P.M., Ngugi, G.W., Kabuye, C.H.S., 1999. Traditional food plants of Kenya Nairobi: National Museums of Kenya.

Mikkelsen, M.D., Hansen, C.H., Wittstock, U., Halkier, B.A., 2000. Cytochrome P450 CYP79B2 from Arabidopsis catalyzes the conversion of tryptophan to indole-3-acetaldoxime, a precursor of indole glucosinolates and indole-3-acetic acid. *J. Biol. Chem* 275, 33712-33717.

Mishra, G.P., Singh, R., Bhoyar, M., Singh, S.B., 2009. *Capparis spinosa*: unconventional potential food source in cold arid deserts of Ladakh. *Curr. Sci* 96, 1563-64.

Nebel, S., Pieroni, A., Heinrich, M., Ta-Chòrta, 2006. Wild edible greens used in the Graecanic area in Calabria, southern Italy. *Appetite* 47, 333-342.

Ogle, B.M., Grivetti, L.E., 1985. Legacy of the chameleon edible wild plants in the Kingdom of Swaziland, South Africa. A cultural, ecological, nutritional study. Parts II-IV, species availability and dietary use, analysis by ecological zone. *Ecology of Food and Nutrition* 17, 1-30.

Olmez, Z., Yahyaoglu, Z., Ucler, A.O., 2004. Effects of H_2SO_4, KNO_3 and GA_3 treatments on germination of caper (*Capparis ovata* Desf.) seeds. *Pak. J. Biological Sci* 7, 879–882.

Panico, A.M., Cardile, V., Garufi, F., Puglia, C., Bonina, F., Ronsisvalle, G., 2005. Protective effect of *Capparis spinosa* on chondrocytes. *Life Sci* 77, 2479–2488.

Pascual, B., San Bautista, A., Imbernon, A., Lopez–Galarza, S., Alagarda, J., Marto, J.V., 2004. Seed treatment for improved germination of caper (*Capparis spinosa* L.). *Seed Sci. Tech.* 32, 637-42.

Pascual, B., San Bautista, A., Imbernon, A., Lopez–Galarza, S., Alagarda, J., Marto, J.V., 2006. Germination behavior after storage of caper seeds. *Seed Sci. Tech* 34, 151–9.

Pascual, B., San Bautista, A., Pascual Seva, N., Garcia Molina, R., Lopez–Galarza, S., Maroto, J.V., 2009. Effect of soaking period and gibberellic acid addition on caper seed germination. *Seed Sci. Tech* 37, 33–41.

Phillips, R., Rix, M., 1998. *Conservatory and Indoor Plants Volumes 1 and 2* Pan Books, London.

Pupalla, N., Fowler, J.I., 2002. Lesquerella seed pre-treatment to improve germination. *Industrial Crops and Products* 17, 61–9.

Ramezani, Z., Aghel, N., Keyghobadi, H., 2008. Rutin from different parts of *Capparis spinosa* growing in Khuzestan/Iran. *Pak. J Biol. Sci* 11, 768-772.

Rodrigo, M., Lazaro, M.J., Alvarruiz, A., Giner, V., 1992. Composition of capers (*Capparis spinosa*): Influence of cultivar, size and harvest date. *J. Food Sci* 57, 1152-1154.

Romeo, V., Ziino, M., Giuffrida, D., Condurso, C., Verzera, A., 2007. Flavour profile of capers (*Capparis spinosa* L.) from the Eolian Archipelago byHS-SPME/GC–MS. *Food Chem* 101, 1272-1278.

Roy Burman, J.J., 2003. Tribal medicine. Mittal Publications, New Delhi.

Sadykov, L., Yu, D., Khodzhimatov, M., 1981. Alkaloids of *Capparis spinosa. Dokl. Akad. Nauktadzh. SSR* 24, 617-620.

Sama, S.K., Krishnamurthy, L., Ramachandran, K., Lal, K., 1976. Efficacy of Liv.52 in acute viral hepatitis. A double-blind study. *Indian J. Med. Res* 64, 738.

Schraudolf, H., 1989. Indole glucosinolates of *Capparis spinosa. Phytochem* 28, 259-260.

Shahina, A.G., 1994. *Handbook of Arabian Medicinal Plants.* USA: CRC Press, Inc. pp 73.

Shaik, S., Dewir, Y.H., Singh, N., Nicholas, A., 2008. Influence of pre–sowing treatments on germination of cancer bush (*Sutherlandia frutescens*), a reputed medicinal plant in arid environments. *Seed Sci. Tech* 36, 795–801.

Sharaf, M., el-Ansari, M.A., Saleh, N.A., 2000. Quercetin triglycoside from *Capparis spinosa. Fitoterapia* 71, 46-49.

Sharma, R., 2003. Medicinal Plants of India – An Encyclopedia, (Daya Publishing House, New Delhi, 42-43.

Singh, N., Singh, R., Kumar, H., Bhoyar, M.S., Singh, S.B., 2008. Sustainable vegetable research for nutritional security and socio-economic upliftment of the tribal's in Indian cold desert. *Advances in Agriculture, Environment and Health.* Edited by Singh *et al.,* SSPH Publication, New Delhi, pp. 43-59.

Smith, J.S.C., Smith, O.S., 1992. Fingerprinting crop varieties. *Advances in Agronomy* 47, 85-140.

Soyler, D., Khawar, K.M., 2007. Seed germination of caper (*Capparis ovate* var. *herbacea*) using α-Napthalene acetic acid and gibberellic acid. *Inter. J. Agril. Biology* 9, 35–37.

Sozzi, G.O., Chiesa, A., 1995. Improvement of caper (*Capparis spinosa* L.) seed germination by breaking seed coat–induced dormancy. *Scientia Horticulturae* 62, 255–61.

Subhose, V., 2005. *Bull Indian Inst Hist Med Hyderba.* 35: 83.

Tesoriere, L., Butera, D., Gentile, C., Livrea, M.A., 2007. Bioactive components of caper (*Capparis spinosa* L.) from Sicily and antioxidant effects in a red meat simulated gastric digestion. *J. Agric Food Chem.* 55, 8465-8471.

Tomas, F., Ferreres, F., 1978. 3-O-Rhamnorutinosyl kaempferol from floral buttons of *Capparis spinosa*. *Rev. Agroquim Tecnol. Ailment* 18, 232-235.

Travlos, I.S., Economou, G., Karamanos, A.I., 2007. Germination and emergence of hard seed coated *Tylosema esculentum* (Bruch) A. Schreib in response to different pre sowing treatments. *J Arid Environ* 68, 501–507.

Trombetta, D., Occhiuto, F., Perri, D., Puglia, C., Santagati, N.A., Pasquale, A.D., Saija, A., Bonina, F., 2005. Anti-allergic and antihistaminic effect of two extracts of *Capparis spinosa* L. flowering buds. *Phytother. Res* 19, 29-33.

USDA, Nutrient database, 2008. http://www.nal.usda.gov/fnic/foodcomp/search/

Zhan, A., 1978. Chemical and Biological characterization of *Capparis spinosa*. *Azerb. Med. Zh* 55, 70-75.

2015, Modern Methods in Phytomedicine *Pages* ***59–67***
Editor: **T. Parimelazhagan**
Published by: **DAYA PUBLISHING HOUSE, NEW DELHI**

5

Direct Shoot Bud Regeneration from Shoot Tip Explants of *Leucas aspera*: An Important Medicinal Plant

T. Bhuvaneawari*[1], *N. Geetha*[2] *and P. Venkatachalam*[1]

[1]Plant Genetic Engineering and Molecular Biology Lab, Department of Biotechnology, Periyar University, Periyar Palkalai Nagar, Salem – 636 011, Tamil Nadu
[2]Department of Biotechnology, Mother Teresa Women's University, Kodaikanal – 624101, Dindugal, Tamil Nadu

1.0 INTRODUCTION

Medicinal plants are an essential source of life saving drugs and play a major role in World health (Constabel, 1990). *Leucas aspera* (Willd) Link is a species of annual branched medicinal herb that belongs to the family Lamiaceae (Labiatae). It is a valuable medicinal plant popularly known as "Thumbai" (Rai, 2005) in Tamil. It grows abundantly in the high land crop fields, roadsides and fallow lands of the wide area of South Asia (India, Bangladesh, Nepal), Malaysia, and Mauritius (Shrestha, 2000). Plants of genus *Leucas* are generally shrubs, sub shrubs, or perennial herbs grows up to 15-45 cm height and flowers are white, sessile or sub sessile in terminal. Whitish hairs are generally present on the outer surface of the upper lip of

* *Corresponding Author.* E-mail: pvenkat67@yahoo.com

the corolla (Ryding, 1998). The genus *Leucas* comprises of about 80 species (Hedge, 1990) and in India, 43 species are available (Mukerjee, 1940). The highest species diversity has been found in East Africa (Ryding, 1998).

There are many *Leucas* species such as *Leucas aspera* (Wild) Spreng, *L. biflora* (Vahl), *L. linifolia* Spreng, *L. lanata* Benth, *L. diffusa* Benth, *L. inflata* Benth, *L. indica* (L.) R.Br, *L. zeylanica.* Among the species, *Leucas indica* (L.) R.Br. is mostly found in the Asian countries (Khanah and Chopra, 2005). It exhibiting a range of biological activities like analgesic-antipyretic, anti-rheumatic, anti-inflammatory, antibacterial, antifungal treatment and its paste is applied topically to inflamed areas (Gani, 2003). *Leucas aspera* is traditionally used as a medicine for coughs, colds, painful swellings and chronic skin eruptions (Chopra, 2002), cobra venom poisoning (Reddy and Viswananthan, 1993), hepatoprotective activity (Gupta, 2005), prostaglandin inhibitory, antioxidant, and tooth infections (Rahmatullah, 2009) and mosquito repellent (Maheswaran, 2008) by the rural people and possesses wound healing properties (Mangathayaru, 2006). Fresh flower of *Leucas aspera* showed significant biphasic RBC membrane stabilization activity against hypotonicity induced hemolysis due to the presence of a flavonoidal glycoside, baicalin (Manivannana and Sukumar, 2007). Due to their medicinal importance, many scientific studies have been carried out on the phytochemical, pharmacological values of *Leucas aspera* which suggested that this genus have immense potential for the discovery of new drugs. Different types of chemicals such as 7-hydroxydotriacontan-2-one, 1 hydroxytetriacontan-4-one, 32-methyl tetratriacontan-8-ol, 5-acetoxytriacontane, dotriacontanol and β-sitosterol also found in *Leucas aspera.*

The pharmacological value of this medicinal herb is mainly due to the presence of major compounds such as triterpenoids (Kamat, 1994) glucosides, tannins, saponins, sterols, oleic acid, linoleic acid, linolenic acid, palmitic acid, stearic acid, oleanolic acid, ursolic acid, nicotin etc. have already been isolated from the leaves, roots, flower and seeds of this plant (Prajapati, 2010; Rahman, 2007 and Srinivasan, 2011). Plant tissue culture is an efficient technique for conservation of endangered plant species, especially those with a limited reproductive capacity and threatened habitats (Ruta and Irene, 2010).

Micropropagation is the well-known method for efficient and large scale production of medicinal and aromatic plants and for commercial utilization of valuable plant-derived pharmaceuticals (Rout, 2002; Faisal, 2005). Explant source is one of the most important factors in the induction of morphogenetic response of *in vitro* culture. There are many *in vitro* studies have been conducted on Lamiaceae species using different explants, like nodal segments, leaf explants, and shoot tips (Begum *et al.*, 2002; Nirmal and Sehgal, 2010; Winthrop and Simon, 2000). Shoot tips provide a preferred source of explant material as they tend to produce genetically similar plants because no intermediate callus phase is involved (Bao *et al.*, 2001; Yookongkaew *et al.*, 2007; Kumar and Chandra, 2009). Over exploitation of the plant for miscellaneous medicinal uses has resulted in its scarcity in nature, which calls for newer approaches for the rapid propagation of this important medicinal plant. Various plant growth regulators play significant roles in the regeneration process of plants under *in vitro* conditions. Plant growth regulators including to the auxin and cytokinin groups are used for *in vitro* morphogenetic control (Ludwig-Müller, 2000; Davey and Anthony, 2010). Lamiaceae is one of the largest and highly evolved angiosperm families having considerable pharmaceutical and culinary interest. The literature on regeneration of *Leucas aspera* and the role of plant growth regulators has not yet been reported, but there are reports on regeneration of plants from different species of this family (Lamiaceae). In the present study, an efficient protocol for *in vitro* plant regeneration of *Leucas aspera* was developed.

2.0 MATERIALS AND METHODS

2.1 Preparation of Explants

Leucas aspera plants were collected from Periyar University Campus, Salem-11. For *in vitro* propagation, the shoot tip explants (1cm) excised from healthy plants were washed in running tap water to remove superficial dust particles. The explants were then treated with 1 per cent (w/v) of Bavistin for 5 min and 5 per cent (v/v) of Teepol for 5 min to get rid off microbes. Then the explants were surface sterilized with 0.1 per cent (w/v) mercuric chloride solution for 2-3 min and finally rinsed 4 times in sterile distilled water to remove traces of mercuric chloride. Eventually, the sterilized shoot tips were placed by inserting their cut-ends in to the MS medium for *in vitro* intiation.

2.2 Culture Medium and Growth Condition

MS media (Murashige and Skoog's, 1962) containing 3 per cent (w/v) sucrose and supplemented with various concentrations of cytokinins such as BAP and KIN were used. The pH of the media was adjusted to 5.8 with 0.1 N NaOH or HCl before adding of 0.7 per cent (w/v) agar. Media (15 mL) were dispensed into 25 mm ×150 mm culture tubes (Borosil, Mumbai) and autoclaved at 121°C for 15 min. In each culture tube one shoot tip explant was implanted. The cultures were incubated at 24±2°C under 16/8 h (light/dark cycle) photoperiod (60μE m^{-2} s^{-1}) and irradiance provided by cool-white fluorescent tubes.

2.3 Initiation of Shoot Buds

Surface sterilized shoot tip explants were cultured on MS medium supplemented with different concentrations of BAP (0.5, 1.0, 1.5 and 2.0 mg/l) in combination with KIN (0.5 mg/l) for shoot bud induction. After ten days of culture, shoot tip explants with newly emerged shoots were subculture onto fresh medium for multiple shoot bud initiation.

2.4 Induction of Multiple Shoots

Shoot buds initiated from the shoot tip explants were transferred to MS medium containing different concentrations of (BAP-1.0, 2.0, 3.0, 4.0, and 5.0 mg/l with KIN-0.5mg/l) as mentioned above for further shoot bud multiplication. Multiple shoot buds were sub cultured on to the same medium for feature growth and elongation.

2.5 Rooting and Acclimatization

The elongated shoots (1-2 cm) were transferred onto half-strength MS medium fortified with different concentrations of IBA (0.5-2.5mg/l) for root induction. Plantlets with well-developed roots were removed from the culture tubes and gently washed under running tap water to remove traces of agar medium. Subsequently, they were transferred to plastic cups containing sterile soil and sand mixture in 2:1 ratios. Plants were covered with transparent polyethylene bags to maintain adequate moisture for a week and shifted to the greenhouse. After two weeks, they were transplanted into the field.

3.0 RESULTS AND DISCUSSION

3.1 Shoot Bud Initiation

In the present study, shoot tip explants collected from mature *Leucas aspera* plants were cultured on MS medium supplemented with different concentrations of BAP (0.5-2.5 mg/l) in combination with 0.5 mg/l KIN. After 15 days of culture, shoot buds were emerged directly from the shoot tip explants. Among the cytokinin combinations tested, highest percent of shoot bud regeneration (97.14 per cent) was noticed on MS medium containing 1.5 mg/l BAP + 0.5 mg/l KIN with 1.98 shoots/explants. Though the shoot bud induction was slightly decreased at higher concentrations of BAP the number of shoots bud induction was increased (Table 5.1). The combined effect of BAP and KIN was found to be essential to induce shoot buds from shoot tip explants of *Leucas aspera.* The promoting effect of BAP in combination with KIN on shoot bud regeneration has been well documented in some medicinal plants. Premkumar *et al.* (2011) reported that the combination of BAP 4.44 µM and KIN 2.32 µM was found to be efficient for direct shoot formation in *Scoparia dulcis.*

Table 5.1: Effect of different Concentrations of BAP in Combination with 0.5 mg/l KIN on Shoot Bud Induction from Shoot Tip Explants of *Leucas aspera*

Cytokinin Concentration (mg/l)		*Percent of Shoot Bud Induction (Mean ± S.E)*	*No of Shoots/ Explant (Mean ± S.E)*
BAP	*KIN*		
0.5	0.5	68.6±2.12	2.47±0.17
1.0	0.5	73.4±3.62	2.67±0.08
1.5	0.5	86.6±4.13	2.82±0.12
2.0	0.5	84.3±3.70	3.12±0.17
2.5	0.5	83.7±2.89	3.31±0.13

3.2 Shoot Bud Multiplication and Elongation

Shoot buds developed from shoot tip explants were subcultured onto MS medium supplemented with different concentrations of BAP (1.0-5.0 mg/l) in combination with 0.5 mg/l.

KIN for shoot bud multiplication. Among the combinations tested, the highest percent of multiple shoot bud formation (100 per cent) was noticed on MS medium fortified with 3.0 mg/l BAP and 0.5 mg/l KIN combination with 5.33 shoots/explants (Table 5.2). By increasing the concentration of BAP, the percent of multiple shoot bud induction was increased up to 3.0 mg/l and the shoot bud regeneration was

Figure 5.1: Direct Shoot Bud Development and Plant Regeneration from Shoot Tip Explants of *Leucas aspera.*

A: Shoot bud initiation; B: Multiple shoot bud development; C: Rooting of elongated shoot and D: Regenerated plant growing in soil.

decreased above the optimum concentration. However the number of multiple shoot bud development was positively correlated with BAP concentrations.

Table 5.2: Effect of different Concentrations of BAP in Combination with 0.5mg/l KIN on Shoot Bud Multiplication from Shoot Tip Explants of *Leucas aspera*

Cytokinin Concentration (mg/l)		*Percent of Multiple Shoot Induction (Mean ± S.E)*	*No of Shoots/ Explant (Mean ± S.E)*
BAP	*KIN*		
1.0	0.5	83.33±2.67	3.50±0.50
2.0	0.5	87.50±2.50	4.90±0.10
3.0	0.5	100.0±1.00	5.33±0.33
4.0	0.5	95.71±11.29	6.15±0.65
5.0	0.5	89.12±2.83	8.41±0.37

Shoot elongation was also noticed on the same media composition upon transfer to fresh medium. Similarly, Palai *et al.* (1997) also reported that the combination of two cytokinins along increased the rate of shoot multiplication in *Zingiber officinale.* Increasing the concentration of BAP (above 1.5 mg/l) in the culture media resulted in reduction of percent of shoot bud multiplication.

3.3 *In vitro* Rooting and Acclimatization

The regenerated shoots were dissected out individually and cultured on half-strength MS medium without hormones and half-strength MS medium augmented with various concentrations of IBA (0.5-2.5 mg/l) alone for root induction. The roots were induced directly from the shoot base without intervening callus phase on media containing IBA and hormone free medium. Among the IBA concentrations used, the maximum percent (100 per cent) of root induction was observed on half-strength MS medium containing 2.0 mg/l IBA with 7.33 roots/shoot (Table 5.3). The frequency of rooting was slightly decreased beyond 2.0 mg/l IBA concentration but number of shoots was increased at high concentration of IBA. In general, IBA showed a strong rooting response in a wide range of plant species.

Table 5.3: Effect of different Concentrations of IBA on Root Induction from Elongated Shoots of *Leucas aspera*

Hormone Concentration (mg/l)	*Percent of Rooting*	*No. of Roots/Shoot (Mean±S.E)*
½ MS	78.4±2.30	3.3±0.21
IBA		
0.5	81.2±2.70	4.0±0.40
1.0	90±0.00	5.33±0.66
1.5	95±2.40	6.66±0.33
2.0	100±0.00	7.33±0.33
2.5	98±0.00	7.5±0.5

Similar observation has also been reported in *Mucuna pruriens* (Faisal *et al.*, 2006) and *Jatropha curcas* (Kumar *et al.*, 2008). It is interesting to note that, the roots were also obtained on half- strength MS medium without hormones, but the formed roots were found to be very thin. Healthy shoots with well developed roots obtained after 20 days of culture were gently removed from the culture tubes and washed the roots with running tap water to remove the traces of agar. Then they were transferred into plastic cups containing sterile soil and sand in the ratio 2:1 and covered with polythene bags to ensure high humidity. The plantlets were kept in the controlled environment for 1 week and polythene bags were gradually removed in order to acclimatize the plantlets and then they were shifted to the green house conditions for further growth. After 2 weeks, they were established in the field condition. The plantlets grew well without showing any phenotypic variation and the survival rate noticed was 87 per cent. The regenerated plants were flowered normally and set seeds.

4.0 CONCLUSION

In conclusion, a reliable and effective protocol for *in vitro* regeneration from shoot tips explants of an important medicinal herb *Leucas aspera* to meet the increasing demand of the herbal medicines instead of chemical drugs. Among the BAP and KIN combinations used for plant regeneration, the combination of BAP (1.5 mg/l) + KIN (0.5 mg/l) was found to be best for direct shoot bud initiation while BAP (3.0 mg/l) + KIN (0.5 mg/l) combination produced maximum percent shooting with highest number of multiple shoot buds. Maximum percent of rooting was noticed on a medium containing 2.0 mg/l IBA. This protocol can be utilized for commercial scale propagation and conservation of this important medicinal plant species.

REFERENCES

Bao, P.H., Granata, S., Castiglione, S., Wang, G., Giordani, C., Cuzzoni, E., Damiani, G., Bandi, C., Datta, S.K., Datta, K., Potrykus, I., Callegarin, A., Sala, F., 2001. Genomic changes in transgenic rice (*Oryza sativa* L.) plants produced by infecting calli with *Agrobacterium tumefaciens*. *Plant Cell Rep.* 20, 325–330.

Begum, F.M., Amin, N., Azad, M.A.K., 2002. *In vitro* rapid clonal propagation of *Ocimum basilicum* L. *Plant Tiss. Cult.* 12, 27-35.

Chopra, R.N., Nayar, S.L., Chopra, I.C., 2002. Glossary of Indian medicinal plants. NISCAIR, CSIR, New Delhi, pp. 153.

Constabel, F., 1990. Medicinal plant biotechnology. *Planta Med.* 56, 421-425.

Davey, M.R., Anthony, P., 2010. Plant cell culture: essential methods. Wiley, Chichester, UK.

Faisal, M., Siddique, I., Anis, M., 2006. *In vitro* rapid regeneration of plantlets from nodal explants of *Mucuna pruriens* -a valuable medicinal plant. *Ann. Appl. Biol.* 148, 1-6.

Premkumar, G., Sankaranarayanan, R., Jeeva, S., Rajarathinam, K., 2011. Cytokinin induced shoot regeneration and flowering of *Scoparia dulcis* L. (Scrophulariaceae)- an ethnomedicinal herb. *Asian Pac J Trop Biomed.* 1(3), 169–172.

Gani, A., 2003. Medicinal plants of Bangladesh: Chemical constituents and uses. Asiatic Society of Bangladesh, pp. 215.

Gupta, Nakul, 2005. Evaluation of hepatoprotective activity of the plant *Leucas aspera* spreng. In Rats. M.Pham Thesis.

Hedge, I.C., 1990. Labiatae, In: Flora of Pakistan. Ali, S.I., Nasir, Y.J. (Ed). University of Karachi, Department of Botany, Karachi, pp. 192.

Kamat, M., Singh, T.P., 1994. Preliminary chemical examination of some compounds in different parts of genus *Leucas*. *Geobios* 21, 31–33.

Kumar, N., Pamidimarri, S.D.V.N., Kaur, M., Boricha, G., Reddy, M.P., 2008. Effects of NaCl on growth, ion accumulation, protein, proline contents, and antioxidant enzymes activity in callus cultures of *Jatropha curcas*. *Biologia* 63, 378-382.

Kumar, S., Chandra, A., 2009. Direct plant regeneration via multiple shoot induction in *Stylosanthes seabrana*. *Cytologia* 74, 391–399.

Ludwig-Müller, J., 2000. Indole-3-butyric acid in plant growth and development. *Plant Growth Reg.* 32, 219–230.

Maheswaran, R., Sathish, S., Ignacimuthu, S., 2008. Larvicidal activity of *Leucas aspera* (Willd.) against the larvae of *Culex quinquefasciatus* Say. and *Aedes aegypti* L. *Int. J. Integr. Biol.* 2, 214-217.

Mangathayaru, K., Thirumurugan, D., Patel, P.S., Pratap, D.V., David D.J., Karthikeyan, J., 2006. Isolation and identification of nicotine from *Leucas aspera* (Willd) Link. *Ind. J. Pharm. Sci.* 68, 88-90.

Manivannan, R., Sukumar, D., 2007. The RBC membrane stablisation in an *in vitro* method by the drug isolated from *Leucas aspera*. *Int. J. Appl. Sci. Eng.* 5, 133-138.

Mukerjee, S.K., 1940. A revision of the Labiatae of the Indian Empire. *Rec. Bot. Surv. India.* 14, 1-205.

Nirmal, K.S., Sehgal, C.B., 2010. Micropropagation of 'Holy Basil' (*Ocimum sanctum* Linn.) from young inflorescences of mature plants. *Plant Growth Regul.* pp 1.

Palai, S.K., Rout, G.R., Das, P., 1997. Micropropagation of giger (*Zingiber officinale* Rosc.)-interaction of growth regulators and culture conditions. In: Ramana S, Sasikumar KV, Nirmal Babu B, Eapen K, editors. Biotechnology of spices, medicinal and aromatic plants of India. Kerala: Indian Society for Spices; pp. 20–24.

Prajapati, M.S., Patel, J.B., Modi, K., Shah, M.B., 2010. *Leucas aspera:* A review. *Pharm. Rev.* 4, 85-87.

Rahman, M.S., Sadhu,S.K., Hasan, C.M., 2007. Preliminary antinociceptive, antioxidant and cytotoxic activities of *Leucas aspera* root. *Fitoterapia* 78, 552-555.

Rahmatullah, M., Das, A.K., Mollik, M.A.H., Jahan, R., Khan, M., Rahman,T., Chowdhury, M.H., 2009. An ethno medicinal Survey of Dhamrai sub-district in Dhaka district, Bangladesh. *American-Eurasian J. Sust. Agric.* 3, 881-888.

Rai, V., Agarwal, M., Agnihotri, A.K., Khatoon, S., Rawat, A.K., Mehrotra, S., 2005. Pharmacognostical evaluation of *Leucas aspera. Nat. Prod. Sci.* 11, 109- 114.

Reddy, M.K., Viswananthan, S.,Thirugnanasambatham, D., Santa, R., Lalitha, K., 1993. Effect of *Leucas aspera* on snake venom poisoning in mice and its possible mechanism of action. *Fitoterapia* 64, 442-446.

Rout, G.R., 2002. Direct plant regeneration from leaf explants of *Plambago* species and its genetic fidelity through RAPD markers. *Ann. Appl. Biol.* 140, 305-313.

Ruta, C., Irene, M.F., 2010. *In vitro* propagation of *Cistus clusii* Dunal, an endangered plant in Italy. *In Vitro Cell. Dev. Biol. Plant* 46, 172-179.

Ryding, O., 1998. Phylogeny of the *Leucas* Group (Lamiaceae). *Syst. Bot.* 23, 235-247.

Shrestha, K.K., Sutton, D.A., 2000. Annotated checklist of the flowering plants of Nepal. Press JR, The Natural History Museum.

Srinivasan, R., 2011. *Leucus aspera* - Medicinal Plant: A Review. *Int. J. Pharma. Bio Sci.* 2(1), 153–159.

Winthrop, B.P., Simon, J.E., 2000. Shoot regeneration of young leaf explants from basil (*Ocimum basilicum* L.). *In Vitro Cell Dev. Biol. Plant* 36, 250-254.

Yookongkaew, N., Srivastanakul, M., Narangajavana, J., 2007. Development of genotype-independent regeneration system for transformation of rice (*Oryza sativa* ssp. *indica*). *J. Plant Res.* 120, 237–245.

2015, Modern Methods in Phytomedicine *Pages* ***69–75***
Editor: **T. Parimelazhagan**
Published by: **DAYA PUBLISHING HOUSE, NEW DELHI**

6

Plant Diversity Assessment at Arulmigu Sri Veeramakali Amman Sacred Grove, Silatur, Pudukkottai District of Tamil Nadu

L. Rajavel[1], B. Muthukumar[1], E. Natarajan[1], S. Soosairaj[2], B. Balaguru[3] and N. Nagamurugan[4]

[1]*P.G and Research Department of Botany, National College, Tiruchirappalli*
[2]*Department of Botany, St. Joseph's College, Tiruchirappalli*
[3]*Department of Botany, Jamal Mohamad College, Tiruchirappalli*
[4]*Department of Biotechnology, Kurinji College of Arts and Science, Tiruchirappalli*

1.0 INTRODUCTION

Biodiversity is responsible for the essential ecosystem services such as regulation of atmospheric gaseous composition, climate, disturbance and water, soil formation, maintenance of soil fertility, waste assimilation and repository of gene pool (Singh, 2003), besides it is the basis of human survival and economic well being since it provides timber, food, fiber, cosmetics, emulsifiers, dyes and natural pesticides (Costanza *et al.,* 1997). The tropical region is rich in biodiversity especially India, which is one of the mega diversity region in the world (Brooks *et al.,* 2006). Due to industrialization, human population explosion and globalization, deforestation is 0.8 per cent per year in tropics (Rosenszweig, 1995) and subsequent loss of species is 2-5 species per hour (Huges *et al.,* 1997). Hence, inventories, conservation and sustainable management of forest are inevitable to prevent further loss of species.

There are numerous inventories from the Western and Eastern ghats such as deciduous forest of Mudumalai (Sukumar *et al.,* 1992), wet evergreen forest of Western ghats (Pascal and Pelissier, 1996), tropical evergreen forest around Sengaltheri (Parthasarathy, 2001) and tropical dry evergreen forest in the Eastern ghats (John Britto *et al.,* 2002). However, floristic and diversity studies in different patches of sacred grove is not much analyzed for various reasons. Hence the present study aims at assessing the plant diversity resources of a sacred grove of Arulmigu Sri Veeramakali Amman, at Silatur, Pudukkottai district, Tamil Nadu.

2.0 MATERIALS AND METHODS

2.1 Study Area

Arulmigu Sri Veeramakali amman sacred grove is situated at Silatur village in Aranthangi Taluk of Pudukkottai district, Tamil Nadu (Figure 6.1). The grove is spread over 2 ha area with the amman grove at the midst. The grove has taken care by the local people and a special festival is celebrated in the month of May every year. The grove is surrounded by agricultural land on west and south, a pond towards the east and the village residence towards the north.

2.2 Methods

The grove was sampled for plant diversity analysis with 10 (10 x 10 m) workable random quadrats. Within each quadrat all individual species of trees (girth > 30) and shrubs (girth > 10) GBH (Girth at Breast Height) were measured and noted in the field note. Herbaceous species and seedlings were gfe died in three 1 x 1 quadrats within each 10 m quadrat. This information is documented in a field note. Herbarium species of all species were collected and dried using standard herbarium techniques and identified at Rapinat Herbarium Tiruchirappalli, St.Joseph's College (Autonomous), Tiruchirappalli, Tamil Nadu. The recorded data were analysed for various biodiversity parameters such as species richness, abundance, density, relative frequency, relative density, basal area, relative basal area, important value index (IVI), AB/per cent F ratio and indices like Shannon Wiener and Simpson.

3.0 RESULTS AND DISCUSSION

Vegetation of Arulmigu Sri Veeramakali amman grove is very dense and well protected on all the sides. The grove is surrounded by village settlement on northern and eastern sides and cultivation on southern and western side. A large pond is located towards the east of the temple. 0.1 ha area was analysed for diversity indices in ten 10 x 10 m quadrats. The present study yield report of 62 species under 48 genera and 38 families contributing 366 individuals. The dominant tree species is *Manilkara hexandara* followed by *Lannea coramandelica*. The shrub layer is dominated by *Memecylon umbellatum* followed by *Cadaba indica*. The family Euphorbiaceae (5 species) followed by Rubiaceae (5 species), Rutaceae (4 species) Papilionaceae (3 species) were the dominant families.

3.1 Tree Diversity

The sacred grove vegetation is comprised of 23 tree species with 111 individuals. The tree stand was dominated by *Lannea coromandelica* (11), *Manilkara hexandra* (10)

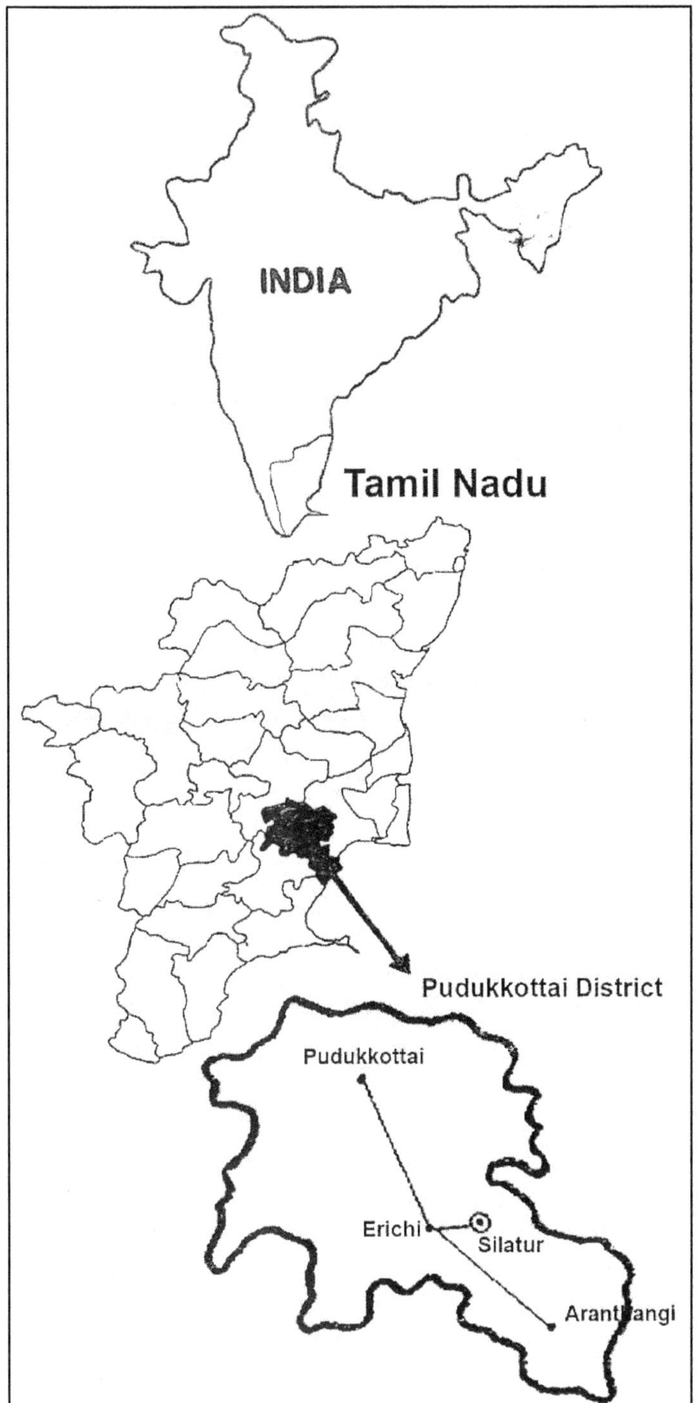

Figure 6.1: Location of Arulmigu Sri Veeramakali Amman Sacred Grove at Silatur, Pudukkottai District of Tamil Nadu.

Table 6.1: Relative Diversity Index of Arulmigu Sri Veeramakali Amman Sacred Grove

Parameters	*Trees*	*Shrub*	*Herb*
Shannon-Wiener index	2.864	2.66	3.57
Simpson index	0.057	0.079	0.029
Individuals	110	126	129
Number of family		36	
Number of genera		58	
Number of species		62	

Table 6.2: Species List and their Habit Recorded from Arulmigu Sri Veeramakali Amman Sacred Grove

Sl.No.	*Species Name*	*Family*	*Habit*
1.	*Abrus precatorious* L.	Fabaceae	Herb
2.	*Acacia caesia* (L.) Willd.	Fabaceae	Straggler
3.	*Acacia leucophloea* Willd.	Fabaceae	Tree
4.	*Achyranthes aspera* L.	Amaranthaceae	Herb
5.	*Adenia wightiana* Wight and Arn.	Passifloraceae	Climber
6.	*Aerva lanata* L. Juss.	Amaranthaceae	Herb
7.	*Albizia amara* (Roxb.) Boivin	Fabaceae	Tree
8.	*Albizia lebbeck* (L.) Benth.	Fabaceae	Tree
9.	*Allophylus serratus* (Hiern) Kurz	Sapindaceae	Shrub
10.	*Aristida setacea* Retz.	Poaceae	Herb
11.	*Aristolochia indica* L.	Aristolochiaceae	Herb
12.	*Atalantia racemosa* Wight and Arn.	Rutaceae	Shrub
13.	*Azadirachta indica* A. Juss.	Meliaceae	Tree
14.	*Benkara malabarica* (Lam.) Tirveng.	Rubiaceae	Shrub
15.	*Cadaba indica* Lam.	Capparaceae	Shrub
16.	*Cadaba trifoliata* (Roxb.) Wight and Arn.	Capparaceae	Shrub
17.	*Capparis brevispina* DC.	Capparaceae	Shrub
18.	*Carissa carandas* L.	Apocynaceae	Shrub
19.	*Carmona retusa* (Vahl) Masam.	Boraginaceae	Shrub
20.	*Cassine glauca* Kuntze	Celastraceae	Tree
21.	*Chloroxylon swietenia* DC.	Rutaceae	Tree
22.	*Cissus vitiginea* L.	Vitaceae	Vine
23.	*Clausena dentata* (Willd.) Roem.	Rutaceae	Shrub
24.	*Clitoria ternatea* L.	Fabaceae	Herb
25.	*Coffea wightiana* Wight and Arn.	Rubiaceae	Shrub

Contd...

Table 6.2–*Contd...*

Sl.No.	*Species Name*	*Family*	*Habit*
26.	*Combretum ovalifolium* Roxb.	Combretaceae	Straggler
27.	*Cyperus rotundus* L.	Cyperaceae	Herb
28.	*Cyrtococcum trigonum* (Retz.) A. Camus	Poaceae	Herb
29.	*Derris scandens* (Roxb.) Benth.	Fabaceae	Straggler
30.	*Desmodium triflorum* (L.) DC.	Fabaceae	Herb
31.	*Dioscorea oppositifolia* L.	Dioscoreaceae	Climber
32.	*Dipteracanthus prostratus* (Poir.) Nees	Acanthaceae	Herb
33.	*Drypetes sepiaria* (Wight and Arn.) Pax and K. Hoffm.	Euphorbiaceae	Tree
34.	*Euphorbia antiquorum* L.	Euphorbiaceae	Tree
35.	*Fluggea leucopyrous* Willd.	Euphorbiaceae	Shrub
36.	*Glycosmis mauritiana* Tanaka	Rutaceae	Shrub
37.	*Grewia bracteata* Roth	Tiliaceae	Straggler
38.	*Gymnema sylvestre* (Retz.) R. Br.	Asclepiadaceae	Climber
39.	*Hemidesmus indicus* (L.) R. Br. ex Schult. var indicus	Asclepiadaceae	Herb
40.	*Hybanthus enneaspermus* (L.) F. Muell.	Violaceae	Herb
41.	*Jasminum angustifolium* Willd.	Oleaceae	Straggler
42.	*Lannea coromandelica* (Houtt.) Merr.	Anacardiaceae	Tree
43.	*Manilkara hexandra* (Roxb.) Dubard	Sapotaceae	Tree
44.	*Maytenus emarginata* Willd.	Celastraceae	Shrub
45.	*Memecylon umbellatum* Burrn. f.	Melastamataceae	Shrub
46.	*Morinda tinctoria* Roxb.	Rubiaceae	Shrub
47.	*Phoenix sylvestris* (L.) Roxb.	Aracaceae	Shrub
48.	*Phyllanthus polyphyllus* Dalzell and Gibson	Euphorbiaceae	Tree
49.	*Phyllanthus reticulates* Poiret	Euphorbiaceae	Shrub
50.	*Premna latifolia* Roxb.	Verbenaceae	Shrub
51.	*Pyrenacantha volubilis* Hook.	Icacinaceae	Herb
52.	*Randia dumetorum* (Retz.) Poiret.	Rubiaceae	Shrub
53.	*Rivea hypocratrifonnis* (Desr.) Choisy	Convolvulaceae	Straggler
54.	*Sansevieria roxburghiana* Schult and Schult. f.	Liliaceae	Herb
55.	*Strychnos lenticellata* Hill.	Loganiaceae	Straggler
56.	*Syzygium cumini* (L.) Skeels	Myrtaceae	Tree
57.	*Tarenna asiatica* (L.) Kuntze ex K. Schum.	Rubiaceae	Shrub
58.	*Tinospora cordifolia* (Willd.) Miers	Menispermaceae	Climber
59.	*Ventilago maderaspatana* Gaertner	Rhamnaceae	Straggler
60.	*Vernonia cinera* L.	Asteraceae	Herb
61.	*Wattakaka volubilis* (L. f.) Stapf	Asclepiadaceae	Straggler
62.	*Zizyphus oenoplia* (L.) Mill.	Rhamnaceae	Straggler

and *Chloroxylon swietenia* (9). In addition to tree species, straggler like *Cissus vitiginea* (14) and *Combretum ovalifolium* (10) were frequently noted. A total of 0.53 m basal area was contributed by 111 individuals of which the maximum contribution was by *Lannea coromandelica* (0.069 m), followed by *Manilkara hexandra* (0.063 m) and *Syzygium cumini* (0.057). Muthukumar *et al.* (2005) report that *Chloroxylon swietenia* and *Wrightia tinctoria* are the dominant tree species in other part of the same district and Ramanujam and Cyril (2003) report that *Aglaia elaeagnoidea* and *Borassus flabellifer* were the dominant species in Pondicherry sacred groves. The tree diversity indices such as Shannon-Wiener and Simpson value were 2.864 and 0.096 respectively. Of the various tree species *Lannea coromandelica* had better regeneration status than the other species.

3.2 Shrub Diversity

The shrub vegetation consists of 20 species in 12 families with 126 individuals. Maximum individuals was contributed by *Glycosmis mauritiana* (49) followed by *Memecylon umbellatum* (16) and minimum individuals was contributed by *Cadaba indica, Morinda tinctoria* and *Randia duematorum* all represented by only one individual. Over all basal area of shrub species is 0.042 m of which the maximum was contributed by *Memecylon umbellatum* (0.009 m) followed by *Glycosmis mauritiana* (0.006 m^2). Similar finding has been reported by John Britto *et al.* (2000) from Vamban sacred grove in Pudukkottai. Shannon-Wiener diversity value of shrub species is 2.661 and the Simpson value is 0.079. Among the various shrub species *Atalantia racemosa* has better regeneration status than the other shrub species.

3.3 Herb Diversity

The herbaceous vegetation consists of 46 species with 129 individuals. Maximum individuals were contributed by seedling of *Strychnos lenticellata* (17) and the herb species *Sansevieria roxburghiana* (9). Diversity indices like Shannon-Wiener value of herbs are 3.576 and the Simpson value is 0.029. Of the three layers the herb vegetation is highly diversified due to presence of seedlings of tree, shrub and climbers.

REFERENCES

Brooks, T.M., Mittermeier, R.A., da Fonseca, G.A.B., Gerlach, J., Hoffmann, M., Lamoreux, J.F., Mittermeier, C.G., Pilgrim, J.D., Rodrigues, A.S.L., 2006. Global Biodiversity Conservation Priorities. *Science* 313 (5783): 58 — 61.

Huges, J.B., Daily, G.C., Ehrlich, P.R., 1997. *Science* 278: 689-692.

John Britto, S., Balaguru, B., Soosairaj, S., Arockiasamy, D.I., 2000. Comparative analysis of species diversity in a sacred grove at Vamban of Pudukkottai district in Tamil Nadu. *J. Swamybot. cl.* 17:79-82.

Muthukumar, B., Dhanasekar, S., Soosairaj, S., Nagamurugan, N., Balaguru, B., 2005. Woody vegetation structure in a sacred grove of Pudukkottai district, Tamil Nadu, South India. Indian *J. Environ. and Ecoplan.* 10(2): 409-412.

Parthasarathy, N., 2001. Changes in forest composition and structure in three sites of tropical evergreen forest around Sengaltheri, Western ghats. *Current Science* 80(3): 389-393.

Pascal, J.P., Pelisser, K., 1996. Stucture and floristic composition of a tropical evergreen forest in Southern India. *Journal of Tropical Ecology* 12: 195-218.

Ramanujam, M.P., Cyril, K.P.K., 2003. Woody species diversity of four sacred groves in the Pondicherry region of South India. *Biodiversity and Conservation* 12: 289-299.

Robert Costanza, Ralph d'Arge, Rudolf de Groot, Stephen Farber, Monica Grasso, Bruce Hannon, Karin Limburg, Shahid Naeem, Robert V. O'Neill, Jose Paruelo, Robert G. Raskin, Paul Sutton and Marjan van den Belt. 1997. The value of the world's ecosystem services and natural capital. *Nature* 387: 253-260.

Rosenszweig, M.L., 1995. Species diversity in space and time, Cambridge University Press, Cambridge.

Singh, J.S., 2002. The biodiversity crisis: A multifaceted review. *Current Science* 82(6): 638-647.

Sukumar R, Dattarja, H.S., Suresh, H.S., Ramakrishna, R., Vasudeva, S., Nirmala, S., Joshi, N.V., 1992. Long term monitoring of vegetation in a tropical deciduous forest in Mudumalai, South India. *Current Science* 62: 608-616.

2015, Modern Methods in Phytomedicine *Pages* ***77–100***
Editor: **T. Parimelazhagan**
Published by: **DAYA PUBLISHING HOUSE, NEW DELHI**

Medicinal Plants with Active Principles in the Treatment of Diabetes Mellitus

Muniappan Ayyanar*

Department of Botany, Pachaiyappa's College, Chennai – 600 030, Tamil Nadu

1.0 INTRODUCTION

Diabetes mellitus is a metabolic disorder in the endocrine system resulting from a variable interaction of hereditary and environmental factors and characterized by abnormal insulin secretion or insulin receptor or post receptor events affecting metabolism involving carbohydrates, proteins and fat metabolism in addition to damaging pancreatic β-cells and also liver and kidney in some cases (Ghosh and Suryawanshi, 2001). It is caused by inherited and/or acquired deficiency in the production of insulin by the pancreas, or by ineffectiveness of the insulin produced and such a deficiency results in increased concentrations of glucose in the blood, which in turn damages many of the body's systems, in particular the blood vessels and nerves (Mukherjee *et al.,* 2006). The disease affects more than 10 million people worldwide (6 per cent of the population) and in the next 10 years it may affect about five times more people than it does now (ADA, 1997). Complications are the major course of morbidity and mortality in diabetes mellitus (DM).

* *Author.* E-mail: asmayyanar@yahoo.com

The National Diabetic Data Group (NDDG) and World Health Organization (WHO) recognized two major forms of diabetes which they termed Insulin dependent diabetes mellitus (IDDM, type-1 diabetes or Immune mediated diabetes) and Non-Insulin dependent diabetes mellitus (NIDDM, Type 2 diabetes). Indigenous people are three to five times more likely to develop the disease than the rest of the population, with most cases (*i.e.*, 90 - 95 per cent) as type 2 or non-insulin dependent diabetes (NIDDM), a disease that is often controllable via diet and exercise (MuCune and Johns, 2002). As a very common chronic disease, diabetes is becoming the third "killer" of the health of mankind along with cancer, cardiovascular and cerebrovascular diseases because of its high prevalence, morbidity and mortality (Li *et al.*, 2004).

Type-I diabetes mellitus is characterized by absolute loss of pancreatic β-cells by autoimmune destruction of pancreatic β-cells. It accounts for 5-10 per cent of diabetes. Insulin itself is the only practical therapy at present. However, after the insulin therapy, transplantation of active islet cells is considered to be only possible way to be employed for type I diabetes mellitus. The Edmonton protocol has already been successfully employed for patients, in which human islets are injected in to the portal vein of patients (Takatsuna and Umezawa, 2004). However, there is a serious problem of the shortage of islets from donors. Bioactive metabolites that induce β-cells differentiation should be useful in preparing large number of islets *ex vivo* (Takatsuna and Umezawa, 2004).

Type 2 diabetes mellitus is characterized by insulin resistance of peripheral tissue and lowered insulin secretion from pancreatic islets. Therefore, restoration of pancreatic β-cells differentiation is should be useful as a regeneration therapy for type 2 diabetes mellitus (Graeme and Kenneth, 2001; Saktiek, 2001). Type 2 diabetes is the most common form of the disease accounting for 90-95 per cent of diabetes mellitus. Type 2 diabetes is nearing epidemic proportions, due to an increased number of elderly people and greater prevalence of obesity and secondary life style (Report of the expert committee on the diagnosis and classification of diabetes mellitus, 2004.). The chronic hyperglycemia of diabetes is associated with long term damage, dysfunction and failure of various organs, especially the eyes, kidneys, nerves and blood vessels (Atkinson and Maclaren, 1994).

At present, treatment of diabetes mellitus is focused on controlling and lowering blood glucose to a normal level and the mechanisms of both western medicines are to stimulate beta cell of pancreatic islet to release insulin; to resist the hormones which rise blood glucose; to increase the number or rise the appetency and sensitivity of insulin receptor site to insulin; to decrease the leading-out of glycogen; to enhance the use of glucose in the tissue and organ; to clear away free radicals, resist lipid peroxidation and correct the metabolic disorder of lipid and protein; to improve microcirculation in the body (Li *et al.*, 2004). Based on these mechanisms, the drugs clinically used to treat diabetes can be mainly divided into insulin, insulin-secretagogues, insulin sensitivity improvement factor, insulin-like growth factor, aldose reductase inhibitor, alpha-glucosidase inhibitors and protein glycation inhibitor, almost all of which are chemical and biochemical drugs. The effect of these drugs is only aimed to lower the level of blood glucose.

2.0 ORAL HYPOGLYCAEMIC AGENTS

Oral medications are initiated when 2-3 months of diet and exercise alone are unable to achieve or maintain their optional plasma glucose levels. The prevalence of type 2 diabetes mellitus has increased rapidly over the past decade. Obesity, which is closely linked to risk of developing the NIDDM. Cardiovascular disease is the major cause of mortality among patients with NIDDM, accounting for 60-80 per cent of deaths in these patients. Good glycaemic control probably does decrease the cardiovascular risk in patients have NIDDM.

2.1 Classes of Oral Hypoglycaemic Agents

The present treatment of diabetes is focused on controlling and lowering blood glucose. The mechanisms of both herbal drugs and herbal formulations in western medicines to decrease blood glucose are i) to stimulate β-cells of pancreatic islet to release insulin; ii) to resist the hormones which rise blood glucose; iii) to increase the number or rise the appetency and sensitivity of insulin receptor site to insulin; iv) to decrease the leading out of glycogen; v) to enhance the use of glucose in tissue and organ; vi) to clear away free radicals, resist lipid peroxidation and correct the metabolic disorder of lipid and protein; and vii) to improve microcirculation in the body (Li *et al.,* 2004).

Based on those mechanisms, the drugs chemically used to treat diabetes can be mainly divided into insulin, insulin secretagogues, insulin sensitivity improvement factors, insulin like growth factor, aldose reductase inhibitors and protein glycation inhibitor almost all of which are chemical and biochemical drugs (Lio and Wang 1996). They are,

- ☆ **Sulfonylureas**-work primarily by stimulating pancreatic insulin secretion, which in turn reduces hepatic glucose output and increases peripheral glucose disposal.
- ☆ **Meglitinides**-stimulate the release of insulin from the pancreatic β- cells.
- ☆ **Biguanides**-works by reducing hepatic glucose output and to a lesser extent, enhancing insulin sensitivity in hepatic and peripheral tissue.
- ☆ **Thiazolidinediones**-increased insulin sensitivity and glucose uptake and to a lesser extent by inhibiting hepatic glucose production.
- ☆ **Alpha-Glucosidase inhibitor**-slows down the breakdown of disaccharides and polysaccharides and other complex carbohydrates into monosaccharide by the inhibition of enzyme α-glucosidase.

3.0 MEDICINAL PLANTS USED IN THE TRADITIONAL MEDICAL SYSTEM FOR DIABETES THERAPY

Plant derivatives with hypoglycaemic properties have been used in folk medicine and traditional healing systems around the world (Yeh *et al.,* 2003) from very ancient time. Medicinal plants used to treat diabetic conditions are of considerable interest and a number of plants have shown varying degree of hypoglycemic and antihyperglycaemic activity. The traditional systems of medicine together with folklore systems continue to serve a large portion of the population, particularly in rural areas, in spite of the advent of the modern medicines (Ayyanar and Ignacimuthu, 2005). Despite the introduction of hypoglycaemic agents from natural and synthetic sources, diabetes and its secondary complications continue to be a major medical problem to people (Ravi *et al.,* 2005). The treatment of diabetic patients with naturally derived agents has the advantage that it does not cause the side effects as do chemical agents such as sulfonylurea (it causes a decreased amount of insulin production by putting too great a strain on the insulin producing beta cells). While treatment with herbal drugs has an effect of protecting beta cells and smoothing out fluctuations in glucose levels (Jia *et al.,* 2003).

The use of plants as therapeutic tools, especially those used to relieve chronic pathologies, have had a remarkable role in the popular medicine of different countries. Since time immemorial, various plants and plant derived compounds have been used in the treatment of diabetes to control the blood sugar of the patients. In the period of 1907 - 1988, antidiabetic activity was reported for about 343 plants for global level and those are mostly used in the indigenous system of medicine as well as scientifically proven plants (Rahman and Zaman, 1989). Plants represent a vast source of potentially useful dietary supplements for improving blood glucose control and preventing long-term complications in type 2 diabetes mellitus. Before the introduction of insulin in 1922 the treatment of diabetes mellitus relied heavily on dietary measures which included the use of traditional plant therapies (Gray and Flatt, 1999).

Before two decades, more than 400 traditional plants were recorded for the treatment of diabetes mellitus with only a small number of scientific and medical evaluations to assess their efficacy (Bailey and Day, 1989). In traditional medicine, diabetes mellitus is treated with diet, physical exercise and medicinal plants, even though, more than 1200 plants are used around the world in the empirical control of diabetes mellitus and approximately 30 per cent of the traditionally used antidiabetic plants were pharmacologically and chemically investigated (Alarcon-Aguilar *et al.,* 2002). On the other hand, potential hypoglycaemic agents have also been detected for more than 100 plants used in antidiabetic therapy. Hypoglycemic action from some treatments has been confirmed in animal models and non-insulin-dependent diabetic patients and now a day a number of hypoglycemic compounds have been identified. Traditional treatments may provide the valuable clues for the development of new oral hypoglycemic agents and simple dietary adjuncts.

Yeh *et al.* (2003) reviewed and analyzed that, a total of 108 trials examining 36 herbs (single or in combination) and 9 vitamin/mineral supplements, involving 4,565 patients with diabetes or impaired glucose tolerance. In their study there were 58 controlled clinical trials involving individuals with diabetes or impaired glucose tolerance and most studies involved patients with type 2 diabetes. Of these 58 trials, the direction of the evidence for improved glucose control was positive in 76 per cent and very few adverse effects were also reported by them. There have been many studies on hypoglycaemic plants and a great variety of compounds have been isolated (alkaloids, glycosides, terpenes, flavonoids, etc.), but the main fact is the further development of such leads into clinically useful medicines and especially phytomedicines or adequate nutritional supplements, which would be of direct benefits to patients (Andrade-Cetto and Heinrich, 2005).

In Trinidad and Tobago, the plants such as *Antigonon leptopus, Bidens alba, Bidens pilosa, Bixa orellana, Bontia daphnoides, Carica papaya, Catharanthus roseus, Cocos nucifera, Gomphrena globosa, Laportea aestuans, Momordica charantia, Morus alba, Phyllanthus urinaria* and *Spiranthes acaulis* are mostly used for the treatment of diabetes among tribal people (Lans, 2006). In south eastern Morocco (Tafilalet region), local traditional herbal healers use 92 plants for the treatment of diabetes mellitus and their complications (Eddouks *et al.,* 2002). Of them most commonly used antidiabetic plants were *Ammi visnaga, Artemesia herba alba, Trigonella foenum-graecum, Marrubium vulgare, Nigella sativa, Globularia alypum, Allium sativum, Olea europaea, Citrullus colocynthis, Aloe succotrina, Artemesia absinthium, Rosmarinus officinalis, Thymus vulgaris, Eucalyptus globulus, Mentha pulegium, Myrtus communis, Linun usitatissimum* and *Carum carvi.* In Mexico, there are 306 plants were recorded as antidiabetic plants and the plants such as *Cecropia obtusifolia, Equisetum myriochaetum, Acosmium panamense, Cucurbita ficifolia, Agarista mexicana, Brickellia veronicaefolia* and *Parmentiera aculeate* are most commonly used for the treatment of diabetes mellitus (Andrade-Cetto *et al.,* 2005).

In the Chinese traditional medical treatment of diabetes, Compound Recipes are often used more than Simple Recipes (prescription with one medicine) due to consideration of integrated effects of different medicines. There are hundreds of prescriptions to aim directly at different symptoms of diabetes, and about 100 of natural medicines and preparations are used in these prescriptions or folk Simple Recipes and diets for diabetes care in China, most of which come from plants (Li *et al.,* 2004). The native people of Canada use 16 plants for the treatment of diabetes from the very ancient period (Leduc *et al.,* 2006). More than 100 medicinal plants are mentioned in the Indian system of medicines including folk medicines for the management of diabetes, which are effective either separately or in combinations (Ajit Kar *et al.,* 2003). A limited number of medicinal plant species have been studied and validated for their hypoglycemic properties using laboratory diabetic animal models and in clinical studies using human subjects; although several medicinal plants and their active principles have been reported in literatures as having been used to control diabetes in the Indian traditional medical system of medicine. In India, the plants such as *Eugenia jambolana, Coccinia indica, Gymnema sylvestre, Momordica charantia* and *Trigonella foenum-graecum* are widely used for the treatment of diabetes in traditional as well as modern medicine.

Many therapeutic agents had been used for the treatment of diabetes mellitus before insulin was discovered and several hundred plants have shown some extent of antidiabetic activity (Helmstadter, 2007). Some of the plants such as *Acosmium panamense, Aegle marmelos, Agarista mexicana, Allium cepa, Allium sativum, Artemisia herba alba, Astragalus membranaceus, Brickellia veronicaefolia, Cecropia obtusifolia, Coccinia indica, Cucurbita ficifolia, Equisetum myriochaetum, Eugenia jambolana* (*Syzygium cumini*), *Globularia alypum, Glycyrrhiza glabra, Gymnema sylvestre, Lupinus albus, Momordica charantia, Nigella sativa, Ocimum sanctum, Opuntia streptacantha, Origanum compactum, Panax ginseng, Parmentiera aculeate, Pueraria lobata, Rehmannia glutinosa, Taraxacum officinale*, *Trigonella foenum-graecum* and *Vitis vinifera* are showed to have promising antidiabetic activity (Ayyanar *et al.*, 2013; Ayyanar and Ignacimuthu, 2011; Ayanar *et al.*, 2008; Mukherjee *et al.*, 2006; Villasenor and Lamadrid, 2006; Andrade-Cetto and Heinrich, 2005; Li *et al.*, 2004; Yeh *et al.*, 2003; Bailey and Dey, 1989; Rahman and Zaman, 1989). *Syzygium cumini* (Jambolan) is one of the widely used plants for the treatment of diabetes by traditional practitioners over many centuries. Clinical and experimental studies of jambolan revealed that different parts of the plant especially fruits, seeds and stem bark possess promising antidiabetic activity (Ayyanar *et al.*, 2013). Any new or previously isolated compound from jambolan can be studied further for its use in diabetic treatments.

Since time immemorial, various medicinal plants and plant derived compounds have been used in the treatment of diabetes to control the blood sugar of the patients. The use of medicinal plants in the management of diabetes has been prevalent in India from very ancient time. Hundreds of medicinal plants have reported to possess potential hypoglycemic activity in Indian system of medicines. There have been several reviews on the hypoglycemic medical plants, more particularly use of Indian botanicals for hypoglycemic activity (Mukherjee *et al.*, 2006). In a recent clinical survey in Senegal, the diabetic patients were ready to use medicinal plants to alleviate health problems; they were aware about the use of medicinal plants in the treatment of diabetes (Dieye *et al.*, 2008).

4.0 ANTIDIABETIC PRINCIPLES ISOLATED FROM MEDICINAL PLANTS

The screening of bioactive metabolites began in 1928 when Alexander Flemming discovered an antibacterial effect of pencillin. In 1940s pencillin was purified and its structure determined. After that huge numbers of antibiotics have been discovered or synthesized. In 1950s the screening of anticancer agents started, since cell culture of cancer cells became available. Late 1960s enzyme inhibitors such as protease inhibitors a tyrosin hydroxylase inhibitors were isolated from the microbial broths. After that many useful enzyme inhibitors including HMG-COA reductase inhibitors have been isolated from microorganism and plants. In the 21st century signal transduction inhibitors will be extended to the compounds for regenerative chemotheraphy and tailor-made medicines. Even environmental modulators may be screened in the same way as signal transduction inhibitors (Takatsuna and Umezawa, 2004).

More than 400 traditional plant treatments for diabetes mellitus have been recorded, but only a small number of these have received scientific and medical evaluation to assess their efficacy. However today it is necessary to provide scientific proof as to whether it is justified to use a plant or its active principles (Singh *et al.*, 2000). Traditional treatments have mostly disappeared in occidental societies, but some are prescribed by practitioners of alternative medicine or taken by patients as supplements to conventional therapy. However, plant remedies are the mainstay of treatment in underdeveloped regions. A hypoglycemic action from some treatments has been confirmed in animal models and non-insulin-dependent diabetic patients, and various hypoglycemic compounds have been identified. A botanical substitute for insulin seems unlikely, but traditional treatments may provide valuable clues for the development of new oral hypoglycemic agents and simple dietary adjuncts (Bailey and Day, 1989).

A wide array of plant derived active principles representing numerous chemical compounds has demonstrated activity consistent with their possible use in the treatment of diabetes mellitus. Many natural products and herbal medicines have been recommended for the treatment of diabetes. The present paper reviews medicinal plants that have shown experimental or clinical antidiabetic activity and that have been used in traditional systems of medicine; the review also covers natural products (active natural components and crude extracts) isolated from the medicinal plants. Traditional lifestyles in which diet, exercise and possibly antioxidant and hypoglycemic medicines played an important role, may have masked people in a pre-diabetic state in the past (MuCune and Johns, 2002).

There are many kinds of natural products, such as alkaloids, glycosides, polysaccharides, peptidoglycans, flavonoids, tannins, steroids, glycopeptides, terpenoids and inorganic ions. The introduction of these indigenous herbal compounds in the management of diabetes mellitus will greatly simplify the management and make it less expensive. Particularly, schulzeines A, B, and C, radicamines A and B, 2,5-imino-1,2,5-trideoxy-L-glucitol, beta-homofuconojirimycin, myrciacitrin IV, dehydro-trametenolic acid, corosolic acid (Glucosol), 4-(alpha-rhamnopyranosyl) ellagic acid and 1,2,3,4,6-pentagalloylglucose have shown significant antidiabetic activities. Among medicinal herbs *Momordica charantia, Pterocarpus marsupium* and *Trigonella foenum- graecum* have been reported as beneficial in the treatment of type 2 diabetes (Jung *et al.,* 2006).

Antioxidants are important in diabetes mellitus, with low levels of plasma antioxidants implicated as a risk factor for the development of the disease and circulating levels of radical scavengers impaired throughout the progression of diabetes. Many of the complications of diabetes, including retinopathy and atherosclerotic vascular disease, the leading cause of mortality in diabetics, have been linked to oxidative stress and antioxidants have been considered as treatments (MuCune and Johns, 2002).

4.1 Flavonoids

Many kinds of flavonoids have been isolated from traditional medicines for antidiabetic effect. Flavonoids are naturally occurring phenolic compounds that are widely distributed in plants. Over 5000 different flavonoids have been described to date and they are classified into at least 10 chemical groups (Harbone, 1993). Among them, flavones, flavonols, flavanols, flavanones, anthocyanins and isoflavones are particularly common in the diet. Flavonols are the most abundant flavonoids in foods, with quercetin, kaempferol and myricetin being the three most common flavonols. Flavanones are mainly found in citrus fruit and flavones in celery. Catechins are present in large amounts in green and black tea and in red wine, whereas anthocyanins are found in strawberries and other berries. Isoflavones are almost exclusively found in soy foods.

Table 7.1: Active Principles Isolated from Medicinal Plants in the Treatment of Diabetes Mellitus

Name of the Plant	*Active Principles Isolated*	*Effect Studied*
Abelmoschus moschatus	Myricetin	Antidiabetic (Liu *et al.*, 2005)
Anacardium occidentale	Stigmast-4-en-3-ol and stigmast-4-en-3-one	Hypoglycemic (Alexander-Lindo *et al.*, 2004)
Astragalus membranaceus and *Pueraria thomsonii*	Isoflavones, especially biochanin A	Hypolipidemic effect and target on PPAR-γ (Shen *et al.*, 2006)
Bauhinia megalandra	Quercetin, kaempferol, astilbin, quercetin 3-*O*-alpha-rhamnoside, kaempferol 3-O-alpha-rhamnoside, quercetin 3-*O*-alpha-arabinoside, quercetin 3-*O*-alpha-(2"-galloyl) rhamnoside and kaempferol 3-O-alpha-(2"galloyl) rhamnoside.	Inhibitors of glucose-6-phosphatase system (Estrada *et al.*, 2005)
Cassia auriculata	Flavonoids, anthracene derivatives, dimeric procyanidins, myristyl alcohol, β-*D*-glucoside, quercetin 3-*O*-glycoside, rutin (Nageswara Rao *et al.*, 2000)	Antidiabetic and antihyperlipidaemic (Pari and Latha, 2002)
Cornus mas	Anthocyanins and ursolic acid	Prohibitors of glucose intolerance and obesity (Jayaprakasam *et al.*, 2006)
Cuminum cyminum	Cinnamaldehyde	Antidiabetic (Lee, 2005)
Curcuma longa	Curcuminoids and sesquiterpenoids (Nishiyama *et al.*, 2005) Curcumin, desmethoxy curcumin, bisdemethoxy curcumin, dihydrocurcumin, α and β-turmerones, eugenol, campesterol, stigmasterol (Rastogi and Mehrotra, 1993)	Antidiabetic (Nishiyama *et al.*, 2005)
Curcuma longa	Curcuminoids and sesquiterpenoids	Antidiabetic (Nishiyama *et al.*, 2005)
Emblica officinalis	Phyllemblin, gallic acid, ellagic acid, phyllantidine, phyllantine, lupeol, emblicanin A and B (Bhatia and Bajaj, 1975)	Antihyperlipidaemic (Mathur *et al.*, 1996)
Enicostemma littorale	Vanillic acid, ferulic acid, p-coumaric acid, apigenin, genkwanin, isovitexin, swertisin, saponarin (Murali *et al.*, 2002)	Antidiabetic (Murali *et al.*, 2002)
Eucommia ulmoides	Quercetin 3-*O*-alpha-L-arabinopyranosyl-(1-2)-beta-*D*-gluco-pyranoside, kaempferol 3-*O*-beta-*D*-glucopyranoside (astragalin), quercetin 3-*O*-beta-D-glucopyranoside (isoquercitrin)	Glycation inhibitory (Kim *et al.*, 2004)

Contd...

Table 7.1–*Contd...*

Name of the Plant	*Active Principles Isolated*	*Effect Studied*
Eugenia jambolana	Gallic acid, ellagic acid, corilagin, ellagitannins, quercetin (Kelkar, 1996)	Antidiabetic (Bhattacharya *et al.*, 1991) and antihyperlipidaemic (Sharma *et al.*, 2003)
Gardenia jaminoides	Crocetin	Prevents insulin resistance (Xi *et al.*, 2005)
Gentiana olivieri	Isoorientin and C-glycosylflavone	Antihyperglycemic (Sezik *et al.*, 2005)
Gymnema sylvestre	Gymnemic acids, saponins, stigmasterol, quercitol, betaine, choline, trimethylamine (Kapoor, 1990)	Antidiabetic (Baskaran *et al.*, 1990) and antihyperlipidaemic (Shigematsu *et al.*, 2001)
Hintonia standleyana	3-*O*-beta- D-glucopyranosyl-23,24-dihydrocucurbitacin; 5-*O*-beta- D-glucopyranosyl-7-methoxy-3',4'-dihydroxy-4-phenyl-coumarin and 5- O-(beta- D-apiofuranosyl-(1-6)- beta- D-gluco-pyranosyl)-7-methoxy-3',4'-dihydroxy-4-phenylcoumarin (4-phenylcoumarins and cucurbitacin glycosides)	Antihyperglycemic (Guerrero-Analco *et al.*, 2005)
Jasonia montana	Diterpenes, namely jasonin-a, jasonin-b and jasonin-c	Antidiabetic (Al-Howiriny *et al.*, 2005)
Llareta *(Azorella compacta)*	Diterpenic compounds mulinolic acid, azorellanol, and mulin-11,13-dien-20-oic acid	Antihyperglycemic (Fuentes *et al.*, 2005)
Melia azedarach	β-carotene, nimbin, azadirachtin, nimbidiol, quercetin, nimbidin and nimbatiktam	Antioxidant (Govindachari, 1992)
Momordica charantia	Charantin, momordicosides A and B, acylglucosyl sterols, P-insulin, V-insulin, stigmasterol (Raman and Lau, 1996)	Antidiabetic (Raman and Lau, 1996) and antihyperlipidaemic (Ahmed *et al.*, 2001)
Mucuna pruriens	Oligocyclitols	Antidiabetic (Donati *et al.*, 2006)
Ocimum sanctum	Eugenol	Antidiabetic (Prakash and Gupta, 2005)
Olea europaea Alhamdani, 2006)	Oleuropein	Hypoglycemic and antioxidant (Al-Azzawie and
Peucedanum japonicum	Coumarin and a cyclitol that is, peucedanol 7-O-beta-D-gluco-pyranoside and myo-inositol	Antidiabetic (Lee *et al.*, 2004)
Pterocarpus marsupium	Pteroside, pteroisoauroside, marsuposide, vijayosin, sesquiterpene (Maurya *et al.*, 2004); pterosupin, pterostilbene, marsupin (Manickam *et al.*, 1997)	Antidiabetic (Manickam *et al.*, 1997) and antihyperlipidaemic (Jahromi and Ray, 1993)

Contd...

Table 7.1–*Contd...*

Name of the Plant	*Active Principles Isolated*	*Effect Studied*
Rheum undulatum	Stilbene, desoxyrhapontigenin, anthraquinones, emodin and chrysophanol.	Antidiabetic (Choi *et al.,* 2005)
Rumex bucephalophorus	trans-resveratrol, piceid and trans-stilbene	Antioxidant (Kerem *et al.,* 2006)
Swertia chirata	Polyoxygenated xanthones, mangiferin, swertinin, swertianin, swerchirin, chiratin, chirataini (Rastogi and Mehrotra, 1993)	Antioxidant (Scartezzini and Speroni, 2000)
Syzygium samarangense	2',4'-Dihydroxy-3',5'-dimethyl-6'-methoxychalcone, flavanone 5-O-methyl-4'-desmethoxymatteucinol, and 2'4'-dihydroxy-6'-methoxy-3'-methylchalcone	Antihyperglycemic effect (Resurreccion-Magno *et al.,* 2005)
Tinospora cordifolia	Tinosporin, isocolumbin, palmatine, tinocordiside (X), tinocordifolioside (XI), cordioside, β-sitosterol (Singh *et al.,* 2003)	Antidiabetic and antihyperlipidaemic (Prince and Menon, 2003), antioxidant (Prince and Menon, 1999)
Trigonella foenum-graecum	4-hydroxyisoleucine an unusual amino acid	Antihyperglycemic and antidyslipidemic (Narender *et al.,* 2006)

4.2 Alkaloids

A few compounds were isolated toward diabetes, which performed excellent effect. Such as conophylline, berberine, anisodamine, vindoline, vindolinine, leurosine, aconitine, hanfangchin A, an multiflorin

4.3 Terpenoids

Terpenoids are many compounds differentiated by triterpenoids, diterpenoids, sesquiterpenoids and monoterpenoids. Triterpenoids and saponins are the promosing compounds with potential to be developing new drug for anti-diabetes.

4.4 Polysaccharides

Many kinds of polysaccharides have been isolated from medicinal plants against diabetes. Most of which performed as good effect. Examples are panaxan, laminaran, coixan, pachymaran, anemarn, moran, lithosperman, trichosan, saciharan, ephedran, abelmosan, atractin. (Li *et al.,* 2004)

4.5 Insulin like Compounds, Polypeptides and Amino Acids

These substances performed excellent an effect for the treatment of diabetes. Examples include p-insulin (*Momordica charantia*), ginseng glycopeptides, α-methylenecyclopropylglycin and S-allyl cystein sulfoxide.

4.6 Sterols

The hypoglycemic effect is similar to sulfonylurea like medicine. An example is charantin (*Momordica charantia*).

4.7 Unsaturated Fatty Acids

The efficacy of anti-hyperglycemia is strong but the effectiveness is shown slowly. Examples are linoleic acid and trihydroxyljecoric

4.8 Miscellaneous

Compounds with a sulpher bond, such as allicin and allylpropyl disulfide, 3-hydroxy-3- methylglutaric acid, sodium oxaloacetate, edyson are active compounds with a varied structure.

Due to the presence of aromatic hydroxyl groups, flavonoids have strong antioxidant properties. They are scavengers of reactive oxygen and nitrogen species and, therefore, inhibit peroxidation reactions. They also protect macrophages from oxidative stress by keeping glutathione in its reduced form (du Thie and Crozier, 2000; Fuhrman and Aviram, 2001). Flavonoids have the capacity to inhibit enzymes such as cyclooxygenases and protein kinases involved in cell proliferation and apoptosis (Formica and Regelson, 1995). It was reported that a flavonoid, (γ)-epicatechin, protects normal rat islets from alloxan, normalizes blood glucose levels and promotes β-cell regeneration in islets of alloxan-treated rats (Chakravarthy *et al.,* 1981, 1982a,b). Tritiated thymidine incorporation into islet cell DNA was also enhanced by this flavonoid in an *in vitro* study (Hii and Howell, 1984). Most of them showed a mechanism to improve the function of β-cells of pancreatic islet. Examples

Figure 7.1: Active Principles Isolated from Medicinal Plants in the Treatment of Diabetes Mellitus.

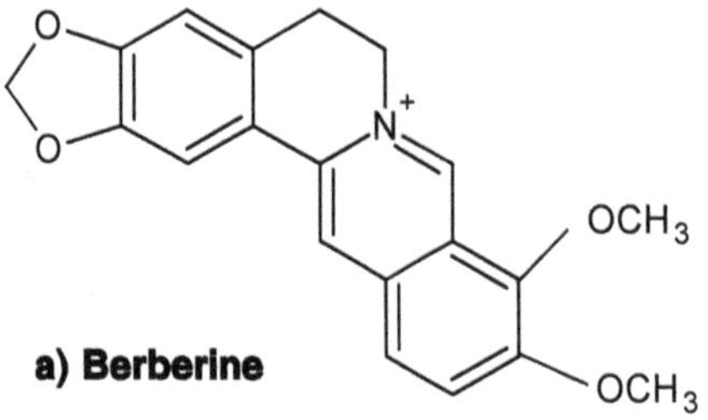

a) Berberine

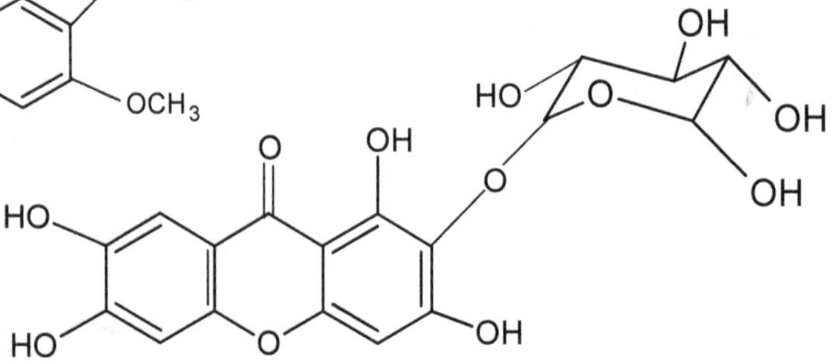

b) Mangiferin

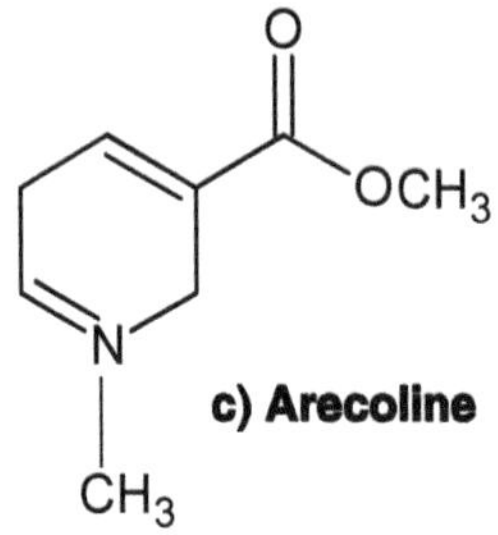

c) Arecoline

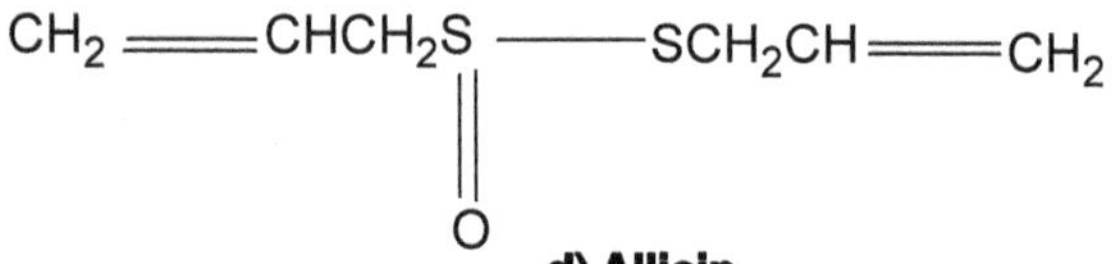

d) Allicin

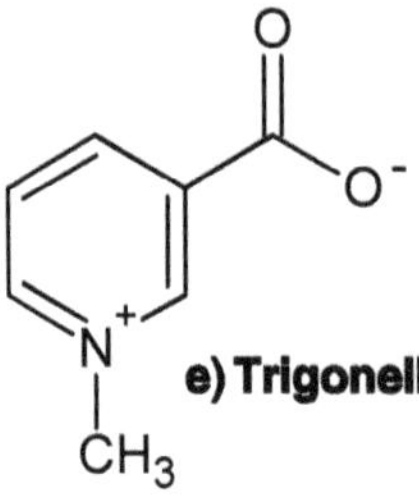

e) Trigonelline

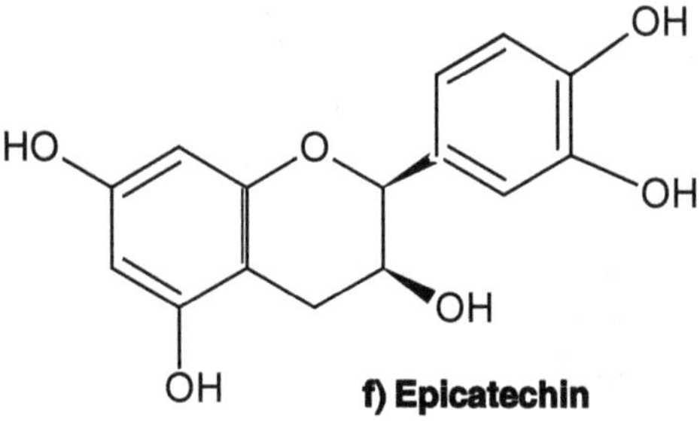

f) Epicatechin

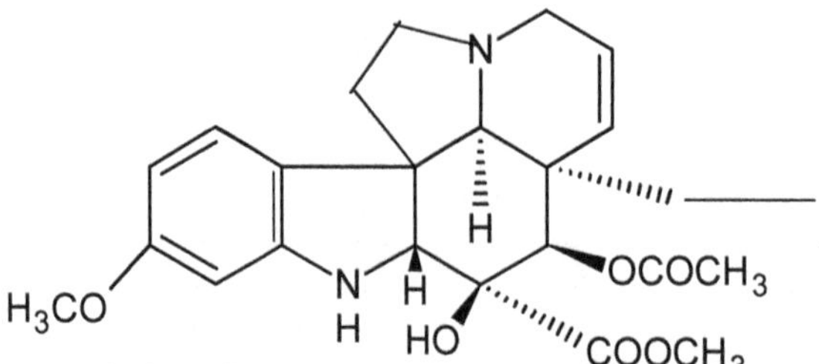

g) Vindoline

Contd...

Figure 7.1–*Contd...*

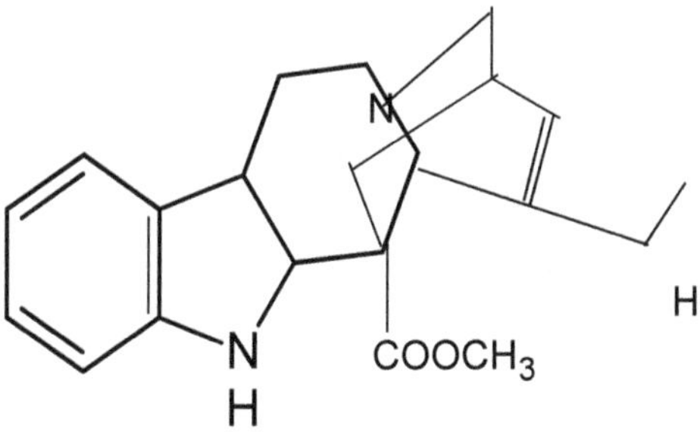

h) Catharanthine

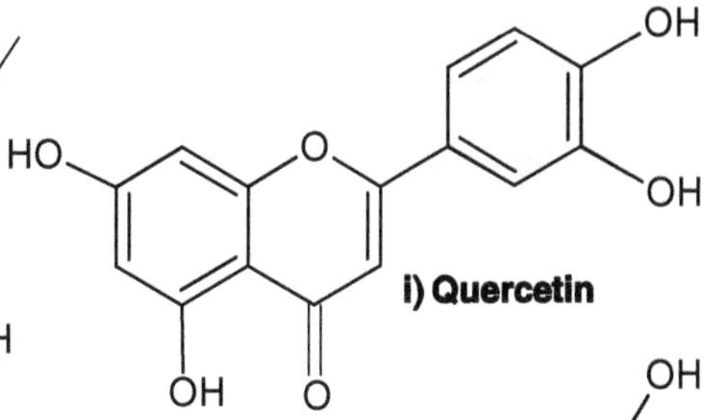

i) Quercetin

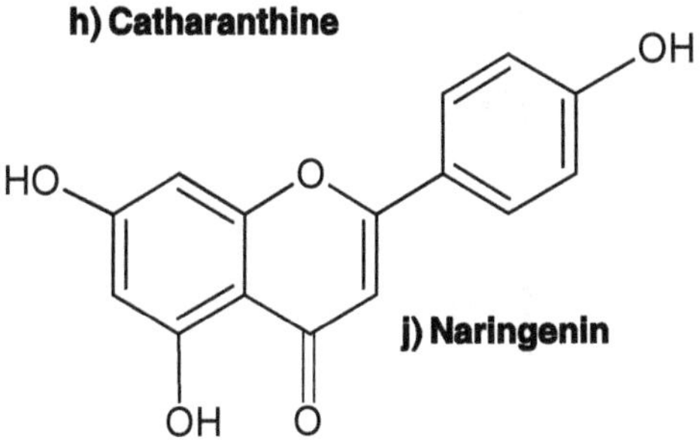

j) Naringenin

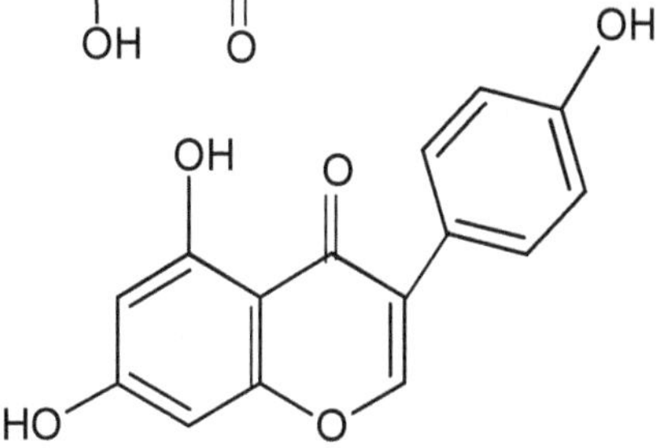

k) Genistein

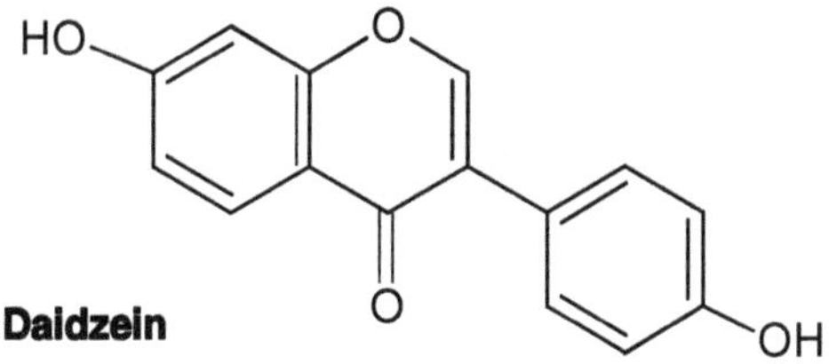

l) Daidzein

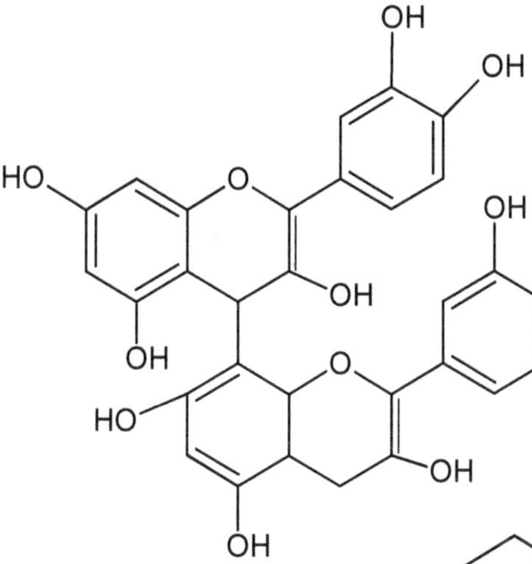

m) Proanthocyanidin

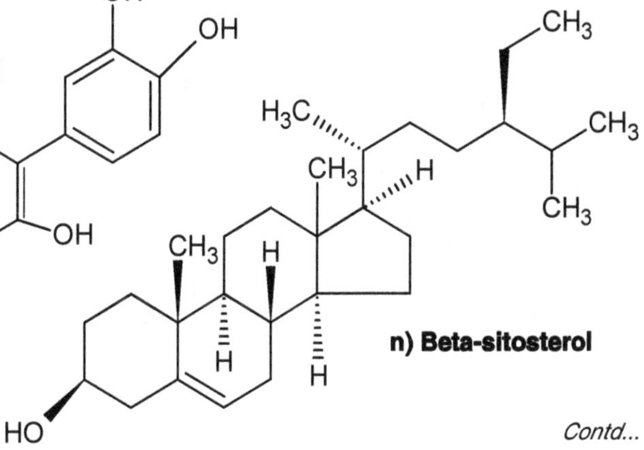

n) Beta-sitosterol

Contd...

Figure 7.1–*Contd...*

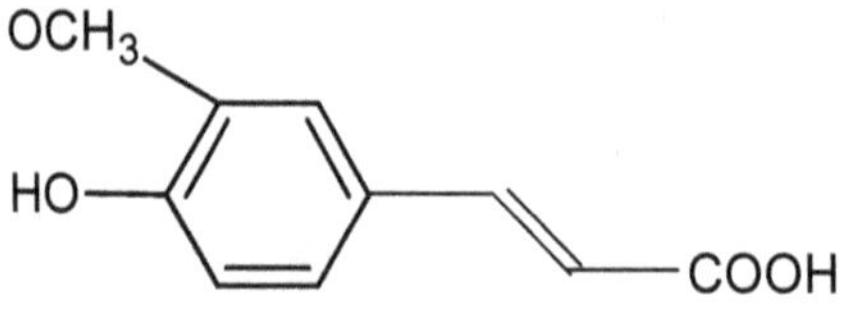

o) Ferulic Acid

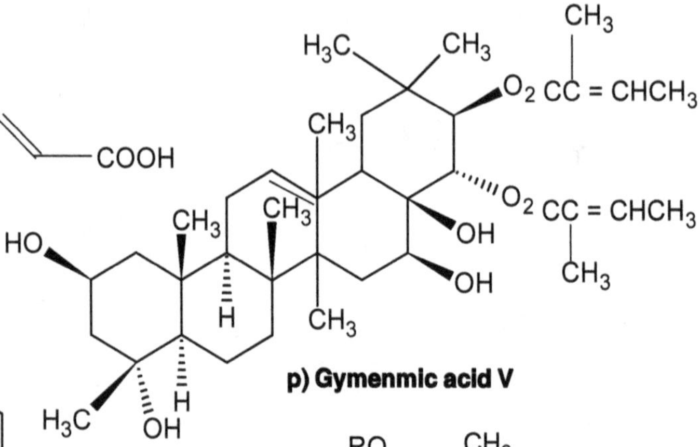

p) Gymenmic acid V

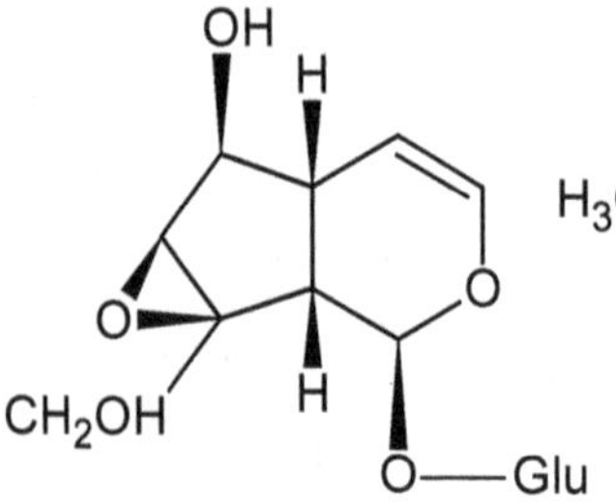

q) Catalpol

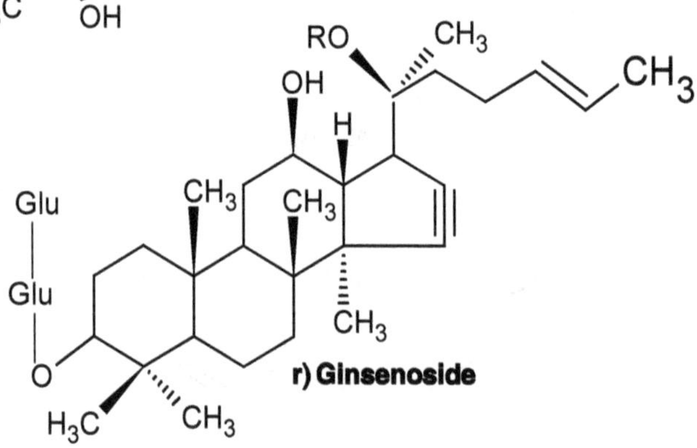

r) Ginsenoside

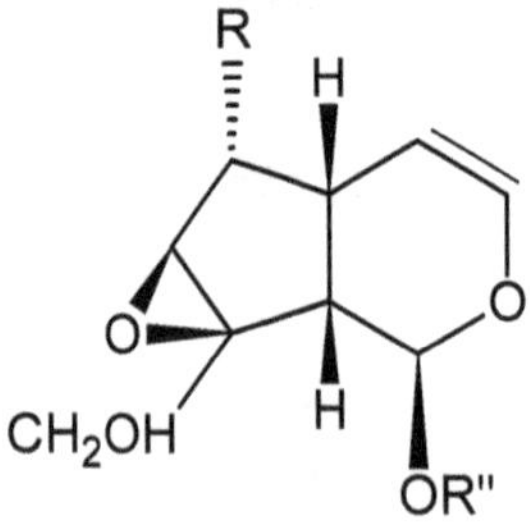

s) Rehmannioside A, B A- R¹= OH;R¹¹ = Glu-GluB- R¹=O-Glu; R¹¹ = Glu

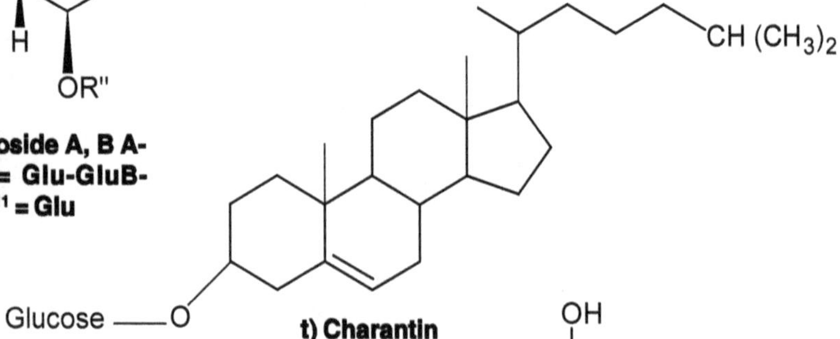

t) Charantin

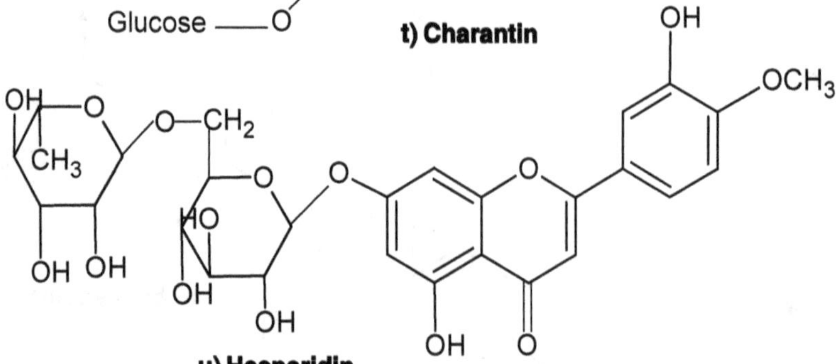

u) Hesperidin

include kakonein, flavone-c-glycoside, icariin, neomyrtillin, sapanchlcone, caesalpin P, 3-deoxysappanone, protosappanin A, brazilin, swerchirin and hyperin.

5.0 POLYHERBAL FORMULATIONS IN COMBINATION WITH JAMBOLAN

A number of herbal formulations prepared in combination (Table 7.2) with this plant available in market showed potential antidiabetic activity and are used regularly by diabetic patients on the advice of the physicians. Different formulations made from the aqueous extract of seeds of Jambolan showed promising hypoglycaemic effect in rabbits (Indira *et al.,* 1954). Some of the polyherbal formulations have been reported to have significant antidiabetic effect in the tested animals.

5.1 Dianex

It is a polyherbal formulation prepared from the mixture of the aqueous extracts of this plant with several other plants showed significant hypoglycaemic activity in both normal and diabetic animals (Mutalik *et al.,* 2005). The elevated triglycerides, cholesterol, ALT, AST, urea and creatinine levels in diabetic mice were significantly; liver glycogen and protein levels were both significantly increased in diabetic mice by the treatment with Dianex. It also increased significant glucose tolerance and produced significant free radical scavenging activity against. With a nonrandomized open labeled clinical trial, dianex was is found to be an effective adjuvant drug with either oral antidiabetic agents or insulin and it can be used in the control of blood sugars in diabetic patients and it is also a safe drug that does not cause any clinical, hematological or biochemical alteration in major organ systems (Sudha *et al.,* 2005).

5.2 DCBT 2345

It an ayurvedic formulation, prepared from jambolan and two more herbal ingredients such as *Gymnema sylvestre* and *Cephalandra indica* showed effective reduction of hyperglycaemia in type 2 diabetes and this study suggested that the drug helps to improve or maintain insulin secretion providing optimal stimulation of pancreatic beta cells (Mohan *et al.,* 2001).

5.3 Diabecon

It is a combination of antidiabetic plants manufactured by 'Himalaya' and reported to increase peripheral utilization of glucose, increase hepatic and muscle glucagon contents, promote β-cells repair and regeneration and increase c peptide level. It has antioxidant properties and protects β-cells from oxidative stress (Modak *et al.,* 2007). It also exerts insulin like action by reducing the glycated haemoglobin levels, normalizing the microalbuminurea and modulating the lipid profile. It minimizes long term diabetic complications. Shri Kant *et al.* (2002) investigated among 30 patients with NIDDM and IDDM and all the patients were given with 2 tablets of Diabecon thrice daily for 12 week and 'Diabecon' inhibited the proliferative changes in retina and controlled progressive retinal damage and visual acuity of patients also showed improvement.

Table 7.2: Herbal Formulations Used in the Treatment of Diabetes Mellitus

Name of the Drug	*Other Plants Used as Ingredients*	*Reported Effect*
Dianex	*Gymnema sylvestre*, *Momordica charantia*, *Azadirachta indica*, *Cassia auriculata*, *Aegle marmelose*, *Syzygium cumini*, *Withania somnifera* and *Curcuma longa*	Hypoglycemic activity
DCBT 2345	*Syzygium cumini*, *Gymnema sylvestre* and *Cephalandra indica*	Reduction of hyperglycaemia
Diabecon (Himalaya)	*Syzygium cumini, Gymnema sylvestre, Pterocarpus marsupium, Glycyrrhiza glabra, Casearia esculenta, Asparagus racemosus, Boerhavia diffusa, Sphaeranthus indicus, Tinospora cordifolia, Swertia chirata, Tribulus terrestris, Phyllanthus amarus, Gmelina arborea, Gossypium herbaceum, Berberis aristata, Aloe vera,* Triphala, *Commiphora wightii, Momordica charantia, Piper nigrum, Ocimum sanctum, Abutilon indicum, Curcuma longa* and *Rumex maritimus*	Minimizes long term diabetic complications
Diabeta (Ayurvedic Herbal Health Products)	*Gymnema sylvestre, Vinca rosea*, *Curcuma longa*, *Azadirachta indica*, *Pterocarpus marsupium*, *Momordica chara*ntia, *Acacia arabica*, *Syzygium cumini*, *Tinospora cordifolia* and *Zingiber officinale*	Hypoglycemic activity
Hyponidd	*Momordica charantia*, *Melia azedarach*, *Pterocarpus marsupium*, *Tinospora cordifolia*, *Syzygium cumini*, *Gymnema sylvestre*, *Enicostemma littorale*, *Emblica officinalis*, *Cassia auriculata* and *Curcuma longa*	Antihyperglycaemic and antioxidant activity
Diasulin	*Cassia auriculata, Coccinia indica, Curcuma longa, Emblica officinalis, Gymnema sylvestre, Syzygium cumini, Momordica charantia, Scoparia dulcis, Tinospora cordifolia* and *Trigonella foenum-graecum*	Hypoglycemic activity
Pancreatic tonic 180 cp (Ayurvedic herbal supplement)	*Pterocarpus marsupium, Gymnema sylvestre, Momordica charantia, Syzygium cumini, Trigonella foenum graceum, Azadirachta indica, Ficus racemosa, Aegle marmelos* and *Cinnamomum tamala*	Initiates insulin release from pancreatic beta cells
Ayurveda alternative herbal formula to Diabetes	*Syzygium cumini*, *Gymnema sylvestre*, *Momordica charantia*, *Inula racemosa*, *Azadirachta indica*, *Trigonella foenum gracecum* and *Tinospora cordifolia*	Hypoglycemic activity
MTEC	*Musa paradisiaca, Syzygium cumini*, *Tamarindus indica* and *Coccinia indica*	Protection in fasting blood glucose and serum insulin levels

5.4 Diabeta

It is an antidiabetic capsule, prepared from the combination of several antidiabetic plants including jambolan and recommended for more than 5000 years in ayurveda and it is formulated based on the ancient ayurvedic references. Diabeta helps to act on different sites in different ways to effectively control factors and pathways leading to diabetes mellitus (Ayurvediccure.com). This drug is safe and effective in managing diabetic conditions as a single agent supplement to synthetic anti-diabetic drugs and also helps overcome resistance to oral hypoglycemic drugs when used as adjuvant to cases of uncontrolled diabetes.

5.5 Hyponidd

Subash Babu and Prince (2004) reported that 'Hyponidd', a herbomineral formulation composed of the extracts of jambolan and nine other medicinal plants exhibited antihyperglycaemic and antioxidant activity in STZ-induced diabetic rats.

5.6 MTEC

Treatment of herbal formulated drug named as MTEC consist of aqueous-methanol extract of four plants including Jambolan to streptozotocin induced diabetic rat at the ratio of 2:2:1:1 at the dose of 60 mg/d for two times a day for 14 d resulted a significant protection in fasting blood glucose and serum insulin levels along with correction of testicular above parameters towards the control level. Also, the herbal formulated drug has no general toxic effects on the body weight, as well as on the activities of serum glutamate and pyruvate transaminases in serum (Mallick *et al.*, 2007).

Modak *et al.* (2007) stated that, one of the major problems with the herbal formulation is that the active ingredients are not well defined and it is important to know the active component and their molecular interaction, which will help to analyse therapeutic efficacy of the product and also to standardize the product. Currently there are number of laboratories have engaged to investigate the mechanism of action of some of these plants using model systems.

6.0 CONCLUSION

Plants have yielded directly or indirectly many important medicines in the past. For example, the discovery of the widely used hypoglycemic drug, metformin, came from the traditional approach of using *Galega officinalis.* Traditional systems of medicine reveal a strong history of use to support antidiabetic action of plants and reproducibility of their safety and efficacy remains questionable. Hence future research aimed at the identification of active molecules is needed for supporting efficacy claims. Our preliminary investigation on the survey of medicinal plants among the tribal people of Tamil Nadu has given some clue in the use of medicinal plants as an antidiabetic plants from very ancient time (Ayyanar and Ignacimuthu, 2005, 2011; Ayyanar *et al.*, 2008; Ayyanar *et al.*, 2013). From an ethnopharmacological perspective, it is important to understand that this disease is one at the interface of conventional biomedical and local (or traditional) treatment. Dieye *et al.* (2008) recommended that in order to demonstrate the efficacy and innocuousness of herbal medicines, scientists in developing countries must work in medicinal plants field in collaboration with traditional therapists and national health authorities to formulate new significant drugs in the treatment of various diseases including diabetes. Instead of trying to identify the active components of herbs through massive collection of plants from natural sources, it is better to start investigating the efficacy of the natural product from the traditional use by patients in randomized clinical trials.

REFERENCES

ADA, 1997. Clinical practice recommendations 1997, screening for diabetes. *Diabetes Care* 20(1), 22-24.

Ahmed, I., Lakhani, M.S., Gillet, M., John, A., Raza, H., 2001. Hypotriglyceridemia and hypocholestrolaemic effects of antidiabetic *Momardica charantia* (Karela) fruit extract in streptozotocin induced diabetic rats. *Diab Res Clin Prac.* 51(3), 155-161.

Kar, A., Choudhary, B.K., Bandyopadhyay, N.G., 2003. Comparative evaluation of hypoglycaemic activity of some Indian medicinal plants in alloxan diabetic rats. *J Ethnopharmacol* 84, 105-108.

Alarcon-Aguilar, F.J., Roman-Ramos, R., Flores-Saenz, J.L., Aguirre-Garcia, F., 2002. Investigation on the hypoglycaemic effects of extracts of four Mexican medicinal plants in normal and alloxan-diabetic mice. *Phytotherapy Research* 16, 383 – 386.

Al-Azzawie, H.F., Alhamdani, M.S., 2006. Hypoglycemic and antioxidant effect of oleuropein in alloxan-diabetic rabbits. *Life Sci.* 16; 78(12), 1371-7.

Alexander-Lindo, R.L., Morrison, E.Y., Nair, M.G., 2004. Hypoglycaemic effect of stigmast-4-en-3-one and its corresponding alcohol from the bark of *Anacardium occidentale* (cashew). *Phytother Res.* 18(5), 403-7.

Al-Howiriny, T.A., Al-Rehaily, A.J., Polsc, J.R., Porter, J.R., Mossa, J.S., Ahmed, B., 2005. Three new diterpenes and the biological activity of different extracts of *Jasonia montana*. *Nat Prod Res.*19(3), 253-65.

Andrade-Cetto, A., Heinrich, M., 2005. Mexican plants with hypoglycaemic effect used in the treatment of diabetes. *J Ethnopharmacol* 99, 325-348.

Atkinson, M.A., Maclaren, N.K., 1994. The pathogenesis of insulin dependent diabetes mellitus. *New Eng J Med.* 331, 1428-1436.

www.ayurvediccure.com. Diabeta – an antidiabetic capsule.

Ayyanar, M., Ignacimuthu, S., 2005. Traditional knowledge of Kani tribals in Kouthalai of Tirunelveli hills, Tamil Nadu, India. *Journal of Ethnopharmacology* 102, 246 - 255.

Ayyanar, M., Sankarasivaraman, K., Ignacimuthu, S., 2008. Traditional herbal medicines used for the treatment of Diabetes among two major tribal groups in south Tamil Nadu, India. *Ethnobotanical Leaflets* 12, 276-280.

Ayyanar, M., Ignacimuthu, S., 2011. Ethnobotanical survey of common medicinal plants used by kani tribals in Tirunelveli hills of Western Ghats, India. *J Ethnopharmacol* 134, 851-864.

Ayyanar, M., Subash-Babu, P., 2012. *Syzygium cumini* (L.) Skeels. A review of its phytochemical constituents and traditional uses. *Asian Pac J Trop Biomed* 2, 240-246.

Ayyanar, M., Subash-Babu, P., Ignacimuthu, S., 2013. *Eugenia jambolana* Lam., A Novel Therapeutic Agent for Diabetes: Folk medicinal and Pharmacological evidences. *Complementary Therapies in Medicine* 21(3), 232-243.

Ayyanar, M., 2013. Indian Medicinal Plants as a source of Therapeutic agents: a review. *International Journal of Bioscience Research* 1(1), 1-24.

Babu, P.S., Prince, P.S.M., 2004. Antihyperglycaemic and antioxidant effect of hyponidd, an ayurvedic herbomineral formulation in streptozotocin-induced diabetic rats. *J Pharm Pharmacol* 56, 1435-1442.

Bailey, C.J., Day, C., 1989. Traditional plant medicines as treatments for diabetes. *Diabetes Care* 12(8), 553-564.

Baskaran, K.M., Ahamath, B.K., Shanmugasundaram, K.R., Shanmugasundaram, E.R.B., 1990. Antidiabetic effect of a leaf extract from *Gymnema sylvestre* in non-insulin dependent diabetes mellitus patients. *J Ethnopharmacol.* 30, 295-305.

Bhatia, I.S., Bajaj, K.L., 1975. Chemical constituents of the seeds and bark of *Syzigium cumini. Planta Med.* 28, 348-352.

Bhattacharya, A., Chatterjee, A., Ghosal, S., Bhattacharya, S.K., 1999. Antioxidant activity of tannoid principles of *Emblica officinalis* (Amla). *Indian J Exp Biol.* 37, 676-680.

Chakravarthy, B.K., Gupta, S., Gambhir, S.S., Gode, K.D., 1981. Pancreatic beta-cell regeneration in rats by (y)- epicatechin. *Lancet* ii 1, 759–760.

Chakravarthy, B.K., Gupta, S., Gode, K.D., 1982a. Antidiabetic effect of (y) epicatechin. *Lancet* ii 1, 272–273.

Chakravarthy, B.K., Gupta, S., Gode, K.D., 1982b. Functional beta cell regeneration in the islets of pancreas in alloxan induced diabetic rats by (y)-epicatechin. *Life Sci.* 31

Choi, S.Z., Lee, S.O., Jang, K.U., Chung, S.H., Park, S.H., Kang, H.C., Yang, E.Y., Cho, H.J., Lee, K.R., 2005. Antidiabetic stilbene and anthraquinone derivatives from *Rheum undulatum*. *Arch Pharm Res.* 28(9), 1027-30.

Dieye, A.M., Sarr, A., Diop, S.N., Ndiaye, M., Yoro Sy, G., Diarra, M., Gaffary, I.R., Sy, A.N., Fayea, B., 2008. Medicinal plants and the treatment of diabetes in Senegal: survey with patients. *Fund Clin Pharmacol* 22, 211-216.

Donati, D., Lampariello, L.R., Pagani, R., Guerranti, R., Cinci, G., Marinello, E., 2005. Antidiabetic oligocyclitols in seeds of *Mucuna pruriens*. *Phytother Res.* 19(12), 1057-60.

Du Thie, G., Crozier, A., 2000. Plant derived phenolic antioxidants. *Curr. Opin. Clin. Nutr. Metab.* 3, 447–451.

Eddouks, M., Maghrani, M., Lemhadri, A., Ouahidi, M.L., Jouad, H., 2002. Ethnopharmacological survey of medicinal plants used for the treatment of diabetes mellitus, hypertension and cardiac diseases in the south-east region of Morocco (Tafilalet). *J Ethnopharmacol* 82, 97-103.

Estrada, O., Hasegawa, M., Gonzalez-Mujica, F., Motta, N., Perdomo, E., Solorzano, A., Mendez, J., Mendez, B., Zea, E.G., 2005. Evaluation of flavonoids from *Bauhinia megalandra* leaves as inhibitors of glucose-6-phosphatase system. *Phytother Res.* 19(10), 859-63.

Formica, J.V., Regelson, W., 1995. Review of the biology of quercetin and related bioflavonoids. *Food Chem. Toxicol.* 33.

Fuentes, N.L., Sagua, H., Morales, G., Borquez, J., San Martin, A., Soto, J., Loyola, L.A., 2005. Experimental antihyperglycemic effect of diterpenoids of llareta *Azorella compacta* (Umbelliferae) Phil in rats. *Phytother Res.* 19(8), 713-6.

Fuhrman, B., Aviram, M., 2001. Flavonoids protect LDL from oxidation and attenuate atherosclerosis. *Curr. Opin. Lipidol.* 12, 41–48.

Ghosh, S., Suryawanshi, S.A., 2001. Effect of *Vinca rosea* extracts in treatment of alloxan diabetes in male albino rats. *Indian J Exp Biol.* 39, 748-759.

Govindachari, T.R., 1992. Chemical and biological investigations on *Azadirachta indica* (the neem tree). *Current Sci.* 63, 117-122.

Graeme, I.B., Kenneth, S.P., 2001. Diabetes mellitus and genetically programmed defects in β-cell function. *Nature* 414, 788-91.

Gray, A.M., Flatt, P.R., 1999. Insulin secreting activity of the traditional antidiabetic plant *Viscum album* (mistletoe). *Journal of Endocrinology* 160, 409–414.

Guerrero-Analco, J.A., Hersch-Martinez, P., Pedraza-Chaverri, J., Navarrete, A., Mata, R., 2005. Antihyperglycemic effect of constituents from *Hintonia standleyana* in streptozotocin-induced diabetic rats. *Planta Med.* 71(12), 1099-105.

Harbone, J.B., 1993. editor. The flavonoids. Advance in research since 1986. London: Chapman and Hall.

Helmstadter, A., 2007. Antidiabetic drugs used in Europe prior to the discovery of insulin. *Pharmazie* 62, 717-720.

Indira, G., Rao, A.S., Devi, M.V., 1954. Studies on hypoglycemic activity of aqueous extract of seeds of *Eugenia jambolana* in different formulations in rabbits. *Indian J Pharmacol* 11, 65.

Jahromi, M.A., Ray, A.B., 1993. Antihyperlipidemic effect of flavonoids from *Pterocarpus marsupium*. *J Nat Prod.* 56(7), 989-994.

Jayaprakasam, B., Olson, L.K., Schutzki, R.E., Tai, M.H., Nair, M.G., 2006. Amelioration of obesity and glucose intolerance in high-fat-fed C57BL/6 mice by anthocyanins and ursolic acid in Cornelian cherry (*Cornus mas*). *J Agric Food Chem.* 11; 54(1), 243-8.

Jia, W., Gao, W., Tang, L., 2003. Antidiabetic herbal drugs officially approved in China. *Phytother Res* 17, 1127-1134.

Jung, M., Park, M., Lee, H.C., Kang, Y.H., Kang, E.S., Kim, S.K., 2006. Antidiabetic agents from medicinal plants. *Curr Med Chem.* 13(10), 1203-18.

Kapoor, L.S., 1990. In: *Handbook of Ayurvedic Medicinal Plants*, CRC Press, Boca Raton FL. pp 200-2201.

Kerem, Z., Bilkis, I., Flaishman, M.A., Sivan, L., 2006. Antioxidant activity and inhibition of alpha-glucosidase by trans-resveratrol, piceid, and a novel trans-stilbene from the roots of Israeli *Rumex bucephalophorus* L. *J Agric Food Chem.* 22; 54(4), 1243-7.

Kim, H.Y., Moon, B.H., Lee, H.J., Choi, D.H., 2004. Flavonol glycosides from the leaves of *Eucommia ulmoides* O. with glycation inhibitory activity. *Journal of Ethnopharmacol.* 93, 227-230.

Lans, C.A., 2006. Ethnomedicines used in Trinidad and Tobago for urinary problems and diabetes mellitus. *Journal of Ethnobiology and Ethnomedicine* 2:45 doi:10.1186/1746-4269-2-45.

Leduc, C., Coonishish, J., Haddad, P., Cuerrier A., 2006. Plants used by the Cree Nation of Eeyou Istchee (Quebec, Canada) for the treatment of diabetes: A novel approach in quantitative ethnobotany. *J Ethnopharmacol* 105, 55-63.

Lee, S.O., Choi, S.Z., Lee, J.H., Chung, S.H., Park, S.H., Kang, H.C., Yang, E.Y., Cho, H.J., Lee, K.R., 2004. Antidiabetic coumarin and cyclitol compounds from *Peucedanum japonicum*. *Arch Pharm Res.* 27(12), 1207-10.

Lee, H.S., 2005. Cuminaldehyde: Aldose Reductase and alpha-Glucosidase Inhibitor Derived from *Cuminum cyminum* L. Seeds. *J Agric Food Chem.* 6; 53(7), 2446-50.

Li, W.L., Zheng, H.C., Bukuru, J., Kimpe, D., 2004. Natural medicine used in the traditional Chinese medical system for therapy of diabetes mellitus. *Journal of Ethnopharmacology* 92, 1-21.

Lio, R.Y., Wang, G.O., 1996. A survey on drugs synthesized for anti-diabetes. *Journal of Shenyang Pharmaceutical University* 13, 148-153.

Liu, I.M., Liou, S.S., Lan, T.W., Hsu, F.L., Cheng, J.T., 2005. Myricetin as the active principle of *Abelmoschus moschatus* to lower plasma glucose in streptozotocin-induced diabetic rats. *Planta Med.* 71(7), 617-21.

Mallick, C., Mandal, S., Barik, B., Bhattacharya, A., Ghosh, D., 2007. Protection of testicular dysfunctions by MTEC, a formulated herbal drug, in streptozotocin induced diabetic rat. *Biol Pharm Bull* 30, 84-90.

Manickam, M., Ramanathan, M., Farboodinary Jahromi, M.A., Chanosuria, J.P.N., Ray, A.B., 1997. Antihyperglycaemic activity of phenolics from *Pterocarpus marsupium*. *J Nat Prod.* 60, 609-610.

Mathur, R., Sharma, A., Dixit, V.P., Varma, M., 1996. Hypolipidaemic effect of fruit juice of *Emblica officinalis* in cholesterol-fed rabbits. *J Ethnopharmacol.* 50, 61-68.

Maurya, R., Sing, R., Deepak, M., Handa, S.S., Yadav, P.P., Mishra, P.K., 2004. Constituents of *Pterocarpus marsupium*: an Ayurvedic crude drug. *Phytochemistry* 65, 915-920.

Mohan, V., Poongothai, S., Deepa, R., Subramanian, S.L., Nalini, K., Murali, P.M., 2001. Efficacy of DCBT 2345 - An Ayurvedic compound in Treatment of Type 2 Diabetic patients with Secondary Failure to Oral Drugs-Randomized Double Blind Placebo Control Study. *Int J Diab Dev Count* 21, 176-183.

Modak, M., Dixit, P., Londhe, J., Ghaskadbi, S., Paul, A.D.T., 2007. Indian herbs and herbal drugs used for the treatment of diabetes. *J Clin Biochem Nutr* 40, 163-173.

McCune, L.M., Johns, T., 2002. Antioxidant activity in medicinal plants associated with the symptoms of diabetes mellitus used by the Indigenous Peoples of the North American boreal forest. *Journal of Ethnopharmacology* 82, 197 -/205.

Mukherjee, P.K., Maiti, K., Mukherjee, K., Houghton, P.J., 2006. Leads from Indian medicinal plants with hypoglycemic activity. *Journal of Ethnopharmacology* 106, 1 – 28.

Murali, B., Upadhyaya, U.M., Goyal, R.K., 2002. Effect of chronic treatment with *Enicostemma littorale* in non-insulin dependent diabetic (NIDDM) rats. *J Ethnopharmacol.* 81, 199-204.

Mutalik, S., Chetana, M., Sulochana, B., Devi, P.U., Udupa, N., 2005. Effect of Dianex, a herbal formulation on experimentally induced diabetes mellitus. *Phytother Res* 19, 409-415.

Nageswara Rao, G., Mahesh Kumar, P., Dhandapani, V.S., Ramakrishna, T., Hayashi, T., 2000. Constituents of *Cassia auriculata*. *Fitotherapia* 71, 82-83.

Narender, T., Puri, A., Shweta Khaliq, T., Saxena, R., Bhatia, G., Chandra, R., 2006. 4-hydroxyisoleucine an unusual amino acid as antidyslipidemic and antihyperglycemic agent. *Bioorg Med Chem Lett.* 15;16(2), 293-6.

Nishiyama, T., Mae, T., Kishida, H., Tsukagawa, M., Mimaki, Y., Kuroda, M., Sashida, Y., Takahashi, K., Kawada, T., Nakagawa, K., Kitahara, M., 2005. Curcuminoids and sesquiterpenoids in turmeric (*Curcuma longa* L.) suppress an increase in blood glucose level in type 2 diabetic KK-Ay mice. *J Agric Food Chem.* 23;53(4), 959-63.

Pari, L., Latha, M., 2002. Effect of *Cassia auriculata* flowers on blood sugar levels, serum and tissue lipids in Streptozotocin Diabetic Rats. *Singapore Med J.* 43, 617-621.

Prakash, P., Gupta, N., 2005. Therapeutic uses of *Ocimum sanctum* Linn (Tulsi) with a note on eugenol and its pharmacological actions: a short review. *Indian J Physiol Pharmacol.* 49(2), 125-31.

Prince, P.S.M., Menon, V.P., 2003. Hypoglycaemic and hypolipidaemic action of alcohol extract of *Tinospora cordifolia* roots in chemical induced diabetes in rats. *Phytother Res.* 17, 410-413.

Prince, P.S.M., Menon, V.P., 1999. Antioxidant activity of *Tinospora cordifolia* roots in experimental diabetes. *J Ethnopharmacol.* 65, 277-281.

Rahman, A.U., Zaman, K., 1989. Medicinal plants with hypoglycemic activity. *J Ethnopharmacol* 26, 1-55.

Raman, A., Lau, C., 1996. Anti-diabetic properties and phytochemistry of *Momordica charantia* L. (Cucurbitaceae). *Phyto Med.* 2, 349-362.

Rastogi, R.P., Mehrotra, B.N., 1993. Compendium of Indian Medicinal Plants. CDRI, Lucknow and Publications and Information Directorate, New Delhi.

Ravi, K., Rajasekaran, S., Subramanian, S., 2005. Antihyperlipidemic effect of *Eugenia jambolana* seed kernel on streptozotocin-induced diabetes in rats. *Food Chem Toxicol* 43, 1433-1439.

Report of the expert committee on the diagnosis and classification of diabetes mellitus. Feb'2004. Vol 27; S5-S10.

Resurreccion-Magno, M.H., Villasenor, I.M., Harada, N., Monde, K., 2005. Antihyperglycaemic flavonoids from *Syzygium samarangense* (Blume) Merr. and Perry. *Phytother Res.* 19(3), 246-51.

Saktiek., 2001. New perspective into molecular pathogenesis and treatment of type 2 diabetes (Review). *Cell* 104, 517-29.

Scartezzini, P., Speroni, E., 2000. Review on some plants of Indian traditional medicine with antioxidant activity. *J Ethnopharmacol.* 71(1-2), 23-43.

Sezik, E., Aslan, M., Yesilada, E., Ito, S., 2005. Hypoglycaemic activity of *Gentiana olivieri* and isolation of the active constituent through bioassay-directed fractionation techniques. *Life Sci.* 28; 76(11), 1223-38.

Sharma, S.B., Nasir, A., Prabhu, K.M., Murthy, P.S., Dev, G., 2003. Hypoglycaemic and hypolipidemic effect of ethanolic extract of seeds of *Eugenia jambolana* in alloxan-induced diabetic ratbbits. *J Ethanopharmacol.* 85, 201-206.

Shen, P., Liu, M.H., Ng, T.Y., Chan, Y.H., Yong, E.L., 2006. Differential effects of isoflavones, from *Astragalus membranaceus* and *Pueraria thomsonii*, on the activation of PPARalpha, PPARgamma, and adipocyte differentiation *in vitro*. *J Nutr.* 136(4), 899-905.

Shigematsu, N., Asano, R., Shimosaka, M., Okazaki, M., 2001. Effect of Administration with the Extract of *Gymnema sylvestre* R. Br Leaves on Lipid Metabolism in Rats. *Biol Pharm Bull.* 24(6), 713-717.

Kant, S., Sahu, M., Sharma, S., 2001. Effect of Diabecon (D-400), an Ayurvedic herbomineral formulation on diabetic retinopathy. *Indian J Clin Pract* 12, 49-56.

Singh, R.P., Padmavathi, B., Rao, A.R., 2000. Modulatory influence of *Adhatoda vasica* (*Justica adhatoda*) leaf extract on the enzyme of xenobiotic metabolism, antioxidant status and lipid peroxidation in mice. *Molecular and Cellular Biochemistry* 213, 99–109.

Singh, S.S., Pandey, S.C., Srivastava, S., Gupta, V.S., Parto, B., Ghosh, A.C., 2003. Chemistry and medicinal properties of *Tinospora cordifolia* (Guduchi). *Indian J Pharmacol.* 35, 83-91.

Sudha, V., Bairy, K.L., Shashikiran, U., Sachidananda, A., Jayaprakash, B., Shalini, S., 2005. Efficacy and tolerability of Dianex in Type 2 diabetes mellitus: a non randomized, open label non-comparative study. *Med J Malaysia* 60, 204-211.

Takatsuna, H., Umezawa, K., 2004. Screening of bioactive metabolites for pancreatic regeneration chemotheraphy. *Biomedicine and Pharmacotheraphy* 58, 610-613.

Villasenor, I.M., Lamadrid, M.R., 2006. Comparative anti-hyperglycemic potentials of medicinal plants. *J Ethnopharmacol* 104, 129-131.

Xi, L., Qian, Z., Shen, X., Wen, N., Zhang, Y., 2005, Crocetin prevents dexamethasone-induced insulin resistance in rats. *Planta Med.* 71(10), 917-22.

Yeh, G.Y., Eisenberg, D.M., Kaptchuk, T.J., Phillips, R.S., 2003. Systematic Review of Herbs and Dietary Supplements for Glycemic Control in Diabetes. *Diab Care* 26, 1277-1294.

2015, Modern Methods in Phytomedicine *Pages* ***101–141***
Editor: **T. Parimelazhagan**
Published by: **DAYA PUBLISHING HOUSE, NEW DELHI**

8

Medicinal Plants Phytochemistry and Biopotentials of Seaweeds: A Review

M. Johnson, G. Sahaya Anthony Xavier and D. Patric Raja*

Department of Botany, St. Xavier's College (Autonomous), Palayamkottai – 627 002, Tamil Nadu

1.0 INTRODUCTION

The marine environment is a rich source of biological and chemical diversity. The diversity has been a unique source of chemical compounds of potential for pharmaceuticals, cosmetics, dietary supplements and agrochemicals (Chau Van Minh *et al.*, 2005). The sea source includes various organisms such as sponges, tunicates, seaweed, marine microorganisms and symbionts etc. (Vallinayagam *et al.*, 2009). Seaweeds belong to a group of plants known as algae. Seaweeds are considered as a rich source of bioactive compounds as they are able to produce a great variety of secondary metabolites characterized by a broad spectrum of biological activities (Rajasulochana, 2009). Seaweeds are classified as Rhodophyta (red algae), Phaeophyta (brown algae) and Chlorophyta (green algae) depending on their nutrient and chemical composition (Cox *et al.*, 2010). In recent years, a significant number of novel metabolites with potent pharmacological properties have been discovered from the marine organisms. The red and brown seaweeds are rich sources of bioactive

* *Corresponding Author.* E-mail: ptcjohnson@gmail.com

secondary metabolites. Numerous studies have focused on their nutraceutical and pharmaceutical properties (Blunt *et al.,* 2011; Cabrita *et al.,* 2010; Narisnh *et al.,* 2005). The algae synthetize a variety of compounds such as carotenoids, terpenoids, xanthophylls, chlorophylls, vitamins, saturated and polyunsaturated fatty acids, amino acids, acetogenins, halogenated compounds such as haloforms, halogenated alkanes and alkenes, alcohols, aldehydes, hydroquinones and ketones (Kandhasamy and Arunachalam, 2008, Lincoln *et al.,* 1999). In the present review, the available information on phytochemistry, bioactivities and pharmacology of seaweeds are presented. Focus is placed on the antibacterial, antioxidant, antifungal, cytotoxic and larvicidal properties of seaweeds and pharmaceutical value. In addition, the emphasis is placed on Seaweed Liquid Fertillizer/extracts and silver nanoparticles synthesis from seaweeds.

2.0 SEAWEEDS

Seaweeds (Macroalgae – Thallophyta) are primitive non-flowering plants without true roots, stem and leaves. They are a commercially important, marine, renewable resource. Seaweeds are the major producers of the oceanic plant community which are distributed widely and recognized for their chemical defence against many biotic factors. Seaweeds have been used as food stuff in the Asian diet for centuries as it contains carotenoids, dietary fibres, proteins, essential fatty acids, vitamins and minerals. Fresh and dry seaweeds are extensively consumed by people especially living in the coastal areas. From the literature, it is observed that the edible seaweeds contain a significant amount of the protein, vitamins and minerals, which are essential nutrition for human. Seaweeds offer a wide range of therapeutic possibilities both internally and externally. Seaweeds have an extensive profile source of secondary metabolites. More than 600 secondary metabolites have been isolated from marine algae (Faulkner, 1986). Compounds with cytostatic, antiviral, antihelmintic, antifungal and antibacterial activities have been detected in green, brown and red algae (Jha, 2004).

The bio-stimulant properties of seaweeds have been explored for use in agriculture and the antimicrobial properties for the development of novel antibiotics. Seaweeds have some valuable medicinal components such as antibiotics, laxatives, anti-coagulants, anti-ulcer products and suspending agents in radiological preparations. Seaweeds have recently received significant attention for their potential as natural antioxidants. Most of the compounds of marine algae show anti-bacterial activities. Many metabolites isolated from marine algae have bioactive efforts (Oh *et al.,* 2008, Gopala *et al.,* 2010). Many of these compounds are bioactive and have been extensively studied using bioassays and pharmacological assays (Paul and Fenical, 1987). The potential antitumor promoting properties of 36 edible/common marine algae from Maozuru, Kyoto, Japan were examined and strong inhibitory activities were found in *Undaria pinnatifida, Laminaria* and *Sargassum* species (Ohigashi *et al.,* 1992). The production of inhibitory substances from seaweeds was noted as early as in 1917 (Harder and Oppermann, 1953). Like other plants, seaweeds contain various inorganic and organic substances which can benefit human health (Kuda *et al.,* 2002). Seaweeds are considered as a source of bioactive compounds as they are able to produce a great variety of secondary metabolites characterized by a broad spectrum of biological activities. Compounds with antioxidant, antiviral, antifungal and antimicrobial activities have been detected in brown, red and green algae (Yuan *et al.*, 2005; Bansemir *et al.*, 2006; Chew *et al.*, 2008). The environment in which seaweeds grow is harsh as they are exposed to a combination of light and high oxygen concentrations. These factors can lead to the formation of free radicals and other strong oxidizing agents but seaweeds seldom suffer any serious photodynamic damage during metabolism. This fact implies that seaweed cells have some protective mechanisms and compounds (Matasukawa *et al.*, 1997).

3.0 BIOCHEMICAL COMPOSITION OF SEAWEEDS

Manivannan *et al.* (2009) analyzed the biochemical composition of *Ulva reticulata, Enteromorpha compressa, Cladophora glomerata, Halimeda macroloba, Halimeda tuna, Dictyota dichotoma, Turbinaria ornate, Padina pavonica, Gelidiella acerosa, Gracilaria crassa* and *Hypnea musciformis* from Vedalai coastal waters, Southeast coast of India. The protein content was recorded to be maximum in *G. acerosa* and minimum in *D. dichotoma*; carbohydrate level was observed to be maximum in *T. ornata* and minimum in *P. pavonica*. The lipid content was maximum in *H. tuna* and minimum in *H. macroloba.*

Manivannan *et al.* (2008) estimated the biochemical composition of *Enteromorpha intestinalis, Enteromorpha clathrata, Ulva lactuca, Codium tomentosum, Padina gymnospora, Colpomenia sinuosa, Sargassum tenerimum, Sargassum wightii, Turbinaria conoides, Gracilaria folifera, Hypnea valentiae* and *Acanthophora spififera* collected from Mandapam on the southeast coast of India. The reported the protein content to vary from 3.25±0.36 to 17.08±0.28 per cent; the maximum protein content was recorded in *P. gymnospora* (17.08±0.28 per cent) followed by *E. intestinalis* (16.38±0.50 per cent) and *S. tenerimum* (12.42±0.63 per cent). The minimum protein content was observed in *U. lactuca* (3.25±0.36 per cent) followed by *C. tomentosum* (6.13±0.23 per cent), *G. folifera* (6.98±0.08 per cent) and *H. valentiae* (8.34±0.30 per cent). The carbohydrate content varied from 20.47±0.50 to 23.9±0.19 per cent. The maximum carbohydrate concentration was recorded from *T. conoides* (23.9±0.19 per cent) followed by *E. intestinalis* (23.84±0.14 per cent), *H. valentiae* (23.60±0.33 per cent), *S. tenerimum* (23.55±0.44 per cent), *A. spicifera* (23.54±0.10 per cent) and *S. wightii* (23.50±0.65 per cent). The minimum carbohydrate content was observed from *C. tomentosum* (20.47±0.50 per cent) followed by *P. gymnospora* (21.88±1.22 per cent), *G. folifera* (22.32±1.40 per cent) and *C. sinuosa* (22.46 ± 1.79 per cent). The lipid content of seaweeds varied from 1.33 ± 0.20 to 4.6 ± 0.17; the maximum lipid content was observed from *E. clathrata* (4.6 ± 0.17 per cent) followed by *G. folifera* (3.23 ± 0.13 per cent), *C. tomentosum* (2.53 ± 0.27 per cent), *C. sinuosa* (2.33±0.37 per cent) and *S. wightii* (2.33 ± 0.37 per cent). The minimum lipid concentration was recorded from *E. intestinalis* (1.33 ± 0.20 per cent) followed by *P. gymnospora* (1.4 ± 0.30 per cent), *S. tenerimum* (1.46±0.20 per cent) and *U. lactuca* (1.6 ± 0.17 per cent).

Johnson *et al.* (2012) identified the phytochemical properties of *Sargassum wightii* and identified the functional constituents present in the crude extracts by spectroscopic and chromatographic analysis. Thillaikkannu *et al.* (2012) evaluated the qualitative and quantitative parameters of six seaweeds namely *Ulva lactuca, Caulerpa racemosa, Sargassum wightii, Padina tetratomatica, Gracilaria corticata* and *Acanthophora spicifera* collected from the Gulf of Mannar and revealed that saponin and polyphenol were absent in the acetone extract of the selected seaweeds and other extracts possess all the phytoconstituents.

Hebsibah *et al.* (2010) evaluated the secondary metabolites in two types of seaweeds (*Sargassam wightii* and *Gracillaria edilus*). Secondary metabolites like phenolic compounds, terpeniods, glycosides, proteins and glycoproteins were analyzed by three solvent extracts: ethanol, methanol and acetone. Leonel *et al.* (2013) identified a variety of polysaccharides present in *Kappaphycus alvarezii, C. jubata* and *C. crispus*-Gigartinales, Rhodophyta; *G. corneum* and *P. Capillacea* - Gelidiales, Rhodophyta; *L. obtusa* - Ceramiales, Rhodophyta; *H. elongata, U. pinnatifida, S. polyschides, S. vulgare* and *P. pavonica* - Phaeophyceae) using FTIR-ATR, FT-Raman analysis and spectroscopic techniques.

Sharma *et al.* (2012) identified and characterized the composition of five seaweed species extracts and commercial formulations (*Ascophyllum nodosum, Fucus serratus, Fucus vesiculosus, Laminaria hyperborea* and *Sargassum muticum*) using thermogravimetry (TGA), energy dispersive X-ray microanalysis (EDX), Fourier-transform infrared spectroscopy (FTIR) and pyrolysis gas chromatography/mass spectrometry (Py-GC/MS). These analyses provided information on the proportions of algal cell wall, inorganic fractions and minerals. Leonel *et al.* (2013) identified the polysaccharides present in several seaweeds (*Kappaphycus alvarezii, Calliblepharis jubata,* and *Chondrus crispus* - Gigartinales, Rhodophyta; *Gelidium corneum* and *Pterocladiella capillacea* - Gelidiales, Rhodophyta; *Laurencia obtuse* - Ceramiales, Rhodophyta; *Himanthalia elongata, Undaria pinnatifida, Saccorhiza polyschides, Sargassum vulgare* and *Padina pavonica* - Phaeophyceae, Ochrophyta) using spectroscopic techniques.

Johnson *et al.* (2012) explored the phytochemical constituents, UV-VIS and HPLC spectrum profile for *Dictyota bartayresiana*. Their phytochemical results showed the presence of alkaloids, steroids, phenolic groups, saponins, tannins, glycosides and sugars. The UV-VIS profile of methanolic, petroleum ether, chloroform, isopropanol extracts of *D. bartayresiana* showed various peaks with different functional groups. The HPLC profile of petroleum ether, chloroform and benzene extracts of *D. bartayresiana* showed some prominent and moderate peaks with different retention time. Based on their results they suggested that *D. bartayresiana* may be a rich source of phytoconstituents.

Fayaz *et al.* (2005) suggested the utility of *Kappaphycus alvarezii* for various nutritional products including antioxidants for use as health foods or nutraceutical supplements. Sanchez-Machado *et al.* (2004) found that the predominant sterol was desmosterol in red seaweeds (87-93 per cent of total sterol content).

Tasende (2000) confirmed that the fatty acids and sterols of algae are characteristic to particular taxa and could be useful as chemotaxonomic tools upto the class, family and sometimes even species levels. Plant sterols have been quantified by gas chromatography (Govindan and Hodge, 1993; Jeong and Lachance, 2001) or by HPLC with UV detection (Indyk, 1990) or evaporative light scattering and detection. However, few studies have presented techniques for parallel determination of different sterols. Further, it is observed that gas chromatography/mass spectrometry techniques are widely employed for identification of sterols.

Ghada *et al.* (2011) isolated the bioactive compounds spatane diterpene tetraol, fucosterol and linoleic acid from the Egyptian brown alga *Sargassum subrepandum*. In addition to this, four hydrocarbons were detected by using GC-MS *viz.*, heptadecane, 2, 6, 10, 14-tetramethyl-hexadecane, nonadecane and heneicosane. The chemical structure of spatane diterpene tetraol was formulated for the first time by spectroscopic analyses including mass spectrometry (EI-MS, HR/EIMS), 1D and 2D NMR experiments. The phytochemical study of the unsaponified fraction of the algal extract and GC-MS analysis confirmed the existence of nine compounds.

Sharma *et al.* (2012) characterised the composition of five seaweed species *viz.*, *Ascophyllum nodosum, Fucus serratus, Fucus vesiculosus, Laminaria hyperborea* and *Sargassum muticum* using FTIR and pyrolysis GC/MS. The main carbohydrate constituents of the five species and their extracts were identified by their pyrolysis products *viz.*, 1-(2-furanyl) ethanone, 1,6-anhydromannopyranose, 5-methyl- 2-furcarboxaldehyde, 2-hydroxy-3-methyl-2-cyclopenten-1-one, diannhydromannitol, and 1,6-anhy dromannofuranose using Py-GC/MS. The differences in relative intensities of the infrared bands of the five species were enhanced especially after acid extraction compared with alkaline or neutral treatments, resulting in improved understanding of the compositional changes.

Microspora floccosa, a filamentous green alga collected from the River Indus was extracted with methanol and it displayed strong antimicrobial activity against 14 bacterial and 20 fungal species. Eleven saturated and eleven unsaturated fatty acids were found in the extracts. The saturated fatty acids were lesser in quantity (47.4 per cent) than the unsaturated ones (52.6 per cent). Palmitic acid was present in the highest amount (12.3 per cent), while oleic acid was detected in an appreciable proportion (6.1 per cent). Two sterols and two terpenes were also obtained from the extract which was chemically elucidated as cholesterol, 24-isopropyl-5-cholesten-3β-ol, trans-phytol and cyclopterospermol (Khalid *et al.,* 2011).

The marine red alga *Laurencia brandenii* collected from the southwest coast of India was extracted and fractioned using column chromatography. It was found that the fraction eluted using petroleum ether: chloroform (6:4) exhibited broader biological activities. GC-MS profile of the active fraction revealed that the main constituent was octadecadienoic acid (49.75 per cent) followed by n-hexadecanoic acid (14.24 per cent), which might have a functional role in the biological activities (Manilal *et al.,* 2011).

Siddiqui *et al.* (1994) isolated four sterols and 19 fatty acids from the methanolic extract green alga *Bryopsis pennata* and it was identified by 1H-NMR, El-MS and GC-MS techniques. The sterol having cholesta skeleton was found as major constituent (78.12 per cent), while other three with ergosta skeleton occurred in traces (5-9 per cent). Seven saturated fatty acids were present in greater quantity (72.58 per cent) than 12 unsaturated ones (27.41 per cent). The latter included 7 monoenoic, 2 dimwit,

2 trienoic and 1 pentaenoic acids. Tricosanoic acid was found in the highest amount (30.24 per cent), clopentenyl undecanoic and heneicosapentaenoic acids were the unique and rare fatty acids detected.

Chlorophylls, carotenoids and degradation products from *Caulerpa prolifera, Jania rubens* and *Padina pavonica* were separated by reversed-phase HPLC using an elution gradient of methanol, acetone and ammonium acetate solution to obtain high resolution peaks. Eighteen photosynthetic pigments were separated from *C. prolifera*, 16 from *J. rubens* and 14 from *P. pavonica*. Chlorophyll b, micronone, microxanthin, neoxanthin, siphonein and siphonoxanthin were the most typical and characteristic pigments of *C. prolifera*, while chlorophyll c_1, c_2, fucoxanthin, fucoxanthol, avoxanthin, diatoxanthin were the most typical pigments in *P. pavonica*. In *J. rubens*, chlorophyll d, α-cryptoxanthin, β-cryptoxanthin and fucoxanthin were the most common pigments (Hegazi *et al.*, 1998). Kumar *et al.* (2013) determined the GC-MS profile and antibacterial activity of *Sargassum tenerrimum*. The presence of bioactive functional groups was revealed by FT-IR analysis. Twelve compounds were identified in GC-MS of which two compounds 1,2-Benzoldicarbonsaeure and Cyclopropanepentanoic acid showed maximum intensity of peak. The methanolic extract showed good antibacterial activity against five pathogenic strains at different concentrations.

4.0 PHYTOCHEMISTRY AND BIOLOGICAL ACTIVITY OF SEAWEEDS

El-Baroty *et al.* (2010) extracted the glycolipids in five species of marine algae: two species of Rhodophyta (*Laurencia popillose, Galaxoura cylindriea*); one species of Chlorophyta (*Ulva fasciata*) and two species of Phaeophyta (*Dilophys fasciola, Taonia atomaria*) collected from Red and Mediterranean Sea. The extracted glycolipids were purified on silica gel column and identified by liquid chromatography MS/MS. Total glycolipid contents (GL) (as per cent of total lipid) were found in between 10.9 to 28.7 per cent. *T. atomaria* had the highest level (28.7 per cent) followed by *L. popillose* (22.5 per cent). GL groups were analyzed for their sugars and fatty acids composition. The results showed that the highest carbohydrate content of GLs were found in *U. fasciata* (6.05 per cent) and *L. popillose* (5.8 per cent) and characterized by high content of monosaccharide: mannouronic acid, galactose and rhamnose. Amongst of the glycolipids of algal species, the most predominate fatty acid identified by GC were palmatic (C16:0 19.20 - 65.89 per cent of total fatty acid) and eicosatrinoic (C20:3 7.52 - 54.41 per cent). GL analysis by LC/MS/MS, revealed the peak at m/z 956 corresponding to the molecular formula of $C_{51}H_{104}O_{17}$ which was the most abundant molecular ion among all GLs of algal species and its fragment peaks at m/z 617($C_{37}H_{58}O_4$) and m/z 337 ($C_{21}H_{58}O_3$) were tentatively identified as digalactosyldiacylglycerol (DGDG). In addition, the *in vitro* anticancer, antimicrobial and antiviral activities of algal glycolipids were evaluated. GL of all algae species showed a remarkable antiviral activity in dose dependent manner. GL from *D. fasciola* has shown the most potent effect against HSV1 (IC_{50} of 10 µg/mL), comparable to that of the current antiviral drug acyclovir (IC_{50} 55 µg/mL). On the other hand, GL of all algal species possessed a moderate antimicrobial activity. GL of *T. atomaria* exhibited a high inhibition effects against all test microorganisms, with MIC value ranging from 60 to 80 µg/mL. Moreover, all algal GL exhibited remarkable anticancer activities against both breast (MCF7) and liver human (HepG2) cancer cells, with an IC_{50} values ranging from 0.47 to 2.89 µg/mL.

Kajal *et al.* (2010) chromatographically purified the two new guaiane sesquiterpene derivatives, guai-2-en-10a-ol (1) and guai-2-en-10a-methanol (2) as major constituents of the $CHCl_3/CH_3OH$ (1:1, v/v) soluble fraction of *Ulva fasciata.* The structures of the compounds were elucidated using one and two-dimensional NMR and mass spectrometric analysis. They observed that the compounds 2 and 3 exhibited significant inhibition against *Vibrio parahaemolyticus* with minimum inhibitory concentrations of 25 and 35 mg/mL, respectively.

Sasidharan *et al.* (2008) isolated the active fraction from crude extract of *Gracilaria changii* and determined its in vitro antifungal activity. They observed that the active fraction was effective as a fungicide against *C. albicans* and showed a dose-dependent antifungal activity. Khanzada *et al.* (2007) screened the various fractions of ethanolic extract of *Solieria robusta* (Greville) Kylin (*Rhodophyta*) for antifungal activity against 5 fruit spoiling fungi isolated from fruits. All fractions were able to inhibit fungal growth. The aqueous fraction showed maximum inhibition ratio followed by methanol, ethyl acetate, chloroform and ethanol. Prabha *et al.* (2013) identified the phytoconstituents and antimicrobial activity in *Kappaphycus alvarezii* using three solvent extracts. Ponnuchamy Kumar *et al.* (2013) studied the presence of various phytoconstituents in the methanolic extract of *S. tenerrimum* by GC-MS analysis and their antibacterial activity was evaluated against five pathogenic strains by disc diffusion method at different concentrations (25, 50, 75 and 100 mg/mL). The bioactive functional groups were revealed using FT-IR analysis. Jeyaseelan *et al.* (2012) reported the phytochemical constituents and antibacterial activity of *S. polycystum, S. tenerrimum, T. ornata, G. crassa* and *C. fragile* against *Escherichia coli* (ATCC 25922) and *Staphylococcus aureus* (NCTC 6571) by agar well diffusion method.

Krishnaveni and Johnson (2012) investigated the phytoconstituents of *Gracilaria corticata* J Ag. In addition they studied UV-VIS, HPLC profiling and the antibacterial potentials of *Gracilaria corticata* J. Ag extracts against the Gram positive and Gram negative bacteria by using agar disc diffusion method. Akbar Esmaeili and Marjan Khakpoor (2012) evaluated the chemical composition of chloroform and methanolic extract of *Stoechospermum marginatum* C. Agardh from Iran and also analysed the chemical composition of the essential oil using Gas chromatography (GC) and gas chromatography-mass spectroscopy (GC-MS). They carried out the antibacterial activity of the methanolic, chloroform and hexane as well as an ethanolic extract against seven Gram-positive and Gram-negative bacteria. Among these ethanolic extract showed the highest antibacterial activity and the hexane extract showed the lowest activity.

Shyamala and Thangaraju (2013) revealed the presence of the secondary metabolites *viz.*, alkaloids, carbohydrates, saponins, glycosides, protein, amino acids, phytosterols, phenolic compounds, flavonoids, terpenoids and tannins present in *C. racemosa, S. marginatum* and *H. musciformis* collected from Gulf of Mannar. They also studied the antibacterial activity against several human pathogenic microbes.

Among these maximum activity was recorded in methanolic extracts of *H. musciformis* against *B. subtilis* and minimum activity was noted in ethanolic extract of *C. racemosa* against *E. coli* among the two different extracts. Nadine *et al.* (2013) isolated the polysaccharides from the brown seaweed *Dictyo pterispolypodioides* grown on the Lebanese coast and reported the percentages of the main polysaccharide alginic acid to be 4.6 per cent in May and 6.25 per cent in July. They also performed the infrared spectroscopy to reveal the functional groups of alginate and mannuronan, anticoagulant activity, activated partial thromboplastintime (APTT) clotting assay and antioxidant activity. Premalatha *et al.* (2011) studied the preliminary phytochemical analysis, antimicrobial and DPPH scavenging activity of *Ulva fasciata* and *Chaetomorpha antennia.* Maximum phytochemical components are present in *Ulva fasciata* when compared to the *Chaetomorpha*. In the DPPH scavenging assay, *U. fasciata* showed more antioxidant activity compared to *C. antennia* and the results of antimicrobial activity clearly expressed that *U. fasciata* has high concentration of active principles when compared to the *C. antennia.*

4.1 Antibacterial and Antifungal Activity of Seaweeds

The discovery and development of antibiotics is one the most powerful and successful achievements of modern science and technology for the control of infectious diseases. The prolonged usage of broad spectrum antibiotics has led to the emergence of drug resistance. There is a tremendous need for novel antimicrobial agents from different sources. The biodiversity of marine ecosystem provides an important source of chemical compounds which have many therapeutic applications. Seaweeds or marine algae have been reported to contain many important compounds which act as antibiotics, laxatives, anticoagulants, anti-ulcer products and suspending agents in radiological preparations. Many substances obtained from marine algae such as alginates, carrageenan and agar as phycocolliods have been used for decades in medicine and pharmacy. More and more chemists and biologists are paying attention to the constituents of the algae; the exploration of the natural constituents may lead to an efficient lead for the discovery of new drug molecules against several pathogens causing infectious diseases (Chanda *et al.*, 2010).

Manivannan *et al.* (2011) evaluated the antimicrobial activity of *Turbinaria conoides, Padina gymnospora* and *Sargassum tenerrimum* against human bacterial and fungal pathogens. The methanol extracts of *Padina gymnospora* showed highest antibacterial activity against *B. subtilus*. The acetone extracts of *Padina gymnospora* showed highest antifungal activity against *Cryptococcus neoformans.* Lavanya and Veerappan (2011) reported the *in vitro* antibacterial activity of six selected marine algae *Codium decorticatum, Caulerpa scalpelliformis, Gracilaria crassa, Acanthophora spicifera, Sargassum wightii* and *Turbinaria conoides*. The acetone, methanol, chloroform, diethyl ether, ethyl acetate, hexane and aqueous extracts of selected marine algae were tested against the selected human pathogens such as *Vibrio parahaemolyticus, Salmonella* sp., *Shewanella* sp., *Escherichia coli, Klebsiella pneumoniae, Streptococcus pyogenes, Staphylococcus aureus, Enterococcus faecalis, Pseudomonas aeruginosa* and *Proteus mirabilis*. All the seaweeds extracts have shown moderate antibacterial activity <10mm of zone of inhibition, out of which only the methanolic extract has shown significant activity. The results showed that high antibacterial activity was found in *Acanthophora spicifera* and low activity was found in *Codium decorticatum*.

Vijayabaskar and Shiyamala (2011) studied the antibacterial activities of brown marine algae (*Sargassum wightii* and *Turbinaria ornata*) from the Gulf of Mannar Biosphere Reserve. The methanol extracts of *Turbinaria ornata* showed the highest activity against the growth of *Bacillus subtilis* (20 ± 0.62mm) and *E. coli* (16 ± 0.58mm). Besides this, the extract showed moderate activity against the growth of *Shigella flexnerii* (14 ± 0.49mm), *Staphylococcus aureus* (15 ± 0.53mm) and moderate activity towards all other pathogens. The methanol extract of *Sargassum wightii* strongly inhibited the growth of *E. coli* (18 ± 0.55mm) and *Aeromonas hydrophila* (15 ± 0.78mm) and showed moderate activity against *Bacillus subtilis* (12 ± 0.53mm) and *Pseudomonas aeruginosa* (12 ± 0.53mm). Vallinayagam *et al.* (2011) screened the antibacterial activities of four important seaweeds namely *Ulva lactuca, Padina gymnospora, Sargassum wightii* and *Gracilaria edulis* against human bacterial pathogens *Staphylococcus aureus, Vibrio cholerae, Shigella dysentriae, Shigella bodii, Salmonella paratyphi, Pseudomonas aeruginosa* and *Klebsiella pneumoniae.* The maximum activity (8.8 mm) was recorded from the extract of *G. edulis* against *S. aureus* and minimum (1.2 mm) by *U. lactuca* against *P. aeruginosa*. The ^{1}H-NMR analysis revealed the presence of signals corresponding to poly unsaturated esters in *Sargassum wightii* and *Gracilaria edulis* and polysaturated alcohols in *Padina gymnospora.*

Divya *et al.* (2011) screened the *in vitro* antimicrobial activity of different extracts of *Sargassum cinereum* against the micro organisms including *Staphylococcus aureus, Pseudomonas auroginosa, Salmonella typhi, Streptococcus and Klebsiella*. They observed that the crude methanol extracts showed high growth inhibitory effect against the selected micro organisms. The chloroform extract did not show marked antimicrobial activity. The methanol extract of *Sargassum cinereum* had more activity against *Aspergillus niger and Candida albicans.*

Ömer Ertürk and Beyhan Ta° (2011) evaluated the antibacterial and antifungal activity of the ethanolic crude extracts from seven marine algal species belonging to Chlorophyceae (*Cladophora glomerata, Enteromorpha linza, Ulva rigida*), Phaeophyceae (*Cystoseira barbata, Padina pavonica*) and Rhodophyceae (*Corallina officinalis, Ceramium ciliatum*) from the coast of Vona against six bacteria and two fungii. They observed that the *Cladophora glomerata* and *Padina pavonica* showed the highest activity against the broadest spectrum of test organisms. In particular, *Enteromorpha linza* and *Padina pavonica* showed the highest antifungal activity against *Aspergillus niger,* while *Cladophora glomerata* showed the highest antibacterial activity against *Staphylococcus aureus.*

Seenivasan *et al.* (2010) tested the *in vitro* antibacterial activity of the acetone, methanol and ethanol extracts of *Ulva fasciata, Enteromorpha intestinalis* and *Chaetomorpha aerea* from Ennore beach near Chennai (coast of Tamil Nadu) against *Escherichia coli, Pseudomonas aeruginosa, Klebsiella pneumoniae* and *Staphylococcus aureus. Ulva fasciata* and *Chaetomorpha aerea* have exhibited average result. Villarreal-Gómez *et al.* (2010) screened the antibacterial and anticancer activity of extracts from the seaweeds *Egregia menziesii, Codium fragile, Sargassum muticum, Endarachne binghamiae,*

Centroceras clavulatum and *Laurencia pacifica* collected from Todos Santos Bay, México. They used the pathogen strains *Staphylococcus aureus, Klebsiella pneumoniae, Proteus mirabilis,* and *Pseudomonas aeruginosa* to test antibacterial activity and HCT-116 colon cancer cells for anticancer activity. Thirty-five bacterial strains were isolated from the surface of seaweeds and identified as belonging to the phyla *Firmicutes, Proteobacteria* and *Actinobacteria* by 16S rDNA sequencing. The strains Cc51 isolated from *Centroceras clavulatum,* Sm36 isolated from *Sargassum muticum,* and Eb46 isolated from *Endarachne binghamiae* showed anticancer activity, with IC_{50} values of 6.492, 5.531, and 2.843 μg mL-1 respectively. Likewise, the extracts from the seaweed-associated bacteria inhibited the growth of the Gram negative bacterium *Proteus mirabilis.*

Renuka Bai *et al.* (2010) evaluated the antimicrobial properties of *Valoniopsis pachynema* extracted in different organic solvents (acetone, chloroform, diethyl ether, ethanol and methanol) against bacteria like *Bacillus subtilis, Escherichia coli, Enterobacter aerogenes, Klebsiella pneumoniae, Pseudomonas aeruginosa, Staphylococcus aureus* and fungal species like *Aspergillus niger* and *Candida albicans*. The chloroform extract displayed remarkable antibacterial activity in terms of inhibiting growth in *E. coli.* There was inhibition of growth of *B. subtilis, P. aeruginosa* and *S. aureus.*

Renuka Bai (2010a) screened the phytochemical constituents and antibacterial activity of different solvent extracts of *Gracilaria fergusonii.* Coumerins, phenols, quinones and steroids were present and tannin was absent in the *Gracilaria fergusonii.* Different organic solvent extracts *viz.*, acetone, chloroform, diethyl ether, ethanol and methanol were evaluated for antibacterial activity, employing gram negative (*Klebsiella pneumoniae* and *Pseudomonas aeruginosa*) and gram positive (*Bacillus subtilis* and *Staphylococcus aureus*) bacteria. The ethanol extract of the alga was found to be active against *Pseudomonas aeruginosa* and *Bacillus subtilis.* Rhimou *et al.* (2010) studied the antibacterial activity of extracts from 26 marine *Rhodophyceae* (8 Ceramiales, 7 Gelidiales, 9 Gigartinales, 1 Bonnemaisoniales and 1 Rhodymeniales) to assess their potential in the pharmaceutical industry. They observed that 96 per cent of extracts were active against at least one of the five test microorganisms. *Staphylococcus aureus* was the most susceptible microorganism. The methanolic extracts of all seaweed extracts tested in the present study exhibited a broad spectrum of antibacterial activity with inhibition diameters ranging from 10 to 35 mm. An extract of *Hypnea musciformis* exhibited high antibacterial activity against all the bacteria tested.

Ibtissam *et al.* (2009) evaluated the antibacterial activity of methanolic extracts from 32 macroalgae (13 Chlorophyta and 19 Phaeophyta) from the Atlantic and Mediterranean coast of Morocco against *Escherichia coli* ATCC 25922, *Staphylococcus aureus* ATCC 25923, *Enterococcus faecalis* ATCC 29212, *Klebsiella pnomeuniae* ATCC 700603 and *E. faecalis* ATCC 29213. In their study, the majority of algal extracts were active against one or two microorganisms. Most of them, 28 (87.5 per cent) showed activity against *S. aureus,* 22 (68.75 per cent) showed activity against *E. coli* and 16 (50 per cent), 9 (28.12 per cent) and 3 (9.37 per cent) presented inhibition activity against *K. pnomeuniae, E. faecalis* ATCC 29212 and *E. faecalis* ATCC 29213, respectively. They indicated that these species of seaweed collected from the Atlantic and Mediterranean coast of Morocco presented a significant capacity for antibacterial activities, which makes them interesting for screening for natural products.

Hebsibah Elsie and Dhanarajan (2010) studied antibacterial and antifungal activity of three different solvents *viz.*, ethanol, methanol and acetone of *Gelidium acerosa* against bacteria like *Staphylococcus aureus, Bacillus cereus, Micrococcus luteus, Klebsiella pneumonia, Pseudomonas aeruginosa,* fungi like *Aspergillus flavus, Aspergillus niger, Aspergillus fumigates, Candida albicans* and *C. tropicalis*. They observed varying degrees of inhibition to the growth of tested organisms in ethanolic extract treated plate, than acetone and methanolic extracts. Osman *et al.* (2010) evaluated the antimicrobial activity of the ethanol, methanol and acetone extracts of nine marine macroalgae (Rhodophyta, Chlorophyta and Phaeophyta) from Abu-Qir bay (Alexandria, Egypt) against pathogenic microbes (*Bacillus subtilis, Staphylococcus aureus* and *Streptococcus aureus* as gram-positive bacteria, and (*Escherichia coli, Salmonella typhi* and *Klebsiella pneumoniae* as gram-negative bacteria) and one yeast strain *Candida albicans*. They observed the best inhibition activity in acetone extracts inoculated plates with inhibition activity (36.7 per cent), followed by the methanol extracts (32.9 per cent) and ethanolic extracts (30.2 per cent) for all tested microorganisms.

Manilal *et al.* (2009) evaluated the antibacterial property of the red algae, *Falkenbergia hillebrandii* (Born) collected from the southwest coast of India (Indian Ocean) against three multidrug resistant human pathogens. They used four different solvents: ethyl acetate, dichloromethane, methanol and phosphate buffer saline (PBS) for this purpose. They observed that highest antimicrobial activity when compared to other solvents in the dried samples extracted with methanol. However, they failed to observe the antibacterial activity in PBS extract. Rajasulochana *et al.* (2009) tested the antibacterial activities of *Kappaphycus* against different types of bacteria using disc diffusion method. The methanol extracts of *Kappaphycus* showed the maximum activity against *Pseudomonas flouresences, Staphylococcus aureus* and less inhibition on *Vibrio chloera* and *Proteus mirabilis*. Karthikaidevi *et al.* (2009) evaluated the antibacterial activity of *Codium adherens, Ulva reticulata* and *Halimeda tuna* by agar diffusion method. Seven different solvents namely acetone, methanol, chloroform, diethyl ether, ethyl acetate, ethanol and petroleum ether were used for extraction. The ethanol extract shows the best result when compared with the other extracts. Some extracts were found to be more effective than the commercial medicine. The maximum antibacterial activity was noted in ethanol extracts which showed activity against *Staphylococcus* sp. (13 mm) and the minimum was recorded in methanol extracts against *Escherichia coli, Staphylococcus* sp., *Proteus* sp. (2 mm), *Streptococcus* sp. (2 mm) and *Enterococci* sp. (3mm).

Kolanjinatha *et al.* (2009) studied the antibacterial activity of crude extracts of *Gracilaria edulis, Calorpha peltada* and *Hydroclothres* sp. against six bacterial pathogens *viz., Escherichia coli, Enterobacter aerogenes, Staphylococcus aureus, Pseudomonas aeruginosa, Streptococcus faecalis* and *Bacillus cereus*. The ethanol extract of *Gracilaria edulis* inhibited growth of all the test organisms except *Bacillus cereus* and *Enterobacter aerogenes*. The seaweed extract of *Calorpha peltada* was found effective against a number of Gram negative and Gram positive bacteria such as *Escherichia coli, Staphylococcus aureus* and *Streptococcus faecalis*. *Hydroclothres* sp. extract inhibited the growth of *Pseudomonas aeruginosa* only out of the six tested pathogens.

Shanmughapriya *et al.* (2008) tested the antimicrobial activity of fourteen seaweeds collected from the intertidal zone of Southwest coast of India against ten human pathogenic bacteria and one human pathogen fungus using the well diffusion test in the casitone agar medium. The species used in their study included five Chlorophyta (*Bryopsis plumosa, Ulva fasciata, Acrosiphonia orientalis, Chaetomorpha antennina, Grateloupia filicina*), five Rhodophyta (*Hypnea pannosa, Gracilaria corticata, Centroceras clavulatum, Portieria hornemannii, Cheilosporum spectabile*) and four Phaeophyta (*Padina tetrastromatica, Sargassum wightii, Stocheospermum marginatum, Chnoospora bicanaliculata*). Of these, seven species were determined to be highly bioactive and screened on the multiresistant pathogens.

Kandhasamy and Arunachalam (2008) studied the *in vitro* antibacterial activities of seaweeds belong to Chlorophyceae (*Caulerpa racemosa* and *Ulva lactuca*), Rhodophyceae (*Gracillaria folifera* and *Hypneme muciformis*) and Phaeophyceae (*Sargassum myricocystum, Sargassum tenneerimum* and *Padina tetrastomatica*) against gram negative and gram-positive pathogenic bacteria. They observed that methanolic extracts of all seaweed extracts tested in the present study exhibited broad spectrum of antibacterial activity of which Chlorophyceae members showed high antibacterial activity.

Patra *et al.* (2009) screened the antibacterial activity of organic solvent extracts of three marine macroalgae *viz., Chaetomorpha linum* (Mell) Kuetzing, *Enteromorpha compressa* (L) Greville and *Polysiphonia subtilissima* Mont. against *Shigella flexneri, Vibrio cholerae, Escherichia coli, Bacillus subtilis* and *Bacillus brevis*. They revealed that the chloroform and ethyl acetate extracts were active against most of the pathogens whereas methanol and ethanol extracts were active only against *S. flexneri*. Salvador *et al.* (2007) studied the antibacterial and antifungal activity of 82 marine macroalgae (18 Chlorophyceae, 25 Phaeophyceae and 39 Rhodophyceae) to evaluate their potential for being used as natural preservatives in the cosmetic industry. The bioactivity was analysed from crude extracts of fresh and lyophilised samples against three gram-positive bacteria, two gram-negative bacteria and yeast using the agar diffusion technique. The samples were collected seasonally from Mediterranean and Atlantic coasts of the Iberian Peninsula. Of the macroalgae analysed, 67 per cent were active against at least one of the six test microorganisms. The highest percentage of active taxa was found in Phaeophyceae (84 per cent), followed by Rhodophyceae (67 per cent) and Chlorophyceae (44 per cent). Nevertheless, red algae had both the highest values and the broadest spectrum of bioactivity. In particular, *Bonnemaisonia asparagoides, Bonnemaisonia hamifera, Asparagopsis armata* and *Falkenbergia rufolanosa* (Bonnemaisoniales) were the most active taxa. *Bacillus cereus* was the most sensitive test microorganism and *Pseudomonas aeruginosa* was the most resistant. The highest percentages of active taxa from Phaeophyceae and Rhodophyceae were found in autumn, whereas they were found in summer for Chlorophyceae.

Ünci T. Ney *et al.* (2006) tested the antimicrobial activities of the methanol, acetone, diethyl ether and ethanol extracts of 11 seaweed species from the coast of Urla against *Candida sp., Enterococcus faecalis, Staphylococcus aureus, Streptococcus epidermidis, Pseudomonas aeruginosa* and *Escherichia coli* with the disc diffusion method. They reported that the diethyl ether was the best solution for extracting the effective antimicrobial materials from the algal species, with the exception of *D. linearis*, for which ethanol was the most effective extraction solution. They observed that the fresh diethyl ether extracts of *Cystoseira mediterranea, Enteromorpha linza, Ulva rigida, Gracilaria gracilis* and *Ectocarpus siliculosus* showed effective results against all test organisms. They found all test organisms were more sensitive to fresh extracts of the algae when compared to the dried extracts. Kayalvizhi *et al.* (2012) studied the antibacterial activity of four seaweeds (*Sargassum wightii, Stocheospermum marginatum, Gracilaria foliifera* and *Padina boergesenii*). They extracted the bioactive constituents in the algae using acetone, methanol, chloroform and diethyl ether as solvents. They tested for their antimicrobial activity against 12 bacterial pathogens (*Klebsiella pneumoniae, Escherichia coli, Staphylococcus aureus, Enterococci sp., Proteus sp., Streptococcus sp., Pseudomonas aeruginosa, Vibrio parahaemolyticus, Salmonella sp., Shewanella sp., Vibrio flurialis* and *Vibrio splendidus*) and five fungal pathogens (*Aspergillus niger, Candida albicans, Penicilium sp., Aspergillus flavus* and *Aspergillus tetreus*).

Seenivasan *et al.* (2012) reported the antibacterial activity of three species of seaweeds *viz., Codium adhaerens* (green algae) *Sargassum wightii* (brown algae), *A. spicifera* (red algae) against human pathogenic bacteria namely *S. aureus, V. cholerae, S. dysentriae, S. bodii, S. paratyphi, P. aeuroginosa* and *K. pneumoniae.* They also investigated and estimated the phytochemical constituents, photosynthetic pigments and mineral composition. Among the three seaweeds screened for their antibacterial activity the brown alga *S. wightii* is more superior to the red alga *A. spicefera* and green alga *C. adharens* in controlling the growth of most of the pathogens tested. The highest zone of inhibition (13 mm) was recorded in methanolic extract of the red alga against *Vibrio cholerae.* Jeyanthi Rebecca *et al.* (2013) represented the antibacterial and phenolic activity of *G. cortica, E. flexuosa* and *E. clathrata*. Kajal *et al.* (2013) evaluated the antioxidant activities and total phenolic contents of brown seaweeds belonging to *Turbinaria conoides* and *Turbinaria cornate* collected form Gulf of southeastern coast of India. Arunkumar *et al.* (2013) identified the antibacterial potential of 23 red, 9 brown and 15 green against two plant pathogenic bacteria such as *Xanthomonas axonopodi* spv. *Citri* and *X. campestris* pv. *Malvacearum.*

Rabia *et al.* (2013) identified antibacterial activity of 19 marine algal species (6 Chlorophyta, 8 Phaeophyta and 5 Rhodophyta) collected from the western coast of Libya against pathogenic bacteria (4 Gram-positive, 4 Gram-negative). The extracts showed a significant antibacterial activity against Gram+^{ve} as well as Gram–^{ve} bacteria. *Cystoseira crinite* exhibited the highest antibacterial activity among tested bacterial species. Seenivasan *et al.* (2012) reported the antibacterial activity of three species of seaweeds *viz., Codium adhaerens* (green algae) *Sargassum wightii* (brown algae), *A. spicifera* (red algae) against human pathogenic bacteria namely *S. aureus, V. Cholerae, S. dysentriae, S. bodii, S. paratyphi, P. aeuroginosa* and *K. pneumoniae.* They also investigated and estimated the phytochemical constituents, photosynthetic pigments and mineral composition. Among the three seaweeds screened for their antibacterial activity the brown alga *S. wightii* was superior to the red alga *A. spicefera* and green alga *C. adharens* in controlling the growth of most of the pathogens tested. The highest zone of inhibition (13 mm) was recorded in methanolic extract of the red alga against *Vibrio cholerae.* Nanthini Devi *et al.* (2012) studied the antibacterial activity on three different extracts of *Sargassum wightii* against human bacterial pathogens.

The crude extracts were purified by silica gel column chromatography and five fractions obtained from each solvent were collected separately and tested for activity. The results revealed that the potential fraction (100 µg/mL) exhibited higher antioxidant activity as compared with standard ascorbic acid with the equivalent concentration between 60-80 µg/mL.

Kayalvizhi *et al.* (2012) studied the antibacterial activity on four extracts in four seaweeds (*Sargassum wightii, Stocheospermum marginatum, Gracilaria foliifera* and *Padina boergesenii*) against 12 bacterial pathogens (*K. pneumoniae, E. coli, S. aureus, Enterococci sp., Proteus sp., Streptococcus sp,. P. aeruginosa, V. parahaemolyticus, Salmonella* sp., *Shewanella sp,. Vibrio flurialis* and *V. splendidus*) and also against five fungal pathogens (*Aspergillus niger, C. albicans, Penicilium* sp., *A. flavus* and *A. tetreus*). Xavier *et al.* (2012) screened the antibacterial efficacy of various solvent extracts of *Sargassum wightii, Chaetomorpha linum* and *Padina gymnospora* against some selected Gram positive and Gram negative human pathogenic bacteria using disc diffusion method and reported that the acetone extracts of marine algae *S. wightii, C. linum* and *P. gymnospora* exhibited good antimicrobial activity. However, the acetone extracts of *S. wightii* possessed highest antibacterial activity than others.

Chong *et al.* (2011) studied the antibacterial activity of n-hexane, dichloromethane and methanolic extracts of brown seaweeds (Phaeophyceae), *Sargassum polycystum* C. Agardh and *Padina australis* Hauck, using the disc diffusion and broth micro dilution methods. The bioactivity of the seaweed extracts was expressed as minimum inhibitory concentration (MIC) and minimum bactericidal concentration (MBC). The antibacterial activity against Gram-negative bacteria (beta-lactamase positive and negative *Escherichia coli, Pseudomonas aeruginosa*) and Gram-positive bacteria (*Staphylococcus aureus, Bacillus cereus*) was discussed. Gram-positive bacteria especially *B. cereus* was more susceptible to the seaweed extracts (MIC = 0.130 to 0.065 mg/mL). Habsah *et al.* (2011) identified the potential antibacterial properties of *Sargassum granuliferum* extract. Dried sample of *S. granuliferum* were ground in powdered form and extracted in methanol by maceration method. All the extracts were tested for antibacterial activity by using disc diffusion method against adhesive bacteria. The methanolic extracts of *S. granuliferum* were further tested for brine shrimp toxicity test and exhibited a nontoxicity (LC_{50}=1.75 mg/mL) against *Artemia salina.*

Johnsi *et al.* (2010) identified aqueous extract of seven species of marine macroalgae which were screened for their antimicrobial potency against ten pathogenic bacterial strains. *Ulva fasciata, Gracilaria corticata, Sargassum wightii* and *Padina tetrastromatica* showed significantly higher activity against 70 per cent of the tested bacterial isolates. The maximum zone of inhibition was noted for the red alga *G. corticata* against *Proteus mirabilis* (17 mm) and brown alga *P. tetrastromatica* against the pathogens *Staphylococcus aureus* and *Vibrio harveyi* (15 mm). The general trend of inhibitory activity was higher towards Gram negative bacteria. Indira *et al.* (2013) examined the antibacterial and antifungal activity of *Halimeda tuna* against 10 bacterial strains (*S. aureus, S. typhimurium, S. paratyphi, K. oxytoca, E. coli, P. mirabillis, L. vulgaris, Pseudomonas* sp., *K. pneumonia* and *V. cholerae*) and nine fungal strains (*Aspergillus niger, A. flavus, A. alternaria, C. albicans, E. floccossum, T. mentagrophytes, T. rubrum, Pencillium* sp. and *Rhizopus* sp.)

Arputha Bibiana *et al.* (2012) studied the antimicrobial effect of four different extracts of two seaweeds *Sargassum wightii* (brown algae) and *Kappaphycus alwarezii* (red algae). The analysis showed the presence of phytochemical constituents like alkaloids, phenols and sugars. Among the tested extracts, the maximum activity of 10 mm was observed with acetic acid extract of *S. wightii.* Samy (2012) reported antimicrobial activities of marine algae *Halimeda opuntia* and *Sarconema filiforme.* Among these *Halimeda* extract exhibited antibacterial activity against six species of microrganisms, with significant inhibition against *Staphylococcus aureus,* while *Sarconema filiforme* extract was better potent as antifungal against *Candida albicans.* Vijayabaskar and Shiyamala (2011) studied the bioactive potential of brown algae, *Sargassum wightii* and *Turbinaria ornate* collected from Gulf of Mannar. The methanolic extracts of both the seaweeds (*S. wightii, T. ornata*) were tested against various Gram $+^{ve}$ and Gram $-^{ve}$ human pathogenic microbes. Thoudam *et al.* (2011) carried out the preliminary phytochemical screening, free radical scavenging activity and total antioxidant activity of various extracts of *Sargassum muticum.* The methanolic extracts displayed highest DPPH scavenging activity and antioxidant activity.

Periasamy *et al.* (2010) studied the phytochemical analysis of the aqueous extracts of some commonly occurring green seaweed *Cladophora glomerata, Ulva lactuca* and *Ulva reticulata,* the red seaweed *Gracilaria corticata* and *Kappaphycus alvarezii* and the brown seaweed *Sargassum wightii* and their antibacterial activity were evaluated by well diffusion assay using two different solvents namely aqueous and methanol. The maximum activity (45 mm) was recorded from 200 mg of aqueous extract of *Ulva reticulata* against *Salmonella typhi* and minimum (9 mm) by *Ulva lactuca* against *Streptococcus pyogenes* at 50 mg level whereas, the methanolic extract showed the maximum activity (40 mm) from 200 mg of *U. reticulata* against *Escherichia coli* and *Streptococcus pyogenes* and *Cladophora glomerata* against *Pseudomonas aeruginosa* and minimum (6 mm) by 50 mg of *Kappaphycus alvarezii* against *Staphylococcus epidermis.*

The petroleum ether extracts and unsaponified fractions of red and green seaweeds, methanolic extracts, lipophilic fractions and unsaponified fractions of brown seaweeds were separated on TLC for their efficacy against *Xanthomonas oryzae* pv. *oryzae.* Two active zones from red seaweeds, one to five from brown seaweeds and two to three from green seaweeds were isolated through TLC profiles. Among eleven seaweeds *Gracilaria edulis, Sargassum wightii* and *Enteromorpha flexuosa* showed highest antibacterial activity. The R_f value 0.30 substance obtained from unsaponified fractions of *E. flexuosa* showed the maximum antibacterial activity against the test bacterium (Arun Kumar and Rengasamy, 2000).

Salvador *et al.* (2007) analysed the antimicrobial activity of 82 marine macroalgae (18 Chlorophyceae, 25 Phaeophyceae and 39 Rhodophyceae) to evaluate their potential for being used as natural preservatives in the cosmetic industry. The bioactivity was analysed from crude extracts of fresh and lyophilised samples against different pathogens using the agar diffusion technique. The samples were collected seasonally from Mediterranean and Atlantic coasts of the Iberian Peninsula. Of the macroalgae analysed, 67 per cent were active against at least one of the six test microorganisms. Highest percentage of active taxa was found in Phaeophyceae (84 per cent), followed by Rhodophyceae (67 per cent) and Chlorophyceae (44 per cent).

Bacillus cereus was the most sensitive test microorganism and *Pseudomonas aeruginosa* was the most resistant. The highest percentages of active taxa from Phaeophyceae and Rhodophyceae were found in autumn and Chlorophyceae were found in summer.

In vitro antibacterial activity of organic solvent extracts of three marine macroalgae *viz.*, *Chaetomorpha linum*, *Enteromorpha compressa* and *Polysiphonia subtilissima* showed specific activity in inhibiting the growth of three Gram-negative bacteria (*Shigella flexneri, Vibrio cholerae* and *Escherichia coli*) and two Gram positive bacteria (*Bacillus subtilis* and *Bacillus brevis*). The results revealed that the chloroform and ethyl acetate extracts were active against most of the pathogens whereas methanolic and ethanolic extracts were active only against *S. flexneri* (Patra *et al.*, 2009).

Antibacterial activities of *Ulva lactuca, Padina gymnospora, Sargassum wightii* and *Gracilaria edulis* were screened against human bacterial pathogens *viz.*, *Staphylococcus aureus, Vibrio cholerae, Shigella dysentriae, Shigella bodii, Salmonella paratyphi, Pseudomonas aeruginosa* and *Klebsiella pneumoniae*. Maximum activity (8.8 mm) was recorded from the extract of *G. edulis* against *S. aureus* and minimum (1.2 mm) by *U. lactuca* against *P. aeruginosa*. The ^{1}H-NMR analysis revealed the presence of signals corresponding to poly unsaturated esters in *S. wightii* and *G. edulis* and poly saturated alcohols in *P. gymnospora* (Vallinayagam *et al.*, 2009).

Srivastava *et al.* (2010) evaluated the *in vitro* antimicrobial activity of *Caulerpa racemosa* and *Grateloupia lithophila* as an alternative to commonly used antibiotics. Extracts of methanol, ethanol, butanol, acetone, chloroform and dichloromethane were tested against selected human pathogens. Both the seaweeds had shown moderate antibacterial activity with <15 mm of zone of inhibition. Out of which only butanolic extract has shown significant activity. Phytochemical screening revealed the presence of alkaloids and phenolic compounds in both the seaweeds whereas flavonoids and steroids were found to be present in only *C. racemosa*.

Ethanolic, methanolic and acetone extracts of nine marine macroalgae (Rhodophyta, Chlorophyta and Phaeophyta) from Abu-Qir bay (Alexandria, Egypt) were evaluated for antimicrobial activity by agar well diffusion methods against pathogenic microbes (*Bacillus subtilis, Staphylococcus aureus, Streptococcus aureus, Escherichia coli, Salmonella typhi* and *Klebsiella pneumoniae* and one yeast strain *Candida albicans*. The best results were obtained by acetone extracts with inhibition activity (36.7 per cent), followed by the methanolic extracts (32.9 per cent), and ethanolic extracts (30.2 per cent) for all tested microorganisms. The tested species of Chlorophyta were the most active followed by Rhodophyta and Phaeophyta. The most active seaweed was *Ulva fasciata* (chlorophyceae) against all tested microorganisms (Osman *et al.*, 2010).

Seenivasan *et al.* (2010) tested the *in vitro* antibacterial activity of the acetone, methanolic and ethanolic extracts of three marine green algae from Ennore beach near Chennai using agar well diffusion technique. The test pathogens include *Escherichia coli, Pseudomonas aeruginosa, Klebsiella pneumoniae* and *Staphylococcus aureus. Ulva fasciata* and *Chaetomorpha aerea* have exhibited average result. *E. coli* in all the solvents have shown significant results for the seaweed *U. fasciata* in selective media.

Bhagavathy *et al.* (2011) tested the existence of bioactive phytochemicals and the antimicrobial role of green algae *Chlorococcum humicola* against the harmful pathogens *Escherichia coli, Pseudomonas aeruginosa, Salmonella typhimurium, Klebsiella pneumoniae, Vibreo cholerae, Staphylococcus aureus, Bacillus subtilis, Candida albicans, Aspergillus niger* and *Aspergillus flavus.* The chemical analysis showed the presence of carotenoids, alkaloids, favanoids, fattyacids, saponins, aminoacids and carbohydrates. Depending upon the existence of bioactive compounds, the different extracts showed difference in their inhibitory zone against the microbes. Out of all the organic extracts tested, benzene and ethyl acetate extracts showed excellent effect of 80 per cent microbial growth inhibition.

Antibacterial activities of petroleum ether, diethyl ether, ethyl acetate and methanolic extracts of marine algae collected from the southern coast of Jeddah, Saudi Arabia during summer and autumn were tested against different pathogenic bacteria. All marine algae extracts tested exhibited a broad spectrum of antibacterial activity. Maximum inhibition activities were shown for extracts of *Padina pavonica* and *Turbinaria triquetra.* The growth inhibitions of bacteria by *Sargassum portieriatum* extracts were higher in samples collected during autumn than in summer. Maximum inhibitory effect of *Gracilaria multipartita* was observed in the petroleum ether extract against *B. subtilis* and *E. coli.* Ethyl acetate and petroleum ether extract of *Enteromorpha prolifera* and *Ulva reticulata* showed strong activity against the tested bacteria (Omar *et al.,* 2012).

Crude methanolic and aqueous extracts of 19 marine algal species (6 Chlorophyta, 8 Phaeophyta and 5 Rhodophyta) collected from the western coast of Libya were evaluated for antibacterial activity against different pathogenic bacteria. The extracts showed a significant antibacterial activity against Gram positive (*Staphylococcus aureus, Bacillus subtilis, Bacillus* spp. and *Staphylococcus epidermidis*) as well as Gram negative bacteria (*Escherichia coli, Salmonella typhi, Klebsiella* spp. and *Pseudomonas aeruginosa*). Methanolic extracts showed higher antibacterial activity than aqueous extracts. Among the tested algae, brown algae *Cystoseira crinita* exhibited the highest antibacterial activity (Alghazeer *et al.,* 2013).

Manilal *et al.* (2010) subjected the ten seaweed species collected from the Kollam coast (Indian Ocean) to antifouling assays against the common fouling organisms such as *Balanus amphitrite, Mytilus edulis* and three biofilm forming bacteria *Vibrio* sp., *Colwellia* sp. and *Pseudoalteromonas* sp. Of all the seaweeds tested, the red algae, *Laurencia brandenii* displayed broadest spectrum of activity. Aqueous and ethanolic extracts of *Sargassum binderi, Amphiroa* sp., *Turbinaria conoides* and *Halimeda macroloba* from the east cost of the Gulf of Thailand were screened for antibacterial and antifungal activities. The test organisms include *Staphylococcus aureus, S. epidermidis, Propionibacterium acnes, Proteus mirabilis* and *Candida albicans.* Aqueous extracts of *T. conoides* and ethanolic extracts of *H. macroloba* were most effective and demonstrated a broad-spectrum antimicrobial activity against all gram-positive, gram-negative bacteria and pathogenic fungus (Boonchum *et al.,* 2011).

Antibacterial effect of the crude methanolic extracts and purified fractions of *Cladophora glomerata* demonstrated appreciable activity against the human pathogen

A. baumannii and fish pathogens *Vibrio fischeri, V. vulnificus, V. anguillarum, V. parahaemolyticus, E. coli* and *B. cereus.* TLC Purified fractions III and V of green seaweed *C. glomerata* inhibited the human pathogen *A. baumannii* and fish pathogens *V. fischeri* and *V. vulnificus.* Purified fraction II of the same seaweed inhibited only *V. fischeri* and *V. vulnificus.* Methanolic extract of *C. glomerata* inhibited *E. coli* and *B. cereus* growth at a minimum inhibitory concentration of 75 µg/mL and other species at 100 µg/mL. Whereas, *V. vulnificus* growth was inhibited at a minimum concentration of 125 µg/mL. GC-MS analysis revealed the presence of hydrocarbon compounds in active fractions II, III and V of *C. glomerata* (Yuvaraj *et al.,* 2011).

4.1 Antioxidant, Cytotoxic and Anticancer Properties

Premalatha *et al.* (2011) studied the preliminary phytochemical analysis in *Ulva fasciata* and *Chaetomorpha antennina.* In the DPPH scavenging assay, both the seaweed extracts showed high antioxidant activity. The *Ulva fasciata* samples have more effective antioxidant activity when compared to the *Chaetomorpha antennina* and the percentage of scavenging was found to be about 83.95 per cent *for U. fasciata* and 63.77 per cent for *C. antennina* sample. The rapid TLC assay is considered as the rapid test to evaluate the antioxidant activity of natural compounds. The compounds showing the bands at hRf = >10, 25 and 94 of both the seaweed extracts and hRf = 52 in *Ulva* sp. alone were proved to be having antioxidant activity. The results of antimicrobial activity by the well diffusion assay also clearly expressed that *Ulva fasciata* has high concentration of active principles when compared to the *Chaetomorpha antennina.*

Siva Kumar and Rajagopal (2011) studied the antioxidant, free radical scavenging activity, total phenolics, total carotenoids, vitamin-C and vitamin-E content of eight green algal species *Chaetomorpha antennina, Cladophora socialis, Acrosiphonia orientalis, Bodlea struveoides, Ulva fasciata, Enteromorpha compresasa, Caulerpa racemosa and Caulerpa taxifolia* to expand their utilization in pharmaceutical and food industry. They extracted fractions rich in phenolics from eight green algal species using methanol as solvent. They studied the free radical scavenging activity using DPPH photometric assay. They observed that *Caulerpa* species exhibited higher levels of antioxidant activity, total phenolics, vitamin C and vitamin E contents compared to other algal species studied.

Carolina Babosa Brito da Matta *et al.* (2011) evaluated the antinociceptive activity of the methanolic (ME), acetate (AE), hexanic (HE) and chloroform (CE) extracts obtained from *Caulerpa mexicana,* and ME, CE and HE obtained from *Caulerpa sertularioides.* They observed that all extracts evaluated were able to significantly inhibit leukocyte migration into the peritoneal cavity in comparison with carrageenan. These data demonstrated that extracts from *Caulerpa* species elicit pronounced antinociceptive and anti-inflamatory activity against several nociception models. Ayesha *et al.* (2010) screened the ethanol extracts of seaweeds *Dictyota dichotoma* var. *velutricata, D. hauckiana, D. indica, Iyengaria stellata, Jolyna laminarioides, Melanothamnus afaqhusainii, Sargassum ilicifolium, S. lanceolatum* and *Ulva fasciata* occurring at Karachi coast for the cytotoxic activity using brine shrimp lethality for larvae (nauplii). They observed that out of 9 seaweeds tested, ethanol extract of eight species showed significant cytotoxicity (LC_{50} <1000µg) on brine shrimp. *Dictyota indica* showed highest cytotoxic activity (LC_{50} =143µg).

Cox *et al.* (2010) assessed the antioxidant and antimicrobial activity of six species of edible Irish seaweeds *viz., Laminaria digitata, Laminaria saccharina, Himanthalia elongata, Palmaria palmata, Chondrus crispus* and *Enteromorpha spirulina*. Extraction of secondary metabolites was carried out using different solvents to determine antioxidant and antimicrobial properties of the dried extracts. The total phenolic contents of dried methanolic extracts were significantly different ($p < 0.05$). *H. elongata* exhibited highest phenolic content at 151.3 mg GAE/g of seaweed extract and also had the highest DPPH scavenging activity ($p < 0.05$) with a 50 per cent inhibition (EC50) level at 0.125µg/mL of extract. *H. elongata* also had the highest total tannin and total flavonoid contents ($p < 0.05$) of 38.34 mg CE/g and 42.5 mg QE/g, respectively. Antimicrobial activity was determined using a microtitre method which allowed detection of bacterial growth inhibition at low levels. All methanolic seaweed extracts inhibited the food spoilage and food pathogenic bacteria tested; *Listeria monocytogenes, Salmonella abony, Enterococcus faecalis* and *Pseudomonas aeruginosa,* except *C. crispus* extracts. It was found that dried methanolic extracts of red and green seaweeds had significantly lower antimicrobial activity than the brown species; *H. elongata* had the highest antimicrobial activity with up to 100 per cent inhibition.

Rahila Najam *et al.* (2010) studied the pharmacological activity of the methanol extracts of *H. musciformis* on rabbit and mice. They observed that the methanolic extracts of *H. musciformis* significantly decreased the serum total cholesterol, triglyceride and low-density lipoprotein cholesterol levels of rabbits. Jorge Mancini-Filhoa *et al.* (2009) evaluated the antioxidant activity of the free phenolic acids (FPA) fraction from the seaweed *Halimeda monile,* and its activity to protect the expression of hepatic enzymes in rats, under experimental CCl_4 injury. The antioxidant activity was measured by the DPPH method. The FPA fraction (80 mg/kg, p.o.) was administered during 20 consecutive days to rats. The peroxidation was performed by thiobarbituric acid reactive substances (TBARS). The SOD and CAT enzymatic expressions were measured by RT/PCR. The histology technique was used to evaluate liver injuries. The expression of both, CAT and SOD genes was more preserved by FPA. Only partial injury could be observed by histology in the liver of rats receiving FPA as compared with the control group; and CCl_4 administration induced 60 per cent more peroxidation as compared with the rats receiving FPA. These data suggest that FPA could modulate the antioxidant enzymes and oxidative status in the liver through protection against adverse effects induced by chemical agents.

Zeliha Demirel *et al.* (2009) evaluated the antioxidant and antimicrobial activity of methanol, dichloromethane and hexane extracts, as well as the essential oils of brown algae (Phaeophyta) *Colpomenia sinuosa, Dictyota dichotoma, Dictyota dichotoma* var. *implexa, Petalonia fascia* and *Scytosiphon lomentaria*. The essential oil of the macroalgae was obtained by steam distillation and analyzed by GC and GC/MS. The antioxidant activity of the algal extracts was determined using the procedures of inhibition of α-carotene bleaching and $ABTS^+$ methods. The antioxidant effects of the extracts were compared with those of commercial antioxidants, such as butylated hydroxytoluene (BHT), butylated hydroxyanisol (BHA) and α-tocopherol. The hexane extracts of *D. dichofoma* var. *implexa* had a higher phenolic content than the other extracts. The dichloromethane extract of *S. lomentaria* was found to be more active in the decolorization of $ABTS^+$ than the other extracts and generally the dichloromethane extracts were more active than the methanol and hexane extracts. Antimicrobial activities of the extracts were assessed against Gram (+) and Gram (–) bacteria and one yeast strain by the disk diffusion method. According to the results, the dichloromethane extracts generally showed more potent antimicrobial activity than the methanol and hexane extracts at concentrations 1.5 and 1.0 mg/disk.

Souza *et al.* (2009) attempted to identify the possible antinociceptive actions of *n*-butanolic phase, chloroformic phase, ethyl acetate phase and crude methanolic extract obtained from *Caulerpa racemosa*. The *n*-butanolic, chloroformic, ethyl acetate phases and crude methanolic extract, all administered orally in the concentration of 100 mg/kg, reduced the nociception produced by acetic acid by 47.39 per cent, 70.51 per cent, 76.11 per cent and 72.24 per cent, respectively. In the neurogenic phase on formalin test, were observed that crude methanolic extract (51.77 per cent), *n*-butanolic phase (35.12 per cent), chloroformic phase (32.70 per cent) and indomethacin (32.06 per cent) were effective in inhibit the nociceptive response. In the inflammatory phase, only the ethyl acetate phase (75.43 per cent) and indomethacin (47.83 per cent) inhibited significantly the nociceptive response.

Patra *et al.* (2008) investigated the free radical scavenging potentials (DPPH radical and hydroxyl radical), inhibition of lipid peroxidation, and glutathione-S-transferase and antimicrobial properties of Sargassum sp. extract. They observed that the tested extract exhibited a dose-dependent free radical scavenging action against DPPH radical and hydroxyl radical and antimicrobial activity. In addition, they observed inhibition of lipid peroxidation and glutathione-S-transferase activities.

Fabíola Dutra Rocha *et al.* (2007) performed the crude extracts of some Brazilian coastal seaweeds for cytotoxic activity against a cultured human melanoma cancer cell line using the sulphorhodamine B assay. The crude dichloromethane: chloroform extract of *Stypopodium zonale* showed good cytotoxic activity against the C32 cell line. Yuan and Walsh (2006) evaluated the effect of red alga, dulse (*Palmaria palmata*) and three kelp (*Laminaria setchellii, Macrocystis integrifolia, Nereocystis leutkeana*) extracts on human cervical adenocarcinoma cell line (HeLa cells) proliferation using the MTT (3-(4,5-dimethylthiazol-2-yl)-2,5-diphenyl tetrazolium bromide) assay. The 1-butanol soluble fractions from the methanol extracts of these algae were also evaluated for reducing activity and total polyphenol content. After 72 h incubation, HeLa cell proliferation was inhibited ($p < 0.05$) between 0 per cent and 78 per cent by *P. palmata*; 0 per cent and 55 per cent by *L. setchellii* and 0 per cent and 69 per cent by *M. integrifolia* and *N. leutkeana* at 0.5–5 mg/mL algal extract. Algal extract reducing activities were as follows: *P. palmata* > *M. integrifolia* > L. *setchellii* > *N. leutkeana*; and total polyphenol contents were: *P. palmata* > *M. integrifolia* = *N. leutkeana* > *L. setchellii*. The antiproliferative efficacy of these algal extracts were positively correlated with the total polyphenol contents ($p < 0.05$), suggesting a causal link related to extract content of kelp phlorotannins and dulse polyphenols including mycosporine-like amino acids and phenolic acids.

Kim *et al.* (2005) assessed the hepatoprotective activity of the ethanolic extracts of 18 seaweed variants against tacrine-induced cytotoxicity in Hep G2 cells. They observed that the *Ecklonia stolonifera* Okamura (Laminariaceae), exhibited promising hepatoprotective activity. Bioassay-guided fractionation of the active ethyl acetate (EtOAc) soluble fraction obtained from the ethanolic extract of *E. stolonifera*, resulted in the isolation of several phlorotannins [phloroglucinol (1), eckstolonol (2), eckol (3), phlorofucofuroeckol A (4), and dieckol (5)]. Compounds 2 and 4 were determined to protect Hep G2 cells against the cytotoxic effects of tacrine, with EC50 values of 62.0 and 79.2 μg/mL, respectively. Silybin, a well characterized hepatoprotective agent, was used as a positive control.

Selvin and Lipton (2004) tested the secondary metabolites of seaweeds *Ulva fasciata* and *Hypnea musciformis*, collected from southeast and southwest coast of India, for biotoxicity potential. Both species showed potent activity in antibacterial, brine shrimp cytotoxicity, larvicidal, antifouling and ichthyotoxicity assays. The green alga *U. fasciata* exhibited broad-spectrum antibacterial activity whereas the red alga *H. musciformis* showed narrow spectrum antibacterial activity. The brine shrimp cytotoxicity profile indicated that the seaweeds were moderately toxic. The overall activity profile indicated that *U. fasciata* contained more biological potency than *H. musciformis*.

Raghavendran *et al.* (2004) examined the effect of pre-treatment with hot water extract of marine brown alga *Sargassum polycystum* C.Ag. (100 mg/kg body wt, orally for period of 15 days) on HCl-ethanol (150 mM of HCl-ethanol mixture containing 0.15 N HCl in 70 per cent v/v ethanol given orally) induced gastric mucosal injury in rats with respect to lipid peroxides, antioxidant enzyme status, acid/pepsin and glycoproteins in the gastric mucosa. They observed that the levels of lipid peroxides of gastric mucosa and volume, acidity of the gastric juice were increased with decreased levels of antioxidant enzymes and glycoproteins in HCl-ethanol induced rats.

Rashmi *et al.* (2010) evaluated the cytotoxic and antioxidant properties of methanolic extracts (MEs) of seven brown seaweeds occurring in the Indian coastal waters were screened for their following various assays. The methanolic extracts of seaweeds in the order of *Dictyo pterisaustralis, Spatoglossum variabile, Stoechospermum marginatum, Spatoglossuma spermum* showed significant cytotoxic activity. A very high DPPH radical scavenging activity was exhibited by the methanolic extracts prepared from *S. marginatum, P. tetrastromatica, D. delicatula* and *S. aspermum*. The antioxidant compounds found in brown seaweeds scavenge free radicals through effective intervention. This decisively promotes them as a potential source of natural antioxidants.

Aseer *et al.* (2009) evaluated for brine shrimp cytotoxicity and hatchability assay using *Artemia salina*. The fraction eluted with petroleum: chloroform (6:4) exhibited excellent activity in both assays was subjected to GC-MS (Hewlett Packard) analysis. At a dose level 200 μg/mL was the active fraction of algae elicited 100 per cent hatching inhibition, whereas in toxicity assay it showed a LD value of 93 μg/mL which might have cytotoxic activity. Lakshmana *et al.* (2013) evaluated α-amylase inhibitory activity, antioxidant activity and toxic effects of ten seaweeds *viz.*, *S. duplicatum, S. wightii, S. tenerrimum, T. conoids, T. ornata* and *P. gymnospora* (brown seaweed), *G. gracilis, C. hornemanni, G. edulis* (red seaweed) and *C. racemosa* (green seaweed) from the southeast coastal area of India. Corpuz *et al.* (2013) identified the total phenolic (TPC) and flavonoid contents (TFC) of methanolic extract of *S.*

*siliquosum*to prevent the initiation of free radicals to cause cellular damage. Nadine *et al.* (2013) isolated the polysaccharides from the brown seaweed *Dictyopteris polypodioides* grown on the Lebanese coast and reported the percentages of the main polysaccharide alginic acid 4.6 per cent in May and 6.25 per cent in July. They also performed the infrared spectroscopy to reveal the functional groups of alginate and mannuronan, anticoagulant activity, activated partial thromboplastintime (APTT) clotting assay and antioxidant activity.

Paola *et al.* (2013) identified antioxidant and trace element content (vitamin C, total polyphenols, zinc, iron, cooper, selenium, cadmium and lead) of eight macroalgae species, three red (*Hypnea spinella, Gracilaria textorii* and *G. vermicullophyla*), four green (*Caulerpa sertularioides, Codium simulans, C. amplivesiculatum* and *Ulva lactuca*) and one brown (*Dictyota flabellata*) macroalgae.

Vijayabaskar and Vaseela (2012) studied the physico-chemical characteristics, total antioxidant capacity (TAC), reducing power and the free radical scavenging potentials (DPPH radical, ABTS, H2O2 radical) of sulfated polysaccharide from marine brown algae *Sargassum tenerrimum* and characterized the sulfated polysaccharide FT-IR spectrum showing the presence of carboxyl, hydroxyl and sulfate groups. Rabia *et al.* (2013) identified antibacterial activity of 19 marine algal species (6 Chlorophyta, 8 Phaeophyta and 5 Rhodophyta) collected from the western coast of Libya against patho-genic bacteria (4 Gram-positive, 4 Gram-negative). The extracts showed a significant antibacterial activity against Gram+ve as well as Gram–ve bacteria whereas *Cystoseira crinite* exhibited the highest antibacterial activity among tested bacterial species.

Rashmi *et al.* (2011) evaluated the cytotoxic and antioxidant properties of methanolic extracts (MEs) of seven brown seaweeds occurring in the Indian coastal waters. Finally the methanolic extracts of seaweeds showed significant cytotoxic activity in the order of *Dictyopteris australis, Spatoglossum variabile, Stoechospermum marginatum, Spatoglossum aspermum.* Weon *et al.* (2008) studied the anti-inflammatory activity of brown alga *D. dichotoma* in murine macrophage RAW 264.7 cells. In their assay, the CH_2Cl_2, fraction of *D. dichotoma* showed decrease in the expression of TNF-α, IL - 1β and IL-6 mRNA. Based on this they suggested that *D. dichotoma* extracts may be considered possible anti-inflammatory candidates for human health.

Ayesha *et al.* (2010) screened the *in vitro* cytotoxicity of seaweeds *viz., Dictyota dichotoma* var. *velutricata, D. hauckiana, D. indica, Iyengaria stellata, Jolyna laminarioides, Melanothamnus afaqhusainii, Sargassum ilicifolium, S. lanceolatum* and *Ulva fasciata*from Karachi coast on brine shrimp. Out of 9 seaweeds tested, ethanolic extract of eight species showed significant cytotoxicity (LC_{50} <1000µg) on brine shrimp. *Dictyota indica* showed highest cytotoxic activity (LC_{50} =143µg). Sahar *et al.* (2003) studied the cytotoxic hydroazulene diterpenes from the brown alga *Dictyota dichotoma.* Two new hydroazulenoid (prenyl guaiane) diterpenes, dictyone acetate (**2**) and 3, 4-epoxy 13-hydroxy pachydictyol A (**4**) were isolated from the petroleum ether fraction of the alcoholic extract of the brown alga. Together with the above four compounds, the steroidal compound (dictyol *E*) (**5**) and stigmasta-5, (*E*)-24(28)-dien-3-_-ol (fucosterol) (**6**) was also isolated. The structures of the isolated compounds have been determined on the basis of spectroscopic evidences as well as physical and chemical correlation with known compounds. Based on their result, they suggested that compounds **1**, **2**, **3** and **5** showed moderate cytotoxic activity. Paola *et al.* (2013) identified antioxidant and trace element content (vitamin C, total polyphenols, zinc, iron, cooper, selenium, cadmium and lead) of eight macroalgae species, three red (*Hypnea spinella, Gracilaria textorii* and *G. vermicullophyla*), four green (*Caulerpa sertularioides, Codiumsimulans, C. amplivesiculatum* and *Ulva lactuca*) and one brown (*Dictyota flabellata*) macroalgae.

Vijayabaskar and Vaseela (2012) studied the physico-chemical characteristics, total antioxidant capacity (TAC), reducing power and the free radical scavenging potentials (DPPH radical, ABTS, H_2O_2 radical) of sulfated polysaccharide from marine brown algae *Sargassum tenerrimum* and characterized the sulfated polysaccharide FT-IR spectrum showing the presence of carboxyl, hydroxyl and sulfate groups. Radhika *et al.* (2012) estimated the antioxidant activity of the selected seaweeds, *Hypnea* and *Amphoria.* Among these, highest activity was found in *Hypnea* extracts where 86.13 per cent of DPPH was scavenged and *Amphiroa* showed no activity. Walailuck *et al.* (2011) carried out the phenolic compounds and antioxidant activities of aqueous and ethanolic extract of four seaweeds *viz.*, *Sargassum binderi* Sonder, *Amphiroa* sp., *Turbinaria conoides* (J. Agardh) Küzting and *Halimeda macroloba* collected from the Gulf of Thailand. In general, the aqueous extracts showed higher antioxidant activities and phenolic contents than ethanolic extracts.

Airanthi *et al.* (2011) assayed the total phenolic content (TPC), fucoxanthin content, radical scavenging activities (DPPH, peroxyl radical, ABTS, and nitric oxide), and antioxidant activity in a liposome system of some Japanese edible brown seaweeds, *Eisenia bicyclis, Kjellmaniella crassifolia, Alaria crassifolia, Sargassum horneri,* and *Cystoseira hakodatensis.* Among the solvents used for extraction, methanolic extract was the most effective to extract total phenolics (TPC) from brown seaweeds. The high antioxidant activity of the extract was based not only on the high content of phenolics, but on the presence of fucoxanthin. Hong-Yu Luo *et al.* (2010) studied antioxidant activities of methanol/chloroform (MC) extracts and fractions of five brown algae (*Sargassum fusiforme, Sargassum kjellmanianum, Sargassum pallidum, Sargassum thunbergii* and *Sargassum horneri*) from China DPPH/hydroxyl radical-scavenging activity and reducing power. Total phenolic content was investigated using Folin–Ciocalteau reagent. The MC extract of *S. kjellmanianum* showed higher antioxidant activity than other seaweeds. Jayanta *et al.* (2008) studied the free radical scavenging potentials (DPPH radical and hydroxyl radical), inhibition of lipid peroxidation, glutathione-S-transferase and antimicrobial properties of *Sargassum* sp. extract. The tested extract exhibited a dose-dependent free radical scavenging action against DPPH radical and hydroxyl radical and antimicrobial activity. In addition, inhibition of lipid peroxidation and glutathione-S-transferase activities were also observed.

4.2 Larvicidal Activity of Seaweeds

Everson *et al.* (2013) studied the larvicidal activity on crude extracts of selected 15 seaweeds collected from north eastern Brazil. Among these *Canistro carpuscervicornis, Laurencia dendroidea, Hypnea musciformis* and *Chaetomorpha antennina* showed e"50 per cent mortality against fourth instar larvae of *A. aegypti* at concentrations of 300 ppm and they isolated a novel agent elatol from seaweed which act against the dengue mosquito. Josmin and Poonguzhali (2013) studied the mosquito larvicidal activity in two different seaweeds *C. antennina* (Bory de Saint-Vincent) Kutzing and *S. wightii* Greville against the mosquito vector *Culex quinquefasciatus.* Of the two algae screened acetone extract of *S. wightii* was found to be effective larvicidal activity. Poonguzhali and Josmin (2012) reported the larvicidal activity of methanol, acetone and benzene extract of seaweeds (*Ulva fasciata* and *Grateloupia lithophila*) against *Culex* larva. Of the two algae screened *G. lithophila* was found to effective against the larva *Culex* in all the 3 extracts.

Mahnaz *et al.* (2011) evaluated the larvicidal activity on various extracts of *Sargassum swartzii* and *Chondria dasyphylla* against malaria vector *Anopheles stephensi.* The results proposed that the larvicidal activity of ethanolic fraction is related to the presence of semi-polar compounds. Margaret Beula *et al.* (2011) carried out larvicidal activity on the seaweed extracts of *Enteromorpha intestinalis, Dictyota dichotoma* and *Acanthopora.* Ethanolic extracts of *D. dichotoma* possess active compounds for development of larvicidal activity.

Cetin *et al.* (2010) identified larvicidal efficacy of the acetone extract of the thalli of *Caulerpa scalpelliformis* var. *denticulata* was determined against late 2nd to early 3rd instars of *Culex pipiens* at concentrations ranging from 100 to 2,000 parts per million (ppm). At 1,200 ppm, the extract caused >70 per cent larval mortality at 24, 48 and 72 h exposure. The LC_{50} (lethal concentration) and LC_{90} values of *C. scalpelliformis* were 338.91 and 1,891.31 ppm respectively. Huseyin *et al.* (2010) studied larvicidal efficacy of the acetone extract of the thalli of *Caulerpa scalpelliformis* var. *denticulata* against late 2nd to early 3rd instars of *Culex pipiens* at concentrations ranging from 100 to 2,000 parts per million (ppm). At 1,200 ppm, the extract caused >70 per cent larval mortality at 24, 48, and 72 h exposure. The LC_{50} (lethal concentration) and LC_{90} values of *C. scalpelliformis* were 338.91 and 1,891.31 ppm, respectively. Our data showed that this species of seaweed contains components with larvicidal properties against mosquitoes.

Mahnaz *et al.* (2011) evaluated the larvicidal activity of native marine algae against main malarial vector *Anopheles stephensi.* The total 70 per cent of the methanolic extract and partition fractions of chloroform ($CHCl_3$), ethylacetate (EtAc), and MeOH from two algae, *Sargassum swartzii* and *Chondria dasyphylla* were investigated for their larvicidal activity against late III and early IV instars larvae of malaria vector *A. stephensi.* Finally they proposed that the larvicidal activity of EtOAc fraction is related to the presence of semi-polar compounds. Manilal *et al.* (2010) studied the weedicidal activity of 29 seaweeds (chlorophyta- 8, Phaeophyta- 6, Rhodophyta- 6) of Kollam coast. Finally, the methanolic extract of seaweeds was subjected for bio activity. Among these *S. marginatum* exhibited broadest and highest mode of bioactivity and they also investigated the phycochemical composition of *S. marginatum* was using GC-MS analysis.

Huseyin *et al.* (2010) studied larvicidal efficacy of the acetone extract of the thalli of *Caulerpa scalpelliformis* var. *denticulata* against late 2nd to early 3rd instars of *Culex pipiens* at concentrations ranging from 100 to 2,000 parts per million (ppm). LC_{50} (lethal concentration) and LC_{90} values of *C. scalpelliformis* were 338.91 and 1,891.31 ppm respectively.

5.0 SEAWEEDS LIQUID FERTILLIZERS

Seaweeds have been used as manure, cattle feed, food for human consumption and as a source of phycocolloids such as agar, alginic acid and carrageenan. Besides their application as farmyard manure (FYM), liquid extracts obtained from seaweeds (LSF/SLF) have recently gained importance as foliar sprays for several crops (Thivy, 1961; Metha *et al.,* 1974) because the extract contains growth promoting hormones (IAA and IBA), cytokinins, trace elements (Fe, Cu, Zn, Co, Mo, Mn, Ni), vitamins and amino acids. Thus, these extracts when applied to seeds or when added to the soil, stimulate growth of the plants (Blunden, 1977). Booth (1965) observed that the value of seaweeds as fertilizers was not only due to nitrogen, phosphorus and potash content, but also because of the presence of trace elements and metabolites. Aqueous extract of *Sargassum wightii* when applied as a foliar spray on *Zizyphus mauritiana* showed an increased yield and quality of fruits (Rama Rao, 1991). Seaweed fertilizer was found to be superior to chemical fertilizer because of the high level of organic matter aids in retaining moisture and minerals in the upper soil level available to the roots. Seaweed extracts are now available commercially under the names, such as Maxicrop (Sea born), Algifert (marinure), Goemar GA14, Kelpak 66, Seaspray, Seasol, SM3, Cytex and Seacrop 16. Recently researchers proved that seaweed fertilizers are better than other fertilizers and are very economical (Gandhiyappan and Perumal, 2001). Any improvement in agricultural system that results in higher production should reduce the negative environmental impact of agriculture and enhance the sustainability of the system. One such approach is the use of biostimulants, which can enhance the effectiveness of conventional mineral fertilizers. Marine bioactive substances extracted from marine algae are used in agricultural and horticultural crops, and many beneficial effects, in the terms of enhancement of yield and quality have been reported. Liquid extracts obtained from seaweeds have recently gained importance as foliar sprays for many crops including various grasses, cereals, flowers and vegetable species (Crouch and Van Staden, 1993).

Bio-stimulant properties of seaweeds are explored for use in agriculture and the antimicrobial activities for the development of novel antibiotics. Seaweeds have some valuable medicinal components such as antibiotics, laxatives, anti-coagulants, anti-ulcer products and suspending agents in radiological preparations. Seaweeds have recently received significant attention for their potential as natural antioxidants. Most of the compounds of marine algae show anti-bacterial activities. Many metabolites isolated from marine algae have bioactive efforts (Oh *et al.,* 2008, Gopala *et al.,* 2010). Many of these compounds are bioactive and have been extensively studied using bioassays and pharmacological assays (Paul and Fenical, 1987). Potential antitumor promoting properties of 36 edible/common marine algae from sea near Maozuru, Kyoto, Japan were examined and strong inhibitory activities were found in *Undaria pinnatifida, Laminaria* and *Sargassum* species (Ohigashi *et al.,* 1992). The production of inhibitory substances from seaweeds was noted as early as in 1917 (Harder and Oppermann, 1953). Like other plants, seaweeds contain various inorganic and organic substances which can benefit human health (Kuda *et al.*, 2002). Seaweeds are considered as a source of bioactive compounds as they are able to produce a great variety of secondary metabolites characterized by a broad spectrum of biological activities. Compounds with antioxidant, antiviral, antifungal and antimicrobial activities have been detected in brown, red and green algae (Yuan *et al.*, 2005; Bansemir *et al.*, 2006; Chew *et al.*, 2008). The environment in which seaweeds grow is harsh as they are exposed to a combination of light and high oxygen concentrations. These factors can lead to the formation of free radicals and other strong oxidizing agents but seaweeds seldom suffer any serious photodynamic damage during metabolism. This fact implies that seaweed cells have some protective mechanisms and compounds (Matasukawa *et al.*, 1997).

Senn (1960) reported that seaweed extracts hasten germination of beet seeds by 25 per cent more. The effect of seaweed extract on germination creeping red fescue has been reported by Button and Noyes (1964), the dilute solution increases the rate of germination but stronger solutions are detrimental. The effect of seaweed extracts on green chillies and turnip shows that the low concentrations of an SLF enhance the rate of seed germination (Dhargalkar and Untawale, 1980). Featonby-Smith and Van Staden (1983) reported that the application of the SLF *Ecklonia maxima* on *Arachis hypogea* increases the protein content. The SLF treatment increased the phosphorus content in the cucumber leaves. The seaweed extract has induced the uptake of unavailable nutrients and improved the efficiency of the utilization of available nutrients (Nelson and Van Staden, 1984).

Application of seaweeds would be beneficial for increasing growth parameters. The soil application of SLF of *Chaetomorpha linnum* and *Hypnea musciformis* has increased the growth characteristics of *Vigna radiata* (Kannan and Tamilselvan, 1990). Cassan *et al.* (1992) have reported the application of seaweed extract GA14 on two varieties of *Spinacia oleracea* which enhanced total fresh matter production of the leaves. The commercial SLF 'Algifert' has promoted the seed germination and seedling growth of *Vigna mungo* and *Vigna radiata* (Mohan and Venkataraman Kumar, 1993). Mohan *et al.* (1994) reported the seeds of *Cajanus cajan* soaked in SLF of *Padina* sp., *Sargassum* sp., *Champia* sp. and *Turbinaria* sp. have achieved 100 per cent germination and increased their growth. The SLF of *Ulva lactuca* has enhanced the growth parameters such as fresh weight, dry weight, root and shoot length, number of lateral roots and leaf area of *Vigna unguiculata* (Sekar *et al.*, 1995).

Rajkumar Immanuel and Subramanian (1999) have reported that the seaweed liquid fertilizers (SLF) from *Ulva lactuca, Sargassum wightii* and *Gelidella acerosa* on maize, cholam, ragi and kambu induce germination percentage, root length and shoot length. The seaweed concentrate 'Kelpak' amended to the *in vitro* culture medium of potato when applied as a leaf/soil drench immediately after transplanting, improved plantlet quality and led to better establishment in the green house (Kowalski *et al.*, 1999). The treatment of the extracts prepared from green seaweeds such as *Cladophora dalmatica, Enteromorpha intestinalis, Ulva lactuca* and red seaweeds such as *Corallina mediterrenia, Jania rubens* and *Pterocladia pinnata* increases protein content in both root and shoot system and soluble sugars and chlorophyll content in leaves.

The amount of cytokinin in the extracts of green seaweeds was found higher than that of red algae, among the seaweeds studied (El- Sheekh and El-Saied, 2000).

Anantharaj and Venkatesalu (2001, 2002) have reported the positive effect of SLF *Caulerpa racemosa* and *Gracilaria edulis* on *Vigna catajung* and *Dolichos biflorus* in seed germination, seedling growth, fresh and dry weight. Murugalakshmikumari *et al.* (2002) have reported that the treatment of *Gracilaria corticata* fertilizer on black gram and cumbu has increased the growth parameters such as shoot length and root length with optimum concentration of seaweed extract. The seaweed liquid fertilizer from *Ulva lactuca* on *Spirulina platensis* enhances the total protein content at lower concentration (Sridhar and Rengasamy, 2002). Thirumal Thangam *et al.* (2003) have reported that the growth parameters, shoot length, root length fresh weight of shoot and root of vegetable plant *Cyamposis tetragonoloba* increased with the optimum concentration of seaweed liquid fertilizer. Venkataraman Kumar and Mohan (2003) reported that the SLF treated plants showed a marked increase in soluble protein and soluble sugar contents.

The effect of SLF extracts of *Sargassum wightii* exhibited better responses than *Caulerpa chemnitzia* on growth and biochemical constituents of *Vigna sinensis* (Sivasankari *et al.,* 2006). Gradual increase in seedling height of *Cajanus cajan* and *Cajanus indicus* was observed with increasing seaweed extracts of *Sargassum wightii, Gracilaria corticata* var. *corticata* and *Caulerpa scalpelliformis* (Kamaladhasan and Subramanian, 2009). The effect of *Rosenvigea intricata* SLF on seed germination, growth, yield, pigment content and soil profile of *Cyamopsis tetragonoloba* was found to be highest at 20 per cent concentration (Thirumaran *et al.,* 2009). *Arachis hypogaea* applied with 1.0 per cent of *Sargassum wightii* and *Ulva lactuca* SLF plus 50 per cent chemical fertilizers showed an increased yield up to ca. 4.1 kg fresh weight which was more than 11 per cent to that of the plants received with 100 per cent recommended rate of chemical fertilizers (Sridhar and Rengasamy, 2010).

The extracts of *Sargassum wightii* showed maximum activity at 1.0 per cent SLF on *Tagetes erecta* (Sridhar and Rengasamy, 2010). The potential of utilizing panchagavya biofertilizer in combination with the aqueous extract of *Sargassum wightii* showed better results than the panchagavya biofertilizer tested on the pulses *Vigna radiata, Vigna mungo, Arachis hypogea, C. tetragonoloba, Lablab purpureus, Cicer arietinum* and the cereal *Oryza sativa* var. *ponni* (Sangeetha and Thevanathan, 2010). Sasikumar *et al.* (2011) determined the effect of Seaweed Liquid Fertilizer of *Dictyota dichotoma* at different concentrations (12.5 per cent, 25 per cent, 50 per cent, 75 per cent and 100 per cent) on growth and yield of *Abelmoschus esculantus.* The seaweed extract was found effective in increasing the biomass, growth of roots and shoots, number of roots, leaves, flowers, and fruits, leaf area index, fruits length, fresh and dry weight of fruits, maturity time and yield. Thambiraj *et al.* (2012) evaluated the effect of seaweed liquid fertilizer prepared from *Sargassum wightii* and *Hypnea musciformis* on the growth and biochemical constituents of the pulse, *Cyamopsis tetragonoloba* (L). The seeds of *C. tetragonoloba* soaked in SLF performed better when compared to the water soaked controls in terms of growth and certain biochemical attributes.

The effect of crude seaweed extracts from *Ulva lactuca* and *Sargassum wightii* was studied on germination and protein profile of five different crops *viz., Amaranthus roxburghinus, Amaranthus tricolor, Arachis hypogea, Capsicum annum* and *Tagetes erecta*. Among the five different crops treated with 1.0 per cent SLF of both seaweeds, *A. roxburghinus* did not show any additional bands in contrast to the rest of crops. A maximum of five additional bands appeared in *C. annum* under *U. lactuca* SLF. Further, one or two bands appeared in *A. tricolor, A. hypogea* and *T. erecta* (Sridhar and Rengasamy, 2011).

6.0 SILVER NANOPARTICLES SYNTHESIS FROM SEAWEEDS

Linga Rao and Savithramma (2012) synthesised the silver nanoparticles using stem extract of *Svensonia hyderobadensis* and confirmed the silver nanoparticles formation by the colour change of plant extracts (SNPs) and UV-Vis spectroscopy. These phytosynthesized silver nanoparticles were tested for antibacterial and antifungal activities using disc diffusion method. Dhanalakshmi *et al.* (2012) reported the Synthesis of silver nanomaterials from the extracts of *S. plagiophyllum, U. reticulata* and *E. compressa.* The AgNPs characterized using UV-Visible Spectroscopy (UV-Vis), Fourier-Transform Infra-red Spectroscopy (FT-IR), Scanning Electron Microscopy (SEM) and X-ray Diffraction (XRD) showed that *Sargassum plagiophyllum* showed higher phytochemicals than the other studied seaweeds.

Shanmugam *et al.* (2012) reported the synthesis of silver nanoparticles extract of marine brown seaweed *Turbinaria conoides* and characterized using UV-Visible Spectrophotometer, Fourier transmittance infrared spectroscopy (FT-IR) and X-Ray diffraction (XRD) data. Scanning electron microscope (SEM) shows spherical shaped with the average size of 96 nm. These silver nanoparticles were found to be highly toxic against gram positive bacteria *Bacillus subtilis* and gram negative bacteria *Klebsiella planticola.* Kumara *et al.* (2012) studied the extracellular synthesis of AgNP's from *Sargassum ilicifolium* and characterized using UV-Vis spectra at 414 nm corresponding to the Plasmon resonance of AgNP's. SEM and TEM analysis showed formation of well dispersed AgNP's in the range of 33-40 nm. Antibacterial activity against five clinical pathogens at various concentrations shows impediment in growth at various nanomolar concentrations. Further, the toxicity of biosynthesized AgNP's were tested against *Artemia salina* to evaluate the cytotoxic effect which displayed LC_{50} value of 10 nM/mL. Arockiya Aarthi Rajathi *et al.* (2012) synthesised the gold nano particles from *S. marginatum* and characterized the photoluminescence spectra using Transform Infrared spectroscopy, Scanning electron microscopy, TEM, FTIR, X-ray diffraction to analyse the particle size of gold nano particle showing absorption peak at 550 nm using UV-Vis spectrophotometer. TEM image was displayed in the spherical shape with the size ranging from 18.7- 93.7 nm and the WD-XRF was 45.92 per cent.

7.0 CONCLUSION

The above review on the phytochemistry, bioactivities and pharmacology of various seaweeds demonstrated that these primitive groups of plant marine algae progressed as successful organism in the marine environment. To survive with the different kinds of environment, they have also devloped different physiological, biochemical and metabolic pathways to synthesize various defense metabolites, which play an important role in the survival of the species. In India, the existence of such metabolites in seaweeds is studied superficially and its biopotentials are not explored.

REFERENCES

Airanthi, M.K., Hosokawa, M., Miyashita, K., 2011. Comparative Antioxidant Activity of Edible Japanese Brown Seaweeds. *Journal of Food Science* 76, 104–111.

Alghazeer, R., Whida, F., Abduelrhman, E., Gammoudi, F., Azwa, S., 2013. Screening of antibacterial activity in marine green, red and brown macroalgae from the western coast of Libya. *Natural Science* 5(1), 7-14.

Anantharaj, M., Venkatesalu, V., 2001. Effect of seaweed liquid fetilizer on Vigna catajung. Seaweed Research Utilisation 23 (1 and 2), 33-39.

Arputha, B.M., Nithya, K., Manikandan, M.S., Selvamani, P., Latha, S., 2012. Antimicrobial evaluation of the organic extracts of *Sargassum wightii* (brown algae) and *Kappaphycus alwarezii* (red algae) collected from the coast of meemesal, Tamilnadu. *IJPCBS,* 2(4), 439-446.

Arunkumar, K., Sivakumar, S.R., Shanthi, N., 2013. Antibacterial potential of Gulf of Mannar seaweeds extracts against two plant pathogenic bacteria *Xanthomonas axonopodispv.* Citri (Hasse) Vauterin *et al.,* and *Xanthomonas campestris* pv *malvacearum* (Smith 1901) Dye 1978b. *IJAPBC* 2(1), 25-31.

Arun Kumar, K., Rengasamy, R., 2000. Antibacterial activities of seaweed extracts/ fractions obtained through a TLC Profile against the Phytopathogenic Bacterium *Xanthomonas oryzae* pv *Oryzae. Botanica Marina* 43, 417- 421.

Arockiya Aarthi Rajathi, F., Parthiban, C., Ganesh Kumar, V., Anantharaman, P., 2012. Biosynthesis of antibacterial gold nanoparticles using brown alga, *Stoechospermum marginatum* (kützing). Spectrochimica Acta Part A: *Molecular and Biomolecular Spectroscopy* 99, 166–173.

Ayesha., Hira., Sultana, V., Jehan Ara., Syed Ehteshamul-Haque., 2010. *In vitro* cytotoxicity of seaweeds from Karachi coast on brine shrimp. *Pak. J. Bot.* 42(5), 3555-3560.

Bansemir, A., Blume, M., Schroder, S., Lindequist, U., 2006. Screening of cultivated seaweeds for antibacterial activity against fish pathogenic bacteria. *Aquaculture* 252, 79-84.

Bhagavathy, S., Sumathi, P., Jancy Sherene Bell I., 2011. Green algae *Chlorococcum humicola* a new source of bioactive compounds with antimicrobial activity. *Asian Pacific Journal of Tropical Biomedicine* S1-S7.

Blunden, G., 1977. Cytokinin activity of seaweed extracts. In: *Marine Natural Products Chemistry.* (D.J. Faulknes and W.H. Fenical, Eds) Plenum Publishing Corporation: New York 337-343.

Blunt, J.W., Copp, B.R., Hu, W.P., Munro, M.H.G., Northcote, P.T., Prinsep, M.R., 2007. Marine natural products. *Nat. Prod. Rep.* 24, 31-86.

Blunt, J.W., Copp, B.R., Munro, M.H.G., Northcote P.T., Prinsep, M.R., 2006. Marine natural products. *Natur. Prod. Report* 23, 26-78.

Blunt, J.W., Copp, B.R., Munso, M.H, Northcote, P.T., Prinsep, M.R., 2011. Marine natural products. *Natural Product Report* 28, 196-268.

Boonchum, W., Peerapornipisal, Y., Kanjanapothi, D., Pekkoh, J., Pumas. C., Jamjai, U., Doungporon, A.D., Noriaksar, T., Vacharapiyasophon. P., 2011. Antioxidant Activity of some seaweed from the Gulf of Thailand, *International Joural Agricultural biology* 13 (1), 95-99.

Booth, E., 1965. The manorial value of seaweed. *Botanica Marina* 8, 138-143.

Booth, E., 1969. The manufacture and properties of liquid seaweed extracts. In: *Proc. Inc. Seaweed Symp* 6, 655- 662.

Button, E.F., Noyes, C.F., 1964. Effect of seaweed extract on emergence and survival of seedlings of creeping red fescue (*Festuca rubra*). *Agronomy Journal* 56, 444-445.

Cabrita, L., Quintela, J.M., Vilalta, R., 2010. *In vitro* activities of three selected brown seaweeds of India. *An. Quim. Ser. C* 8, 113–115.

Carolina Babosa Brito da Matta., Éverton Tenório de Souza., Aline Cavalcanti de Queiroz., Daysianne Pereira de Lira., Morgana Vital de Araújo., Luiz Henrique Agra Cavalcante-Silva., George Emmanuel C. de Miranda., João Xavier de Araújo-Júnior, José Maria Barbosa-Filho., Bárbara Viviana de Oliveira Santos., Magna Suzana Alexandre-Moreira., 2011. Antinociceptive and Anti-Inflammatory Activity from Algae of the Genus *Caulerpa. Mar. Drugs* 9, 307-318.

Cassan, L., Jennin, I., Lamaze, T., 1992. The effect of the *Ascophyllum nodosum* extract Goemar G. A. 14 on the growth of spinach. *Botanica Marina* 35, 437-43.

Cetin, H., Gokoglu, M., Oz, E., 2010. Larvicidal activity of the extract of seaweed, *Caulerpa scalpelliformis*, against *Culex pipiens. J Am Mosq Control Assoc.* 26(4), 433-435.

Chanda, S., Dave, R., Kaneria, M., Nagani, K., 2010. Seaweeds: A novel, untapped source of drugs from sea to combat Infectious diseases. Current Research, technology and Education Topics in Applied Microbiology and Biotechnology A. Medez- Vilas (Ed). 473-480.

Chau Van, M., Phan Van, K., Ngyen Hai, D., 2005. Marine natural products and their potential application in the future. *AJST* 22(4), 297-311.

Chew, Y.L., Lim, Y.Y., Omar, M., Khoo, K.S., 2008. Antioxidant activity of three edible seaweeds from two areas in South East Asia. *LWT* 41, 1067-1072.

Chong Chiao-Wei., Hii Siew-Ling., Wong Ching., 2011. Antibacterial activity of *Sargassum polycystum* C. Agardh and *Padina australis* Hauck (Phaeophyceae) Lee. *African Journal of Biotechnology* 10(64), 14125-14131.

Corpuz, M.J.A.T., Osi, M.O., Santiago, L.A., 2013. Free radical scavenging activity of *Sargassum siliquosum* J. G. Agardh. *International Food Research Journal* 20(1), 291-297.

Cox, S., Abu-Ghannam, N., Gupta, S., 2010. An assessment of the antioxidant and antimicrobial activity of six species of edible Irish seaweeds. *International Food Research Journal* 17, 205-220.

Crouch, I.J., Van Staden. J., 1993. Evidence for the presence of plant growth regulators in commercial seaweed product. *Plant growth Regulator* 13, 21-29.

Dhanalakshmi, P.K., Riyazulla Azeez., Rekha R., Poonkodi S., 2012. Thangaraju Nallamuthu. Synthesis of silver nanoparticles using green and brown seaweeds. *Phykos* 42 (2), 39-45.

Dhargalkar, V.K., Untawale, A.G., 1980. Some observations on the effect of seaweed liquid fertilizers on the higher plants. Proc. Nat. Work. Algae System, Indian Society of Biotechnology. IIT: New Delhi.

Divya, C.V., Devika, V., Asham. K.R.T., Bharat, G., 2011. Antimicrobial Screening of the Brown Algae *Sargassum cinereum. Journal of Pharmacy Research* 4(2), 420-421.

El-Sheekh, M.M., El-Saiedh, A.E.F., 1999. Effect of seaweed extracts on seed germination, seedling growth and some metabolic processes of faba beans (*Vicia faba* L.). *Phykos* 38, 55-64.

El-Baroty, G.S, Moussa, M.Y, Shallan, M.A., Ali, M.A, Sabh, A.Z., Shalaby, E.A., 2007. Contribution to the aroma, biological activities, minerals, protein, pigments and lipid contents of the red alga: *Asparagopsis taxiformis* (Delile) Trevisan. *J. Appl. Sci. Res.*, 1825-1834.

Fabíola Dutra Rocha., Angélica Ribeiro Soares., Peter John Houghton., Renato Crespo Pereira., Maria Auxiliadora Coelho Kaplan., 2007. Valéria Laneuville Teixe Potential Cytotoxic Activity of Some Brazilian Seaweeds on Human Melanoma Cells. *Phytother. Res.* 21, 170–175.

Faulkner, D.J., 1986. Marine natural product. *Nat. Prod.Rep.*, 3, 2-33.

Fayaz, M., Namitha, K.K., Chidambara Murthy, K.N., Mahadeva Swamy, M., Sarada, R., Salma Khanam, Subbarao, P.V., Ravishankar, G.A., 2005. Chemical composition, Iron Vioavailability and Antioxidant Activity of *Kappaphycus alvarezzi* (Doty). *J. Agric. Food Chem* 53, 792-797.

Featonby-Smith, B.C., Van Staden, J.C., 1983. The effect of seaweed concentration on the growth of tomato plants in nematode infested soil. *Scientia Horticulture* 20, 137-146.

Gandhiyappan, K., and Perumal, P., 2001. Growth promoting effect of seaweed liquid fertilizer [*Enteromorpha intestinalis]* on the seame crop plant [*Sesamum indicum*.L] *Seaweed Research and Utilisation* 23 (1 and 2), 23-25.

Ghada, S., Abou-El-Wafaa, E., Khaled Shaabanb, A., Mohamed, E., El-Naggara, E., Mohamed Shaabanb C., 2011. Bioactive constituents and biochemical composition of the egyptian brown alga *Sargassum subrepandum* (Forsk). *Revista Latinoamericana de Química* 39, 1-2.

Gopala, S.V.K., Panchagnula Aditya, L., Gottumukkala Gottumukkala, V., Subbaraju., 2007. Synthesis, structural revision, and biological activities of 42-chloroaurone, a metabolite of marine brown alga Spatoglossum variabile. *Tetrahedron* 63(29), 6909-14.

Habsah, M., Kamariah, B., Aisha, M.R.S., Julius, Y.F.S., Desy, F.S., Asnulizawati, A., Faizah, S., 2011. The Potential of Local *Sargassum granuliferum* crude extract as antibacterial and antifouling properties. *UMTAS* 112, 721-726.

Harder, R., Oppermann, A., 1953. Antibiotische Stoffe bie den Grunalgen *Stichoccus bacillaris* and *Protosiphon bomyoides. Archives of Microbiology* 19, 98-401.

Hebsibah Elsie, B., Dhanarajan, M.S., 2010. Evaluation of antimicrobial activity and phytochemical screening of *Gelidium acerosa. J. Pharm. Sci. and Res.* 2 (11), 704-707.

Hegazia, M., Angel Pe´rez-Ruzafaa., Luis Almelab., Mar´ýa-Emilia Candelac., 1998. Separation and identification of chlorophylls and carotenoids from *Caulerpa prolifera, Jania rubens* and *Padina pavonica* by reversedphase high-performance liquid chromatography. *Journal of Chromatography A* 829, 153–159.

Hong-Yu Luo., Bin Wang., Chun-Guang Yu., You-le Qu and Chuan-ling Su., 2010. Evaluation of antioxidant activities of five selected brown seaweeds from China. *Journal of Medicinal Plants Research* 4(18), 2557-2565.

Huseyin, C., Mehmet, G., Emre, O., 2010. Larvicidal Activity of the Extract of Seaweed, *Caulerpa scalpelliformis,* Against *Culex pipiens Journal of the American Mosquito Control Association* 26(4), 433-435.

Ibtissam, C., Hassane, R., José, M.L., Francisco, D.S.J., Antonio, G.V.J., Hassan, B., Mohamed, K., 2009. Screening of antibacterial activity in marine green and brown macroalgae from the coast of Morocco. *African Journal of Biotechnology* 8 (7), 1258-1262.

Indira, K., Balakrishnan, S., Srinivasan, M., Bragadeeswaran, S., Balasubramanian, T., 2013. Evaluation of in vitro antimicrobial property of seaweed *(Halimeda tuna*) from Tuticorin coast, Tamil Nadu, Southeast coast of India. *African Journal of Biotechnology* 12(3), 284-289.

Jayanta, K.P., Mohapatra, A.K., Rath, S.K., Dhal, N.K., Thaoti, H., 2009. Screening of antioxidant and antifilarial activity of leaf extracts of *Excoecaria agallocha* L. *International J of Integrative Biology* 7 (1), 9-15.

Jeyanthi Rebecca, L., Dhanalakshmi, V., Avinash Kumar., Smity Priyadarshini., Shivani., 2013. Research Journal of Pharmaceutical, Biological and Chemical Sciences Isolation of Phenolic Compounds from Marine Algal Extracts. *RJPBCS* 4(1), 38 -42.

Jeyaseelan, E.C, Kothai, S., Kavitha, R., Tharmila, S., Thavaranjit, A.C. 2012. Antibacterial activity of some selected algae present in the Costal Lines of Jaffna Peninsula. *Journal of Pharmaceutical and Biological Archives* 3(2), 352-356.

Jha, R, K., Zi-rong, X., 2004. Biomedical compounds from marine organisms. *Mar. Drugs* 2, 123-146.

Johnsi christobel, G., Lipton, A.P., Aishwarya, M.S., Sarika, A.R., Udayakumar, A., 2010. Antibacterial activity of aqueous extract from selected macroalgae of southwest coast of India Seaweed *Res. Util* 33(1): 67-75.

Johnson, M., Krishnaveni, E., 2012. UV- VIS Spectroscopic and HPLC Studies on *Dictyota bartayresiana* Lamour. *Asian Pacific Journal of Tropical Biomedicine* 3(6): 514-518.

Johnson, M., Petchiammal, E., Janakiraman, N., Babu, A., Renisheya Joy Jeba Malar, T., Sivaraman, A., 2012. Phytochemical characterization of brown seaweed *Sargassum wightii. Asian Pacific Journal of Tropical Disease* 109-113.

Jorge Mancini-Filhoa., Alexis Vidal Novoab., Ana Elsa Batista Gonzálezb., Elma Regina S de Andrade-Warthaa., Ana Mara de O e Silvaa., José Ricardo Pintoc., and Dalva Assunção Portari Mancinic., 2009. Free Phenolic Acids from the Seaweed *Halimeda monile* with Antioxidant Effect Protecting against Liver Injury Z. *Naturforsch* 64 c, 657 – 663.

Josminlaalinisha, L.L., Poonguzhali, T.V., 2013. Effect of two seaweeds *Chaetomorpha antennina* (Bory de Saint-Vincent) Kützing and *Sargassum wightii* Greville as larvicide against mosquito vector. *Culex quinquefasciatus IJBPAS* 2(3): 705-711.

Kajal, C., Lipton, A.P., Paulraj, R., Rekha, D.C., 2010. Guaiane sesquiterpenes from seaweed *Ulva fasciata* Delile and their antibacterial properties. *European Journal of Medicinal Chemistry* 45, 2237–2244.

Kajal, C., Krishnankartha, P., Kodayan, K., Gonugontla, S.R., 2013. Evaluation of phenolic contents and antioxidant activities of brown seaweeds belonging to *Turbinaria* spp. (Phaeophyta, Sargassaceae) collected from Gulf of Mannar. *Asian Pac J Trop Biomed* 3(1): 8-16.

Kamaladhasan, N., Subramanian, S.K., 2009. Influence of seaweed liquid fertilizers on legume crop, red gram. *Journal of Basic and Applied Biology* 3(1 and 2): 21-24.

Kandhasamy, M., Arunachalam, K.D., 2008. Evaluation of *in vitro* antibacterial property of seaweeds of southeast coast of India. *African Journal of Biotechnology* 7 (12):1958-1961.

Kannan, L., Tamilselvan, C., 1990. Effect of seaweed manures on *Vigna radiata* L. (Green gram). In: *Perpectives in Phycology* (Prof. M.O.P. Iyengar Centenary Celebration, Volume Ed. V.N. Raja Rao), Today and Tomorrows Printers and Publishe: New Delhi. 427-430.

Karthikaidevi, G., Manivannan, K., Thirumaran, G., Anantharaman, P., Balasubaramanian, T., Antibacterial Properties of Selected Green Seaweeds from Vedalai Coastal Waters; *Gulf of Mannar Marine Biosphere Reserve Global Journal of Pharmacology* 3 (2): 107-112.

Kayalvizhi, K., Vasuki Subramanian., Anantharaman, P., Kathiresan, K., 2012. Antimicrobial activity of seaweeds from the Gulf of Mannar. *International Journal of Pharmaceutical Applications* 3(2): 306-314.

Khalid, M.N., Mustafa, S., Ahmad, V.U., 2011. Bioactivity and phycochemical studies on *Microspora floccose* (Chlorophycota) from Sindh. *Pakistan Journal of Botany* 43(5): 2557-2560.

Khanzada, A.K., Wazir sheikh, Kazi, T.G., Samina kabir., and Shahzadi Soofia., 2007. Antifungal activity, elemental analysis and determination of total protein of seaweed, *Solieria robusta* (greville) Kylin from The coast of Karachi, *Pak. J. Bot.,* 39(3): 931-937.

Kim, Y.C., An R.O., Yoon, N.Y., Nam, T.J., Choi, J.S., 2005. Hepatoprotective Constituents of the Edible Brown Alga *Ecklonia stolonifera* on Tacrine-induced Cytotoxicity in Hep G2 Cells. *Arch Pharm Res* 28(12): 1376-1380.

Kolanjinathan, K., Ganesh, P., Govindarajan, M., 2009. Antibacterial activity of ethanol extracts of seaweeds against fish bacterial pathogens. European Review for Medical and Pharmacological Sciences 13, 173-177.

Kowalski, B., Jager, A.K., Van Staden, J., 1999. Influence of cutliver season, seaweed explant type and seaweed concentrate on potato plantlet quality. *Potato Research* 42: 181-188.

Krishnaveni, E., Johnson Marimuthu A., 2012. Preliminary Phytochemical, UV-VIS, HPLC and Anti-bacterial Studies on *Gracilaria corticata* J. Ag *Asian Pacific Journal of Tropical Biomedicine* 6 (2): 1-5.

Kuda, T., Kunii, T., Goto, H., Suzuki, T., and Yano, T., 2007. Varieties of antioxidant and antibacterial properties of *Ecklonia stolonifera* and *Ecklonia kurome* products harvested and processed in the Noto peninsula, Japan. Food Chemistry 103: 900-905.

Kumar, P., Senthamilselvi, S., Govindaraju, M., 2013. GC-MS profiling and antibacterial activity of *Sargassum tenerrimum. Journal of Pharmacy Research* 6, 88 -92.

Kumara, P., Senthamilselvi, S., Lakshmi prabha, A., Selvaraja, M., Macklin rania, L., Suganthi, P., Sarojinidevi, B., Govindaraju, M., 2012. Antibacterial activity and *in-vitro* cytotoxicity assay against brine shrimp using silver nanoparticles synthesized from *sargassum ilicifolium. Digest Journal of Nanomaterials and Biostructures* 7(4): 1447-1455.

Lakshmana, S., Vinothkumar, T., Geetharamani, D., Maruthupandi, T., 2013. Screening of seaweeds collected from southeast coastal area of India for α-amylase inhibitory activity, antioxidant activity and biocompatibility, *Int J Pharm PharmSci,* 5(1): 240-244.

Lavanya, R., Veerappan, N., 2011. Antibacterial Potential of Six Seaweeds Collected from Gulf of Mannar of Southeast Coast of India. *Advances in Biological Research* 5 (1): 38-44.

Leonel Pereira., Saly, Gheda, Paulo, J.A., Ribeiro-Claro, 1999. Analysis by Vibrational Spectroscopy of seaweed polysaccharides with potential use in Food, Pharmaceutical, and Cosmetic Industries. *International Journal of Carbohydrate Chemistry* 2013; 537202- 537207.Lincoln MM. Plants products as antimicrobial agents. *Clinical Microbiology Review* 12, 564-582.

Linga Rao, M., and Savithramma, N., 2012. Antimicrobial activity of silver nanoparticles synthesized by using stem extract of *Svensonia hyderobadensis* (Walp.) Mold – A rare medicinal plant. *Research in Biotechnology* 3(3): 41-47.

Mahnaz Khanavi., Pouyan Bagheri Toulabi., Mohammad, R.A., Nargess Sadati., Farzaneh, H., Abbas, H., Hassan, V., 2011. Larvicidal activity of marine algae, *Sargassum swartzii* and *Chondria dasyphylla,* against malaria vector *Anopheles stephensi. J Vector Borne Dis* 48: 241–244.

Manilal, A., Sugathan Sujith., George Seghal Kiran., Joseph Selvin., Chippu Shakir., Ramakrishnan Gandhimathi., 2009. Aaron Premnath Lipton. Antimicrobial potential and seasonality of red algae collected from the southwest coast of India tested against shrimp, human and phytopathogens, *Annals of Microbiology* 59 (2): 207-219.

Manilal, A., Sujith, S., Selvin, J., Shakir, C., Kiran, G.S., 2009. Antibacterial activity of *Falkenbergia hillebrandii* (Born) from the Indian coast against human pathogens. FYTON 78: 161-166.

Manilal, A., Sujith, S., Selvin, J., Kiran, G.S., Shakir, C., Lipton, A.P., 2010. Antimicrobial potential of marine organisms collected from the southwest coast of India against multiresistant human and shrimp pathogens. *Scientia Marina* 74(2): 287-296.

Manilal, A., Sujith, S., Kiran, G.S., Selvin, J., Panikkar, M.V.N., 2011. Evaluation of seaweed bioactives on common aquatic floral and faunal weeds of shrimp ponds. *Thalassas An international Journal of Marine Sciences* 27 (1): 47–55.

Manivannan, K., Karthikai devi, G., Anantharaman, P., Balasubaramanian, T., 2011. Antimicrobial potential of selected brown seaweeds from Vedalai coastal waters, Gulf of Mannar. *Asian Pacific Journal of Tropical Biomedicine* 117-123.

Manivannan, K., Karthikai devi, G., Anantharaman, P., Balasubramanian, T., 2011. Antimicrobial potential of selected brown seaweeds from Vedalai coastal waters, Gulf of Mannar. *Asian Pacific Journal of Tropical Biomedicine* 114-120.

Manivannan, K., Thirumaran, G., Karthikai Devi, G., Anantharaman, P., Balasubramanian, T., 2009. Proximate Composition of Different Group of Seaweeds from Vedalai Coastal Waters (Gulf of Mannar): Southeast Coast of India. Middle-East Journal of Scientific Research 4 (2): 72-77.

Manivannan, K., Thirumaran, G., Karthikai Devi, G., 2008. Biochemical Composition of Seaweeds from Mandapam Coastal Regions along Southeast Coast of India. *American-Eurasian Journal of Botany* 1 (2): 32-37.

Margaret Beula, J., Ravikumar, S., Syed Ali, M., 2011. Mosquito larvicidal efficacy of seaweed extracts against dengue vector of *Aedesaegypti Asian Pacific Journal of Tropical Biomedicine* 1(2):143–146.

Matsukawa, R., Dubinsky, Z., Kishimoto, E., Masaki, K.F.Y., Takeuchi, T., 1997. A comparison of screening methods for antioxidant activity in seaweeds. *Journal of Applied Phycology* 9: 29-35.

Metha, V.C., Trivedi, B.S., Bokil, K.K., and Narayanan, M.R., 1974. Seaweed as manure I. Studies on nitrification. *Proceedings on the Seminar sea salt and plant, CSMCRI Bhavanagar* 357-365.

Mohan, V.R., Venkataraman Kumar., 1993. Effect of seaweed extract Algifert on seed ermination and seedling growth in black gram and green gram. *Seaweed Research and Utilisation* 16, 53- 55.

Mohan,V.R., Venkataraman kumar., Murugewari, R., Muthuswami, S., 1994. Effect of crude and commercial seaweed extract on seed germination and seeding growth in *Cajanus cajan*.L. *Phykos* 33(1 and 2): 47-51.

Murugalakshmi Kumari, R., Ramasubramanian, V., Muthuchezhian, K., 2002. Studies on the utilization of seaweed as an organic fertilizer on the growth and some biochemical characteristics of black gram and cumbu. *Seaweed Research and Utilisation* 24(1):125-128.

Nadine Karaki., Carine Sebaaly., Nathalie Chahine., Tarek Faour., Alexandre Zinchenko., Samar Rachid., Hussein Kanaan., 2013. The antioxidant and anticoagulant activities of polysaccharides isolated from the brown algae *Dictyopteris polypodioides* growing on the Lebanese Coast. *Journal of Applied Pharmaceutical Science* 3(02): 043-051.

Nanthini Devi, K. Ajithkumar, T.T., Dhaneesh, K.V., Marudhupandi, T., Balasubramanian, T., 2012. Evaluation of antibacterial and antioxidant properties from brown seaweed, *Sargassum wightii* (Greville, 1848) against human bacterial pathogens. *Int J Pharm PharmSci* 4(3): 143-149.

Narisnh, L., Archana, N., Werner, *E.G.*, Muller., 2005. Marine natural products on drug discovery. *Natural product radiance* 4(6): 28-48.

Nelson, W.R., Van Staden, J., 1984. The effect of seaweed concentrate on growth of nutrientstressed greenhouse cucumbers. *Hortscience* 19: 81–82.

Oh, K.B, Lee, J.H., Chung, S.C., Shin, J., Shin, H.J., Kim. H.K., *et al.*, 2008. Antimicrobial activities of the bromophenols from the red alga *Odonthalia corymbifera* and some synthetic derivatives, Bioorganic and Med Chem Lett 18: 104-108.

Ohigashi, H., Sakai, Y., Yamguchi, K., Umezaki, I., Koshimizu, K., 1992. Possible anti-tumor promoting properties of marine algae and *in vivo* activity of wakame seaweed extract. *Bioscience, Biotechnology, and Biochemistry* 56, 994–995.

Omar, H.H., Shiekh, H.M., Gumgumjee, N.M., El-Kazan M.M., El-Gendy, A.M., 2012. Antibacterial activity of extracts of marine algae from the Red Sea of Jeddah, Saudi Arabia. *African Journal of Biotechnology* 11(71): 13576-13585.

Ömer Ertürk, Beyhan Ta°., 2011. Antibacterial and Antifungal Eff ects of Some Marine Algae. *Kafkas Univ Vet Fak Derg* 17 (Suppl A): S121-S124.

Osman, M.E.H., Abushady, A.M., Elshobary, M.E., 2010. *In vitro* screening of antimicrobial activity of extracts of some *macroalgae* collected from Abu-Qir bay Alexandria, Egypt.*African Journal of Biotechnology* 9(12): 7203-7208.

Paola, A., Tenorio Rodriguez., Mendez-Rodriguez, L.C., Serviere-Zaragoza, E., Hara, T, O., Zenteno-Savín, T., 2013. Antioxidant substances and trace element content in macroalgae from a subtropical lagoon in the West Coast of the Baja California Peninsula *Vitam Trace Elem* 2: 1-13.

Patra, J.K., Patra, A.P., Mahapatra, N.K., Thatoi, H.N., Das, S., Sahu, R.K., Swain, G.C., 2009. Antimicrobial activity of organic solvent extracts of three marine macroalgae from Chilika Lake, Orissa, India. *Malaysian Journal of Microbiology* 5(2):128-131.

Patra, J.K., Rath, S.K., Jena, K., Rathod, V.K., Thatoi, H., 2008. Evaluation of Antioxidant and Antimicrobial Activity of Seaweed (Sargassum sp.) Extract: A Study on Inhibition of Glutathione-S-Transferase Activity. *Turk J Biol* 32: 119-125.

Paul, V.J., and Fenical, W., 1987. Natural Products Chemistry and Chemical Defense in Tropical Marine Algae of The Phytum Chlorophyta. In: Bioorganic Marine Chemistry. P.J. Scheuer, (Ed.). *Spring-Verag. Berlin.* 1-29.

Periasamy Mansuya., Pandurangan Aruna., Sekaran Sridhar., Jebamalai Suresh Kumar Sarangam Babu., 2010. Antibacterial activity and qualitative phytochemical analysis of selected seaweeds from Gulf of Mannar Region. *J ExpSci* 1(8): 23-26.

Ponnuchamy Kumar, Singaravelu Senthamil Selvi, Munisamy. GC-MS profiling and antibacterial activity of *Sargassum tenerrimum* Govindaraju. *Journal of pharmacy research* 2013; **6**: 88-92.

Poonguzhali, T.V., JosminLaali Nisha, L.L., 2012. Larvicidal activity of two seaweeds, *Ulvafasciata* and *Grateloupialithophila* against mosquito vector, *Culex quinquefasciatus. INT J CURR SCI* 8(3): 163-168.

Prabha, V., Prakash, D.J., Sudha, P.N., 2013. Analysis of bioactivecompounds and antimicrobial activity of marine algae *Kappaphycus alvarezii* using three solvent extracts. *IJPSR* 4(1): 306-310.

Premalatha, M., Dhasarathan, P., Theriappan, P., 2011. Phytochemical characterization and antimicrobial efficiency of seaweed sample *Ulva fasciata* and *Chaetomorpha antennina. International Journal of Pharma and Bio Sciences* 2(1), 288- 293.

Rabia Alghazeer., Fauzi W., Entesar Abduelrhman., Fatiem Gammoudi., Salah Azwai., 2013. Screening of antibacterial activity in marine green, red and brown macroalgae from the western coast of Libya. *Natural Science* 5(1): 7-14

Radhika, D., Veerabahu, C., Priya, R., 2012. *In vitro* studies on antioxidant and haemagglutination activity of some selected seaweeds. *Int J Pharm PharmSci* 5(1): 152-155.

Raghavendran, H.R.B., Sathivel, A., Devaki, T., 2004. Efficacy of Brown Seaweed Hot Water Extract Against HCl-ethanol Induced Gastric Mucosal Injury in Rats. *Arch Pharm Res* 27 (4): 449-453.

Rahila Najam., Shahida, P.A., Azhar, I., 2010. Pharmacological Activities of *Hypnea musci*formis. *Afr. J. Biomed. Res.* 13: 69 – 74.

Rajasulochana, P., Dhamotharan, R., Krishnamoorthy, P., Murugesan, S., 2009. Antibacterial Activity of the Extracts of Marine Red and Brown Algae. *Journal of American Science* 5(3): 20-25.

Rajasulochana, P., Dhamotharan, R., Krishnamoorthy, P., Murugesan, S., 2009. Antibacterial Activity of the Extracts of Marine Red and Brown Algae. *Journal of American Science* 5(3): 20-25.

Rajkumar Immanuel, S., and Subramanian, S.K., 1999. Effect of fresh extracts and seaweed liquid fertilizers on some cereals and millets. *Seaweed Res. Utiln.* 21: 91-94.

Rama Rao, K., 1991. Effect of aqueous seaweed extract on *Ziziphus mauritioana* Lam. *J.India. Botanical Society* 71, 19-21.

Rashmi, C., Vinayak., Sabu, A.S., Anil Chatterji., 2010. Bio-Prospecting of a Few Brown Seaweeds for their cytotoxic and antioxidant activitiese *CAM advance access published* 3(8): 1-9.

Rashmi, C., Vinayak., Sabu, A.S., and Anil Chatterji., 2011. Bio-Prospecting of a Few Brown Seaweeds for Their Cytotoxic and Antioxidant Activities, Evidence-Based Complementary and Alternative Medicine Article ID 673083, http://dx.doi.org/10.1093/ecam/neq024

Renuka Bai, N., Mary Christi, R., Christy Kala, T., 2010. Antimicrobial potency of the marine alga, *Valoniopsis pachynema* (MAR.) BOERY. *Plant Archives* 10 (2): 699-701.

Renuka Bai, N., 2010a. Evaluation of *Gracilaria fergusonii* for phytochemical analysis and antibacterial activity. *Plant Archives* 10 (2): 711-713.

Rhimou, B., Hassane, R., José, M., and Nathalie, B. 2010. The antibacterial potential of the seaweeds (Rhodophyceae) of the Strait of Gibraltar and the Mediterranean Coast of Morocco. *African Journal of Biotechnology* 9(38), 6365-6372.

Sahar, R.G., Abdel-Halima, O.B., El-Sharkawya, S.H., Salamaa, O.M., Shierb, T.W., and Halim, A.F., Cytotoxic Hydroazulene Diterpenes from the Brown Alga *Dictyota dichotoma. http://www.znaturforsch.com/ac/v58c/s58c0017.pdf*

Salvador, N., Gomez-Garreta, A., Lavelli, L., Ribera, L., 2007. Antimicrobial activity of Iberian macroalgae. *Sci Mar* 71: 101-113.

Samy, A., Selim., 2012. Antimicrobial, antiplasmid and cytotoxicity potentials of marine algae *Halimeda opuntia* and *Sarconema filiforme* collected from Red Sea Coast *World Academy of Science, Engineering and Technology* 61: 1154- 1159.

Sanchez-Machado, D.I., Lopez-Hernandez, J., Paseiro-Losada, P., Lopez-Cervantes, J., 2004. An HPLC method for the quantification of sterols in edible seaweeds. *Biomed. Chromatogr.* 18, 183-190.

Sangeetha, V., and Thevanathan, R, 2010. Biofertilizer Potential of Traditional and Panchagavya Amended with Seaweed Extract. *J Amer Sci* 6: 61-67.

Sasidharan, S., Darah, I., Noordin, M.K.M.J., 2010. *In vitro* antimicrobial activity against *Pseudomonas aeruginosa* and acute oral toxicity of marine algae *Gracilaria changii. New Biotechnology* 27: 390-396.

Sasikumar, K., Govindan, T., Anuradha, C., 2011. Effect of Seaweed Liquid Fertilizer of *Dictyota dichotoma* on growth and yield of *Abelmoschus esculantus* L. *European Journal of Experimental Biology* 1 (3): 223-227.

Seenivasan, R., Indu, H., Archana, G., Geetha, S., 2010. The antibacterial activity of some marine algae from south east coast of India. *Journal of Pharmacy Research* 3(8): 1907-1912.

Seenivasan, R., Indu, H., Archana, G., Geetha, S., 2010. The Antibacterial Activity of Some Marine Algae from South East Coast of India. *American-Eurasian J Agric Environ Sci* 9 (5): 480-489.

Seenivasan, R., Rekha, M., Indu, H., and Geetha, S., 2012. Antibacterial Activity and Phytochemical Analysis of Selected Seaweeds from Mandapam Coast, *India Journal of Applied Pharmaceutical Science* 2 (10), 159-169.

Sekar, R., Suhasini, R., Raghavarao, R., 1995. Evolution of plasma bubbles in the equatorial F region with different seeding conditions. *Geophysical Research Letters* 22: 45-52.

Selvin, J., Lipton, A.P., 2004. Biopotentials of *Ulva fasciata* and *Hypnea musciformis* collected from the peninsular coast of India. *Journal of Marine Science and Technology* 12 (1): 1-6.

Senn, T.L., 1960. Seaweed extracts hasten germination of beet seeds. *Science News Letter* Feb. 2: 14-23.

Shanmugam, R., Malarkodi, C., Gnanajobitha, G., Paulkumar, K., Vanaja, M., Kannan, C., Annadurai, G., 2013. Seaweed-mediated synthesis of gold nanoparticles using *Turbinaria conoides* and its characterization, *Journal of Nanostructure in Chemistry* 3, 44.

Shanmughapriya, S., Aseer Manilal., Sugathan Sujith., Joseph Selvin., George Seghal Kiran., Kalimuthusamy Natarajaseenivasan., 2008. Antimicrobial activity of seaweeds extracts against multiresistant pathogens. *Annals of Microbiology* 58 (3): 535-541.

Shyamala, V., Thangaraju, N., 2013. Antibacterial and antioxidant activity of red seaweeds from Kilakarai, Rameswaram, Tamil Nadu, India. *Journal of Pharmaceutical and Biomedical Sciences* 32(32):1386 - 1395.

Siddiqui, S., Usmanghani, K., Shameel, M., 1994. Sterol and Fatty Acid compositions of a marine alga *Bryopsis pennata* (Bryopsidophyceae, Chlorophyta). Pakistan *Journal of Pharmaceutical Sciences* 7(1):73-82.

SivaKumar, K., Rajagopal, S.V., 2011. Radical scavenging activity of green algal species, *Journal of Pharmacy Research* 4(3): 723–725.

Sivasankari, S., Chandrasekaran, M., Kannathasan, K., Vengateslu, V., 2006a. Studies on the biochemical constituents of *Vigna radiata* L. treated with seaweed liquid fertilizer. *Seaweed Research and Utilization* 28(1): 151-158.

Sivasankari, S., Venkatesalu, V., Anantharaj, M., Chandrasekaran, M., 2006. Effect of seaweed extracts on the growth and biochemical constituents of *Vigna sinensis. Bioresource Technology* 97: 1745–1751.

Souza, E.T., Aline C de Queiroz, Emmanuel, G., de Miranda C., Lorenzo, P.V., Evandro, F da Silva, Thays, L. M. Freire-Dias, Yolanda K. Cupertino-Silva, Gabriela Muniz de A. Melo, Bárbara V.O.Santos, Maria Célia de O. Chaves, Magna S. Alexandre-Moreira, 2009. Antinociceptive activities of crude methanolic extract and phases, n-butanolic, chloroformic and ethyl acetate from *Caulerpa racemosa* (Caulerpaceae) *Brazilian Journal of Pharmacognosy* 19(1A): 115-120.

Sridhar, S., Rengasamy, R., 2002. Effects of seaweed liquid fertilizer obtained from *Ulva lactuca* on biomass, pigments and protein content of *Spinulina platensis. Seaweed Research and Utilisation Association* 24(1): 145-149.

Sridhar, S., Rengasamy, R., 2011. Potential of seaweed liquid fertilizers (SLFs) on some agricultural crop with special reference to protein profile of seedlings. *International Journal of Development Research* 1(7): 055-057.

Sridhar, S., Rengasamy, R., 2010. Significance of seaweed liquid fertilizers for minimizing chemical fertilizers and improving yield of *Arachis hypogaea* under field trial. *Recent Research in Science and Technology* 2(5): 73-80.

Sridhar, S., Rengasamy, R., 2010. Studies on the Effect of seaweed liquid fertilizer on the flowering Plant *Tagetes erecta* in Field Trial. *Advances in Bioresearch* 1 (2): 29 – 34.

Srivastava, N., Saurav, K., Mohanasrinivasan, V., Kannabiran, K., Singh, M., 2010. Antibacterial potential of macroalgae collected from the Madappam coast, India. *British Journal of Pharmacology and Toxicology* 1(2): 72-76.

Tasende, M.G., 2000. Fatty acid and sterol composition of gametophytes and saprophytes of *Chondrusb crispus. Scientia Marina* 64(4): 421-426.

Thambiraj, J., Lingakumar, K., Paulsamy, S., 2012. Effect of seaweed liquid fertilizer (SLF) prepared from *Sargassum wightii* and *Hypnea musciformis* on the growth and biochemical constituents of the pulse, *Cyamopsis tetragonoloba* (L). *Journal of Research in Agriculture* 1: 065-070.

Thillaikkannu Thinakaran, Mohan Balamurugan., Kathiresan Sivakumar, 2012. Screening of phytochemical constituents qualitatively and quantitatively certain seaweeds from Gulf of mannar Biosphere Reserve. *IRJP* 3 (7): 261-265.

Thirumal Thangam, R., Maria Victorial Rani, S., Peter Marian, M., 2003. Effect of seaweed liquid fertilizer on the growth and biochemical constituents of *Cyamopsis tetragonoloba* [h.] Taub. *Seaweed Research and Utilisation* 25: 99-104.

Thirumaran, G., Anantharaman, P., 2006a. Antibacterial activity and antifungal activities of marine macro alga (*Hydroclathrus clathratus*) from the Gulf of Mannar Biosphere Reserve. *Environ and Ecol* 24 S(1): 55-8.

Thirumaran, G.P., Baskar, V., Anantharaman, P., 2006b. Antibacterial and antifungal activities of seaweed (*Dictyota dichotoma*) from the Gulf of Mannar Biosphere Reserve. *J Ecotoxicol Environ and Ecol* 24 S (1): 37-40.

Thirumaran, G., Arumugam, M., Arumugam, R., Anantharaman, P., 2009. Effect of seaweed liquid fertilizer on growth and pigment concentration of *Cyamopsis tetrogonolaba* (L.) Taub. *American-Eurasian Journal of Agronomy* 2(2): 50-56.

Thivy, F., 1961. Seaweed ulitization in India. Proceedings of the Symposium on Algae, New Delhi. Seaweed maure for perfect soil and smiling fields salt. *Res. Indust* 1:1-4.

Ünci, T., Ney, Bilge Hilal. Cadirci, Dilek, Nal, Atakan Sukatar, 2006. Antimicrobial activities of the Extracts of Marine Algae from the Coast of Urla (Üzmir, Turkey). *Turk J Biol* 30: 171-175.

Vallinayagam, K., Arumugam, R., Ragupathi Raja Kannan, R., Thirumaran, G., Anantharaman, P., 2009. Antibacterial Activity of Some Selected Seaweeds from Pudumadam Coastal Regions Global. *Journal of Pharmacology* 3(1): 50-52.

Venkataraman Kumar, Mohan, V.R., 2003. Effect of seaweed liquid fertilizer on drought stressed ragi, (*Eleusine coracana* L. Gaertn.). *Seaweed Research and Utilisation* 25(102): 105-107.

Vijayabaskar, P., Shiyamala, V., 2011. Antibacterial Activities of Brown Marine Algae (*Sargassum wightii* and *Turbinaria ornata*) from the Gulf of Mannar Biosphere Reserve. *Advances in Biological Research* 5 (2): 99-102.

Vijayabaskar, P., Vaseela, N., 2012. In vitro antioxidant properties of sulfated polysaccharide from brown marine algae *Sargassum tenerrimum. Asian Pacific Journal of Tropical Disease* 4: 890-896.

Villarreal-Gómez, L.J., Soria-Mercado, *I.E.*, Graciela Guerra-Rivas., Ayala-Sánchez, N.E., 2010. Antibacterial and anticancer activity of seaweeds and bacteria associated with their surface. *Revista de Biología Marina Oceanografía* 45(2): 267-275.

Walailuck, B., Yuwadeepeera, P., Duangtakan, J., Jeereporn, Chayakornpumas, Utanjamjai, Doung, P., Thidarat, Panmu, k., 2011. Antioxidant activity of some seaweed from the Gulf of Thailand, *Int. J. Agric. Biol.,* 13 (1): 32-40.

Xavier Devanya, R., Shanmugavel, S., Kuppu Rajendran, Sundaram, J., 2012. Screening of selected marine algae from the coastal Tamil Nadu, South India for antibacterial activity, *Asian Pacific Journal of Tropical Biomedicine* 7(2): 1-9.

Yuan, Y.V., Walsh, N.A., 2006. Antioxidant and antiproliferative activities of extracts from a variety of edible seaweeds, *Food and Chemical Toxicology* 44: 1144–1150.

Yuan, Y.V., Carrington, M.F., Walsh, N.A., 2005. Extracts from dulse (*Palmaria palmata*) are effective antioxidants and inhibitors of cell proliferation *in vitro. Food and Chemical Toxicology* 43: 1073-1081.

Yuvaraj, N., Kanmani, P., Satishkumar, R., Paari, K.A., Pattukumar, V., Arul, V., 2011. Extraction, purification and partial characterization of *Cladophora glomerata* against multidrug resistant human pathogen *Acinetobacter baumannii* and fish pathogens. *World Journal of Fish and Marine Sciences* 3(1): 51-57.

Zeliha Demirel, Ferda, F., Yilmaz-Koz., Ulku, N., Karabay-Yavasoglu, Guven Ozdemir, Atakan Sukatar, 2009. Antimicrobial and antioxidant activity of brown algae from the Aegean Sea. *J. Serb. Chem. Soc.* 74 (6): 619–628.

2015, Modern Methods in Phytomedicine *Pages* ***143–153***
Editor: **T. Parimelazhagan**
Published by: **DAYA PUBLISHING HOUSE, NEW DELHI**

9

Traditional and Therapeutic Uses of *Withania somnifera*: A Siddha Perspective

G.S. Murugesan[1] and M. Sudalai Das[2]

[1]Department of Biotechnology; [2]Department of Yoga,
Bannari Amman Institute of Technology, Sathyamangalam, Tamil Nadu

1.0 INTRODUCTION

Withania somnifera, Dunal (ashwagandha, WS) is widely used in Siddha medicine, the traditional medical system of India. It is an ingredient in many formulations prescribed for a variety of musculoskeletal conditions (*e.g.*, arthritis, rheumatism), and as a general tonic to increase energy, improve overall health and longevity, and prevent disease in athletes, the elderly, and during pregnancy (Chatterjee and Pakrashi, 1995; Bone, 1996). Numerous studies indicated that ashwagandha possesses antioxidant, antitumor, antistress, anti-inflammatory, immunomodulatory, hematopoetic, anti-ageing, anxiolytic, antidepressive rejuvenating properties and also influences various neurotransmitter receptors in the central nervous system. Many pharmacological studies have been conducted to investigate the properties of ashwagandha in an attempt to authenticate its use as a multi-purpose medicinal agent. For example, anti-inflammatory properties have been investigated to validate its use in inflammatory arthritis (Anbalagan and Sadique, 1981; Somasundaram *et al.*, 1983), and animal stress studies have been performed to investigate its use as an antistress agent (Dadkar *et al.*, 1987; Singh *et al.*, 1982).

Several studies have examined the antitumor and radiosensitizing effect of WS (Singh *et al.*, 1986; Sharad *et al.*, 1996).

Withania somnifera Dunal belongs to the family solanaceae. It is a xerophytic plant, found in the drier parts of India, Sri Lanka, Afghanistan, Baluchistan and Sind and is distributed in the Mediterranean regions, the Canaries and Cape of Good Hope. It is found in high altitude ascending to 5,500 feet in the Himalayas. This shrub is common in Bombay and Western India, occasionally met within Bengal. It grows wildly throughout India particularly in hotter parts, on waste places and on road sides. It is also cultivated for medicinal purposes in fields and open grounds throughout India. It is widely cultivated in Bikaner and Pilani areas of Rajasthan, Rajputana, Punjab and Manasa (M.P.) (Anonymous, 2007, Chopra *et al.*, 1980, Dey *et al.*, 1973, Dymock *et al.*, 1976, Kirtikar *et al.*, 1980, Nadkarni, 1982). In Unani system of medicine, roots of *Withania somnifera* commonly known as Asgand are used for the medicinal properties. However, leaves of the plant are also reported to be used medicinally (Anonymous, 1982). The fresh roots are collected during January to March and dried under shade for several days. The drug retains its therapeutic efficacy for less than 2 years. It is prone to decomposition and loses its potentials within 2 years. So the fresh dried roots are preferred for medicinal uses.

According to the World health organization, traditional medicines are widely used in India. Approximately 80 per cent of the population of developing countries relies on traditional medicines for their primary health care needs. Medicinal plants continue to play a central role in the health care system of large proportions of the world's populations. Recognition and development of the medicinal and economic benefits of these plants are on the increase in both developing and industrialised nations. The medicinal plants contain several phytochemicals such as Vitamins (A, C, E, and K), Carotenoids, Terpenoids, Flavonoids, Polyphenols, Alkaloids, Tannins, Saponins, Enzymes, and Minerals etc. These phytochemicals possess antioxidant activities, which can be used in the treatment of multiple ailments. Most often the medicinal plants are collected from the wild. This uncontrolled harvesting has resulted in the extinction of many plants and created huge issues related to the potency and quality of medicinal products derived from these plants.

The researchers revealed that a specific extract from the plant, Withaferin A, was more effective in the inhibition than the common cancer chemotherapy drug, doxorubicin. Studies revealed that the anti-inflammatory and immunomodulatory properties of *Withania somnifera* (WS) root extracts are likely to contribute to the chemo preventive action. The roots of the plant are categorised as rasayanas, which are reputed to promote health and longevity by augmenting defence against disease, arresting the ageing process, revitalising the body in debilitate conditions, increasing the capability of the individual to resist adverse environmental factors and creating a sense of mental wellbeing. It is in use for a very long time for all groups and both sexes and even during pregnancy without any side effects. In this review, we have attempted to summarize briefly the information available on the potency of WS because of its immense therapeutic potential.

2.0 MORPHOLOGY

Withania somnifera is an evergreen, erect, branching, tomentose shrub, 30-150 cm in height. Leaves are simple, ovate, glabrous, and up to 10 cm long. Flowers are greenish or lurid yellow, small about 1 cm long; few flowers (usually about 5) born together in axillary, umbellate cymes (short axillary clusters). Fruits are globose berries, 6 mm in diameter, orange red when mature, enclosed in the inflated and membranous persistent calyx. Seeds are yellow, reniform and 2.5 mm in diameter (Qamar Uddin *et al.,* 2012).

2.1 Vernacular Names

- ☆ **Arabic:** Kaknaj-e-Hindi
- ☆ **Bengali:** Ashvaganda, Asvagandha
- ☆ **English:** Winter cherry
- ☆ **Gujarati:** Asan, Asana, Asoda, Asundha, Ghodaasoda
- ☆ **Hindi:** Asgandh, Punir
- ☆ **Malayalam:** Amukkiram, Pevetti
- ☆ **Marathi:** Askandha, Kanchuki, Tilli
- ☆ **Odiya:** Asugandha
- ☆ **Persian:** Kaknaj-e-Hindi, Asgand Nagaori
- ☆ **Sanskrit:** Ashvagandha,Ashvakandika, Gandhapatri, Palashaparni
- ☆ **Tamil:** Amukkira, Asubam, Asuvagandi
- ☆ **Telugu:** Asvagandhi, Penneru, Pennerugadda, Dommadolu
- ☆ **Urdu:** Asgand, Asgand Nagori (Anonymous, 2007, Chopra *et al.,* 1980, Kirtikar *et al.,* 1980).

3.0 CHEMISTRY

Since many of ashwagandha's uses have not been scientifically validated, skepticism can naturally be expected when presented with an herb purportedly useful in so many ailments. In Ayurvedic medicine there is a class of herbs, including WS, known as adaptogens or vitalizers. Adaptogens cause adaptive reactions to disease, are useful in many unrelated illnesses, and appear to produce a state of nonspecific increased resistance (SNIR) to adverse effects of physical, chemical, and biological agents. They are relatively innocuous, have no known specific mechanism of action, normalize pathological effects, and are usually glycosides or alkaloids of a plant.17,18 The chemistry of WS has been extensively studied and over 35 chemical constituents have been identified, extracted, and isolated. 19 The biologically active chemical constituents are alkaloids (isopelletierine, anaferine), steroidal lactones (withanolides, withaferins), saponins containing an additional acyl group (sitoindoside VII and VIII), and withanolides with a glucose at carbon 27 (sitoindoside IX and X). WS is also rich in iron.

4.0 THERAPEUTIC USES

Aswagand (*Withania somnifera*) has been recommended for the treatment of various ailments which include polyarthritis (*Waja-ul-Mafasil*), rheumatoid arthritis (*Hudar*), lumbago (*Wajaul- Qutn*), painful swellings (*Tawwarum-e-Alami*), spermatorrhoea (*Jaryan-e-Mani*), asthma (*Zeeq-un-Nafas*), leucoderma (*Bars*), general debility (*Zof-e-Aam*), sexual debility (*Zof-e-Bah*), (Ali, 1997, Anonymous, 2007, Ghani, 1920, Kabiruddin, 1955, Khare, 2007, Kirtikar *et al.*, 1980, Nadkarni, 1982), amnesia (*Nisyan*) (Ali, 1997, Ghani, 1920), anxiety neurosis (*Qalaq-e-Usabi*), (Ali, 1997, Khare, 2007), scabies (*Jarb*), ulcers (*Qurooh*), marasmus (*Saghal*) and leucorrhoea (*Sailan-ur-Rahem*), anti-inflammatory (*Mohallil-e-Warm*), sedative (*Musakkin*), hypnotic (*Munawim*), narcotic (*Munashshi*), general tonic (*Muqawwi-e-Aam*), diuretic (*Mudir-e-Baul*) (Fruits and Seeds), aphrodisiac (*Muqawwi-e-Bah*), alterative (*Muaddil*), deobstruent (*Mufatteh Sudad*), (Ali, 1997, Anonymous, 2007, Chopra *et al.*, 1980, Ghani, 1920, Kabiruddin, 1955, Khare, 2007, Kirtikar *et al.*, 1980, Nadkarni, 1982), uterine tonic (*Muqawwi-e-Rahem*) and increases production of semen (*Muwallid-e-Mani*) (Anonymous, 2007, Ghani, 1920, Kabiruddin, 1955).

5.0 PHYTOCHEMICAL STUDIES

A review of literature reveals the presence of various chemical constituents in the different parts of the plant which are as follows:

5.1 Root

The roots are reported to contain alkaloids, amino acids, steroids, volatile oil, starch, reducing sugars, glycosides, hentriacontane, dulcitol, withaniol, an acid (m.p. 280-283? decomp.), and a neutral compound (m.p. 294-296?). The total alkaloidal content of the Indian roots has been reported to vary between 0.13 and 0.31 percent, though much higher yields (up to 4.3 per cent) have been recorded elsewhere (Anonymous, 1982, Anonymous, 2007).

Many biochemically heterogeneous alkaloids have been reported in the roots. Basic alkaloids include cuscohygrine, anahygrine, tropine, pseudotropine, anaferine, isopelletierine, withananine, withananinine, pseudo-withanine, somnine, somniferine, somniferinine. Neutral alkaloids include 3- tropyltigloate and an unidentified alkaloid. Other alkaloids include withanine, withasomnine, and visamine. Withanine is sedative and hypnotic (Khare, 2007). Withasomnine has been separated from the roots of the plant grown in West Germany. Visamine is a new alkaloid which has been separated from the roots of the plant grown in Soviet Union. It prolonged hexanal-induced sleeping time and showed hypothermic and nicotinolytic effects in mice (Rastogi *et al.*, 1998).The free amino acids identified in the root include aspartic acid, glycine, tyrosine, alanine, proline, tryptophan, glutamic acid, and cystine (Khare, 2007).

5.2 Leaf

The leaves of the plant (Indian chemotype) are reported to contain 12 withanolides, 5 unidentified alkaloids (yield, 0.09 per cent), many free amino acids, chlorogenic acid, glycosides, glucose, condensed tannins, and flavonoids (Khare,

2007). The leaves of the plant from different habitats contain different withanolides– a group of C28 steroids characterized by a 6-membered lactone ring in the 9-carbon atom side chain. Withaferin A, a steroidal lactone is the most important withanolide isolated from the extract of the leaves and dried roots of *Withania somnifera*. It is thermostable and slowly inactivated at pH 7.2. It is insoluble in water and is administered in the form of suspension. For its separation, the leaves are extracted with cold alcohol; the extract is purified and dried, and finally crystallized from aqueous alcohol (yield, 0.18 per cent air dry basis). The yield of this compound from the South-African plants is reported to be as high as 0.86 percent. The curative properties of the leaves and roots are attributed to Withaferin A (Anonymous, 1982).

5.3 Fruit

The green berries contain amino acids, a proteolytic enzyme, condensed tannins, and flavonoids. They contain a high proportion of free amino acids which include proline, valine, tyrosine, alanine, glycine, hydroxyproline, aspartic acid, glutamic acid, cystine and cysteine. The presence of a proteolytic enzyme, chamase, in the berries may be responsible for the high content of the amino acid.

5.4 Shoots

The tender shoots are rich in crude protein, calcium and phosphorous, and are not fibrous. They are reported to contain scopoletin.

5.5 Stem

The stem of the plant contains condensed tannins and flavonoids.

5.6 Bark

The bark contains a number of free amino acids (Anonymous, 1982).

6.0 PHARMACOLOGICAL STUDIES

The drug consists of the dried roots of *Withania somnifera* which is official as a sedative in the pharmacopoeia of India. The pharmacological activity of the roots is attributed to the presence of several alkaloids. The total extract (70 per cent alcoholic) of the roots possesses the same properties as the total alkaloids, but is nearly half as potent (Anonymous, 1982).

6.1 Anti-inflammatory Activity

Withaferin A exhibits fairly potent anti-arthritic and anti-inflammatory activities. Anti-inflammatory activity has been attributed to biologically active steroids, of which Withaferin A is a major component. It is as effective as hydrocortisone sodium succinate dose for dose (Khare, 2007). It was found to suppress effectively arthritic syndrome without any toxic effect. Unlike hydrocortisone-treated animals which lost weight, the animals treated with Withaferin A showed gain in weight in arthritic syndrome. It is interesting that Withaferin A seems to be more potent than hydrocortisone in adjuvant-induced arthritis in rats, a close experimental approximation to human rheumatoid arthritis. In its oedema inhibiting activity, the compound gave a good dose response in the dose range of 12-25 mg/kg body weight of Albino rats intraperitoneally and a single dose had a good duration of action, as it could effectively suppress the inflammation after 4 hours of its administration (Anonymous, 1982, Rastogi *et al.*, 1998). Asgand (*Withania somnifera*) has been shown to possess anti-inflammatory property in many animal models of inflammations like carrageenan-induced inflammation. The experiments showed interesting results as most of the APR were influenced in a very short duration and also suppressed the degree of inflammation (Anabalagan *et al.*, 1985).

6.2 Antitumour Activity

Withaferin A, withanolide D and E exhibited significant antitumour activity *in vitro* against cells derived from human epidermoid carcinoma of nasopharynx (KB) and *in vivo* against Ehrlich ascites carcinoma, Sarcoma 180, Sarcoma Black (SBL), and E 0771 mammary adenocarcinoma in mice in doses of 10, 12, 15 mg/kg body-weight. Growth of Ehrlich ascites carcinoma was completely inhibited in more than half the mice which survived for 100 days without the evidence of growth of the tumour. They also acted as a mitotic poison arresting the division of cultured human larynx carcinoma cells at metaphase and in HeLa cultures similar to star metaphase. Withaferin A caused mitotic arrest in embryonic chicken fibroblast cells. Methylthiodeacetyl colchicine potentiated the effect of Withaferin A. The presence of an unsaturated lactone in the side-chain to which an allelic primary alcohol group is attached at C25 and the highly oxygenated rings at the other end of the molecule may well suggest specific chemical systems possessing carcinostatic properties (Anonymous, 1982, Rastogi *et al.*, 1998, Khare, 2007). Withaferin A has been shown to possess growth inhibitory and radio-sensitizing effects on experimental mouse tumours (Ganasoundary *et al.*, 1997). Administration of Withaferin A in mice inoculated with Ehrlich ascites carcinoma cells was found to inhibit tumour growth and increase tumour-free animal survival in a dose dependent manner (Devi *et al.*, 1995, Sharada *et al.*, 1996). The alcoholic extract of the dried roots of the plant as well as the active component Withaferin A isolated from the extract showed significant antitumour and radio-sensitizing effects in experimental tumours *in vivo*, without any noticeable systemic toxicity. One-hour treatment with Withaferin A in a nontoxic dose of 2.1 μM before irradiation significantly enhanced cell killing. Withaferin A gave a sensitizer enhancement ratio (SER) of 1.5 for *in vitro* cell killing of V79 Chinese hamster cells at a nontoxic concentration of approximately 2 μM. SER increased with drug dose (Devi *et al.*, 1996).

6.3 Immunomodulatory Activity

Asgand showed a significant modulation of immune reactivity in animal models. Administration of Asgand was found to prevent myelo-suppression in mice treated with three immunosuppressive drugs *viz.* cyclophosphamide, azathioprin, and prednisolone. Treatment with Asgand was found to significantly increase Hb concentration, RBC count, platelet count, and body weight in mice (Ziauddin *et al.*, 1996). Administration of Asgand extract was found to significantly reduce leucopenia induced by cyclophosphamide (CTX) treatment. Administration of Asgand extract increased the number of α-esterase positive cells in the bone marrow of CTX treated animals, compared to the CTX alone treated group (Davis *et al.*, 1998). Administration of Asgand extract was found to significantly reduce leucopenia induced by sub-lethal dose of gamma radiation (Kuttan, 1996). Withaferin A and Withanolide E exhibited specific immunosuppressive effect on human B and T lymphocytes and on mice thymocytes. Withanolide E had specific effect on T lymphocytes whereas Withaferin A affected both B and T lymphocytes (Aggarwal *et al.*, 1999, Davis *et al.*, 2000, Gautam *et al.*, 2004, Rasool *et al.*, 2006, Rastogi *et al.*, 1998).

7.0 PURIFICATION OF THE ASWAGANDHA ROOT (SUTHI METHOD)

The Aswagandha root was boiled with milk and the roots were dried and powdered. This process will neutralize the ill effects of the compounds and reduces the incompatibility.

8.0 TRADITIONAL USES OF ASWAGANDHA

8.1. Traditional Use of Aswagandha as Internal Medicine

Table 9.1: Effects of Aswagandha Root

Aswagandha Root Powder (1-2g) with the Following Ingredient	*Effect*
Warm water	Body strength, reduces obesity
Honey	Body strength, dry cough, wheezing
Ghee	Body strength, White discharge, increase biomass
Milk	Body strength, dry cough, wheezing, toning up of nerves, erectile dysfunction, increase sperm count

8.2 Traditional Use of Aswagandha as Topical Medicine

The Aswagandha root crushed with water can be applied on skin for reducing inflammation and the Aswagandha root crushed with water and mixed with *Zingiber officinalis* (Sukku) can be applied on skin for inflammation in knee.

8.3 Traditional Use of Aswagandha Sooranam (Powder) (Kannusamy Pillai, 2010)

Composition

- ☆ *Withania somnifera* (root) – 64 g
- ☆ *Zingiber officinalis* – 32 g
- ☆ *Piper longum* – 16 g
- ☆ *Piper nigrum* – 8 g
- ☆ *Coriandrum sativum* (seeds) – 7 g
- ☆ *Cuminum cyminum* – 6 g
- ☆ *Cinnamomum tamala* – 5 g
- ☆ *Cinnamomum aromaticum* – 4 g
- ☆ *Elettaria cardamomum* – 3 g

- ☆ *Mesua ferrea* – 2 g
- ☆ *Myrtus caryophyllus* – 1 g
- ☆ *Saccharum officinarum* (sugar) – 148 g

Preparation

1-2 g of the above formulated powder was taken and mixed with honey/milk/ghee depending upon the Vatha, Pitha or Kapa conditions of the individual twice daily after food.

Uses

Wheezing, severe cold and cough, anemia, nervous disorder/weakness, impotency, sexual disorders, skin itching and bone rigidity.

8.4 Traditional use of Aswagandha Legiyam (Gel) (Kannusamy Pillai, 2011)

Composition

- ☆ *Withania somnifera* (root) – 64 g
- ☆ *Zingerber officinalis* – 32 g
- ☆ *Piper longum* – 16 g
- ☆ *Piper nigrum* – 8 g
- ☆ *Coriandrum sativum* (seeds) – 7 g
- ☆ *Cuminum cyminum* – 6 g
- ☆ *Cinnamum tamala* – 5 g
- ☆ *Cinnamomum aromaticum* – 4 g
- ☆ *Elettaria cardomomum* – 3 g
- ☆ *Mesua ferrea* – 2 g
- ☆ *Myrtus caryophyllus* – 1 g
- ☆ *Saccharum officinarum* (sugar) – 148 g

Preparation

1-2 g of the above formulated powder was taken and mixed with honey/milk/ghee depending upon the Vatha, Pitha or Kapa conditions of the individual twice daily after food.

Uses

Wheezing, severe cold and cough, anemia, nervous disorder/weakness, impotency, sexual disorders, skin itching and bone rigidity.

9.0 CONCLUSION

Medicinal plants maintain the health and vitality of individuals and also cure disease, without causing toxicity. As *Withania somnifera* possess good immunomodulatory anti-inflammatory, antitumor, antioxidant, anticancer properties and many pharmacologically and medicinally important chemicals, such as withaferins, sitoindosides and various alkaloids, they protect the cells from oxidative damage and diseases. Thus consume a good diet, rich in antioxidant plant foods (eg. fruits and vegetables) will provide health- protective effects. In conclusion, this article provides the therapeutic knowledge about *Withania somnifera*, which is used by the people all over the world. Also, it is of significance to exploit novel medicines from *Withania somnifera.*

REFERENCES

Aggarwal, R., Diwanay, S., Patki, P., Patwardhan, B., 1999. Studies on immunomodulatory activity of *Withania somnifera* (Ashwagandha) extracts in experimental immune inflammation. *J Ethnopharmacol.,* 97, 27-35.

Anbalagan, K., and Sadique, J., 1981. Influence of an Indian medicine (Ashwagandha) on acutephase reactants in inflammation. *Indian J ExpBiol,* 19, 245-249.

Anabalagan, K., Sadique, J., 1985. *Withania somnifera,* a rejuvenating herbal drug which controls alpha-2 macroglobulin synthesis during inflammation. *Intl J Crude Drug Res.* 23, 177-183.

Anonymous, 1982. The Wealth of India. Vol. X (Sp-W), Publications and Information Directorate, Council of Scientific and Industrial Research (CSIR), New Delhi, 580-585.

Anonymous, 2007. Standardisation of Single Drugs of Unani Medicine. Part III, 1st ed. Central Council for Research in Unani Medicine (CCRUM), New Delhi, 9-14.

Almeida, C.F.C.B.R., de Amorium, E.L.C., de Albuquerque, U.P., Maia, M.B., 2006. Medicinal plants popularly used in the Xingó region – a semi-arid location in Northeastern Brazil. *Journal of Ethnobiology and Ethnomedicine,* 2: http://www.ethnobiomed.com/content/2/1/15.

Bone, K., 1996. *Clinical Applications of Ayurvedic and Chinese Herbs. Monographs for the Western Herbal Practitioner.* Australia: *Phytotherapy Press,* 137-141.

Chatterjee, A., Pakrashi, S.C., 1995. *The Treatise on Indian Medicinal Plants* 4, 208-212.

Chopra, R.N., Nayar, S.L., Chopra, I.C., 1980. Glossary of Indian Medicinal Plants. Council of Scientific and Industrial Research, New Delhi, 191, 258.

Dadkar, V.N., Ranadive, N.U., Dhar, H.L., 1987. Evaluation of antistress (adaptogen) activity of *Withania somnifera* (Ashwagandha). *Ind J Clin Biochem,* 2, 101-108.

Davis, L., Kuttan, G., 1998. Suppressive effect of cyclophosphamide induced toxicity by *Withania somnifera* extract in mice. *J. Ethnopharmacol.* 62 (3), 209-214.

Davis, L., Kuttan, G., 2000. Immunomodulatory activity of *Withania somnifera. J Ethnopharmacol.* 71 (1-2), 193-200.

Devi, P.U., Sharada, A.C., Soloman, F.E., 1995. *In vivo* growth inhibitory and radio-sensitizing effects of Withaferin A on mouse Ehrlich ascites carcinoma. *Cancer Letters* 95 (1-2), 189-193.

Devi, P.U., Akagi, K., Ostapenko, V., Tanaka, Y., Sugahara, T., 1996. Withaferin A: A new radiosensitizer from Indian medicinal plant *Withania somnifera. Intl J Radiation Biol.* 69 (2), 193-197.

Dey, K.L., Bahadur R., 1973. Indigenous Drugs of India. Prime Lane, Chronica Botanica, New Delhi, 670.

Dymock, W., Warden, C.J.H., Hooper, D., 1976. Pharmacographia Indica. Vol. II, M/s Bishen Singh Mahendra Pal Singh, Dehradun and M/s Periodical Experts, New Delhi, 566-572.

Ganasoundari, A., Zare, S.M., Devi, P.U., 1997. Modification of bone marrow radiosensitivity by medicinal plant extracts. *British J Radiology* 70 (834), 599-602.

Gautam, M., 2004. Immune response modulation to DPT vaccine by aqueous extract of *Withania somnifera* in experimental system. *Int. Immunopharmacol.* 4, 841–849.

Kannusamy Pillai, C., 2010. Aswagandhi. *MATERIA MEDICA (Vegetable Kingdom),* B.Rathina Nayagar and Sons, Chennai, India, 9-10.

Kannusamy Pillai, C., 2011. Aswagandhi Legiyam. *Sigicha Rathna Deepam,* B.Rathina Nayagar and Sons, Chennai, India, 180-181.

Khare, C.P., 2007. Indian Medicinal Plants–An Illustrated Dictionary. First Indian Reprint, Springer (India) Pvt. Ltd., New Delhi, 717- 718.

Kirtikar, K.R., Basu, B.D., 1980. Indian Medicinal Plants. 2nd ed. Vol. III, Lalit Mohan Basu, Allahabad, India, 1774-1777.

Kuttan, G., 1996. Use of *Withania somnifera* Dunal as an adjuvant during radiation therapy. *Indian J Exp. Biol.* 34 (9), 854-856.

Mishra, L.C., Singh, B.B., 2000. Scientific Basis for the Therapeutic Use of *Withania somnifera* (Ashwagandha): A Review. *Alternative Medicine Review* 5, 4.

Nadkarni, K.M., 1982. Indian Materia Medica. 3rd ed. Vol. I, Popular Prakashan Pvt. Ltd., Bombay, 1292-1294.

Qamar Uddin, L., Samiulla, V.K., Singh Jamil, S.S., 2012. *Phytochemical and Pharmaceutical Science* 02 (01), 170-175.

Rasool, M., Varalakshmi, P., 2006. Immunomodulatory role of *Withania somnifera* root powder on experimental induced inflammation: An *in vivo* and *in vitro* study. *Vascul Pharmacol.* 44 (6), 406-410.

Rastogi, R.P., Mehrotra, B.N., 1998. Compendium of Indian Medicinal Plants. 2nd Reprint, Central Drug Research Institute, Lucknow and National Institute of Science Communication, Council of Scientific and Industrial Research, New Delhi Vol. 1: 434-436; Vol. 2: 708-710; Vol. 3: 682-684; Vol. 4: 765-766; Vol. 5: 889-891; Vol. 6: 148.

Sharad, A.C., Solomon, F.E., Devi, P.U., 1996. Antitumor and radiosensitizing effects of withaferin A on mouse Ehrlich ascites carcinoma *in vivo. Acta Oncol,* 35, 95-100.

Singh, N., Nath, R., Lata, A., 1982. *Withania somnifera* (ashwagandha), a rejuvenating herbal drug which enhances survival during stress (an adaptogen). *Int J Crude Drug Res* 20, 29-35.

Singh, N., Singh, S.P., Nath, R., 1986. Prevention of urethane-induced lung adenomas by *Withania somnifera* (L.) Dunal in albino mice. *Int J Crude Drug Res,* 24, 90-100.

Somasundaram, S., Sadique, J., Subramoniam, A., 1983. *In vitro* absorption of [14C] leucine during inflammation and the effect of anti-inflammatory drugs in the jejunum of rats. *Biochem Med* 29, 259-264.

Sharada, A.C., Solomon, F.E., Devi, P.U., Udupa, N., Srinivasan, K.K., 1996. Antitumour and radiosensitizing effects of withaferin A on mouse Ehrlich ascites carcinoma *in vivo. Acta. Oncology* 35 (1), 95-100.

Veena, S., Sadhana, S., Pracheta, Ritu, P., 2011. *Withania somnifera*: A Rejuvenating Ayurvedic Medicinal Herb for the Treatment of various Human ailments. *International Journal of Pharm. Tech. Research* 3(1), 187-192.

Ziauddin, M., Phansalkar, N., Patki, P., Diwanay, S., Patwardhan, B., 1996. Studies on the immunomodulatory effect of Asgandh. *J Ethnopharmacol.* 50 (2), 69-76.

2015, Modern Methods in Phytomedicine *Pages* ***155–169***
Editor: **T. Parimelazhagan**
Published by: **DAYA PUBLISHING HOUSE, NEW DELHI**

10

Evaluation of DNA Barcodes for the Effective Discrimination of Zingiberaceae Species

Dhivya Selvaraj, Rajeevkumar Sarma, Dhivya Shanmughanandhan, Saravanan Mohanasundaram and Sathishkumar Ramalingam*

Plant Genetic Engineering Laboratory, Department of Biotechnology, Bharathiar University, Coimbatore – 641 046, Tamil Nadu

1.0 INTRODUCTION

DNA barcoding is a technique that helps for the rapid identification of species, with the standardized region or a tag (Hebert and Barrett, 2005). The cytochrome oxidase 1 (*COX1*) genehas been widely used as the barcode candidate for the animal kingdom (Omnia *et al.*, 2010). In plants, chloroplast genes like *rbcL* and *psbA-trnH,* (Kress and Erickson, 2006, Newmaster *et al.*, 2006), the combination of internal transcribed spacer (*ITS*) and intergenic spacer *psbA-trnH,* (Chase *et al.*, 2005, Kress *et al.*, 2005, Lahaye *et al.*, 2008) *matK, atpF-H* and *psbI-psbK,* (Kim *et al.*, 2004) intron *trnL* from degraded samples (Taberlet *et al.*, 2007) and *trnT-trnL, ITS, psbA-trnH* (Edwards *et al.*, 2008), have been proved as an effective barcodes. In some studies, the two noncoding plastid regions like *psbA-trnH* and *trnL-trnF* were proposed as a

* *Corresponding Author.* E-mail: rsathish@buc.edu.in

multimarker system that can be used throughout the plant kingdom for species identification (Gianmarco *et al.*, 2009). The *psbA-trnH* region has been proved to be effective barcode for identification of 17 *Dendrobium* species and also in differentiating the *Dendrobium* species from other adulterating species (Hui *et al.*, 2009). Among these proposed barcodes candidates the most important genes like *matK, rbcL, psbA-trnH* and *ITS2* were taken for the present study to analyze and identify the medicinally important plant species of Zingiberaceae.

Zingiberaceae family includes flowering plants comprising 52 genera and more than 1200 species (Kress, 2009). This family consists of many Indian traditional medicinal plants like *Alpinia, Curcuma, Kaempferia* and also plants that can yield commercially valuable products like dyes, perfumes, aesthetics etc., This family also includes a vital group of rhizomatous medicinal and aromatic plants characterized by the presence of volatile oils and oleoresins which has high export value. In south India nearly eight species of the genus *Zingiber* including 1 cultivar and 7 wild species were distributed and also the two species *Z. neesanum, Z. nimnoni* are endemic to Western Ghats (Kavitha *et al.*, 2010). Most medicinal plants in the genus *Curcuma, Alpinia, Globba, Kaempferia, Etingera, Amomum* and *Zingiber* were found to be highly similar in their morphological characters and this complexity leads to misidentification of plant species by a non-taxonomist (Leong-Škornièková *et al.*, 2007).

In Ayurvedic herbal formulations, sometimes knowingly or unknowingly the potential herbal plant species were substituted or adulterated by other species or closely related species. The species *Hedychium spicatum* is well known for drug "Karchura", and it is being substituted by *H. coronarium* and *H. coronarium* var. *flavum* species. The plants belong to the genus *Curcuma i.e. C. sattayasai, C. zeodoaroides* are being morphologically similar like *C. longa*. The species of *Aromatica galangal* is used in the ayurvedic preparation such as "Rasnadichoornam" which is used in the treatment of rheumatoid, arthritis and neuromuscular pains (Gamble, 1987). From the species of *Alpinia*, ayurvedic decoctions like "*Rasna-saptak-kwath*" and "*Rasna-adikamath*" were made to cure anti-inflammatory disease (Thakur *et al.*, 1989).

Due to lack of accurate identification, the herbal products lose its efficacy. Molecular markers like RAPD, RFLP and AFLP are employed to distinguish the medicinal plant species from their adulterants for example, *Amomum villosum* plant can be identified and distinguished from their adulterants using RAPD markers (Wang *et al.*, 2000). Fifteen *Boesenbergia* species have been discriminated by AFLP analysis (Techaprasan *et al.*, 2008). As the members of the Zingiberaceae possess commercial importance, an easy and accurate method for authenticating Zingiberaceae family is vital for ensuring the drug and food safety of internationally traded herbs. In this study, more recent and reliable tool "DNA barcoding" was used for species identification.The present analysis provides sufficient confirmation for the use of DNA sequences as a suitable marker for species identification in the family Zingiberaceae. In this study, four proposed DNA barcodes (*matK, rbcL, psbA-trnH* and *ITS2*) have been analyzed to determine the potential barcode candidate for Zingiberaceae family. The most effective DNA barcode region *ITS2* showed several polymorphic sites that distinguished the medicinally important species of the genus

Alpinia, Aframomum, Amomum, Curcuma, Zingiber, Hedychium, Kaempferia, Globba, Boesenbergia, and *Zingiber.*

2.0 MATERIALS AND METHODS

2.1 Sampling of Plant Materials

The data consists of two sets. In the first set, thirty samples representing 7 genera of Zingiberaceae were collected from the Western Ghats, India (76°45' and 76°55' E longitude, 11°0' N latitude). In the second set, 1124 sequences belonging to 102 genera were downloaded from GenBank database and many other closely related species were also included in the dataset (Table 10.1).

Table 10.1: Total Number of DNA Barcode Sequences Used In the Study

Markers	*No. of Sequences*	*No. of Sequences Belonging to Genera Containing more than One Species*	*No. of Sequences Belonging to Species Containing more than One Sample*	*Percentage of Variance*
matK	3870	195(18)	62(11)	0.3
rbcL	66	51(18)	20(6)	0.2
psbA-trnH	102	56(10)	22(7)	1.1
ITS	1124	102(9)	148(15)	1.4

2.2 DNA Extraction and Amplification

Total genomic DNA was extracted from the fresh leaf tissue according to Suman *et al.* (1999) (Figure 10.1). The *ITS* gene was PCR amplified using the forward primer 5′-GGAAGTTATTGACACAGGAGT-3′ and reverse primer 5′CCGGAAAGGAT CAATTGAAAC-3′. The reaction mixture included 10-15 ng of total DNA,10 pmol of each primer (IDT, USA), 0.2 mM of each dNTP (Fermantas®), 10X Taq Buffer (Fermentas®), 15mM of Magnesium Chloride (Fermentas®) and 0.2 units Taq DNA polymerase (Fermentas®). PCR was performed using Eppendorf Gradient Master Cycler and the amplification was carried out under the following conditions: initial denaturation at 94°C for 5 min, denaturation at 94°C for 30 s, annealing at 57°C for 1 min and elongation at 72°C for 1 min (30 cycles). PCR products were electrophoresed in 0.8 per cent agarose gel electrophoresis and documented.

2.3 Sequence Alignment and Analysis

All the sequences of Zingiberaceae family corresponding to the four marker genes were downloaded from GenBank database and also the sequences generated from our laboratory were used for sequence analysis. Sequences were aligned using the CLUSTAL X 1.81 program (Thompson *et al.,* 1994) and then optimized by using BioEdit (Hall, 1999). The inter and intraspecific variation of each barcoding region were characterized by calculating Kimura 2-parameter (K2P) distances in MEGA 4 (Tamura *et al.,* 2007). The barcoding gaps were analyzed by comparing the distributions of intra and inter-specific divergences of each candidate locus (Chen *et al.,* 2010).

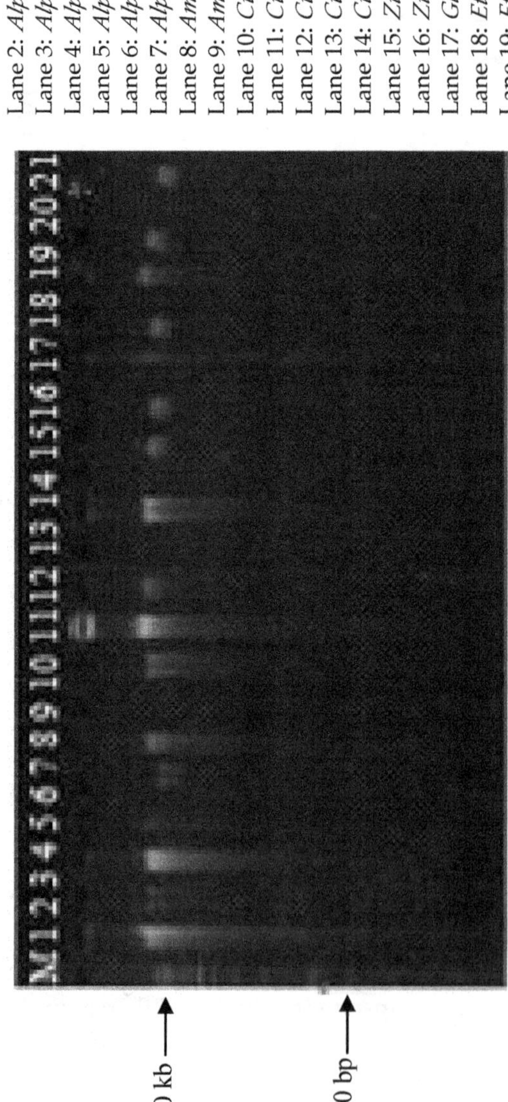

Lane M: Marker
Lane 1: *Hedychium coronarium*
Lane 2: *Alpinia smithiae*
Lane 3: *Alpinia purpurata*
Lane 4: *Alpinia calcarata*
Lane 5: *Alpinia malaccensis*
Lane 6: *Alpinia vittata*
Lane 7: *Alpinia hypoleucum*
Lane 8: *Amomum subulatum*
Lane 9: *Amomum aculeatum*
Lane 10: *Curcuma aromatic*
Lane 11: *Curcuma zeodania*
Lane 12: *Curcuma caesia*
Lane 13: *Curcuma caesia*
Lane 14: *Curcuma ratakonda*
Lane 15: *Zingiber zerumpet*
Lane 16: *Zingiber purpureum*
Lane 17: *Globbas chomburgkii*
Lane 18: *Etlingera elatior*
Lane 19: *Etlingera linguiformis*
Lane 20: *Hedychium flavescens*
Lane 21: *Hedychium coccineum*

Figure 10.1: Genomic DNA Isolation from the Plant Species of Zingiberaceae.

2.4 Species Identification by Genetic Analysis

Phylogenetic tree was constructed by Maximum Likelihood (ML) method using PhyML program (Dereeper *et al.,* 2010). Branch support for ML was assessed with 500 bootstrap replicates (Felsenstein, 1985). The two methods for species identification namely BLAST1 and the nearest genetic distance method were performed (Chase *et al.,* 2007; Ross *et al.,* 2008). The genetic divergence was optimized using Wilcoxon signed- rank test and StatDirect (Bhuchan, 2000).

3. RESULTS

3.1 Analysis of Five Potential Barcoding Regions

From the sequence analysis, it was clear that the universal coding regions namely *rbcL* and *matK* produce highly conserved regions with comparatively constant length, making the alignment easier. Similarly, the non-coding region nuclear internal transcribed spacer region *ITS* was amplified and sequenced (Figure 10.2). *ITS* region was edited for *ITS1* and *ITS2* region. Sequencing was relatively straightforward for all putative loci except for the plastid, the non-coding intergenic spacer gene, *psbA-trnH*. The *matK, rbcL, psbA- trnH* and *ITS2* candidates showed 0.3, 0.2, 1.1 and 1.4 per cent of variations respectively. Henceforth, the nuclear internal transcribed spacer *ITS2* and chloroplast plastid non coding intergenic spacer *psbA-trnH* showed better discrimination at the species level than *rbcL* and *matK.*

3.2 Measurement of DNA Divergence for ITS2

Significant variation in DNA sequences between the genus and less variations within the species are important criteria to identify various species. First, we characterized the intra and inter-specific variation for *ITS* and *psbA-trnH* gene sequences. Within ITS, specifically the *ITS1* region showed no variation and contains highly conserved sequences but the *ITS2* region contains highly polymorphic sites, which makes them as a potential barcode region for the species discrimination. The length for *ITS2* and *psbA-trnH* ranged from 200 to 250 bp and 577 to 689 bp respectively. The inter specific variation between the genus ranged from 9.7 per cent (*Haniffia albiflora* Vs *Hedychium forrestii*) to 26.7 per cent (*Zingiber wrayi* vs. *Amomum villosum*) with an average of 15.2 per cent (*Curcuma aromatic* vs. *Globba radicalis*) and *Aframomum alpinum* vs. *Amomum villosum*, with an average of 2.1 per cent. The intra specific variation among the 1124 Zingiberaceae sequences ranged from 0.09 per cent (*Hedychium forrestii* vs. *Hedychium flavum*, *Boesenbergia rotunda* vs. *Boesenbergia albomaculata*) to 4 per cent (*Zingiber ellipticum* vs. *Zingiber purpureum*). In this study, we used six metrics to characterized the inter versus intra-specific variation among the five candidate loci. Relatively higher levels of intra specific divergence were observed for the gene *ITS2* when compared to other candidate genes. Similarly, the percentages of interspecific differences were greater for *ITS2* (Table 10.2 and Figures 10.3–10.5). The results of Wilcoxon signed-rank test confirmed that *ITS2* has more inter specific variation with the significant p value of 0.5273. Therefore, we have concluded that *ITS2* region of the Zingiberaceae family have optimum levels of genetic divergence between the species, hence this region can be an ideal DNA barcode candidate.

Table 10.2: Analysis of Inter-specific Divergence between Congeneric Species and intra-specific Variation among Candidate DNA Barcodes Tested

Marker	*matK*	*psbA-trnH*	*rbcL*	*ITS2*
All Inter-specific distance	0.0157±0.0308	0.037 ± 0.112	0.0110±0.0202	0.143 ± 0.431
Theta prime	0.0175 ± 0.0212	0.019 ± 0.049	0.0018± 0.0116	0.122 ± 0.146
Minimum Inter-specific distance	0.003 ± 0.0260	0.013 ± 0.085	0.0055 ±0.0183	0.090 ± 0.090
All Intra-specific distance	0.0164 ± 0.079	0.012 ± 0.812	0.0029 ± 0.0057	0.088 ± 0.180
Theta	0.010 ± 0.0352	0.009 ± 0.273	0.002 ± 0.0037	0.023 ±0.098
Coalescent depth	0.044 ± 0.799	0.081 ± 0.202	0.001 ± 0.373	0.078 ±1.876

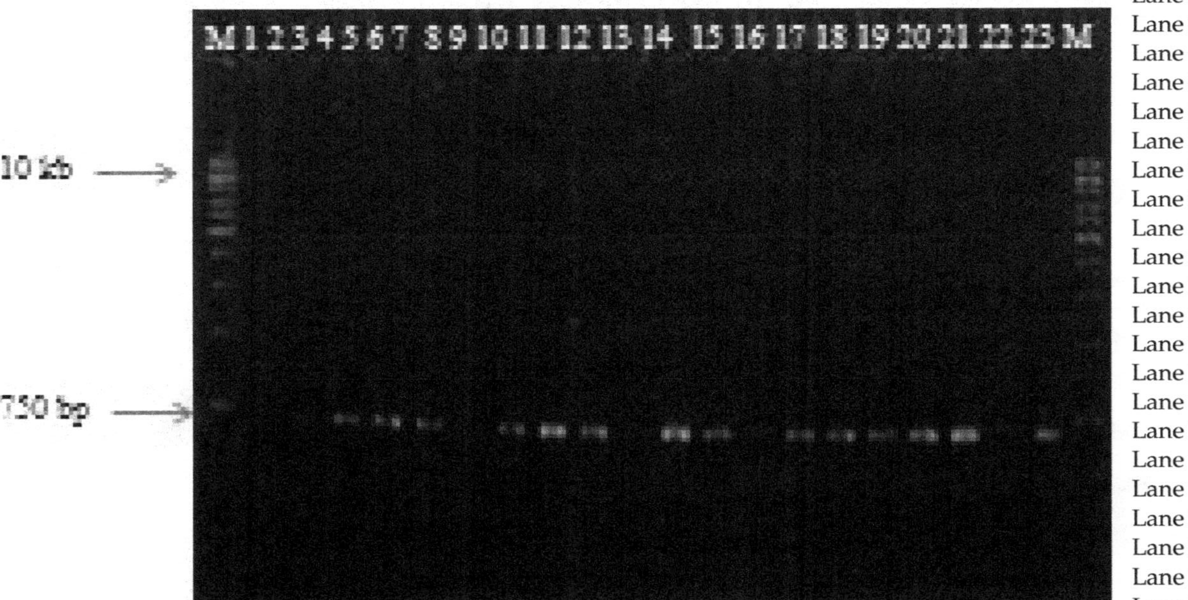

Lane 1: Marker
Lane 2: *Alpinia smithiae*
Lane 3: *Alpinia purpurata*
Lane 4: *Alpinia calcarata*
Lane 5: *Alpinia malaccensis*
Lane 6: *Alpinia vittata*
Lane 7: *Alpinia hypoleucum*
Lane 8: *Amomum subulatum*
Lane 9: *Amomum aculeatum*
Lane 10: *Curcuma aromatic*
Lane 11: *Curcuma zeodania*
Lane 12: *Curcuma caesia*
Lane 13: *Curcuma caesia*
Lane 14: *Curcuma ratakonda*
Lane 15: *Zingiber zerumpet*
Lane 16: *Zingiber purpureum*
Lane 17: *Globba schomburgkii*
Lane 18: *Etlingera elatior*
Lane 19: *Etlingera linguiformis*
Lane 20: *Hedychium flavescens*
Lane 21: *Hedychium coccineum*
Lane 22: *Hedychium coronarium*
Lane 23: *Kaempferia rotunda*
Lane 24: Marker

Figure 10.2: PCR Amplification of the *ITS* Region different Species of *Zingiberaceae*.

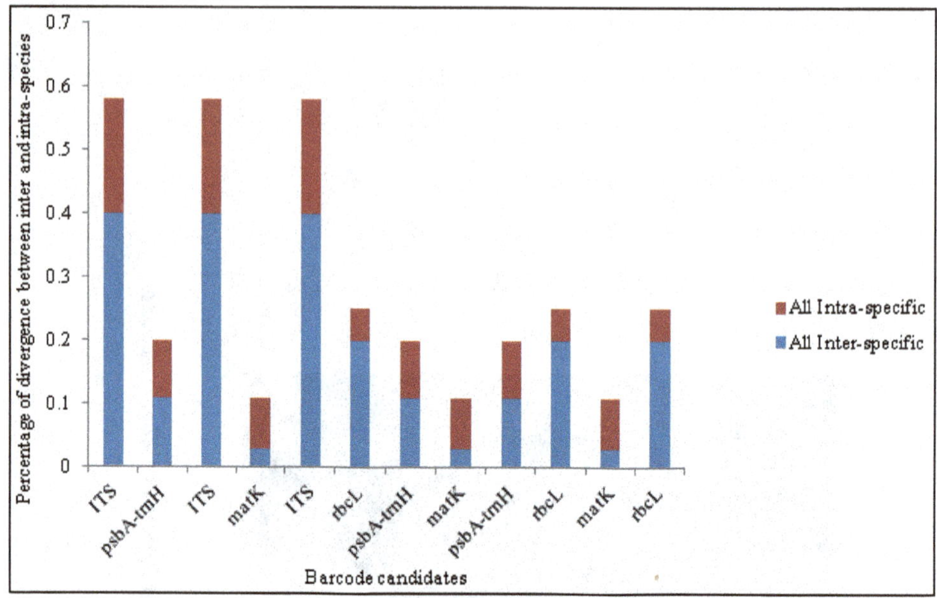

Figure 10.3: Analyses of the Inter-Specific and Intra-Specific Divergence between Congeneric Species for Four Loci.

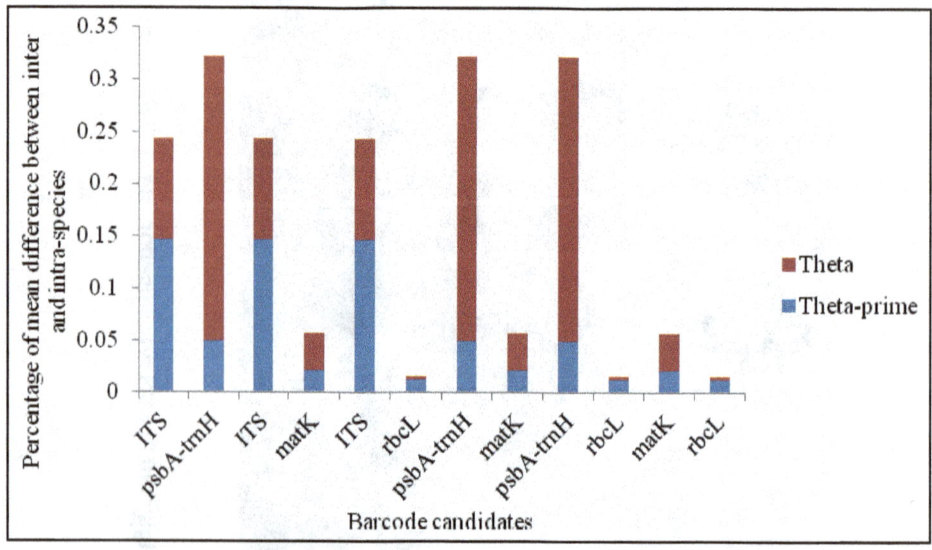

Figure 10.4: Analyses of Theta and Theta Prime for Species at Inter and Intra-species Level Divergence of Four Loci.

3.3 Assessment of DNA Barcoding Gap

In an unknown situation, genetic variation of a DNA barcode should demonstrate separate, non-overlapping distributions between intra and interspecific samples. When the number of closelyrelated species increases, the overlap of genetic variation

Plate 10.1: Plant Species of Zingiberaceae Collected from Bharathiar University Campus, Coimbatore, Tamil Nadu, India.

Alpinia calcarata | *Alpinia purpurata* | *Curcuma zeodaria*

Hedychium coronarium | *Hedychium flavescens* | *Curcuma caesia*

Globba schomburgkii | *Alpinia smithiae* | *Zingiber purpureum*

Hedychium coccineum | *Kaempferia rotunda* | *Kaempferia galangal*

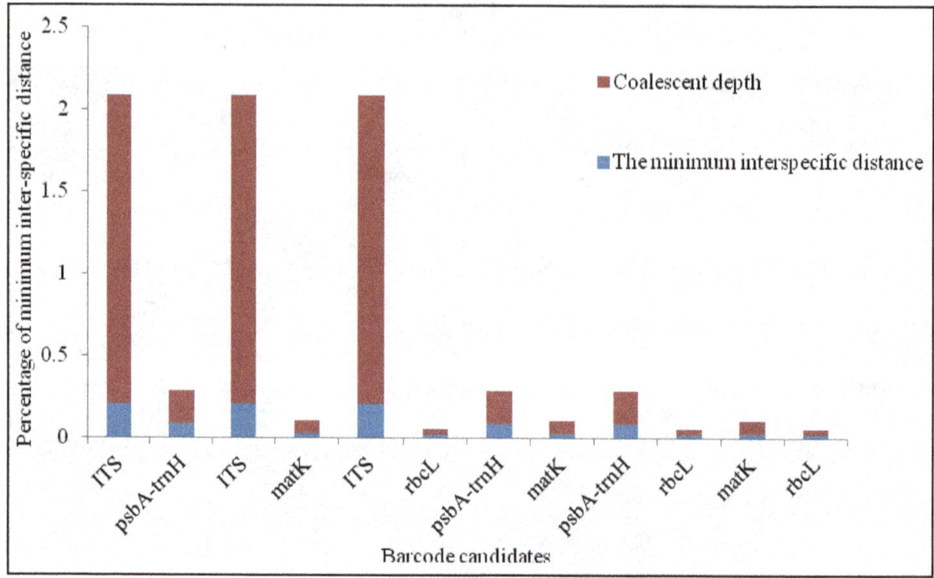

Figure 10.5: Analyses of the Inter-Specific Divergence between Congeneric Species and Intra-Specific Variation in the 4 Loci.

Table 10.3: Identification Efficiency of the Locus *ITS* Locus for the Family and Eight Large Genera using different Methods

Genus	*Length (bp)*	*Number of Species*	*Per cent of Variation*	*Method*	*Success Identification*
Aframomum	605	53	0.3	BLAST 1	89
				Distance	92.2
Boesenbergia	577	3	0.09	BLAST 1	93
				Distance	89.6
Curcuma	602	9	0.06	BLAST 1	90
				Distance	97.6
Globba	572	2	0.06	BLAST 1	92
				Distance	93
Haniffia	575	2	0.0	BLAST 1	100
				Distance	100
Hedychium	572	4	0.03	BLAST 1	98
				Distance	97
Kaempferia	689	3	0.1	BLAST 1	89
				Distance	93.3
Roscoea	569	5	0.04	BLAST 1	95
				Distance	98.8
Zingiber	450	21	0.21	BLAST 1	81.2
				Distance	88.3

without barcoding gaps significantly increases. Our results demonstrated that the distribution of intra and inter-specific variation of *psbA-trnH* and *ITS2* exhibit distinct gaps. When intra-specific variation between non-specific individuals and inter-specific divergence between all species were calculated using *matK* and *rbcL*, there was a significant overlap without gaps (Table 10.2).

3.4 Testing the Efficiency for Identification

The specific identities of the query sequences was determined using BLAST1 and the nearest genetic distance method. Identification efficiency was performed well by *ITS2* by both methods. The variations among eight large genera of zingiberaceae family were individually analyzed (Table 10.3 and Figure 10.2). The percentage of successful identification for the genus *Hedychium, Curcuma, Zingiber* was 98, 90 and 81.2 per cent respectively by BLAST 1 method. By distance method, the genus *Rosaceae, Aframomum* showed 98.8 per cent, 92.2 per cent and 88.3 per cent respectively. The rate of successful identification was around 81-97 per cent at the species level, which is relatively good when compared to the other reports. The results indicated that the BLAST1 and nearest genetic distance methods, which were used for evaluating the barcode candidates revealed that the *ITS2* can be used for a categorical identification at genus and species level when compared to other barcodes. The percentage of identification capacity has also been summarized in Table 10.3. The phylogenetic tree at the species level representing 22 taxa is show in Figure 10.6.

4.0 DISCUSSION

We explored the probability of using the putative barcodes for the medicinally important herbs of Zingiberaceae family. The gene *rbcL* showed very low polymorphic site to discriminate the species. Similarly, the gene *matK* was able to distinguish at the species level but the percentage of variation was less when compared with *psbA-trnH* and *ITS*. The results clearly showed that the nuclear DNA *ITS2* region and the chloroplast *psbA-trnH* region can distinguish all the 28 species of Zingiberaceae family and it was further confirmed by Wilcoxon signed rank test. Out of the selected four DNA barcodes tested the *ITS2* showed perfect species discrimination and hence this will be an ideal tool for the molecular taxonomists and evolutionary biologists. The other tested less effective barcodes could also have many different applications especially where the application is particularly, only for one family/genus like in the case, to trace the smuggled timbers/any plant material that can be very handy to the customs officials, forensic examiners, food processing specialists etc. (Ting *et al.*, 2010). Thus *ITS2* has a strong propensity to group the plant samples into their right genus and has a relatively high accuracy for grouping samples into their exact species. The impressive practical application is to identify the plant specimens by the individuals who don't have enough taxonomic training apart and to study the evolutionary studies (Coleman, 2003; Coleman, 2007). There are previous reports of *ITS2* proved to be effective in discriminating the plant species of Fabaceae (Xiaohui *et al.*, 2010). There was study on evaluation on 1410 plant samples representing 893 species in 96 diverse genera from the Rosaceae. *ITS2* showed discrimination of 78 per cent and 100 per cent at the species and genus levels respectively. Reports also suggest

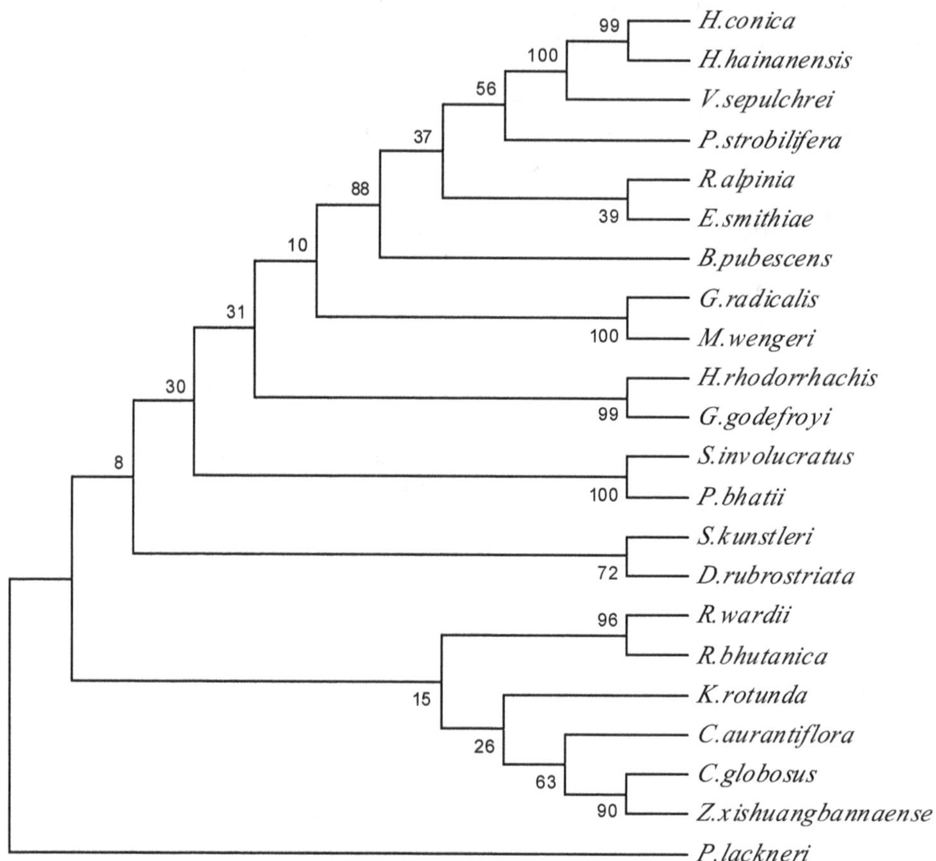

Figure 10.6: Inter-generic Tree for 22 Species of Zingiberaceae using *ITS* Region.
***Note*: Maximum likelihood tree for 21 taxa based on the *ITS* region using substitution model HKY85. Model with bootstrap values are shown above in the relevant branches. Transition/transversion ratio is 6.789 and the proportion of invariant is 0.303.**

that *ITS2* region was used for authenticating Chinese herbal medicine (Coleman, 2009).

From the above analysis, nuclear Internal Transcribed Spacer *ITS2* region and the Chloroplast *psbA-trnH* region are proved to be the most promising universal DNA barcodes for the plant family Zingiberaceae. Some of the other advantages of these candidates are, 1) the size of the *psbA-trnH* and *ITS2*, *i.e.* 500bp and 250bp respectively, which can be easily amplified using the universal primer, 2) evaluation of genetic divergence between inter and intra-species for paired loci have also been distinguished. Analysis of phylogenetic tree supports the concept that, the mean inter specific divergence of *ITS2* and *psbA-trnH* region is higher than that of intra-specific variation, 3). It can be correctly identified by using BLAST1 and nearest genetic distance methods.

In conclusion, the study on DNA barcoding across the medicinally important genus was explored for the family Zingiberaceae. Overall, our results states that *ITS2* is an important locus for differentiating the closely related species and also effective gene for phylogenetic analysis, followed by plastid gene *psbA-trnH* among the five candidate genes tested.This study will have significant impact on plant identification, resolving taxonomic disputes and especially in adulterant identification.

ACKNOWLEDGEMENTS

We wish to thank Ms. Uma, Department of Botany, School of Life Science, Bharathiar University, Coimbatore, India for providing plant materials. The authors thank the Bharathiar University, University Grants Commission- Research Fellowship for Meritorious Students (UGC RFMS), India for financial support to Dhivya Selvaraj, UGC-MRP, UGC-SAP and DST-FIST for funding the Plant Genetic Engineering lab.

REFERENCES

Bhuchan, *I.E.*, 2000. The development of a statistical computer software resource for medical research [Dissertation], University of Liverpool 322p. England.

Barrett, D.H., Hebert, P.D.N., 2005. Identifying spiders through DNA barcodes. *Can. J Zoo* 83: 505–506.

Chase, M.W., Salamin, N., Wilkinson, M., Dunwell, J.M., Kesanakurthi, R.P., Haidar, N., Savolainen, V., 2005. Land plants and DNA barcodes: short-term and long-term goals. *Philos Trans R Soc. Lond Biol. Sci.* 360: 1889–1895.

Chase, M.W., Cowan, R.S., Hollingsworth, P.M., Berg, C.V., Madrinan, S., Petersen, G., Seberg, O., Jorgsensen, T., Cameron, K.M., Carine, M., Pedersen, N., Hedderson, T.A.J., Conrad, F., Salazar, G.A., Richardson, J.E., Hollingsworth, M.L., Barraclough, T.G., Kelly, L., Wilkinson, M., 2007. A proposal for a standardized protocol to barcode allland plants. *Taxon* 56, 295–299.

Chen, S., Yao, H., Han, J., Liu, C., Song, J., Shi, L., Zhu, Y., Ma, X., Gao, T., Pang, X., Luo, K., Li, Y., Li, X., Jia, X., Lin, Y.L., Leon, C., 2010. Validation of the ITS2 Region as a novel DNA barcode for identifying medicinal plantspecies. PLoS One: e8613.

Chiou, S.J., Yen, J.H., Fang, C.L., Chen, H.L., Lin, T., 2007. Authentication of medicinal herbs using PCR- amplified ITS2 with specific primers. *Planta Med* 73, 1421-1426.

Coleman, A.W., 2003. ITS2 is a double-edged tool for eukaryote evolutionary comparisons. *Trends Genet* 19, 370-375.

Coleman, A.W., 2007. Pan-eukaryote ITS2 homologies revealed by RNA secondary structure. *Nucleic Acids Res* 35, 3322-3329.

Coleman, A.W., 2009. Is there a molecular key to the level of biological species in eukaryotes: A DNA guide. *Mol Phylogenet Evol* 50, 197-203.

Dereeper, A., Audic, S., Claverie, J.M., Blanc, G., 2010. BLAST-EXPLORER helps you building datasets for phylogenetic analysis. *BMC Evol. Biol.* 10, 1-8.

Edwards, D., Horn, A., Taylor, D., Savolainen, V., Hawkins, J.A., 2008. DNA barcoding of a large genus *Aspalathus* L. Fabaceae. *Taxon* 57, 1317–1327.

Felsenstein, J., 1985. Confidence limits on phylogenies: an approach using the bootstrap. *Evol* 39, 783–791.

Ferri, G., Alu, M., Corradini, B., Beduschi, G., 2009. Forensic botany: species identification of botanical trace evidence using a multigene barcoding approach. *Int J Legal Med* 123, 395- 401.

Gamble, J.S., 1987. Flora of the presidency of Madras. Bishen Singh Mahendra Pal Singh, Dehra Dun, India 3:1478-1493.

Gao, T., Yao, H., Song, J., Liu, C., Zhu, Y., Ma, X., Pang, X., Xu, H., Chen, S., 2010. Identification of medicinal plants in the family Fabaceae using a potential DNA barcode ITS2. *J Ethnopharmacol* 120, 116-121.

Jianping, H.,Yingjie, Z., Xiaochen, C., Baoshen, L., Hui, Y., Jingyuan, S., Shilin, C., Fanyun, M., 2013. The Short *ITS2* Sequence Serves as an Efficient Taxonomic Sequence Tag in Comparison with the Full-Length ITS. *Bio Med Research International*. http://dx.doi.org/10.1155/2013/741476.

Skornickova, J.L., Sida, O., Jarolimova, V., Sabu, M., Fer, T., Travnicek, P., Suda, J., 2007. Chromosome Numbersand Genome Size Variation in Indian Species of *Curcuma* (Zingiberaceae). *Ann Bot* 100, 505-526.

Kavitha, P.G., Kiran, A.G., Dinesh Raj, R., Sabu, M., Thomas, G., 2010. Amplified fragment length polymorphism analyses unravel a striking difference in the intraspecific genetic diversity of four species of genus *Zingiber* Boehm from the Western Ghats, south India. *Curr Sci* 98, 242 – 247.

Kim, Y.D., Kim, S.H., Landrum, L.R., 2004. Taxonomic and phytogeographic implications from *ITS* phylogeny in Berberis (Berberidaceae). *J Plant Res* 117, 175–182.

Kress, S.G., Erickson, D.L., 2007. A two-locus global DNA barcode for land plants: the coding rbcL gene complements the non-coding trnH-psbA spacer region. PLoS ONE 2: e508.

Kress, W.J., Wurdack, K.J., Zimmer, E.A., Weigt, L.A., Janzen, D.H., 2005. Use of DNA barcodes to identifyflowering plant. *Proc of Natl Acad Sci* 106, 8369–8374.

Lahaye, R., Vander, B.M., Bogarin, D., Warner. J., Pupulin, F., 2008. DNA barcoding the floras of biodiversity hotspots. *Proc of Natl Acad Sci* 8, 2923–2928.

Newmaster, SG, Fazekas AJ, Ragupathy. S., 2006. DNA barcoding in the land plants evaluation of *rbcL* in a multigene tiered approach. *Can J Bot* 84, 335–341.

Kandil, O.M., Mahmoud, M.S., Allam, N.A.T., Namaky, A.H.E., 2010. Mitochonderial Cytochrome C Oxidase Subunit 1 (*cox 1*) Gene Sequence of the *Hymenolepis* Species. *J Am Sci* 6, 1346 -1353.

Pang, X., Song, J., Zhu, Y., Xu, H., Huang, L., Chen, S., 2011.Applying plant DNA barcodes for Rosaceae species identification. *Cladistics* 27, 165–170.

Ross, H.A., Murugan, S., Li, W.L., 2008. Testing the reliability of genetic methods of species identification via simulation. *Syst Biol* 57, 216-230.

Taberlet, P., Coissac, E., Pompanon, F., Gielly, L., Miquel, C., Valentini, A., Vermat, T., Corthier, G., Brochmann, C., Willerslev, E., 2007. Power and limitations of the chloroplast trnL (UAA) intron for plant DNA barcoding. *Nucleic Acids Res* 35, *e*14.

Tamura, K., Dudley, J., Nei, M., Kumar, S., 2007. MEGA 4: Molecular Evolutionary Genetics Analysis (MEGA) software version 4.0. *Mol Biol Evol* 8, 1596-1599.

Techaprasan, J., Ngamriabsakul, C., Klinbunga, S., Chusacultanachai, S., Jenjittikul, T., 2008. Genetic variation species identification of Thai *Boesenbergia* (Zingiberaceae) analyzed by chloroplast DNA polymorphism. *J Bio chem. Mol Biol* 39, 361-370.

Thompson, J.D., Higgins, D.G., Gibson, T.J., 1994. CLUSTAL W: Improving the sensitivity of progressive multiple sequence alignment through sequence weighting, position-specific gap penalties and weight matrixchoice. *Nucleic Acids Res* 22, 4673- 4680.

Wang, P.X., Huang, F., Zhou, L., Cao, L.Y., Liang, R.Y., Xu, H.H., Liu, J.M., 2000. Analysis of *Amomum villosum* species and some adulterants of Zingiberaceae by RAPD. *Zhong Yao Cai* 23, 71-74.

Yao, H., Song, J.Y., Ma, X.Y., Liu, C., Li, Y., Xu, H.X., Han, J.P., Duan, L.S., Chen, S.L., 2009. Identification of *Dendrobium* Species by a Candidate DNA Barcode Sequence: The Chloroplast *psbA-trnH* Intergenic Region. *Planta Med* 75, 667–669.

2015, Modern Methods in Phytomedicine *Pages* ***171–190***
Editor: **T. Parimelazhagan**
Published by: **DAYA PUBLISHING HOUSE, NEW DELHI**

11

Review on Conventional Medicinal Plants for Wound Healing

Sembian Suriyamoorthy and Kalidass Subramaniam*

Department of Biotechnology, School of Biotechnology and Health Sciences, Karunya University, Coimbatore – 641 114, Tamil Nadu

1.0 INTRODUCTION

Higher plants have provided the basic necessities of life to human beings from the very beginning of human civilization. Although there has been considerable development in the areas of synthetic drug chemistry and antibiotics, plant still occupy an important place in modern as well as traditional system of medicine all over the world. Because of the growing realization of the toxic effects of the synthetic drugs and antibiotics, herbal medicine and health foods, derived from plants are becoming more important not only in developing countries, but also in the superpower countries. It is expected that in the future, plant product will play a major role in the health care programmes of all countries in the world.

India is one among the top 12 mega bio-diversity centres of the world, due to immense variety of climatic and altitudinal condition coupled with varied ecological habitat. India, with her unique geographical position is endowed with a variety of medicinal plants. Knowledge of the healing properties of herbs was known to ancient

* *Corresponding Author.* E-mail: kalidass@karunya.edu

Indians and medicinal plants were given a prominent place in the ancient systems of medicine an intensive study of indigenous drug plants and their therapeutic potentials will certainly unearth many more useful remedies and widen the scope of the traditional systems of India (Kokate *et al.*, 1997 and Faraz *et al.*, 2003).

Medicinal plants are rich in secondary metabolites such as alkaloids, glycosides, flavonoids, Steroids, etc. (Rotblatt and Ziment, 2002). The medicinal property is due to the presence of these chemical substances which may have some definite physiological action on the human body (Gupta and Varshneya, 1975).

The screening of plant extracts and plant products for anti-microbial activity has shown that higher plants represent a potential source of novel antibiotic prototypes (Afolayan, 2003). The isolation, characterization of bio active compounds begins with the general screening of plants to identify those with bioactivity against pathogenic organisms (Oyewale *et al.*, 2004).

The genetic ability of pathogenic bacteria to develop resistance against commonly used antibiotics is a major medical problem and challenge worldwide, posing a big threat to human society (Sathish, 2009). This has necessitated a search for novel anti-bacterial substances from various natural sources, including flowering plants. Some plants have shown the ability to overcome resistance in such organisms which led the researchers to isolate active principles and investigate mechanisms. The use of plant extracts and the phytochemicals can be of great significance in therapeutic treatments and could be helpful to curb the problem of these multi-drug resistant microorganisms (Basso *et al.*, 2005).

New medicines have been discovered with traditional, empirical and molecular approaches. The traditional approach makes use of material that has been found by trial and error over many years in different cultures and system of medicine. With the development of molecular biological techniques and the advances in genomics, the majority of drug discovery is currently based on the molecular approach.

The major advantages of natural product for random screening is the structural diversity provided by most available combinatorial approaches based on heterocyclic compounds. Bioactive natural products often occur as a part of family of related molecules so that it is possible to isolate a number of homologues and obtain structure-activity information (Prashant *et al.*, 2011). Of course, lead compounds found from screening of natural products can be optimised by traditional medicinal chemistry or by application of combinatorial approaches. Overall, when faced with molecular targets in screening assays for which there is no information about low molecular weight leads, use of a natural products library seems more likely to provide the chemical diversity to yield a hit than a library of similar numbers of compounds made by combinatorial synthesis.

Plants are viable for medicine in basic 5 ways:

1. Plants are used as sources of direct medicinal agents.
2. Plants serve as raw material base for elaboration of more complex semi-synthetic chemical compounds.

3. Plants chemical structure derived from phytoconstituents can be used as models for new synthetic compounds.
4. Plants can be taxonomic markers for discovery of new therapeutic compounds.
5. Important plant secondary metabolites such as glycosides, flavonoids, lignins, terpenoids and alkaloids have been isolated from plants. Since the use of botanicals and herbal remedies has increased dramatically in the last several years, the numbers of researchers on the isolation and identification of compounds from plants have also been increased.

Wounds have affected humans since pre-historic times and the treatment and healing of wounds is an art as old as humanity (Robson *et al.,* 2001). Due to the increasing life expectancy coupled with a more modern way of life, wounds and particularly chronic wounds increasingly affect a growing number of elderly patients and seriously reduce their quality of life. Research on wound healing drugs is a rapidly developing area in modern biomedical sciences. The progress in this field has allowed the synthesis of large numbers of molecules associated with wound repair process. Delivery of exogenous growth factors in order to mimic the natural microenvironments of tissue formation and repair is believed to be therapeutically effective. Despite finding new methods of stimulation of the wound repair process, wound care has returned to the roots of medicine and is embracing some of the remedies used millennia ago.

Plant-derived natural products are significant as sources of medicinal agents and models for the design of new remedies. As plants are a source of many bioactive compounds and many plant ingredients are traditionally used to accelerate healing. Scientists go back to traditional folk medicines as they are generally characterized by high acceptability and good toleration (Jagetia *et al.,* 2004). The healing potential of phytomedicines is often associated with angiogenesis, which is a critical step of wound healing (Sagar *et al.,* 2006). It is the essential part of the repair process as it enables the nutrient supply to sustain cell metabolism, creates an intact delivery system, and facilitates the clearance of debris. The healing efficacy seen in phytomedicine treated wounds shows great promise although for most natural products no well controlled scientific data are available.

2.0 METHODS OF EXTRACTION OF DRUG PRINCIPLES

Extraction methods used pharmaceutically involves the separation of medicinally active portions of plant tissues from the inactive/inert components by using selective solvents. During extraction, solvents diffuse into the solid plant material and solubilize compounds with similar polarity (Ncube *et al.,* 2008). The purpose of standardized extraction procedures for crude drugs (medicinal plant parts) is to attain the therapeutically desired portions and to eliminate unwanted material by treatment with a selective solvent known as menstrum. The extract thus obtained, after standardization, may be used as medicinal agent as such in the form of tinctures or fluid extracts or further processed to be incorporated in any dosage form such as tablets and capsules. These products contain complex mixture of many medicinal plant metabolites, such as alkaloids, glycosides, terpenoids, flavonoids, phenols and lignans (Handa *et al.,* 2008).

The general techniques of medicinal plant extraction include maceration, infusion, percolation, digestion, decoction, hot continuous extraction (Soxhlet), aqueous-alcoholic extraction by fermentation, counter- current extraction, microwave-assisted extraction, ultrasound extraction (sonication), supercritical fluid extraction, and phytonic extraction (with hydro fluorocarbon solvents). For aromatic plants, hydro distillation techniques (water distillation, steam distillation, water and steam distillation), hydrolytic maceration followed by distillation, expression and enfleurage (cold fat extraction) may be employed. Some of the latest extraction methods for aromatic plants include headspace trapping, solid phase micro- extraction, protoplast extraction, micro distillation, thermo micro distillation and molecular distillation (Handa *et al.,* 2008).

2.1 Choice of Solvents

Successful determination of biologically active compounds from plant material is largely dependent on the type of solvent used in the extraction procedure. Properties of a good solvent in plant extractions includes, low toxicity, ease of evaporation at low heat, promotion of rapid physiologic absorption of the extract, preservative action, inability to cause the extract to complex or dissociate. The factors affecting the choice of solvent are quantity of phytochemicals to be extracted, rate of extraction, diversity of different compounds extracted, diversity of inhibitory compounds extracted, ease of subsequent handling of the extracts, toxicity of the solvent in the bioassay process, potential health hazard of the extractants (Eloff 1998). The choice of solvent is influenced by what is intended with the extract. Since the end product will contain traces of residual solvent, the solvent should be non- toxic and should not interfere with the bioassay. The choice will also depend on the targeted compounds to be extracted (Table 11.1) (Ncube *et al.,* 2008, Das *et al.,* 2010).

Table 11.1: Solvents Used for Phytochemical Extraction

Water	*Ethanol*	*Methanol*	*Chloroform*	*Ether*	*Acetone*
Anthocyanins	Tannins	Anthocyanins	Terpenoids	Alkaloids	Phenol
Starches	Polyphenols	Terpenoids	Flavonoids	Terpenoids	Flavonols
Tannins	Polyacetylenes	Saponins		Coumarins	
Saponons	Flavonol	Tannins		Fatty acids	
Terpenoids	Terpenoids	Xanthoxyllines			
Polypeptides	Sterols	Totarol			
Lectins	Alkaloids	Quassinoids			
		Lactones			
		Flavones			
		Phenones			
		Poyphenols			

3.0 PHYTOCHEMICALS

For many decades, the use of synthetic chemicals as drugs has been effective in the treatment of most diseases. Moreover, from ancient to modern history, many traditional plant based medicines are playing an important role in health care. Phytochemicals are natural bioactive compounds found in vegetables, fruits, medicinal plants, aromatic plants, leaves, flowers and roots which act as a defense system to combat against diseases. Phytochemicals appear to have significant physiological effects in the body. The phytochemicals from natural products cover a diverse range of chemical entities such as polyphenols, flavonoids, steroidal saponins, organosulphur compounds and vitamins. A number of bioactive compounds generally obtained from terrestrial plants such as isoflavones, diosgenin, resveratrol, quercetin, catechin, sulforaphane, tocotrienols and carotenoids are proven to reduce the risk of human diseases especially wounds which is a great menace globally. Some phytochemicals work alone, others work in combination, and some seem to work in conjunction with other nutrients in food, such as vitamins.

The wound healing effects of the various phytochemicals are perhaps due to their anti-oxidative, anti-microbial, inhibition of platelet aggregation and anti-inflammatory activities that reduce the risk of wound disorders. The multi-faceted role of the phytochemicals is mediated by its structure-function relationship and can be considered as leads for wound healing drug design in future. Table 11.2 summarizes some important phytochemicals, their role in human health.

Table 11.2: Mechanism of Action of some Phytochemicals

Phytochemicals	*Activity*	*Mechanism of Action*
Quinones	Anti-microbial	Binds to adhesins, complex with cell wall, inactivates enzymes.
Flavonoids	Anti-microbial	Complex with cell wall, binds to adhesins.
	Anti-diarrhoeal	Inhibits release of autocoids and prostaglandins, Inhibits contractions caused by spasmogens, Stimulates normalization of the deranged water transport across the mucosal cells, Inhibits GI release of acetylcholine.
Polyphenols and Tannins	Anti-microbial	Binds to adhesins, enzyme inhibition, substrate deprivation, complex with cell wall, membrane disruption, metal ioncomplexation.
	Anti-diarrhoeal	Makes intestinal mucosa more resistant and reduces secretion, stimulates normalization of deranged water transport across the mucosal cells and reduction of the intestinal transit, blocks the binding of B subunit of heat-labile enterotoxin to GM 1, resulting in the suppression of heat-labile enterotoxin-induced diarrhea, astringent action.
	Anthelmintic	Increases supply of digestible proteins by animals by forming protein complexes in rumen, interferes with energy generation by uncoupling oxidative phosphorylation, causes a decrease in G.I. metabolism.
Coumarins	Anti-viral	Interaction with eucaryotic DNA.

Contd...

Table 11.2–*Contd...*

Phytochemicals	*Activity*	*Mechanism of Action*
Terpenoids and	Anti-microbial	Membrane disruption.
essential oils	Anti-diarrhoeal	Inhibits release of autocoids and prostaglandins.
Alkaloids	Anti-microbial	Intercalates into cell wall and DNA of parasites.
	Anti-diarrhoeal	Inhibits release of autocoids and prostaglandins.
	Anthelmintic	Possess anti-oxidating effects, thus reduces nitrate generation which is useful for protein synthesis, suppresses transfer of sucrose from stomach to small intestine, diminishing the support of glucose to the helminthes, acts on CNS causing paralysis.
Lectins and Polypeptides	Anti-viral	Blocks viral fusion or adsorption, forms disulfide bridges.
Glycosides	Anti-diarrhoeal	Inhibits release of autocoids and prostaglandins.
Saponins	Anti-diarrhoeal	Inhibits histamine release *in vitro.*
	Anti-cancer	Possesses membrane permeabilizing properties.
	Anthelmintic	Leads to vacuolization and disintegration of teguments.
Steroids	Anti-diarrhoeal	Enhance intestinal absorption of Na^+ and water.

Courtesy: Prashant *et al.* (2011) Phytochemical screening and Extraction: A Review.

3.1 Anti-microbial Agents

Infectious diseases, particularly skin and mucosal infections, are common in most of the tribal inhabitants due to lack of sanitation, potable water and awareness of hygienic food habits. An important group of these skin pathogens are the fungi, among which dermatophytes and *Candida* spp., besides certain pathogenic bacteria are the most frequent (Caceres *et al.,* 1993; Desta, 1993). Furthermore, in the last few years, the number of immuno supressed and immuno compromised patients, who frequently develop opportunistic systemic and superficial mycoses (Rahalison *et al.,* 1994; Li *et al.,* 1995). This is mainly due to the non-availability of effective anti-fungal drugs for systemic fungal infections and toxicity of available drugs like amphotericin-B (Maddux and Brarriere, 1980; Saral, 1991). Thus there is an increased need for the development of alternative anti-pathogenic substances. One possible approach is to screen local medicinal plants in search of suitable chemotherapeutic anti-bacterial and anti-fungal substances, many such anti-microbials have been identified. The herbalists prescribed various preparations of medicinal plants in treating ailments such as itch, eczema, scabies and skin diseases (Chopra *et al.,* 1992, Behl *et al.,* 1993, Iyengar *et al.,* 1997).

4.0 ANTIOXIDANTS FROM PLANTS

An anti-oxidant is any substance that, when present at low concentrations significantly delays or prevents oxidation of cell content. In nature there are a wide variety of naturally occurring anti-oxidants which are different in their composition, physical and chemical properties, mechanisms and site of action. Plants having vitamins, flavonoids, anthocyanins and polyphenols are reported to possess remarkable anti-oxidant activity. Anti-oxidant activity is neither restricted to a particular part of the plant nor the specific families.

5.0 WOUND HEALING

A wound is a break in the epithelial integrity of the skin and may be accompanied by disruption of the structure and function of underlying normal tissue.

5.1 Healing of Acute Wounds

Healing of acute wounds (as seen in primary healing) occurs as a carefully regulated, systemic cascade of overlapping processes that require the coordinated completion of a variety of cellular activities, including phagocytosis, chemotaxis, mitogenesis, and synthesis of components of the extracellular matrix. These activities occur in a cascade that correlates with the appearance of different cell types in the wound during various stages of the healing process. These processes (triggered by tissue injury) involve four overlapping (but well-defined) phases of haemostasis, inflammation, proliferation and remodelling and scar maturation (Figure 11.1). The regulation of these events is multifactorial.

The aim of wound treatment is to keep the wounded surface away from microbial infections and to shorten the period of epithelialization. Table 11.3 shows some of the available anti-biotics for wound healing and their mode of action.

6.0 DRAWBACKS OF ANTIBIOTICS IN WOUND HEALING

Despite the advantages of available antibiotics for wound healing, there are drawbacks in antibiotics used for wound healing. They are:

1. Significant over use of antibiotics in the treatment of acute and chronic wounds.
2. Antibiotics show selectivity only for specific microorganisms. Chemical compounds that either kill or inhibit growth of bacteria not viruses or fungi.
3. Spectrum of action varies from compound to compound.
4. Anti-septics are generally non selective, potentially damage all cells on contact need to carefully evaluate the use of all chemical agents used in wound management.
5. Anti-bacterial agents prevent or treat infections and can aid in wound healing, they do not necessarily take an active physiological part in the wound healing process (Komarcevic, 2000).
6. Much of the information about the negative effects of antibiotics on wound healing involves topical application.
7. Limitations in the topical application of antibiotics:
   - ☆ No penetration of the wound bed by topical application.
   - ☆ Only treats surface infection.
   - ☆ Inhibit contraction and delay reepithelialization.
   - ☆ Not dose controlled.

   Frequent and inappropriate prescription.

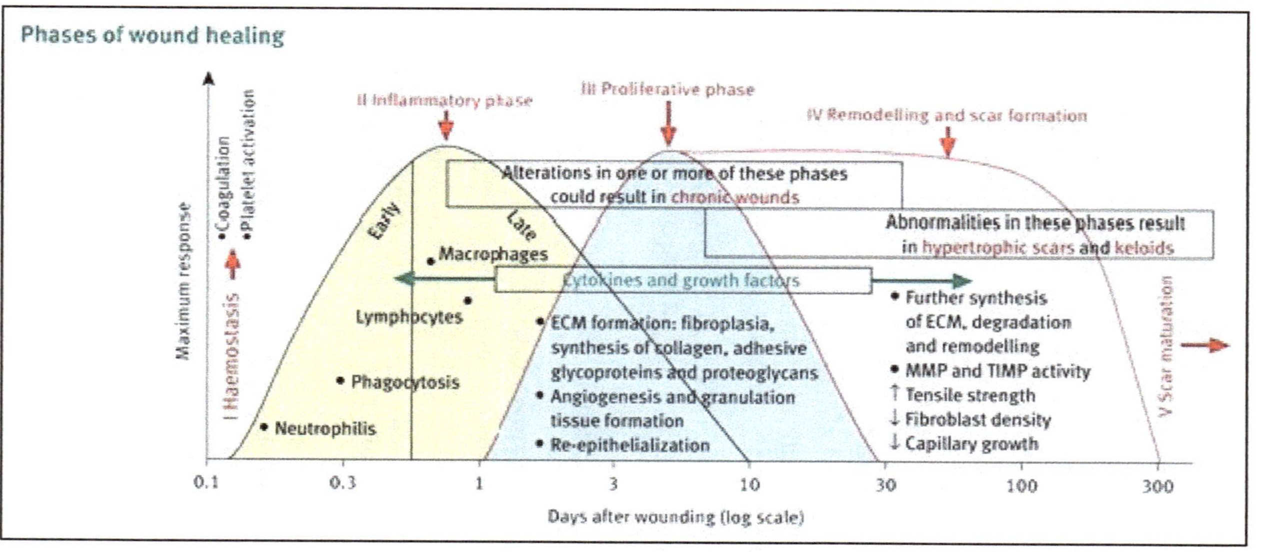

Figure 11.1: Phases of Wound Healing.

ECM: Extracellular matrix; MMP: Metalloproteinases; TIMP: Tissue inhibitors of Metalloproteinases. (Adapted from: Clark, 1991).

Table 11.3: Spectrum of Coverage and Mechanisms of Action of Topical Anti-bacterial Agents/Drugs Used for Wound Care

Name	*Bacterial Coverage*	*Mechanism of Action*	*Origin*
Bacitracin	Bactericidal against Gram (+) and *Neisseria* species	Interferes with bacterial wall synthesis; occurs by inhibition of phospholipid receptors involved in peptidoglycan synthesis	*Licheniformis* group of *Bacillus subtilis* var. Tracey I
Polymyxin B	Bactericidal against Gram (–) bacteria only; effective against *P. aeruginosa*	Increases permeability of bacteria cell membrane; occurs by interacting with phospholipid components of membrane	*Bacillus polymyxa* *Bacillus subtilis*
Neomycin	Bactericidal against Gram (+) and Gram (–) bacteria; good *S. aureus* coverage	Inhibits protein synthesis; occurs by binding to 30s subunit of ribosomal RNA; end result is misreading of bacterial genetic code	Aminoglycoside antibiotic derived from S*treptomyces fradiae*
Mupirocin	Bactericidal against methicillin-resistant *S. aureus*; *S. pyrogenes*	Inhibits bacterial RNA and protein synthesis; occurs by reversibly binding to bacterial isoleucyl transfer RNA synthetase	*P. fluorescens*
Retapamulin	Bacteriostatic against *S. pyogenes,* mupirocin-resistant and methicillin-resistant *S. aureus, anaerobes*	Inhibits bacterial protein synthesis; occurs by binding to protein L3 on 50s ribosomal subunit	Pleuromutilin antibiotic derived from *Clitopilus scyphoides*
Gentamicin	Bactericidal against Gram (+) and Gram (–) organisms; coverage includes *P. aeruginosa*	Inhibits bacterial protein synthesis; occurs by irreversibly binding to 30s ribosomal subunits	Aminoglycoside antibiotic derived from *M. purpurea*
Silver sulfadiazine	Bactericidal against Gram (+) and Gram (–) organisms	Binds to bacteria DNA and inhibits its replication	Synthesized from reaction of silver nitrate and sodium sulfadiazine
Iodoquinol	Active against Gram (+) and Gram (–) organisms	Unknown	Synthetic halogenated derivative of quinolone

On considering the above drawbacks on antibiotics in wound healing, researchers are now in search of new naturally available phytochemicals and anti-oxidants from plants as antibiotics for the treatment of wound healing.

7.0 PLANTS IN WOUND HEALING

The search for "natural remedies" for a commonly occurring disorder such as wounds has drawn attention to herbals. From ancient times, herbs have been routinely used to treat wounds, and in many cultures their use in traditional medicine has persisted to the present. Some time-tested herbal remedies are indeed effective.

There are numerous herbal derivatives that have been tried for their ability to promote wound healing. Most studies are purely observational in nature; a few others have attempted to address the underlying mechanisms (Table 11.4). Plant extracts have been shown to favorably influence collagen metabolism and support wound healing. Topical administration of herbal derivatives were reported to accelerated granuloma maturation, increases fibroblast and endothelial cell growth. The extract increases expression of several components of the adhesion complex and fibronectin by human keratinocytes.

8.0 SUMMARY

Wound healing is a complex process and differs pathologically making it complex to discuss. The readings were obtained for different parameters from different groups. Plant extracts when topically applied on rats that showed convincing wound healing suggesting that they enhances various stages of the healing process. Secondary metabolites that serve as the defensive agents are produced under stressed condition by the plants. Secondary metabolites produced by the plants have been used to find new therapeutic properties thereby serve as drug molecules; these plant origin compounds have historically served as templates for the development of many important classes of drugs (Fakim, 2006). These phytoconstituents include various chemical families like alkaloids, essential oils, flavonoids, tannins, terpenoids, saponins and phenolic compounds. The wound healing action of the plant extracts may be probably due to the synergistic or individual activity of the phytoconstituents present. Earlier literature reports on plant extracts have shown that phytochemical constituents like flavonoids (Tsuchiya *et al.*, 1996), triterpenoids (Scortichini *et al.*, 1991) and tannins (Rane *et al.*, 2003) are known to enhance the wound healing process. It appears that different mechanisms like free radical scavenging, metal chelation as well as immune modulation by plant extracts may act at different levels individually or in combination to bring about the wound healing effects of medicinal plants (Somashekar *et al.*, 2008).

Wound healing is a natural process which could be delayed by reactive oxygen species and/or by microbial infection (Bodeker and Hughes, 1998). In recent years, oxidative stress has been implicated in a variety of degenerative processes and diseases; these include acute and chronic inflammatory conditions such as wounds (Soliman and Ibrahim, 2011). Wound healing process includes cell proliferation, suppression of inûammation and contraction of the collagen tissue (Houghton *et al.*, 2005). Anti-oxidant activity helps in the release of oxygen radicals, thus controlling

Table 11.4: Plants Reported for Wound Healing Activity

Sl.No.	*Name of the Plant*	*Parts Used*	*Solvent Used for Extraction*	*Animal Model*	*Compound Responsible*	*Reference*
1.	*Aristolochia bracteolata*	Leaves	Ethanol	Wistar Rats	Flavonoids, Alkaloids	Shirwaikar *et al.*, 2003
2.	*Tragia involucrata*	Root	Methanol	Wistar Rats	Phenolics, Flavonoids and tannins	Perumal Samy *et al.*, 2006
3.	*Pterocarpus santalinus*	Stem	Ethanol	Charles Foster rats	Santalin A Isoflavones, Triterpenes	Tuhin *et al.*, 2004
4.	*Cassia fistula*	Leaves	Ethanol	Wistar Rats	Not Reported	Muthusamy *et al.*, 2005
5.	*Centella asiatica*	Whole plant	Not Reported	Guinea pigs and Sprague Dawley Rats	Asiaticoside	Shukla *et al.*, 1999
6.	*Butea monosperma*	Bark	Ethanol	Wistar Rats	Not Reported	Miriyala *et al.*, 2005
7.	*Ixora coccinea*	Flowers	Ethanol	Wistar Rats	Not Reported	Nayak *et al.*, 1999
8.	*Celosia argentea*	Leaves	Ethanol	Wistar Rats	Flavanoids and Vitamins	Kulasekaran *et al.*, 2004
9.	*Sphaeranthus amaranthoides*	Whole plant	Diethyl ether, Acetone, Chloroform, Methanol and Water	Wistar Rats	Not Reported	Geethalakshmi *et al.*, 2013
10.	*Ocimum sanctum*	Leaves	Ethanol	Wistar Rats	Flavonoids	Somashekar *et al.*, 2007
11.	*Spathodea campanulata*	Bark	Methanol	Sprague Dawley rats	Not Reported	Kwabena *et al.*, 2011
12.	*Blechnum orientale*	Leaves	Methanol	Sprague Dawley rats	Phenols, Flavonoids and Tannins	How *et al.*, 2011
13.	*Acanthus ebracteatus*	Stem	Ethanol	Balb/c mice	ß-sitosterol	Jutamas *et al.*, 2012
14.	*Elaeis guineensis*	Leaf	Methanol	Sprague Dawley rats	Not Reported	Sreenivasan *et al.*, 2012

Contd...

Table 11.4–*Contd...*

Sl.No.	Name of the Plant	Parts Used	Solvent Used for Extraction	Animal Model	Compound Responsible	Reference
15.	*Jatropha curcas*	Seed	Hexane	Wistar Rats	Oleic acid, Linoleic acid, Palmitic acid, and Stearic acid	Jose *et al.*, 2012
16.	*Ageratum conyzoides*	Roots	Ethanol	Wistar Rats	Alkaloidsand Saponins	Jain *et al.*, 2009
17.	*Ficus religiosa*	Rhizome	Ethanol	Wistar Rats	Alkaloidsand Saponins	Jain *et al.*, 2009
18.	*Curcuma longa*	Stem, Bark	Ethanol	Wistar Rats	Alkaloidsand Saponins	Jain *et al.*, 2009
19.	*Tamarindus indica*	Leaves	Ethanol	Wistar Rats	Alkaloidsand Saponins	Jain *et al.*, 2009
20.	*Ocimum gratissimum*	Leaves	Methanol	Wistar Rats	Not Reported	Chah *et al.*, 2006
21.	*Psidium guajava*	Leaves	Methanol	Wistar Rats	Not Reported	Chah *et al.*, 2006
22.	*Anthocleista djalonensis*	Root	Methanol	Wistar Rats	Not Reported	Chah *et al.*, 2006
23.	*Napoleonaea imperialis*	Leaves	Methanol	Wistar Rats	Not Reported	Chah *et al.*, 2006
24.	*Quercus infectoria*	Galls	Ethanol	Wistarrats	Phenols,Tannins, Gallic acid, Ellagic acid, Syringic acid, ß-sitosterol and Amentoflavone	Umachigi *et al.*, 2008
25.	*Heliotropium indicum*	Whole plant	Ethanol	Wistar rats	Not Reported	Suresh *et al.*, 2002
26.	*Plumbago zeylanicum*	Whole plant	Ethanol	Wistar rats	Not Reported	Suresh *et al.*, 2002
27.	*Acalypha indica*	Whole plant	Ethanol	Wistar rats	Not Reported	Suresh *et al.*, 2002
28.	*Curcuma aromatica*	Rhizome	Ethanol	Swiss albino mice	Not Reported	Amit *et al.*, 2009
29.	*Cordia dichotoma*	Fruit	Ethanol	Wistar rats	Flavonoids	Kuppast and Nayak 2005
30.	*Lantana camara*	Leaves	Ethanol	Sprague Dawley rats	Not Reported	Nayak *et al.*, 2008
31.	*Spilanthes acmella*	Aerial parts	Water	Wistar rats	Flavonoids	Chakraborty *et al.*, 2003
32.	*Pongamia pinnata*	Leaves	Ethanol	Wistar rats	Not Reported	Srinivasan *et al.*, 2001

Contd...

Table 11.4–*Contd...*

Sl.No.	Name of the Plant	Parts Used	Solvent Used for Extraction	Animal Model	Compound Responsible	Reference
33.	*Pongamia pinnata*	Flower	Ethanol	Wistar rats	Flavonoids, Alkaloids and Phenols	Punitha and Manoharan 2006
34.	*Leucas lavandulaefolia*	Whole plant	Methanol	Wistar rats	Glycosides,Terpenoids	Kakali *et al.*, 1997
35.	*Calotropis procera*	Latex		Guinea pigs	Not Reported	Rasik *et al.*, 1999
36.	*Datura alba*	Leaves	Ethanol	Wistar rats	Not Reported	Shanmuga *et al.*, 2002
37.	*Hamelia patens*	Whole plant	Ethanol	Sprague Dawley rats	Not Reported	Alfredo *et al.*, 2004
38.	*Ajuga chia*	Aerial parts	Ethanol	Swiss albino mice	Not Reported	Enam *et al.*, 2007
39.	*Inula viscosa*	Aerial parts	Ethanol	Swiss albino mice	Not Reported	Enam *et al.*, 2007
40.	*Parieteria diffusa*	Leaves	Ethanol	Swiss albino mice	Not Reported	Enam *et al.*, 2007
41.	*Rubia taenifolia*	Aerial parts	Ethanol	Swiss albino mice	Not Reported	Enam *et al.*, 2007
42.	*Laurus nobilis*	Seed	Ethanol	Swiss albino mice	Not Reported	Enam *et al.*, 2007
43.	*Acalypha langinia*	Leaves	Water	Wistar rats	Not Reported	Perez and Vargas 2006
44.	*Plagiochasma appendiculatum*	Whole plant	Ethanol	Sprague Dawley rats	Not Reported	Meenakshi *et al.*, 2006
45.	*Hypericum perforatum*	Aerial parts	Ethanol	Chicken egg	Not Reported	Nilgun *et al.*, 2007

microbial infection and clearing the wound fibrin matrix, thus enhancing the healing process (Grinnell and Zhu, 1994).

The role of anti-oxidants from plant extracts in wound healing has been reported widely (Tran *et al.,* 1996). Many plant extracts and medicinal herbs have shown potent anti-oxidant activity. Flavonoids, the main components of many plant extracts, act as powerful free radical scavengers (Tran *et al.,* 1997). Positive correlation was found between anti-oxidant activity and phenolic content of the plant extracts. Similar results were reported by Elzaawely and Tawata, 2012. The free radical scavenging activity of plant flavonoids help in the healing of wounds (Havsteen, 2009). It is reported that radical-scavenging capacity of plant extract could be responsible for wound healing (Martin, 1996).

Healing of wound starts from the moment of injury and can continue through varying periods of time. Researchers have reported depending on the extent of wounding and the process of wound healing can be broadly categorized into three stages: Inflammatory phase, proliferative phase and finally the remodeling phase which ultimately determine the strength and appearance of the healed tissue. Wound healing process holds several steps which involve hemostasis, coagulation, inflammation, formation of granulation tissue, matrix formation, remodeling of connective tissue, collagenization and acquisition of wound strength (Suresh *et al.,* 2002).

Re-epithelialization is a process of restoring the epidermis and involves proliferation and migration of keratinocytes. Cell proliferation is an essential event during re-epithelialization, so proliferating keratinocytes ensure an adequate supply of cells to migrate and cover the wound. Accelerated dermal and epidermal regeneration in plant extract treated rats also confirmed that the extract had a positive effect towards cellular proliferation, granular tissue formation and epithelialization. Synthesis of ECM is also a key feature of wound healing. Dermal reconstruction is characterized by the formation of granulation tissue, which includes cell proliferation, ECM deposition, wound contraction and angiogenesis (Li *et al.,* 2007).

Earlier reports promised that plant products are potential agents for wound healing and are largely preferred because of their wide spread availability, non-toxicity, effectiveness and absence of unwanted side effects as crude preparations (Sandhya *et al.,* 2011). Topically administered drugs are effective in faster wound contraction because of the larger availability at the wound site (Akkol *et al.,* 2011).

REFERENCES

Afolayan, A.J., 2003. Extracts from the shoots of *Arctotis artotoides* inhibit the growth of bacteria and fungi. *Pharmaceutical Biology* 41, 22-25.

Akkol, E, K., Ac kara, O.B., Suntar, I., Citoglu, G.S., Keles, H., Ergene, B., 2011. Enhancement of wound healing by topical application of *Scorzonera species*: determination of the constituents by HPLC with new validated reverse phase method. *Journal of Ethnopharmacology* 137, 1018–1027.

Alfredo Gomez-Beloz, James, C., Rucinski, Michael. J. Balick, Camille Tipton, 2003. Double incision wound healing bioassay using *Hamelia patens* from El Salvador. *Journal of Ethnopharmacology* 88, 169–173.

Amit Kumar, Rajiv Chomwal, Praveen Kumar and Renu Sawal, 2009. Anti-inflammatory and wound healing activity of *Curcuma aromatica* salisb extract and its formulation. *Journal of Chemical and Pharmaceutical Research* 1 (1), 304-310.

Basso, K., Margolin, A.A., Stolovitzky, G., Klein, U., Dalla-Favera, R., Califano, A., 2005. Reverse engineering of regulatory networks in human B cells. *Nature Genetics* 37, 382–390.

Behl, P.N., Arora, R.B., Srivastava, G., Malhotra, S.C., 1993. *Herbs useful in Dermatological Therapy*, First edition, Delhi, India, 70–134.

Bodeker, G., Hughes, M.A., 1998. Wound healing, traditional treatments and research policy. *Plants for Food and Medicine*, 245–359.

Caceres, A., Fletes, L., Aguilar, L., Ramirez, O., Ligia, F., Tareena, A.M., 1993. Plants used in Guatemala for the treatment of gastrointestinal disorders, confirmation of activity against Enterobacteriaceae. *Journal of Ethnopharmacology* 38, 31-38.

Chaha, K.F., Eze, C.A., Emuelosi, C.E., Esimone, C.O., 2006. Antibacterial and wound healing properties of methanolic extracts of some Nigerian medicinal plants. *Journal of Ethnopharmacology* 104, 164–167.

Chakraborty, A., Devi, R.K.B., Rita S., Kh. Sharatchandra, Th.I. Singh. 2004. Preliminary studies on antiinflammatory and analgesic activities of *Spilanthes acmella* in experimental animal models. *Indian Journal of Pharmacology* 36 (3), 148-150.

Chopra, R.N., Nayar, S.L., Chopra, I.C., 1992. *Glossary of Indian Medicinal Plants*, Third edition, CSIR, New Delhi, 7–246.

Clark, R.A., 1991. In: Goldsmith LA (Editor): Physiology, Biochemistry and molecular biology of the skin. 2nd edition, Volume I. New York: Oxford University Press, 577.

Das, K., Tiwari, R.K.S., Shrivastava, D.K., 2010. Techniques for evaluation of medicinal plant products as antimicrobial agent: Current methods and future trends. *Journal of Medicinal Plants Research* 4(2), 104-111.

Eloff, J.N., 1998. Which extractant should be used for the screening and isolation of antimicrobial components from plants. *Journal of Ethnopharmacology* 60, 1–8.

Elzaawely, A.A., Tawata, S., 2012. Antioxidant activity of phenolic rich fraction obtained from *Convolvulus arvensis* L. leaves grown in Egypt. *Asian Journal of Crop Science*, 4 (1), 32–40.

Enam A. Khalil, Fatma, U., Afifia, Maysa Al-Hussaini, 2007. Evaluation of the wound healing effect of some Jordanian traditional medicinal plants formulated in Pluronic F127 using mice (*Mus musculus*). *Journal of Ethnopharmacology* 109, 104–112.

Fakim, A.G., 2006. Medicinal plants: Tradition of yesterday and drugs of tomorrow. *Molecular Aspects of Medicine* 27, 1-93.

Faraz, M., Mohammad, K., Naysaneh, G., Hamid, R.V., 2003. Phytochemical screening of some species of Iranian plants. *Iranian Journal of pharmaceutical Research*, 77-82.

Geethalakshmi, R., Sakravarthi, C., Kritika, T., Arul Kirubakaran, M., Sarada, D.V.L., 2013. Evaluation of Antioxidant and Wound Healing Potentials of *Sphaeranthus amaranthoides* Burm.f. *BioMed Research International* 1-7.

Grinnell, F., Zhu, M.F., 1994. Identification of neutrophil elastase as the proteinase in burn wound fluid responsible for degradation of fibronectin. *Journal of Investigative Dermatology* 103, 155–161.

Gupta B.L., Varshneya A.K., 1975. Anesthetic accident caused by unusual leakage of leakage. *British Journal of anesthesia* 47, 805.

Handa, S.S., Khanuja, S.P.S., Longo, G., Rakesh, D.D., 2008. Extraction Technologies for Medicinal and Aromatic Plants. International centre for science and high technology, Trieste, 21-25.

Havsteen, B.H., 2002. The biochemistry and medical significance of the flavonoids. *Pharmacology and Therapeutics* 96, 67–72.

Houghton, P.J., Hylands P.J., Mensah A.Y., Hensel, A., Deters A.M., 2005. *In vitro* tests and ethnopharmacological investigations: wound healing as an example. *Journal of Ethnopharmacology* 100: 100–107.

How Yee Lai, Yau Yan Lim and Kah Hwi Kim, 2011. Potential dermal wound healing agent in *Blechnum orientale* Linn. *BMC Complementary and Alternative Medicine* 11, 62.

Iyengar, M.A., Tripathi, M., Sirnivas, C.R., Nayak, S.G.K., 1997. Studies on some recommended ayurvedic herbs for contact dermatitis. *Ancient Science of Life* 17, 111–113.

Jagetia, G.C., Baliga, M.S., Venkatesh, P., Ulloor, J.N., 2003. Inuence of ginger rhizome (*Zingiber ocinale* Rosc.) on survival, glutathione and lipid peroxidation in mice after whole-body exposure to gamma radiation. *Radiation Research* 160, 584–592.

Jain Sachin, Jain Neetesh, Tiwari, A., Balekar, N., Jain, D.K., 2009. Simple Evaluation of Wound Healing Activity of Polyherbal Formulation of Roots of *Ageratum conyzoides* Linn. *Asian Journal of Research in Chemistry* 2(2), 135-138.

Jose Roberto Passarini Junior I, Fernanda Oliveira de Gaspari de Gaspi II, Lia Mara Grosso Neves III, Marcelo Augusto Marreto Esquisatto IV, Gláucia Maria Tech dos Santos V, Fernanda Aparecida Sampaio Mendonça VI. 2012. Application of *Jatropha curcas* L. seed oil (Euphorbiaceae) and microcurrent on the healing of experimental wounds in Wistar rats. *Acta Cirúrgica Brasileira* 2 (7), 441.

Jutamas Somchaichana, Tanom Bunaprasert and Suthiluk Patumraj, 2012. *Acanthus ebracteatus* Vahl. Ethanol Extract Enhancement of the Efficacy of the Collagen Scaffold in Wound Closure: A Study in a Full-Thickness-Wound Mouse Model. *Journal of Biomedicine and Biotechnology* 1-8.

Kakali Saha, Pulok, K., Mukherjee, Das, J., Pal, M., Saha, B.P., 1997. Wound healing activity of *leucas lavandulaefolia* Rees. *Journal of Ethnopharmacology* 56, 139-144.

Kokate, C.K., Purohit, A.P., Gohale, S.B., 1997. *Drugs Containing Tannins*, Text Book of Pharmacognosy, Fifth Edition, Pune, India, 461 -464.

Komarcevic, A., 2000. The modern approach to wound treatment. Medicinski pregled 53, 363–368.

Kulasekaran, S., Priya., Gnanamani Arumugam., Bhuvaneswari Rathinam., Alan Wells, Mary Babu, 2004. *Celosia argentea* Linn. leaf extract improves wound healing in a rat burn wound model. Wound Repair and Regeneration 12, 618–625.

Kuppast, I.J., Vasudeva Nayak, P., 2005. Wound healing activity of *Cordia dichotoma* Forst. f. fruits. *Natural Product Radiance*, 99-102.

Kwabena Ofori-Kwakye, Awo Afi Kwapong, Marcel Tunkumgnen Bayor, 2011. Wound healing potential of methanol extract of *Spathodea campanulata* stem bark formulated into a topical preparation. *African Journal of Traditional, Complementary and Alternative medicines* 8 (3), 218223.

Li, E., Clark, A.M., Hufford, C.D., 1995. Antifungal evaluation of Pseudolaric acid B, a major constituent of *Pseudolarix kaempferi. Journal of Natural Products* 58, 57–67.

Li, J., Chen J., Kirsner, R., 2007. Pathophysiology of acute wound healing. *Clinics in Dermatology* 25, 9–18.

Maddux, M.S., Brarriere, S.L., 1980. A review of complications of amphotericin B therapy: recommendation for prevention and management. *Drug Intellectual Clinical Pharmacology* 14, 177–181.

Martin, A., 1996. The use of antioxidants in healing. *Dermatologic Surgery* 22, 156–160.

Meenakshi Singh, Raghavan Govindarajan, Virendra Nath, Ajay Kumar Singh Rawat, Shanta Mehrotra, 2006. Antimicrobial, wound healing and antioxidant activity of *Plagiochasma appendiculatum* Lehm. et Lind. *Journal of Ethnopharmacology* 107, 67–72.

Miriyala Sumitraa, Panchatcharam Manikandana, Lochin Sugunab, 2005. Efficacy of *Butea monosperma* on dermal wound healing in rats. *The International Journal of Biochemistry and Cell Biology* 37, 566–573.

Muthusamy Senthil Kumar, Ramasamy Sripriya, Harinarayanan Vijaya Raghavan, Praveen Kumar Sehgal, 2006. Wound Healing Potential of *Cassia fistula* on Infected Albino Rat Model. *Journal of Surgical Research* 131, 283–289.

Nayak, B.S., Raju, S.S., Ramsubhag, A., 2008. Investigation of wound healing activity of *Lantana camara L.* in Sprague dawley rats using a burn wound model. *International Journal of Applied Research in Natural Products* 1 (1), 15-19.

Nayaka, B.S., Udupab, A.L., Udupac, S.L., 2009. Effect of *Ixora coccinea* flowers on dead space wound healing in rats. *Fitoterapia* 70, 233-236.

Ncube, N.S., Afolayan, A.J., Okoh, A.I., 2008. Assessment techniques of antimicrobial properties of natural compounds of plant origin: current methods and future trends. *African Journal of Biotechnology* 7(12), 1797-1806.

Nielsen, I.C., 1992. *Flora Malesiana: Mimosaceae,* Vol 1, 45.

Nilgun O Zturk, Seval Korkmaz, Yusuf O Zturk, 2007. Wound-healing activity of St. John's Wort (*Hypericum perforatum* L.) on chicken embryonic fibroblasts. *Journal of Ethnopharmacology* 111, 33–39.

Oyewale, A.O., Audu, O.T., Ayo, R.G., Amupitan, J.O., 2004. Cytotoxic correlation of some traditional medicinal plants using brine shrimp lethality test. *Chem Class Journal* 1, 110–112.

Perez Gutierrez, R.M., Vargas, S.R., 2006. Evaluation of the wound healing properties of *Acalypha langiana* in diabetic rats. *Fitoterapia* 77, 286–289.

Perumal Samy, R., Gopalakrishnakone, P., Sarumathi, M., Ignacimuthu, S., 2006. Wound healing potential of *Tragia involucrata* extract in rats. *Fitoterapia* 77, 300–302.

Prashant Tiwari., Bimlesh Kumar., Mandeep Kaur., Gurpreet Kaur, Harleen Kaur, 2011. Phytochemical screening and Extraction: A Review. *Internationale Pharmaceutica Sciencia* 1(1), 98-106.

Punitha, R., Manoharan, S., 2006. Antihyperglycemic and antilipidperoxidative effects of *Pongamia pinnata* (Linn.) *Pierre* flowers in alloxan induced diabetic rats. *Journal of Ethnopharmacology* 105, 39–46.

Rahalison, L., Hamburger, M., Monod, M., Frenk, E., Hostettaman, K., 1994. Antifungal tests in phytochemical investigations: comparison of bioautographic methods using phytopathogenic and human pathogenic fungi. *Planta Medica* 60, 41–44.

Rane, M., Madhura, Mengi, A., Shusma, 2003. Comparative effect of oral administration and topical application of alcoholic extract of *Terminalia arjuna* bark on incision and excision wounds in rats. *Fitoterapia* 74, 553-558.

Rasik, A.M., Ram Raghubir, Gupta, A, Shukla, A, Dubey, M.P., Srivastava, S., Jain, H.K., Kulshrestha, D.K., 1999. Healing potential of *Calotropis procera* on dermal wounds in Guinea pigs. *Journal of Ethnopharmacology* 68, 261–266.

Robson, M.C., Steed, D.L., Franz, M.G., 2001. Wound Healing: Biologic features and approaches to maximize healing trajectories. *Current Problems in Surgery* 38 (2), 65–140.

Rotblatt, M., Ziment, I., 2002. *Evidence-Based Herbal Medicine,* Hanley and Belfus, Philadelphia, PA, 464.

Sagar, S.M., Yance, D., Wong, R.K., 2006. Natural health products that inhibit angiogenesis: a potential source for investigational new agents to treat cancer. *Current Oncology* 13(1), 14-26.

Sandhya, S., Sai Kumar, P., Vinod, K.R., Banji, D., Kumar, K., 2011. Plants as potent anti-diabetic and wound healing agents – a review. *Hygeia Journal for Drugs and Medicine* 3, 11–19.

Saral, R., 1991. *Candida* and *Aspergillus* infection in immunocompromised patients: An overview. *Revision on Infectious Diseases* 13, 487–492.

Sathishkumar, P., Paulsamy, S., Anandakumar, A.M., Senthilkumar, P., 2009. Effect of habitat variation on the content of certain secondary metabolites of medicinal importance in the leaves of the plant, *Acacia caesia* Willd. *Advances in Plant Sciences* 22 (2), 451-453.

Scortichini, M., Pia, Roosi, M.J., 1991. Preliminary *in vitro* evaluation of the antimicrobial activity of terpenes and terpenoids towards *Erwinia amylovora. Journal of Applied Bacteriology* 71, 109 112.

Shanmuga Priya, K., Gnanamani, A., Radhakrishnan, N., Mary Babu, 2002. Healing potential of *Datura alba* on burn wounds in albino rats. *Journal of Ethnopharmacology* 83, 193-199.

Shirwaikar, A., Somashekar, A.P., Udupa, A.L., Udupa, S.L., Somashekar, S., 2003. Wound healing studies of *Aristolochia bracteolata* Lam. with supportive action of antioxidant enzymes. *Phytomedicine* 10, 558–562.

Shukla, A., Rasik, A.M., Jain, G.K., Shankar, R., Kulshrestha, D.K., Dhawan, B.N., 1999. *In vitro* and *in vivo* wound healing activity of asiaticoside isolated from *Centella asiatica. Journal of Ethnopharmacology* 65, 1–11.

Soliman Yusufoglu, H., Ibrahim Alqasoumi, S., 2011. Anti inflammatory and wound healing activities of herbal gel containing an antioxidant *Tamarix aphylla* leaf extract. *International Journal of Pharmacology* 7, 829–835.

Somashekar Shetty, Saraswati Udupa, Laxminarayana Udupa, 2008. Evaluation of Antioxidant and Wound Healing Effects of Alcoholic and Aqueous Extract of *Ocimum sanctum* Linn in Rats. *Evidence-Based Complementary and Alternative Medicine* 5 (1), 95–101.

Sreenivasan Sasidharan, Selvarasoo Logeswaran, Lachimanan Yoga Latha, 2012. Wound Healing Activity of *Elaeis guineensis* Leaf Extract Ointment. *International Journal of Molecular Sciences* 13, 336-347.

Srinivasan, K., Muruganandan, S., Lal, J., Chandra, S., Tandan, S.K., Ravi Prakash, V., 2001. Evaluation of anti-inflammatory activity of *Pongamia pinnata* leaves in rats. *Journal of Ethnopharmacology* 78, 151–157.

Suresh Reddy, J., Rajeswara Rao, P., Mada Reddy, S., 2002. Wound healing effects of *Heliotropium indicum, Plumbago zeylanicum* and *Acalypha indica* in rats. *Journal of Ethnopharmacology* 79, 249–251.

Tran, V.H., 1997. Antioxidant and free radical scavenging effects of some polyphenolic extracts from medicinal plants on wound healing. Asian European Tissue Repair Society Post-satellite Meeting, Hanoi, Vietnam.

Tran, V.H., Hughes, M.A., Cherry, G.W., 1996. The effects of polyphenolic extract from *Cudrania cochinchinenesis* on cell response to oxidative damage caused by H_2O_2 and xanthine oxidase. First Joint Meeting of Chinese and Europian Tissue Repair Society, Xi'an, China.

Tsuchiya, H., Sato, M., Miyazaki, T., Fujiwara, S., Tanigaki, S., Ohyama, M., Tanaka, T., Linuma, M., 1996. Comparative study on the antibacterial activity of phytochemical flavanones against methicillin resistant *Staphylococcus aureus. Journal of Ethnopharmacology* 50, 27-34.

Tuhin Kanti Biswas, M.D., Lakshmi Narayan Maity, M.D., Biswapati Mukherjee, 2004. Wound Healing Potential of *Pterocarpus santalinus* Linn: A Pharmacological Evaluation. *Lower Extremity Wounds* 3 (3), 143–150.

Umachigia, S.P., Jayaveerab, K.N., Ashok Kumarc, C.K., Kumard, G.S., Vrushabendra swamye, B.M., Kishore Kumarf, D.V., 2008. Studies on Wound Healing Properties of *Quercus infectoria. Tropical Journal of Pharmaceutical Research, March* 7 (1), 913-919.

2015, Modern Methods in Phytomedicine *Pages* ***191–220***
Editor: **T. Parimelazhagan**
Published by: **DAYA PUBLISHING HOUSE, NEW DELHI**

12

Plant Growth Promoting Rhizobacteria: Their Role in Growth Enhancement and Productivity of Medicinal Plants

R. Pemila Edith Chitraselvi and S. Kalidass*

Department of Biotechnology, School of Biotechnology and Health Sciences, Karunya University, Coimbatore – 641 114, Tamil Nadu

1.0 INTRODUCTION

Medicinal plants rich in secondary metabolites are potentially useful to produce natural drugs and are the precious part of the world flora (Savitharamma *et al.*, 2011). India is known to be the home of many important and valuable medicinal plants from ancient times and more than 5000 plant species across the globe are said to possess some specific therapeutic value (Malleshwari and Bagyanarayana, 2013). Almost 83 per cent of the plant species are at the risk of extinction because of human activities mainly due to over exploitation, habitat loss, destruction and loss in the genetic diversity (Tasheva and Kosturkova, 2013).

Biotechnological approaches using plant tissue culture and genetic manipulation have emerged to grow plants with desired traits under *in vitro* conditions. These techniques are applied when the particular species has high medicinal and economical

* *Corresponding Author.* E-mail: kalidass@karunya.edu

value and the plant resources are limited. But to preserve these plants in their natural habitat, a systematic agro technique needs to be developed. Employing bio-fertilizer as a cheap and reliable tool can be effective not only in enhancing the growth of the rare plants without affecting the soil composition and soil characters but also can help preserve plants in their natural habitat or our backyards.

One group of beneficial microorganisms that inhabits the rhizosphere region of plants is referred as Plant Growth Promoting Rhizobacteria (PGPR). These bacteria utilize the photosynthates rich in nutrients that are released from the host by exudation for their growth and they secrete metabolites into the rhizosphere which are plant growth promoting factors. These metabolites can act as signaling compounds that are perceived and taken up by root cells of the host plant (Van Loon, 2007). Plant Growth Promoting Rhizobacteria (PGPRs) can be excellent in bio-fertilization as well as in bio control (Ahmad *et al.*, 2008). Many PGPR belonging to various families have proved to be effective bio-control agents against various bacterial and fungal pathogens. The potential environmental benefits of these PGPR can help reduce use of chemical fertilizers and fungicides and enhance sustainable management practices.

1.1 Need for Protection of Endangered Medicinal Plant Species

About three quarters of the world population rely mainly on plants and plant extracts for health care. The global demand for herbal medicine is not only large, but growing.

Medicinal plants are often the only easily accessible health care alternative for the most of the population and traditional medicines remained a part of our integral health system (Vasudha *et al.*, 2013). The ethnic people and local tribes living in the remote forest areas still depend to a great extent on this indigenous system of medicine.

The number of higher plant species (angiosperms and gymnosperms) is estimated between 215,000 and 500,000. Of these, only about 6 per cent have been screened for biological activity and about 15 per cent have been evaluated for phytochemicals. In India about 1100 species are used in different systems. India is rich in its biological resources and known for its valuable heritage of herbal medicinal knowledge. It possesses an extremely rich plant bio-diversity which is gradually decreasing.

Natural medicines are in great demand also in the developed world for primary health care because of their efficacy, safety and lesser side effects. These medicinal plants hold great potential as an inexhaustible reservoir for the identification and isolation of therapeutics even for cancer and AIDS for which there is no known cure yet. Protecting these resources and integrating them into modern medical practice would bring enormous benefits to mankind.

The large scale deforestation is leading to an accelerated loss of valuable biodiversity, extinction of species and genetic erosion. It has been reported by Botanical Survey of India that around 93 per cent of medicinal plants of India now belong to endangered species. Unsustainable harvesting of these medicinal plants has led to exploitation and decrease of the species.

Most sustainable and environmentally acceptable mode of preserving the medicinal plants in their own habitat can be achieved by using bio-fertilizers and bio-control agents as an effort to reduce the use of agrochemicals and their residues in the environment. Identifying, understanding and utilizing microorganisms or microbial products has become integral part of sustainable agriculture.

1.2 Significance of Plant Growth Promoting Rhizobacteria

In the rhizosphere, that is on the plant root or its close vicinity, bacteria are abundantly present, most often organized in micro-colonies. Some of these rhizobacteria not only benefit from the nutrients secreted by the plant root but also beneficially influence the plant in a direct or indirect way, resulting in a stimulation of its growth. These plant growth- promoting rhizobacteria (PGPRs) can be classified according to their beneficial effects. Biofertilizers can fix nitrogen, which can subsequently be used by the plant, thereby improving plant growth when the amount of nitrogen in the soil is limiting. Phytostimulators can directly promote the growth of plants, usually by the production of hormones. Biocontrol agents are able to protect plants from infection by phytopathogenic organisms.

2.0 GROWTH PROMOTION BY PGPR

2.1 Nitrogen Fixation

At present, biofertilization accounts for approximately 65 per cent of the nitrogen supply to crops worldwide. The most efficient nitrogen fixers are bacterial strains belonging to the genera *Rhizobium, Sinorhizobium, Mesorhizobium, Bradyrhizobium, Azorhizobium* and *Allorhizobium*. These strains have been studied in most detail. All of these bacteria form a host-specific symbiosis with leguminous plants. The symbiosis is initiated by the formation of root or stem nodules in response to the presence of the bacterium.

The bacteria penetrate the cortex, induce root nodules, multiply and subsequently differentiate into bacteroids, which produce the nitrogenase enzyme complex. Within the root nodules, a low oxygen state is formed and the bacterial nitrogenase converts atmospheric nitrogen into ammonia. Plant supplies carbon source through the photosynthates. The gene products of both the *nif* and the *fix* genes, which are involved in nitrogen fixation, have been characterized. However, understanding the bacterial signaling mechanism in the nodule formation is a major challenge.

Free-living nitrogen-fixing rhizobacteria such as *Azospirillum, Herbaspirillum, Acetobacter, Azotobacter* and *Azoarcus* are also able to fix atmospheric nitrogen. They use a nitrogenase complex that functions under low oxygen conditions and that is not as specific in its interaction with the plant as are *Rhizobia*. *Azospirillum* predominantly colonizes the rhizosphere, whereas the other bacteria are predominantly found as endophytes inside roots, stems and leaves. The genes involved in nitrogen fixation, nitrogen assimilation and nitrogen regulation have been described for *Azospirillum*. Several of the *nif* genes have also been described for the other free-living nitrogen fixers, which all have similar nitrogenase complexes, except for *Azoarcus* which possesses three differently encoded nitrogenase complexes (Bloemberg and Lugtenberg, 2001).

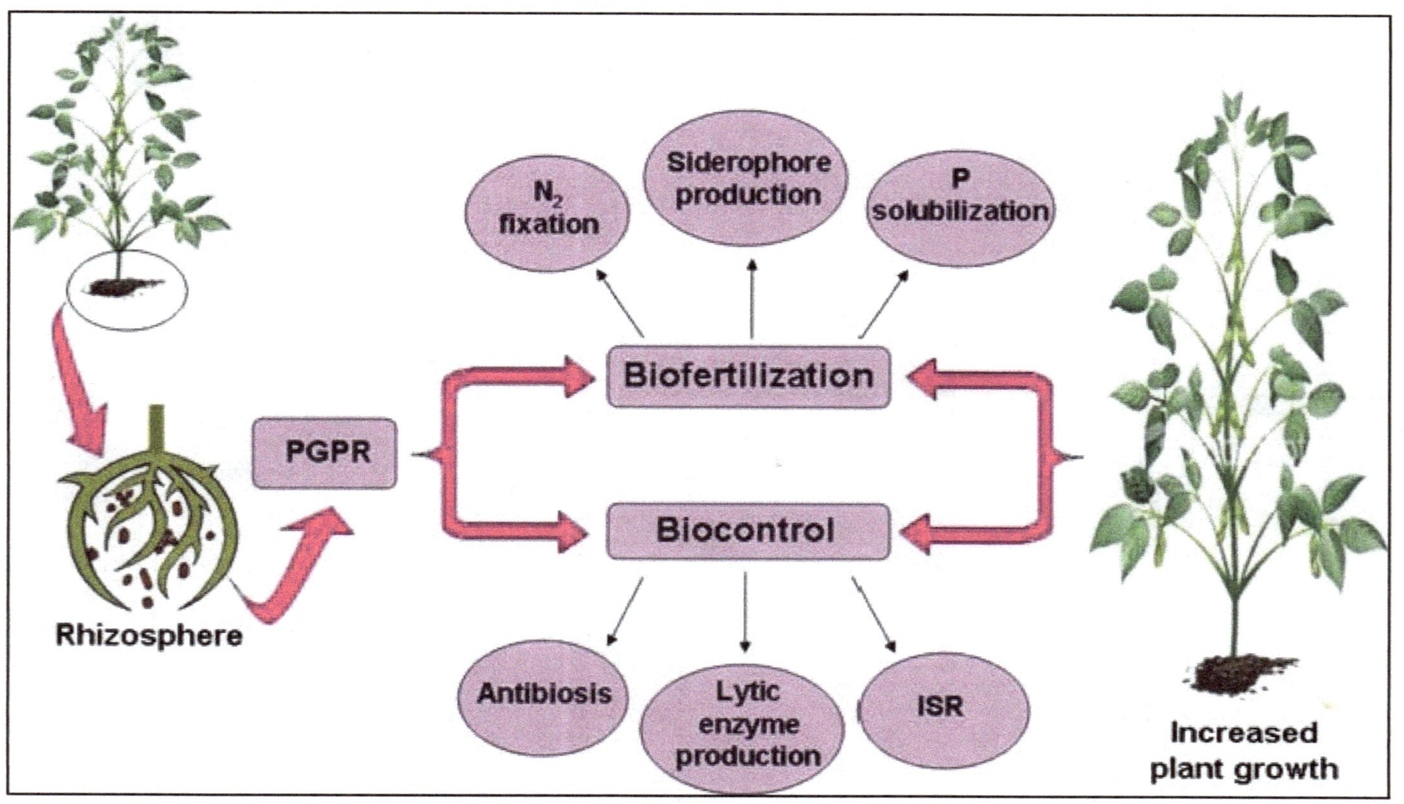

Figure 12.1: Mechanism of Action of PGPR in the Plant Rhizosphere (*Source*: Kumar *et al.*, 2011).

Works are being done to widen the host range of the symbiosis towards non-leguminous crops. Recent studies also showed that the genes in *Sym* (symbiosis) plasmid in *Rhizobium* are homologous to the gene encoding Type III secretion systems which is also identified in plant beneficial *Pseudomonas fluorescens* strains.

2.2 Phytohormone Production

Phytohormones are natural plant growth regulators that influence physiological processes, when at low concentrations. They enhance plant growth in a direct way. Plant growth regulators are classified as auxins (cell differentiation, root and fruit growth, and abscission control), cytokinins (growth regulation, cell differentiation, and plant senescence), gibberellins (cell division and elongation, interruption of dormancy, and increase in fruit growth), abscisic acid (transpiration regulation, interruption of dormancy, and initial seed development), and ethylene (ripening of fruits, promotion of leaf, fruit and leaf abscission, and influence in female sexual expression) (Arshad and Frankenberger, 1998). The production of these growth regulators has been reported in several bacteria including *Gluconacetobacter, Azospirillum, Herbaspirillum, Methylobacterium, Erwinia, Pantoea*, and *Pseudomonas* (Fuentes – Ramirez *et al.*, 1993; Koenig *et al.*, 2002; Lucangeli and Bottini, 1997; Patten and Glick, 1996; Verma *et al.*, 2001) and has been consistently observed in bacteria that live in association with plants.

Besides having nitrogen fixing ability, *Azospirillum* spp. secrete phytohormones such as auxins, cytokinins and gibberellins. Auxins are quantitatively the most abundant phytohormones secreted by *Azospirillum*, and it is generally agreed that auxin production, rather than nitrogen fixation, is the major factor responsible for the stimulation of rooting and, hence, enhanced plant growth.

Among the phytohormones produced by soil microorganisms indole-3 acetic acid (IAA) is an important hormone for plant growth and development. The PGPR that are capable of producing IAA include *Pseudomonas* sp., *Bacillus* sp., *Klebsiella* sp., *Azospirillum* sp., *Enterobacter* and *Serratia* sp. (Martens and Frankenberger, 1991; Frankenberger and Arshad, 1995).

2.3 Phosphate Solubilization and Sulfur Oxidation

The Rhizobacteria increase the availability of essential nutrients, by solubilizing phosphate, oxidizing sulfur. Phosphorus (P) frequently limits crop growth in organic form. Several bacterial species, in association with plant rhizosphere, are capable of increasing availability of Phosphorus to plants either by mineralization of organic phosphate or by solubilization of inorganic phosphate by production of acids (Ekin, 2010).

Bacteria in the groups *Bacillus, Rhizobium* and *Pseudomonas* have proven to be the most powerful phosphate-solubilizing bacteria. When inoculated with seeds these PGPRs colonize plant roots and they provide P in a plant available source (Rodriguez and Fraga, 1999). Table 12.1 contains some phosphate solubilizing bacteria and the substrates they act upon.

Table 12.1: Phosphate Mineralization from P-substrates by some Soil Bacterial Species (Rodriguez and Fraga, 1999)

Phosphate Solubilizing Bacteria (PSB)	*Substrate*	*Enzyme Produced by the Organism*
Pseudomonas fluorescens	Non-specific	Acid Phosphatase
Pseudomonas sp.	Non-specific	Acid Phosphatase
Burkholderia cepacia	Non-specific	Acid Phosphatase
Enterobacter aerogens	Non-specific	Acid Phosphatase
Enterobacter cloacae	Non-specific	Acid Phosphatase
Citrobacter freundi	Non-specific	Acid Phosphatase
Proteus mirabalis	Non-specific	Acid Phosphatase
Serratia marcensens	Non-specific	Acid Phosphatase
Bacillus subtilis	Inositol Phosphate	Phytase
Pseudomonas putida	Inositol Phosphate	Phytase
Pseudomonas mendocina	Inositol Phosphate	Phytase
Pseudomonas fluorescens	Phosphonoacetate	Phosphonoacetate hydrolase
Bacillus licheniformis	D-α- glycerophosphate	D-α- glycerophosphatase
Klebsiella aerogenes	Phosphonates	C-P Lyase

Similarly elemental sulfur, gypsum and other sulfur bearing mined minerals must be transformed (or oxidized) by bacteria into sulphate making it available for plants.

These PGPR increase nutrient uptake from soil, thus reducing the need for fertilizers and preventing accumulation of minerals in soil (Ashrafuzzaman *et al.,* 2009).

2.4 Siderophore Production

Siderophores are low molecular weight biomolecules secreted by microorganisms in response to iron starvation. Some siderophores can chelate other many heavy metal ions, yet their specificity and avidity for iron is the most consistent feature (Sarode *et al.,* 2009) and they are called generally as Microbial Iron Transport Agents. Environmental factors that modulate siderophores synthesis include pH, the level of iron and the form of iron ions, the presence of other trace elements, and an adequate supply of carbon, nitrogen, and phosphorus (Saharan and Nehra, 2011).

Soil bacteria isolates including *Azotobacter vinelandii* strain, *Bacillus megaterium, Bacillus cereus* strains and *Pseudomonas fluorescens-putida* group produce siderophores and these can be used as efficient PGPR to increase the yield of the crop. PGPR that produce siderophores exert their plant growth-promoting activity by depriving native microflora of iron. The extracellular siderophores efficiently complex with environmental iron and make it unavailable to other microbes and pathogenic fungi.

The phenolate siderophores 2, 3 dihydroxybenzoic acid, 3, 5-dihydroxybenzoic acid, pyoverdine, pyochelin, salicylic acid (Sarode *et al.,* 2009), Coelchelin and DFOB

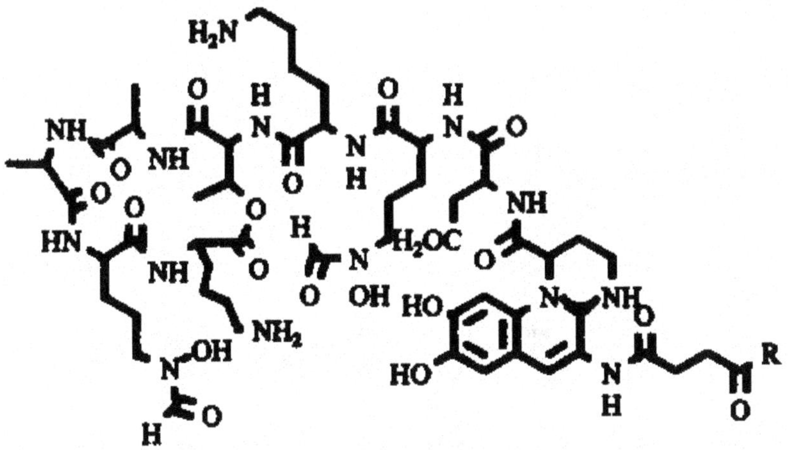

Figure 12.2: Structure of Siderophore Pyoverdin.

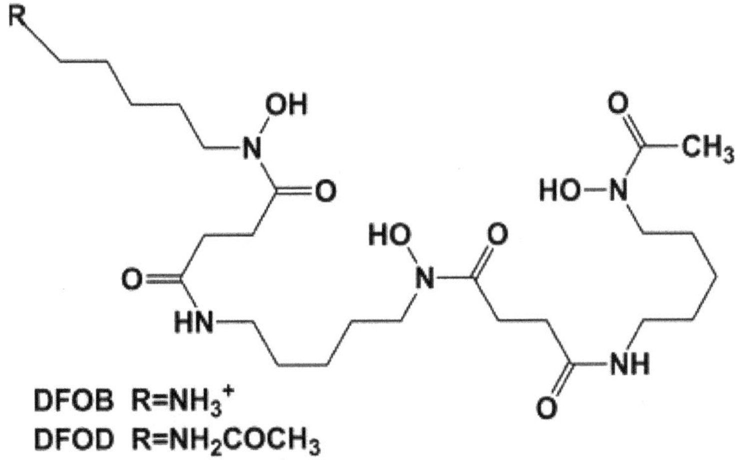

DFOB R=NH_3^+
DFOD R=NH_2COCH_3

Figure 12.3: Structure of Desferrioxamine B.

were reported by under iron-starved conditions *by Azospirillum lipoferum* strain *D-2* and Pseudobactin produced by *Pseudomonas fluorescens* (Ramamoorthy *et al.,* 2001).

These microbial siderophores enhance heavy metal micro nutrient acquisition and supplies the plant with essential minerals even in starved conditions. They also induce Systemic Resistance against plant pathogens.

2.5 Mycorrhizal Association

Mycorrhizal (AM) fungi and bacteria can interact synergistically to stimulate plant growth through a range of mechanisms that include improved nutrient acquisition and inhibition of fungal plant pathogens. These interactions may be of crucial importance within sustainable, low-input agricultural cropping systems.

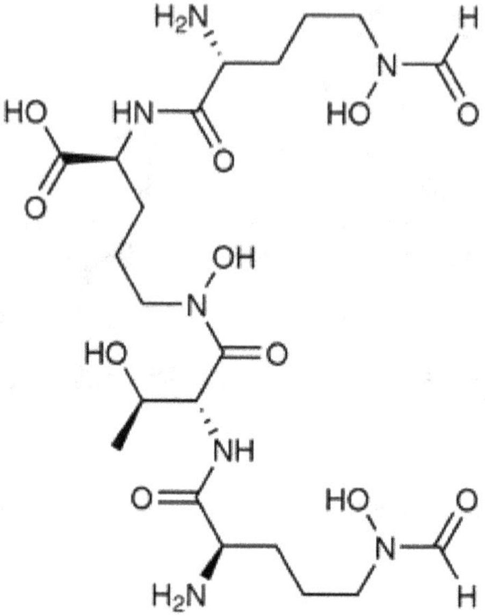

Coelichelin (**3**)

Figure 12.4: Structure of Coelichelin.

Generally most plant roots are colonized by mycorrhizal fungi and their presence stimulates plant growth. The synergistic effects of bacteria and mycorrhizal fungi and their combined beneficial impacts on plants has been studied recently. Both ectomycorrhizal and endomycorrhizal (Meyer and Linderman, 1986) fungi can interact with different bacterial species. These interactions occur in the zone of soil surrounding the roots and fungal hyphae; commonly referred to as the 'mycorrhizosphere' (Rambelli, 1973). The interactions between bacteria and AM fungi are potentially beneficial, especially when PGPR including N_2 fixing bacteria are involved (Meyer and Linderman, 1986; Von Alten *et al.*, 1993; Secilia and Bagyaraj, 1987). Some bacteria directly affect AM fungal germination and growth rate thus establishing a direct synergistic relationship (Mosse, 1959; Daniels and Trappe, 1980; Mayo *et al.*, 1986; Carpenter-Boggs *et al.*, 1995). Bacteria together with AM fungi also create a more indirect synergism that supports plant growth, including nutrient acquisition (Barea *et al.*, 2002), inhibition of plant pathogenic fungi (Budi *et al.*, 1999), and enhancement of root branching (Gamalero *et al.*, 2004).

Mycorrhizal establishment changes the chemical composition of root exudates which are the source of nutrients to associated bacteria in the mycorrhizosphere and hence the composition of the bacterial communities inhabiting the Mycorhizosphere (Harley and Smith, 1983; Linderman *et al.*, 1992; Azcon-Aguilar and Bago, 1994; Barea, 2000; Gryndler, 2000). A few studies (Andrade *et al.*, 1997), suggest a high degree of specificity between bacteria associated with AM fungi. This may be due to the activation of bacteria by species specific fungal exudates.

Some PGPR and AM fungi could be co-inoculated to optimize the formation and functioning of the AM symbiosis. Apart from having effects on AM fungal growth, PGPR have been suggested to possess a variety of other direct mechanisms to support the mycorrhizal symbiosis. In 1989, Azcon reported the enhancement of the growth of the AM fungus *Glomus mosseae.* It is said that the bacterial groups most commonly interact synergistically with AM fungi are mainly Gram-positive bacteria and γ – proteobacteria, maybe due to the fact that some members of these phylogenetic groups are more integrally associated with AM fungi than others. However the association between fluorescent pseudomonad and *G. mosseae* which indicate gram negative bacteria are also capable of forming association with AM fungi.

2.6 Biological Control against Plant Pathogens

Biological control of plant diseases is gaining attention due to increased pollution concerns because of pesticides use for crop protection and development of pathogen resistance (Wisniiewski and Wilson, 1992). With the development of more and more pesticide-resistant strains, the replacement of pesticides by the controlled use of alternative agents and/or products has become the focus of considerable interest in the context of a sustainable, economically profitable agriculture (Benhamou and Nicole, 1999). The use of environmental friendly microorganisms has proved useful in plant-growth promotion and disease control in modern agriculture (Weller, 1988; Fatima *et al.,* 2009).

PGPR may promote plant growth through several mechanisms among which preventing growth of pathogenic microorganisms comes under indirect mechanisms (Ramos-Solano *et al.,* 2008). Different mechanisms of action used by the bio-control agents have great advantage over chemicals which work on a single target (Lucas *et al.,* 2009).

Many PGPR belonging to various families have proved to be effective bio-control agents against various bacterial and fungal pathogens. Even though there is not much success in bio-control with PGPR, proper understanding of their diversity, colonization ability, modes of action, formulation and field application can facilitate their development as reliable tools in the management of sustainable agricultural systems and enhance commercial development.

PGPR hold promise as good bio-control agents against these pathogens. Root colonizing bacteria belonging to the genus *Pseudomonas* and *Bacillus* have been proven to be effective biocontrol agents. *P. flourescens, P. putida, B. polymyxa, B. Coagulans, B. Pumulus, B. lentus, B. cereus, B. circulans, Enterobacter agglomerans* and *Paenibacillus* sp. are some of the most effective bio-control agents that are widely studied (Gnanamanickam, 2009). These PGPR indirectly enhance plant growth via suppression of phytopathogens (Arrebola *et al.,* 2010; Joshi and McSpadden Gardener, 2006).

2.6.1 Antagonism

The PGPR bio-control bacteria act by various mechanisms like competition for nutrition and niches on the plant root, antagonism and by producing metabolites that are toxic to the pathogens. Most of these bio-control agents produce fungal cell wall-lysing enzymes such as chitinase that retards fungal growth as the fungal hyphal wall ruptures, some produce toxic compounds like hydrogen cyanide which are toxic to pathogenic fungi even in low concentrations.

Production of siderophores is yet another strategy in bio-control mechanism. They scavenge and chelate iron and other heavy metal ions so the pathogens are starved of iron and other essential minerals and their growth retards. Some siderophores produced by microbes are pyoverdine, pyochelin and salicylic acid.

2.6.2 Induced Systemic Resistance and Systemic Acquired Resistance

Exposure to PGPR triggers a defense response by the plant as if attacked by pathogenic organism. Plants respond by activating broad-spectrum innate immune responses that can be expressed locally at the site of invasion as well as systemically in the uninfected tissue (Maleck and Lawton, 1998). The crop will thus be armed and prepared to mount successful defense against eventual attacks of the pathogens.

Systemic Acquired Resistance (SAR)

SAR implies the production of translocated signals that are involved in the activation of resistance mechanisms in uninfected parts of the plant. The first infection predisposes the plant to resist further attacks. This is similar to immune response in higher animals (Maunch-Mani and Metraux, 1998).

Salicylic acid produced by PGPR induces Systemic Resistance by endogenous trigger of thermogenesis by increasing the expression of an alternate oxidase enzyme linked to thermogenesis. This SA dependent alternate oxidase production is involved in oxidative burst via NADPH dependent pathway after pathogen infection and lead to the activation of defense response (Maunch-Mani and Metraux, 1998).

Induced Systemic Resistance (ISR)

Induced Systemic Resistance is the process of active resistance dependent on the host plant's physical barriers activated by biotic or abiotic agents (Kloepper *et al.,* 1992).

The phenotypic effects and the biochemical and mechanical changes in the plant after root inoculation with bacteria suggest that the plant exhibits both IR and SAR and the term Induced Systemic Resistance (ISR) is introduced for bacterially induced resistance and systemic acquired resistance for the other forms (Pieterse *et al.,* 1996; Benhamou and Nicole, 1999). The ability of a plant to survive pathogen attack depends both on preformed barriers and on induced active defence mechanisms (Nandakumar *et al.,* 2001). ISR can be triggered independent of SA through Volatile Organic Compounds (Choudhary and Johri, 2009).

Following changes take place in plant roots that exhibit ISR;

- ✰ Strengthening of epidermal and cortical cell walls and deposition of newly formed barriers beyond infection sites including callose, lignin and phenolics.
- ✰ Increasing levels of enzymes such chitinase, peroxidises, polyphenol oxidases and phenylalaninde ammonia lyase.

☆ Enhanced phytolexin production.

☆ Enhanced expression of stress related genes.

Ability of bacteria to colonize root tissue is also considered as one of the factors inducing ISR (Whipps, 2001). The compounds that produce ISR include siderophores and O-antigen of lipopolysaccharide layer (Emmert *et al.,* 2004).

2.6.3 Anti-Fungal Metabolites (Antibiotics)

Numerous Anti - Fungal Metabolites (AFM) are produced by bacteria. These include Ammonia, Butyrolactones, 2,4-Diacetyl Phloroglucinol (phl), Hydrogen Cyanide (HCN), Kanosamine, Oligomycin A, Oomycin A, Phenazine-1-Carboxylic Acid (PCA), Pyoluterin (Plt), Pyrrolnitrin (Pln), Viscosinamide, Xanthobaccin and Zwittermycin A (Whipps, 2001; Emmert *et al.,* 2004; Thomashow, 1996). The various antifungal metabolites produced by biocontrol agents are listed below in Table 12.2.

Table 12.2: Antibiotics Produced by Rhizobacteria

PGPR	*Antibiotics*	*Reference*
Bacillus sp.	Kanosamine	Milner *et al.,* 1996
	Zwittermycin A	Silo-Suh *et al.,* 1994
	Iturin A (Cyclopeptide)	Constantinescu, 2001
	Bacillomycin	Volpon *et al.,* 1999
	Plipastatins A & B	Volpon *et al.,* 2000
Pseudomonas	*Antifungal Antibiotics*	
	Phenazine	Burkhead *et al.,* 1994
	Phenazine-1- carboxylic acid	
	Phenazine-1- carboxamide	Chin-A-Woeng *et al.,* 1998
	Pyrrolnitrin	Thomashow and Weller, 1988
	Pyoluteorin	Howel and Stipanovic, 1980
	2,4-Diacetylphloroglucinol	Shanahan *et al.,* 1992
	Rhamnolipids	
	Oomycin A	
	Ecomycins	Jiao *et al.,* 1996
	DDR	Miller *et al.,* 1998
	Viscosinamide	Hokeberg *et al.,* 1998; Nielsen *et al.,* 1999 Thrane *et al.,* 2000
	Butyrolactone	Gamard *et al.,* 1997
	N-butylbenzene	
	Sulphonamide	Baron and Rowe, 1981
	Pyocyanin	
	Antibacterial Antibiotics	Fuller *et al.,* 1971
	Pseudomonic Acid	
	Azomycin	

The scope of developing these microbial metabolites for commercial pesticides as an alternative to chemical fungicides is gaining importance due to increased concerns on environmental pollution, pathogen resistance and high plant protection costs.

Polyketides

DAPG: 2, 4- Diacetyl phloroglucinol is a secondary metabolite from *Pseudomonas flourescens*. It is a major factor in controlling of plant pathogens. Bacteria producing 2, 4-DAPG play a key role in agricultural environments and their potential for use in sustainable agriculture is promising (Reddy *et al.,* 2009).

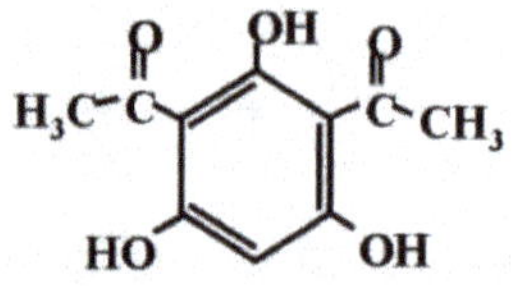

Structure of DAPG

Pyoluteorin: Pyoluteorin (Plt) is a phenolic polyketide with resorcinol ring. The ring is linked to a bichlorinated pyrrole moiety. It was first isolated from *P. aeruginosa* (Takeda, 1958).

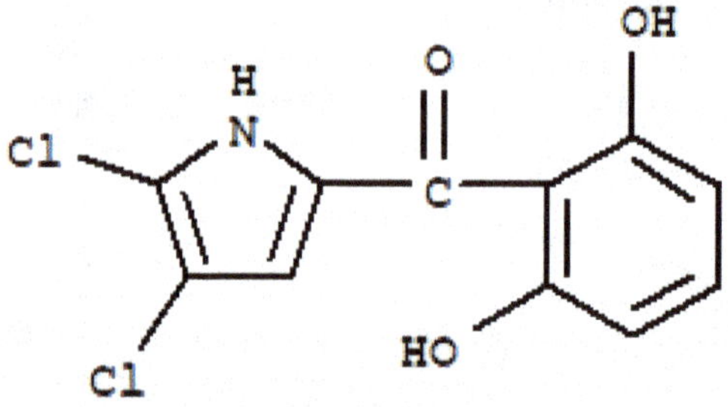

Structure of Pyoluteorin

Mupirocin: Among the major metabolites pseudomonic acid known as mupirocin is also responsible for its bactericidal activity (Fuller *et al.,* 1971). Mupirocin has a unique chemical structure and contains C9 saturated fatty acid (9-hydroxynonanoic acid) linked to monic acid A by an ester linkage. Mupirocin is derived from acetate. The acetate units are incorporated in to monic acid A and 9 - hydroxy nonanoic acid via polyketide synthesis.

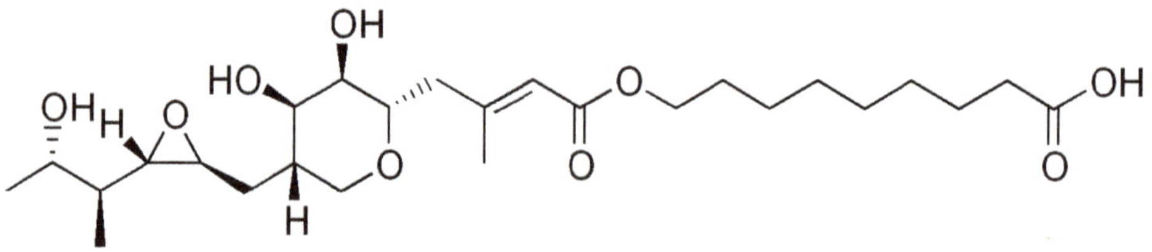

Structure of Mupirocin

Hetrocyclic Nitrogen Compounds

PCA: Phenazines are broad spectrum antibiotics, produced by biocontrol bacteria to kill the phytopathogens in soil. Different Phenazine derivatives originate from Phenazine-1-Carboxylic Acid (PCA) (Upadhyay and Srivastava, 2010). PCA derivatives are formed by modifying on the 1-position of PCA, as Carboxylic acid shows high chemical reactivity with other reagents such as amines and alcohols (Ye *et al.*, 2010). Commonly identified derivatives of phenazine produced by *Pseudomonas* spp. are pyocyanin, PCA, PCN and hydroxy phenazines (Turner and Messenger, 1986). The antimicrobial activity of phenazine depends on the rate of oxidative reductive, transformation of the compound coupled with the accumulation of toxic superoxide radicals in the target cells (Hassett *et al.*, 1992; Hassett *et al.*, 1993).

PCA production by *Pseudomonas* involves a signalling system. The components of this quorum sensing pathway are the pili or fimbriae and other putative proteins. This signalling system involves the gene product of *phzI* gene of the bacteria. Pathogen growth on the root increases root exudation and this results in increase in the *Pseudomonas* population which inturn increases the signal molecule N - Acyl – L – Homoserine Lactone (HSL) produced by the gene *phzI*. In response to the HSL signal molecule the PCA synthesis pathway which is controlled by *phzR* gene is switched on.

Phenyl Pyrrole Antibiotics

Pyrrolnitrin: The antibiotic that belongs to phenylpyrrole group receives much attention due to its broad-spectrum action. The antibiotic pyrrolnitrin belongs to phenylpyrrole group. Pyrrolnitrin (PRN) is a chlorinated phenylpyrrole antibiotic produced by several fluorescent and non-fluorescent pseudomonads. It was first isolated from *Burkholderia pyrrocinia* (Arima *et al.*, 1964).

PRN persists actively in the soil for almost one month without diffusing. But it is released after lysis of host bacterial cell, resulting in the slow release. *P. fluorescens* strains producing PRN are excellent biocontrol agents (Tazawa *et al.*, 2000).

Cyclic Lipopeptides: Cyclic lipopeptides (CLPs) are produced by both gram-positive and gram-negative bacteria. Different kinds of CLP are produced by fluorescent *Pseudomonas* spp. All CLPs have either 9 or 11 amino acids in the peptide ring with a C10 fatty acid at one of the amino acids (Nielsen *et al.*, 2000). Its synthesis is non-ribosomal and catalyzed by large peptide synthetase complexes. CLP has antimicrobial (Takesako *et al.*, 1993; Gerard *et al.*, 1997; Vollenbroich *et al.*, 1997) and biosurfactant properties (Rosenberg and Ron, 1999).

Lipopeptide Antibiotics

Bacillus strains produce a broad spectrum of bioactive peptides. A well-known class of such compounds includes the lipopeptides surfactins, fengycin and the iturins compounds (iturins, mycosubtilins and bacillomycins), which are amphiphilic membrane active biosurfactants and peptide antibiotics with potent antimicrobial activities.

Several strains of *B. subtilis* produce cyclic lipopeptides, which belong to the family Iturin. Iturin A and other antibiotics of their family bacillomycin L, bacillomycin

D, bacillomycin F and mycosubtilins are powerful antifungal agents. Iturin A is a cyclolipopeptide containing seven residues of alpha and one residue of beta amino acid. Iturin A has strong antimicrobial action in suppressing *P. ultimum, R. solani, F. oxysporum, S. sclerotiorum and M. phaseolina* (Constantinescu, 2001).

Aminopolyols

Zwittermicin A: Zwittermicin A is a novel bioactive molecule produced by *Bacillus* sp. It is an aminopolyol antibiotic having structural similarities to polyketide antibiotics with broad spectrum of action against various microbes (Silo-Suh *et al.*, 1998; Stohl *et al.*, 1999). It is produced by *B. cereus* and *B. thuringiensis* (Raffel *et al.*, 1996) and effective against oomycetes and other pathogenic fungi (Silo-Suh *et al.*, 1998).

Volatile Antibiotics

Hydrogen Cyanide (HCN): Cyanide is a secondary metabolite produced by gram-negative *P. fluorescens, P. aeruginosa*, and *Chromobacterium violaceum* (Askeland and Morrison, 1983). Hydrogen cyanide (HCN) and CO_2 are formed from glycine catalyzed by HCN synthase. HCN synthase of *Pseudomonas* sp. oxidize glycine in the presence of electron acceptors, *e.g.*, phenazine methosulfate to form HCN. HCN producing *P. fluorescens* strains are reported to protect several plants from root diseases caused by soil borne fungi (Voisard *et al.*, 1994).

3.0 IDENTIFICATION AND CHARACTERIZATION OF PGPR

3.1 Phenotypic and Chemotaxonomic Approaches

PGPR are distributed widely in many families and orders. They do not belong to a single genus or family. Therefore it is not possible to concise PGPR to a specific family or order. However through taxonomic approaches, can to study the relationships among the bacterial species and classify them into appropriate groups (Coenye *et al.*, 2005). The accurate comparison of organisms depends on a reliable taxonomic system.

Although many new characterization methods have been developed, the basic principle of identification mainly depends on phenotypic characters (traditional biochemical, morphological, and physiological characters) and chemotaxonomic characters (such as polyacrylamide gel electrophoresis (PAGE), and fatty acid methyl ester (FAME) profiles).

Phenotype includes morphological, physiological, and biochemical properties of the microorganism. Traditional phenotypic tests used comprise colony morphology (color, dimensions, form) and microscopic appearance of the cells (shape, endospore, flagella, inclusion bodies), Gram's staining, characteristics of the organism on different growth substrates, growth range of microorganisms on different conditions of salt, pH, and temperature, and susceptibility toward different kinds of antimicrobial agents, etc. Biochemical tests in bacterial identification include the relationship with oxygen, fermentation reactions, and nitrogen metabolism. Other tests may be performed as appropriate, depending on the bacterial strains studied (Heritage *et al.*, 1996).

However, reproducibility of results from phenotypic tests between different laboratories is a great problem, and only standardized procedure should be used during execution of experiment. Other major disadvantage with phenotypic methods is the conditional nature of gene expression wherein the same organism might show different phenotypic characters in different environmental conditions. Therefore, phenotypic data must be compared with similar set of data from type strain of closely related organism(s).

Chemotaxonomic fingerprinting techniques are applied to PGPR identification. The techniques include FAME profiling, PAGE analysis of whole-cell proteins, polar lipid analysis, quinone content, cell wall diamino acid content, pyrolysis mass spectrometry, Fourier transform infrared spectroscopy, Raman spectroscopy, and Matrix-assisted laser desorption/ionization time-of-flight (MALDI-TOF) mass spectrometry. Among these techniques, Pyrolysis mass spectrometry, Fourier transform infrared spectroscopy, and UV resonance Raman spectroscopy are sophisticated analytical techniques which examine the total chemical composition of bacterial cells. These methods have been used for taxonomic studies of particular groups of bacteria, including the members of the family Bacillaceae (Vandamme *et al.*, 1996; Logan *et al.*, 2009).

Fatty acids are the major constituents of lipids and lipopolysaccharides and have been used extensively for taxonomic purposes. FAME analysis is linked to commercial database for identification purposes. Fatty acid profiles showing variability in chain length, positions of double-bond, and substituent groups are excellent tools for taxon description and also for comparative analyses of similar profiles (Suzuki *et al.*, 1993).

SDS-PAGE of whole-cell proteins can be a powerful tool in chemotaxonomy, but it requires standardized conditions of growth, rigorously standardized procedure for analysis, and normalization of the data for computer-assisted comparison of the results (Logan *et al.*, 2009).

Determination of the cell wall composition has traditionally been important in Gram-positive bacteria which contain various peptidoglycan types. The peptidoglycan type of Gram-negative bacteria is rather uniform and provides little information. However in the cytoplasmic membranes of gram negative groups, the major structural units are quinones. They play important role in electron transport, oxidative phosphorylation, and active transport (Collins and Jones, 1981). The large variability of the quinones as Isoprenoid quinones, naphthoquinones and benzoquinones can be used to characterize bacteria into different taxons (Figueiredo *et al.*, 2010).

However a single phenotypic technique will not be suitable for identifying all bacterial species. Therefore, the potentials of chemotaxonomic analyses and studies of nucleic acids have been investigated and a polyphasic approach is introduced for an efficient classification. Polyphasic approach refers to the integration of genotypic, chemotypic and phenotypic information of a microbe in order to perform reliable grouping of the organism (Colwell, 1970).

3.2 Molecular Approaches

Phylogenetic information of bacteria has increased greatly after the advent of molecular methods for the measurement of genetic relatedness. The availability of several sensitive and accurate PCR-based methods has enabled differentiation among closely related bacterial strains and the detection of higher rhizobacterial diversity (Tan *et al.,* 2001). Ribosomal RNA sequence analysis has been extensively used to study phylogenetic relationships between microorganisms and taxon identification (Woese, 1987; Woese *et al.,* 1985). Conserved regions of the 16S rRNA gene (*rrs*) is useful not only for studying phylogenetic relationships but also to design oligonucleotide probes and primers used for identification by hybridizations and PCR-amplifications, respectively (Givannoni, 1991). Variable regions of 16S rRNA provide sequence data for detection of the bacteria (Stahl and Amann, 1991; Ward *et al.,* 1991). The availability and use of PCR based amplification methods and sequencing of the PCR products on automated sequencers has dramatically expanded RNA databases during the past few years. Below are given some primers used to amplify the 16S rRNA gene segment in various studies.

Table 12.3: Primers for 16S rDNA Amplification

Primers	*Sequence*	*Reference*
Universal primer	518F(CCAGCAGCCGCGGTAATACG) 800R (TACCAGGGTATCTAATCC)	Bhromsiri and Bhromsiri, 2010
Universal primer	27F (AGAGTTTGATCMTGGCACAG) 1429R (TACGGYTACCTTGTTACGACTT).	Bhromsiri and Bhromsiri, 2010
Universal primer	FGPS4-281 BIS: AGA GTT TGA TCC TGG CTC AG FGPS1509-153: AAG GAG GTG ATC CAG CCG CA	Mehnaz *et al.,* 2001
Universal primer	FGPS5-255 (5B-TGGAAAGCTTGATCCTGGCT-3B) FGPS1509B- 153 (5B-AAGGAGGGGATCCAGCCGCA-3B)	Bertrand *et al.,* 2001
Universal primer	PA (5'AGAGTTTGATCCTGGCTAG-3') PH (5'AGGAGGTGATCCAGCCGCA-3')	Upadhyay *et al.,* 2009
Actinobacter	243F (5'GGA TGA GCC CGC CGC CTA 3') 1492R (5'TA CGG GTA CCT TGT TAC GAC TT 3')	Conn *et al.,* 2004

The revolutionary technological developments in high throughput DNA sequencing have resulted in the publication of many whole-genome sequences. The sequencing of approximately 35 microbial genomes has been completed and those of another 150 are in progress. Among these are several rhizosphere inhabiting bacteria such as *P. aeruginosa* (Stover *et al.,* 2000), *P. putida, P. fluorescens, P. syringae pathovar tomato, Sinorhizobium meliloti, Mesorhizobium loti* (Kaneko *et al.,* 2000), *Bacillus subtilis* (Kunst *et al.,* 1997) and *Streptomyces coelicolor.*

The genes that are specifically present in (plant beneficial) rhizosphere bacteria, that are involved in the regulation and production of secondary metabolites (*e.g.* anti-fungal metabolites), expressed on roots or in the rhizosphere, whose expression is influenced by other rhizosphere organisms (such as fungi), can be identified by comparative and functional genomics.

The construction of bacterial artificial chromosome (BAC) libraries for the study of gene expression and to identify genes of interest is of great value, especially in the study of bacteria whose genome has not been sequenced (Rondon *et al.,* 1999, Bloemberg and Lugtenberg, 2001).

The nucleotide information and tools available in NCBI and NLM (National Library of Medicine, Bethesda, USA) database are of infinite value in phylogenetic analysis of any bacterial strain. The indispensable tools in molecular phylogeny are BLAST for homology search, CLUSTAL W, CLUSTAL X (Thompson *et al.,* 1997) for data alignment, Neighbour – Joining method (Saitou and Nei, 1987) for analyzing the relationship of the sequences and Bootstrap method (Felsenstein, 1985) to evaluate the tree topology (Zou *et al.,* 2010). Software packages like MEGA (V 4.0) are available for performing phylogenetic analysis of the aligned data.

3.3 Chromatographic Identification

Plant growth excretion is one of the chief plant growth promotion mechanisms of the PGPR bacteria. The instigation of the seed or root with plant growth promoters has been attributed to plant growth enhancement. Rhizosphere microflora are the sources of these materials instigation and many different micro-organisms have the ability to exude different plant growth promoters. *Azotobacter, Pseudomonas, Azosprillium, Rhizobium, Bacillus, Enthrobacter* and *Mycorrhiza* fungus are some of the PGPR organisms capable of produce herbal hormones.

Chromatographic techniques are of great help in identifying these volatiles and plant growth promoting hormones thereby identifying PGPR. TLC is commonly used to detect the production of auxins. Gas Chromatography coupled with Mass Spectroscopy has been effective in some studies to detect auxin production. HPLC method is found to be superior to all of the other methods including TLC and GC. Studies by Crozier and Reeve also suggest that HPLC is a powerful method for simplifying the auxins' identification in comparison with mass spectrophotometer method as in HPLC the IAA separated is measurable and the amount of unexpected substances or impurities in samples is reduced thus giving precise results (Khakipour *et al.,* 2008).

4.0 DEVELOPMENT OF PGPR BIOFERTILIZERS

Characteristics of an ideal formulation as described by Jeyarajan and Nakkeeran (2000) are that it must

- ☆ Have increased shelf life
- ☆ Not be phytotoxic to the crop plants
- ☆ Dissolve well in water and should release the bacteria
- ☆ Tolerate adverse environmental conditions
- ☆ Be cost effective and should give reliable control of plant diseases
- ☆ Be compatible with other agrochemicals
- ☆ Carriers must be cheap and readily available for formulation development

Talc, peat, vermiculate, etc. are some of the inert carriers which are proposed as substrates for bioformulation. In a study they have shown that the bacteria can survive well in talc or peat based formulations more than 8 months. PGPR have already been reported to survive in certain dry formulations. Population of PGPR did not decline in talc mixture with 20 per cent xanthan gum after storage for 2 months at 4°C. However vermicompost and farmyard manure (Chakravarty and Kalita, 2011) when used as substrate carrier in conjunction with carboxymethyl cellulose (CMC) as adhesive in formulations are able to provide better nutrient sources and congenial microenvironment required for proper growth and subsequent longer shelf life compared to inert carriers.

Some reports hold that the performance of the rhizosphere strains that performed well under in vitro conditions are not consistent under field conditions. This inconsistent performance can be overcome by the combined application of several rhizosphere strains that mimic the natural environment (Schisler *et al.,* 1997; Raupach and Kloepper, 1998). Also combined application of PGPR bioformulation, as seed treatment, root and soil application and foliage treatment was the most effective method for reducing the disease. The combined rhizosphere and phyllosphere population of PGPR strain can effectively control onset of fungal plant diseases at early stages Root zone application can be done again to control phytopathogens in later stages (Vidhyasekaran and Muthamilan, 1999).

In various studies, it is proven that root inoculation with combined PGPR can promote significant increase in growth and alkaloid content although the growth response may vary between different rhizobacterial strains. In general it can be concluded that the growth response (in terms of increased plant growth and alkaloid production compared to control) can be enhanced when the PGPR strains are applied in combination. Earlier reports have shown that combined inoculation of *Azospirillum, Azotobacter chrococcum, Pseudomonas fluorescens* and *Bacillus megaterium* significantly increased grain yield in sorghum. The stimulatory effects on the yield and growth of the sorgham crops were attributed to the combined effect of the PGPRs applied *viz.,* N_2 fixation ability, plant growth regulator production and phosphate solubilizing capacity respectively (Cakmakci *et al.,* 2007; Kevinvessy, 2003; Karilag *et al.,* 2007). So it can be drawn to a conclusion that a combination of microbial consortium strains have a great potential for use as bioinoculants to increase production in medicinal plants and other crops (Rajasekar and Elango, 2011).

5.0 LIMITATIONS AND SHORTCOMINGS OF PGPR BIOFERTILIZERS

For the effective management of any soil borne disease, the introduced antagonist should colonize the roots. It is difficult to introduce an antagonist to the established rhizosphere since once a portion of the root is occupied by the native bacteria, the introduced bacteria are unable to displace them. The successful antagonist should colonize the rhizosphere at seed germination stage. The PGPR should move from the spermosphere to the rhizosphere and establish there. Seed treatment with cell suspensions *P. fluorescens* has found effective in controlling several diseases. For commercial exploitation this method will be impractical due to difficulty in handling, transport, and storage.

Plant growth promoting bacterial strains must be rhizospheric competent, able to survive and colonize in the rhizospheric soil (Cattelan *et al.,* 1999). Unfortunately, the interaction between associative PGPR and plants can be unstable and under field conditions results may fluctuate from the results obtained under *in vitro* conditions (Chanway and Holl, 1993; Zhender *et al.,* 1999). Variability in performance of PGPR may be due to various environmental factors that may affect their growth and exert their effects on plant. The environmental factors include climate, weather conditions, soil characteristics or the composition or activity of the indigenous microbial flora of the soil. However these difficulties can be overcome by exploring potential plant growth promoting bacteria from the native population or by employing highly competent rhizosphere strains.

6. FUTURE PROSPECTS

The use of PGPR offers an attractive way to replace chemical fertilizer, pesticides, and supplements; most of the isolates result in a significant increase in plant height, root length, and dry matter production of shoot and root of plants. PGPR help in the disease control in plants. Some PGPR especially if they are inoculated on the seed before planting, are able to establish themselves on the crop roots. PGPR as a component in integrated management systems in which reduced rates of agrochemicals and cultural control practices are used as biocontrol agents.

To achieve the maximum growth promoting interaction between PGPR and nursery seedlings it is important to discover how the rhizobacteria exerting their effects on plant and whether the effects are altered by various environmental factors, including the presence of other microorganisms (Bent *et al.,* 2001). By exploring soil microbial diversity for PGPR having combination of PGP activities and its adaption to particular soil environment, it is possible to develop biofertilizer which is efficient in field conditions.

Rhizosphere isolates, especially those with phosphate solubilizing activity can be applied with complex phosphorus fertilizers to enhance time to time solubilization from the complex source thereby preventing fertilizer run off and inefficiency.

In last few decades a large array of bacteria including species of *Pseudomonas, Azospirillum, Azotobacter, Klebsiella, Enterobacter, Alcaligenes, Arthrobacter, Burkholderia, Bacillus, Rhizobium* and *Serratia* have reported to enhance plant growth. Genetic enhancement of these potent PGPR strains to enhance colonization and effectiveness may involve addition of one or more traits associated with plant growth promotion (Bloemberg and Lugtenberg, 2001; Glick, 1995; Lubeck *et al.,* 2000). Genetic manipulation of host crops for root-associated traits to enhance establishment and proliferation of beneficial microorganisms (Mansouri *et al.,* 2002; Smith and Goodman, 1999) is being pursued. The use of multi-strain inocula of PGPR with known functions is of interest as these formulations may increase consistency in the field (Jetiyanon and Kloepper, 2002; Siddiqui, 2002). They offer the potential to address multiple modes of action, multiple pathogens, and temporal or spatial variability. PGPR offer an environmentally sustainable approach to increase crop production and health. The application of molecular tools is enhancing our ability to understand and manage the rhizosphere and will lead to new products with improved effectiveness (Nelson, 2004).

7.0 SUMMARY

As our understanding of the complexity of the rhizosphere environment, the mechanisms of action of PGPR, and the development of inoculant formulation and delivery systems, we can have new PGPR products available in market. The ultimate success of these biofertilizers greatly depends on the survival and competence of the beneficial organisms in the rhizosphere (Bowen and Rovira, 1999).

In developing countries like India, demand of chemical fertilizers for crop production has increased tremendously due to the release of several high yielding and nutrient demanding varieties of crop plants. The use of chemical fertilizers has resulted not only in the deterioration of soil health but also has led to some major environmental problems, such as soil and water pollution and other health related problems, besides increasing the input cost for crop production especially on the marginal farmers. So, there is an urgent need to recycle available organics and manipulate rhizospheric microflora in a more efficient way and improve and expand their usage. Search for ecologically adaptable Plant Growth Promoting Rhizobacteria with enhanced plant growth promoting properties and their use to enhance crop productivity can improve the socio-economic status of poor farmers.

If proper and efficient delivery system is developed, PGPR based biofertilizers can be used as effective means to reduce chemical fertilizer usage and prevent contamination of water and soil. The process of medicinal plants cultivation thus creates the need for interdisciplinary studies on rhizosphere biology, microbiology, ecology and agricultural technology to develop effective methods of biomass production and obtaining quality material enriched with phytochemicals.

REFERENCES

Ahmad, F., Ahmad, I., Khan, M.S., 2008. Screening of Free-living Rhizospheric Bacteria for their multiple plant growth promoting activities, *Microbiological Research* 163, 173-181.

Andrade, G., Mihara, K.L., Lindermann, R.G., Bethlenfalvay, G.J., 1997. Bacteria from Rhizosphere and Hyphosphere soils of different arbuscular Mycorrhizal fungi, *Plant and Soil* 192, 71 -79.

Arima, K., Imanaka, H., Kousaka, M., Fukuda, A., Tamura, G., 1964. Agric. Biol. Chem. 28, 575-576.

Arrebola, E., Sivakumar, D., Korsten, L., 2010. Effect of volatile compounds produced by *Bacillus* strains on postharvest decay in citrus, *Biol. Control* 53, 122 – 128.

Arshad, M., Frankenberger Jr.W.T., 1998. Plant growth regulating substances in the rhizosphere: Microbial Production and Functions, *Adv. Agron.* 62, 46-151.

Ashrafuzzaman, M., Hossen, F.A., Ismail, M.R., Hoque, M., Islam, Z.M., Shahidullah, S.M., Meon, S., 2009. Efficiency of Plant Growth-Promoting Rhizobacteria (PGPR) for the enhancement of rice growth, *African Journal of Biotechnology* 8(7), 1247-1252.

Askeland, R.A., Morrison, S.M., 1983. Cyanide production by *Pseudomonas fluorescens* and *Pseudomonas aeruginosa, Appl. Environ. Microbiol.* 45, 1802-1807.

Azcon, R., 1989. Selective interaction between free living rhizosphere bacteria and vescicular – arbuscular Mycorrhizal fungi, Soil biology and Biochemistry 21(5), 639 – 644.

Azcón-Aguilar, C., Bago, B., 1994. Physiological characteristics of the host plant promoting an undisturbed functioning of the mycorrhizal symbiosis, in: S. Gianinazzi, H. Schüepp (Eds.), Impact of Arbuscular Mycorrhizas on Sustainable Agriculture and Natural Ecosystems, Birkhäuser-Verlag, Basel, pp. 47–60.

Barea, J.M., Azcón, R., Azcón-Aguilar, C., 2002. Mycorrhizosphere interactions to improve plant fitness and soil quality, Antonie van Leeuwenhoek. *International Journal of General and Molecular Microbiology* 81, 343–351.

Barea, J.M., 2000. Rhizosphere and mycorrhiza of field crops, in: E. Balázs, E. Galante, J.M. Lynch, J.S. Schepers, J.P. Toutant, D. Werner, J. Werry Path (Eds.), *Biological resource management: connecting science and policy.* Berlin, Heidelberg, New York, INRA Editions, Springer-Verlag, pp. 110–125.

Baron, S.S., Rowe, J.J., 1981. Antibiotic Action of pyocyanin, Antimicrobial Agents Ch 20, 814 – 820.

Benhamou, N., Nicole, M., 1999. Cell biology of plant immunization against microbial infection: The potential of induced resistance in controlling plant diseases, *Plant Physiology and Biochemistry* 37(10), 703-719.

Bent, E., Tuzun, S., Chanway, C.P., Enebak, S., 2001. Alterations in plant growth and in root hormone levels of lodgepole pines inoculated with rhizobacteria, *Canadian Journal of Microbiology* 47, 793–800.

Bertrand, H., Nalin, R., Bally, R., Cleyet-Marel J.C., 2001. Isolation and identification of the most efficient plant growth-promoting bacteria associated with canola (*Brassica napus*), *Biol. Fertil. Soils* 33, 152–156

Bhromsiri, C., Bhromsiri, A., 2010. The effect of plant growth-promoting rhizobacteria and arbuscular mycorrhizal fungi on the growth, development and nutrient uptake of different vetiver ecotypes, *Thai J. Agric. Sci.* 43 (4), 239–249.

Bloemberg, G.V., Lugtenberg, B.J.J., 2001. Molecular Basis of Plant Growth Promotion and Biocontrol by Rhizobacteria, *Current Opinion in Plant Biology* 4(4), 343-350.

Bowen, G.D., Rovira, A.D., 1999. The rhizosphere and its management to improve plant growth. *Advances in Agronomy* 66, 1-102.

Budi, S.W., Van Tuinen, D., Martinotti, G., Gianinazzi, S., 1999. Isolation from *Sorghum bicolor* mycorrhizosphere of a bacterium compatible with *Arbuscular mycorrhiza* development and antagonistic towards soil-borne fungal pathogens. *Applied and Environmental Microbiology* 65, 5148–5150.

Burkhead, K.D., Schisler, D.A., Slininger, P.J., 1994. Pyrrolnitrin production by biological control agent *Pseudomonas cepacia* B37W in culture and in colonized wounds of potato, *Appl. and Environ. Microbiol.* 60(6), 2031-2039.

Cakmakci, R., Donmez, M.F., Erdogan, U., 2007. The effect of plant growth promoting rhizobacteira on barley seedling growth, some soil properties and bacterial counts, *Turkish Journal of Agriculture and Forestry* 31, 189-199.

Carpenter-Boggs, L., Loynachan, T.E., Stahl, P.D., 1995. Spore germination of *Gigaspora margarita* stimulated by volatiles of soil-isolated actinomycetes, *Soil Biol. Biochem.* 27, 1445–1451.

Cattelan, A.J., Hartel, P.G., Fuhrmann, J.J., 1999. Screening for Plant Growth–Promoting Rhizobacteria to Promote Early Soybean Growth, *Soil Science Society of America Journal* 63, 1670–1680.

Chakravarty, G., Kalita, M.C., 2011. Comparative evaluation of organic formulations of *Pseudomonas fluorescens* based biopesticides and their application in the management of bacterial wilt of brinjal (*Solanum melongena* L.), *African Journal of Biotechnology* 10(37), 7174-7182.

Chanway, C.P., Holl, F.B., 1993. First year yield performance of spruce seedlings inoculated with plant growth promoting rhizobacteria, *Canadian Journal of Microbiology* 39, 1084–1088.

Chin-A-Woeng, T.F., Bloemberg, G.V., Lugtenberg, B.J., 2003. Phenazines and their role in biocontrol by *Pseudomonas* bacteria, *New Phytologist* 157, 503-523.

Choudhary, D.K., Johri, B.N., 2009. Interactions of *Bacillus* spp. and plants – With special Reference to Induced Systemic Resistance (ISR), *Microbiological Research* 164, 493-513.

Coenye, T., Gevers, D., Van de Peer, Y., Vandamme, P., Swings, J., 2005. Towards a prokaryotic genomic taxonomy, *FEMS Microbiol. Rev.* 29, 147 – 167.

Collins, M.D., Jones, D., 1981. Distribution of isolprenoid quinone structural types in bacteria and their taxonomic implications, *Microbiol. Rev.* 45, 316-354.

Colwell, R.R., 1970. Polyphasic taxonomy of genus *Vibrio*: numerical taxonomy of *Vibrio cholera*, *Vibrio parahaemolyticus* and related *Vibrio* species, *J. Bacteriol.* 104, 410-433.

Conn, V.M., Franco, C.M.M., 2004. Analysis of the endophytic actinobacterial population in the roots of wheat (*Triticum aestivum* L.) by terminal restriction fragment length polymorphism and sequencing of 16S rRNA clones, *Appl. Environ. Microbiol.* 70, 1787-1794.

Constantinescu, F., 2001. Extraction and identification of antifungal metabolites produced by some *B. subtilis* strains, Analele Institutului de Cercetari Pentru Cereale Protectia Plantelor 31, 17-23.

Daniels, B.A., Trappe, J.M., 1980. Factors affecting spore germination of the vesicular arbuscular mycorrhizal fungus, *Glomus epigaeus*. *Mycologia* 72, 457-471.

Ekin, Z., 2010. Performance of phosphate solubilizing bacteria for improving growth and yield of sunflower (*Helianthus annuus* L.) in the presence of phosphorus fertilizer, *African Journal of Biotechnology* 9, 25, 3794-3800.

Emmert, E.A., Klimowicz, A.K., Handelsman, M.G., 2004. Genetics of Zwittermycin A production by *Bacillus cereus, Appl. Environ. Microbiology* 70, 104-113.

Fatima, Z., Saleema, M., Zia, M., Sultan, T., Aslam, M., Riaz-Ur-Rehman, Chaudhary, M.R., 2009. Antifungal activity of plant growth promoting rhizobacteria isolates against *Rhizoctonia solani* in wheat, *African Journal of Biotechnology* 8(2), 219-225.

Felsenstein, J., 1985. Confidence limits on phylogenies: an approach using the bootstrap, *Evolution* 39, 783-789.

Figueiredo, M.V.B., Seldin, L., Araujo, F.F., Mariano, R.R., 2010. Plant Growth Promoting Rhizobacteria: Fundamentals and Applications. In: Maheshwari, D.K. (Eds.), Plant Growth and Health Promoting Bacteria, Microbiology Monographs. Springer-Verlag Berlin Heidelberg, pp. 21-47.

Frankenberger Jr., W.T., Arshad, M., 1995. Phytohormones in Soil: Microbial production and function, New York, Marcel Dekker, pp. 503.

Fuentes-Ramirez, T., Jimenez-Salgade, I.R., Abarceocampo, J. Caballero-Mellado, *Acetobacter diazotrophicus*, an IAA producing bacterium isolated from sugarcane cultivars of Mexico, *Plant Soil* 154, 145-150.

Fuller, A.T., Mellows, G., Woolford, M., Banks, G.T., Barrow, K.D., Chain, E.B., 1971. Pseudomonic acid: an antibiotic produced by *Pseudomonas fluorescens*, Nature 234, 416-417.

Gamalero, E., Lingua, G., Capri, F.G., Fusconi, A., Berta, G., Lemanceau, P., 2004. Colonization pattern of primary tomato roots by *Pseudomonas fluorescens* A6RI characterized by dilution plating, flow cytometry, fluorescence, confocal and scanning electron microscopy, *FEMS Microbiology Ecology* 48, 79-87.

Gamard, P., Sauriol, F., Benhamou, N., Belanger, R.R., Paulitz, T.C., 1997. Novel butyrolactones with antifungal activity produced by *Pseudomonas aureofaciens* strain 63-28, *Journal of Antibiotics* 50, 742-9.

Gerard, J.R., Lloyd, T., Barsby, P., Haden, M., Kelly, T., Andersen, R.J., Massetolides, A-H, 1997. antimycobacterial cyclic depsipeptides produced by two pseudomonads isolated from marine habitats, *J Nat Prod* 60, 223-229.

Givannoni, S., 1991. The polymerase chain reaction. In, E. Stackebrandt, M. Goodfellow, Nucleic acid techniques in bacterial systematics, (Eds.), John Wiley and Sons, New York, pp. 177–203.

Glick, B.R., 1995. The enhancement of plant growth by free living bacteria, *Canadian Journal of Microbiology* 41(2), 109–114.

Gnanamanickam, S.S., 2009. Biological control of Rice diseases, Progress in biological control 8, 1-93.

Gryndler, M., 2000. Interactions of Arbuscular Mycorrhizal Fungi with other Soil Organisms. In: Y. Kapulnik, D.D. Douds Jr. (Eds.), Arbuscular Mycorrhizas:

Physiology and Function, Kluwer Academic Publishers, Dordrecht, Netherlands, 239-262.

Harley, J.L., Smith, S.E., 1983. Mycorrhizal symbiosis. London, New York: Academic Press.

Hassett, D.J., Charniga, L., Bean, K., Ohman, D.E., Cohen, M.S., 1992. Response of *Pseudomonas aeruginosa* to pyocyanin: mechanisms of resistance, antioxidant defenses, and demonstration of manganese - cofactored superoxide dismutase, *Infection and Immunity* 60 (1992) 328 - 336.

Hassett, D.J., Woodruff, W.A., Wozniak, D.J., Vasil, M.L., Cohen, M.S., Ohman, D.E., 1993. Cloning of *sodA* and *sodB* genes encoding manganese and iron superoxide dismutase in *Pseudomonas aeruginosa*: demonstration of increased manganese superoxide dismutase activity in alginate-producing bacteria, *J. Bacteriol.* 175 (1993) 7658–7665.

Heritage, J., Evans, *E.G.*V., Killington, R.A., 1996. Introductory Microbiology, Cambridge University Press, England, 234.

Hokeberg, M., Wright, S.A.I., Svensson, M., Lundgren, L.N., Gerhardson, B., 1998. Mutants of *Pseudomonas chlororaphis* defective in the production of an antifungal metabolite express reduced biocontrol activity, Edinburgh, Scotland, Abstract Proceedings IC, 98.

Howel, C.R., Stipanovic, R.D., 1980. Suppression of *Pythium ultimuminduced* damping-off on cotton seedlings by *Pseudomonas fluorescens* and its antibiotic, pyoluteorin, *Phytopathology* 70, 712-715.

Jetiyanon, J., Kloepper, J.W., 2002. Mixtures of plant growth-promoting rhizobacteria for induction of systemic resistance against multiple plant diseases, *Journal of Biology* 24(3), 285-291.

Jeyarajan, R., Nakkeeran, S., 2000. Exploitation of microorganisms and viruses as biocontrol agents for crop disease mangement. In: Upadhyay *et al.* (Eds.), Biocontrol Potential and their exploitation in Sustainable agriculture, Kluwer Academic/Plenum Publishers, USA, 95-116.

Jiao, Y., Yoshihara, T., Ishikuri, S., Uchino, H., Ichihara, A., 1996. Structural identification of cepaciamide A, a novel fungitoxic compound from *Pseudomonas cepacia* D-202, Tetrahedron Lett. 37, 1039-1042.

Joshi, R., McSpadden Gardener, B., 2006. Identification and characterization of novel genetic markers associated with biological control activities of *Bacillus subtilis*, *Phytopathology* 96, 145-154.

Kaneko, T., Nakamura, Y., Sato, S., Asamizu, E., Kato, T., Sasamoto, S., Watanabe, A., Idesawa, K., Ishikawa, A., Kawashima, K., 2000. Complete genome structure of the nitrogen-fixing symbiotic bacterium *Mesorhizobium loti*, *DNA research* 7, 331-338.

Karlidag, H.A. Esitken, Turan, M., Sahin, F., 2007. Effects of root inoculation of plant growth promoting rhizobiacteria (PGPR) on yield, growth and nutrient element contents of apple, *Scientia Horticulture* 114, 16-20.

Kevinvessey, J., 2003. Plant growth promoting Rhizobacteria on Radishes. Angers (Ed.) Gibert - Clarey, Tours, 879.

Khakipour, N., Khavazi, K., Mojallali, H., Pazira, E., Asadirahmani, H., 2008. Production of auxin hormone by fluorescent Pseudomonads, *American – Eurasian Journal of Agricultural and Environmental Sciences* 4(6), 687-692.

Kloepper, J., Tuzun, S., Kuc, J., 1992. Proposed Definitions Related to Induced Disease Resistance, *Biocontrol Science and Technology* 2, 347-349.

Koenig, R.L., Morris, R.O., Polacco, J.C., 2002. tRNA is the source of low-level *trans*-zeatin production in *Methylobacterium* spp., *J. Bacteriol.* 184, 1832–1842.

Kumar, A., Prakash, A., Johri, B.N., 2011. *Bacillus* as PGPR in crop ecosystem, In: D.K. Maheshwari, (Eds.), Bacteria in Agrobiology: Crop systems, Springer-Verlag Berlin Heidelberg, 37-59.

Kunst, F., Ogasawara, N., Moszer, I., Albertini, A.M., Alloni, G., Azevedo, V., Bertero, M.G., Bessieres, P., Bolotin, A., Borchert, S., 1997. The complete genome sequence of the Gram-positive bacterium *Bacillus subtilis*, *Nature* 390, 249-256.

Linderman, R.G., 1992. In: G.J. Bathlenfalvay, R.G. Linderman, (Eds.), Mycorrhizae in sustainable agriculture, ASA specila publication, Madiso, Wisconsin, 45 – 70.

Logan, N.A., Berge, O., Bishop, A.H., Busse, H.J., De Vos, P., Fritze, D., Heyndrick, M., Kempfer, P., Rabinovitch, L., Salkinoja Salonen, M.S., Seldin, L., Ventosa, A., 2009. Proposed minimal standards for describing new taxa of aerobic, endospore - forming bacteria, *Int. J. Syst. Evol. Microbiol.* 59(8), 2114-2121.

Lubeck, P.S., Hansen, M., Sorensen, J., 2000. Simultaneous detection of the establishment of seed-inoculated *Pseudomonas fluorescens* strain DR54 and native soil bacteria on sugar beet root surfaces using fluorescence antibody and *in situ* hybridization techniques, FEMS Microbiology Ecology 33(1), 11-19.

Lucangeli, C., Bottini, R., 1997. Effects of *Azospirillum* spp. on endogenous gibberellin content and growth of maize (*Zea mays* L.) treated with uniconazol, *Symbiosis* 23, 63-71.

Lucas, J.A., Ramos Solano, B., Montes, F., Ojeda, J., Megias, M., Gutierrez Manero, F.J., 2009. Use of two PGPR strains in the integrated management of blast disease in rice (*Oryza sativa*) in southern Spain, *Field Crops Research* 114, 404 - 410.

Maleck, K., Lawton, K., 1998. Plant strategies for resistance to pathogens, *Current Opinion in Biotechnology* 9, 208-213.

Malleshwari, D., Bagyanarayana, G., 2013. *In vitro* screening of rhizobacteria isolated from the rhizosphere of medicinal and aromatic plants for multiple plant growth promoting activities, *Journal of Microbiological and Biotechnological research* 3(1), 84 – 91.

Mansouri, H., Petit, A., Oger, P., Dessaux, Y., 2002. Engineered rhizosphere: the trophic bias generated by opine-producing plants is independent of the opine type, the soil origin, and the plant species. *Applied and Environmental Microbiology* 68(5), 2562-2566.

Martens, D.A., Frankenberger Jr., W.T., 1991. On-line solid-phase extraction of soil auxins produced from exogenously-applied tryptophan with ion-suppression reverse phase HPLC analysis, *Chromtaographia* 32, 417–422.

Maunch-Mani, B., Metraux, J.P., Salicylic Acid and Systemic Acquired Resistance to pathogen Attack, *Annals of Botany* 82 (5), 535-540.

Mayo, K., Davis, R.E., Motta, J., 1986. Stimulation of germination of spores of *Glomus versiforme* by sporeassociated bacteria, *Mycologia* 78, 426–431.

Mehnaz, S., Mirza, M.S., Haurat, J., Bally, R., Normand, P., Malik, K.A., 2001. Isolation and 16S rRNA sequence analysis of the beneficial bacteria from the rhizosphere of rice, *Can. J. Microbiol.* 47, 110–117.

Meyer, J.R., Linderman, R.G., 1986. Response of subterranean clover to dual inoculation with vesicular-arbuscular fungi and a plant growth-promoting bacterium, *Pseudomonas putida, Soil Biology and Biochemistry* 18, 185-190.

Miller, S.A., Abbasi, P.A., Sahin, F., Al-Dahmani, J., Hoitink, H.A.J., 1998. Control of foliar and fruit diseases of tomato by compost amendments and Actigard, *Phytopath.* 88, 63.

Milner, J.L., Silo-Suh, L., Lee, J.C., He, H., Clardy, J., Handelsman, J., 1996. Production of kanosamine by *Bacillus cereus* UW85, *Appl. Environ. Microbiol.* 62, 3061–3065.

Mosse, B., 1959. The regular germination of resting spores and some observations on the growth requirements of an *Endogone* sp. causing vesicular-arbuscular mycorrhiza, Transactions of the British mycological society 42, 273-286.

Nandakumar, R., Babu, S., Viswanathan, R., Raguchander, T., Samiyappan, R., 2001. Induction of systemic resistance in rice against sheath blight disease by *Pseudomonas fuorescens, Soil Biology and Biochemistry* 33, 603-612.

Nelson, L.M., 2004. Plant Growth Promoting Rhizobacteria (PGPR): Prospects for New Inoculants, Crop Management.

Nielsen, T.H., Thrane, C., Christophersen, C., Anthoni, U., Sorensen, J., 2000. Structure, production characteristics and fungal antagonism of tensin–a new antifungal cyclic lipopeptide from *Pseudomonas fluorescens* strain 96.578, *J. Appl. Microbiol.* 89, 992-1001.

Patten, C.L., Glick, B.R., 1996. Bacterial biosynthesis of IAA, *Canadian Journal of Microbiology* 42, 207-220.

Pieterse, C.M.J., Van Wees, S.C.M., Hoffland, E., Van Pelt, J.A., Van Loon, L.C., 1996. Systemic Resistance in Arabidopsis Induced by biocontrol Bacteria is Independent of Salicylic Acid Accumulation and Pathogenesis – Related Gene Expression, *The Plant Cell* 8, 1225-1237.

Raffel, S.J., Stabb, E.V., Milner, J.L., Handelsman, L., 1996. Genotypic and phenotypic analysis of zwittermicin A-producing strains of *Bacillus cereus, Microbiology* 142, 3425–3436.

Rajasekar, S., Elango, R., 2011. Effect of microbial consortium on plant growth and improvement of alkaloid content in *Withania somnifera* (Ashwagandha), *Current Botany* 2(8), 27-30.

Ramamoorthy, V., Viswanathan, R., Raguchander, T., Prakasam, V., Samiyappan R., 2001. Induction of systemic resistance by plant growth promoting rhizobacteria in crop plants against pests and diseases, *Crop Protection* 20, 1-11.

Rambelli, A., 1973. The rhizosphere of mycorrhizae. In: G.C. Marks, T.T. Kozlowski, (Eds.), Ectomycorrhizae: Their Ecology and Physiology, Academic Press, New York, NY. 299-350.

Ramos Solano, B., Barriuso Maicas, J., Pereyra De La Iglesia, M.T., Domenech, J., Gutierrez Manero, F.J., 2008. Systemic disease protection elicited by plant growth promoting rhizobacteria strains: relationship between metabolic responses, systemic disease protection and biotic elicitors, *Phytopathology* 98, 451–457.

Raupach, G.S., Kloepper, J.W., 1998. Mixtures of Plant growth – promoting rhizobacteria enhance biological control of multiple cucumber pathogens, *Phytopathology* 88(11), 1158-1164.

Reddy, B.P., Reddy, M.S., Krishnakumar, K.V., 2009. Characterization of Antifungal Metabolites of Pseudomonas flourescens and their effect on mycelia growth of *Magnaporthe griseae* and *Rhizoctonia solani, International Journal of Pharm. Tech. Research* 1(4), 1490-1493.

Rodríguez, H., Fraga, R., 1999. Phosphate solubilizing bacteria and their role in plant growth promotion, *Biotechnology Advances* 17, 319–339.

Rondon, M.R., Raffel, S.J., Goodman, R.M., Handelsman, J., 1999. Toward functional genomics in bacteria: analysis of gene expression in *Escherichia coli* from a bacterial artificial chromosome library of *Bacillus cereus, Proc. Natl. Acad. Sci., USA*, 96, 6451-6455.

Rosenberg, E., Ron, E.Z., 1999. High- and low-molecular-mass microbial surfactants, Appl. *Microbiol. Biotechnol.* 52, 154-162.

Saharan, B.S., Nehra, V., 2011. Plant growth promoting rhizobacteria: A critical review, Life Sciences and Medicine research 21, 1 -30.

Saitou, N., Nei, M., 1987. The neighbor-joining method: a new method for reconstructing phylogenetic trees, *Molecular Biology and Evolution* 406-425.

Sarode, P.D., Rane, M.R., Chaudari, B.L., 2009. Chincholkar, S.B., 2009. Siderophorogenic *Acinetobacter calcoaceticus* isolated from Wheat Rhizosphere with Strong PGPR Activity, *Malaysian Journal of Microbiology* 5(1), 6-12.

Savithramma, N., Linga Rao, M., Beenaprabha, 2011. Phytochemical studies of *Dysophylla myosuroides* (*Roth.*) Benth. In. wall. and *Talinum cuneifolium* (Vahl.) *Willd, Res. J. Phyto.* 5(3), 163-169.

Schisler, D.A., Slininger, P.J., Bothast, R.J., 1997. Effects of antagonist cell concentration and two-strain mixtures on biological control of *Fusarium* dry rot of potatoes, *Phytopathology* 87, 177-183.

Secilia, J., Bagyaraj, D.J., 1987. Bacteria and actinomycetyes associated with pot cultures of vesicular arbuscular mycorrhizas, *Can. J. Microbiol.* 33, 1069-1073.

Shanahan, P., O'Sullivan, D.J., Simpson, P., Glennon, J.D., O'Gara, F., 1992. Isolation of 2, 4-diacetylphlorogucinol from a fluorescent pseudomonad and investigation of physiological parameters influencing its production, *Applied Environmental Microbiology* 58, 353–358.

Siddiqui, I.A., Shaukat, S.S., 2002. Resistance against damping-off fungus *Rhizoctonia solani* systematically induced by the plant-growth-promoting rhizobacteria *Pseudomonas aeruginosa* (1E-6S(+)) and *P. fluorescens* (CHAO), *Journal of Phytopathology* 150, 500-506.

Silo-Suh, L.A., Lethbridge, B.J., Raffel, S.J., He, H., Clardy, J., Handelsman, J., 1994. Biological activities of two fungistatic antibiotics produced by *Bacillus cereus* UW85, *Applied Environmental Microbiology* 60, 2023-2030.

Silo-suh, L.A., Stab, V.E., Raffel, S.R., Handelsman, J., 1998. Target range of Zwittermicin A, an Aminopolyol antibiotic from *Bacillus cereus, Curr. Microbiol.* 37, 6-11.

Smith, K.P., Goodman, R.M., 1999. Host variation for interactions with beneficial plant-associated microbes, *Annual Review of Phytopathology* 37, 473-491.

Stahl, D.A., Amann, R., 1991. Development and application of nucleic acid probes. In: E. Stackebrandt, M. Goodfellow, (Eds.), Nucleic Acid Techniques in Bacterial Systematics. John Wiley and Sons Ltd., Chichester, UK, 205- 248.

Stohl, E.A., Milner, J.L., Handelsman, J., 1999. Zwittermicin A biosynthetic cluster, Gene 237, 430-411.

Stover, C.K., Pham, X.Q., Erwin, A.L., Mizoguchi, S.D., Warrener, P., Hickey, M.J., Brinkman, F.S., Hufnagle, W.O., Kowalik, D.J., Lagrou, M., 2000. Complete genome sequence of *Pseudomonas aeruginosa* PAO1, an opportunistic pathogen, *Nature* 406, 959-964.

Suzuki, H., Hashimoto, W., Kumagai, H., 1993. *Escherichia coli* K-12 can utilize an exogenous γ-gluatamyl peptide as an amino acid scource for which γ- glutamyl - transpeptidase is essential, *Journal of Bacteriology* 175, 6038-6040.

Takeda, R, 1958. *Pseudomonas* pigments. I. Pyoluteorin, a new chlorine-containing pigment produced by *Pseudomonas aeruginosa,* Hako Kogaku Zasshi 36, 281-290.

Takesako, K., Kuroda, H., Inoue, T., Haruna, F., Yoshikawa, Y., Kato, I., Uchida, K., Hiratani, T., Yamaguchi, H., 1993. Biological properties of aureobasidin A, a cyclic depsipeptide antifungal antibiotic, *J. Antibiot.* 46, 1414 - 1420.

Tan, B.C., Cline, K., McCarthy, D.R., 2001. Localization and targeting of the VP14 epoxy-carotenoid dioxygenase to chloroplast membranes, *Plant J.* 27, 373-382.

Tasheva, K., Kosturkova, G., 2013. Role of Biotechnology for protection of endangered medicinal plants, Environmental Biotechnology – New Approaches and Prospective Applications 235-285.

Tazawa, J., Watanabe, K., Yoshida, H., Sato, M., Homma, Y., 2000. Simple method of detection of the strains of fluorescent Pseudomonas spp. producing antibiotics, pyrrolnitrin and phloroglucinol, *Soil Microorg.* 54, 61-67.

Thomashow, L.S., 1996. Biological control of plant root pathogens, *Current Opinion in Biotechnology* 7, 343-347.

Thomashow, L.S., Weller, D.M., 1988. Role of phenazine antibiotic from *Pseudomonas fluorescens* in biological control of *Gaeumannomyces graminis* var. *tritici*, *J. Bacteriol.* 170, 3499-3508.

Thompson, J.D., Gibson, T.J. Plewniak, F., Jeanmougin, F., Higgins, D.G., 1997. The CLUSTAL_X windows interface: flexible strategies for multiple sequence alignment aided by quality analysis tools, *Nucleic Acids Res.* 25, 4876-4882.

Thrane, C., Nielsen, T.H., Nielsen, M.N., Olsson, S., Sorensen, J., 2000. Viscosinamide producing *Pseudomonas fluorescens* DR54 exerts biocontrol effect on *Pythium ultimum* in sugar beet rhizosphere, *FEMS Microbiol. Ecol.* 33, 139-146.

Turner, J.M., Messenger, A.J., 1986. Occurrence, biochemistry and physiology of phenazine pigment production, *Adv. Microbial Physiol.* 27, 211-275.

Upadhyay, A., Srivastava, S., 2010. Phenazine-1-carboxylic acid is a more important contributor to biocontrol *Fusarium oxysporum* than pyrrolnitrin in *Pseudomonas fluorescens* strain Psd, *Microbiological Research* 2, 53 - 67.

Upadhyay, V.P., Malviya, H.S., Sudha Behura, Rout, D.K., 2009. Ecological methods for biodiversity assessment in EIA, *Indian J. Env. Ecoplan.* 16, 157-168.

Van Loon, L.C., 2007. Plant responses to plant growth-promoting rhizobacteria, *European Journal of Plant Pathology* 119, 243-254.

Vandamme, P., Pot, B., Gillis, M., de Vos, P., Kersters, K., Swings, J., 1996. Polyphasic taxonomy, a consensus approach to bacterial systematics, *Microbiol. Rev.* 60, 407–438.

Vasudha, S., Shivesh, S., Prasad, S.K., 2013. Harnessing PGPR from rhizosphere of prevalent medicinal plants in tribal area of central India, *Research Journal of Biotechnology* 8(5), 76 – 85.

Verma, S.C., Ladha, J.K., Tripathi, A.K., 2001. Evaluation of plant growth promotion and colonization ability of endophytic diazotrophs from deep water rice, *J. Biotechnol.* 91, 127–141.

Vidhyasekaran, P., Muthamilan, M., 1999. Evaluation of powder formulation of *Pseudomonas fluorescens* Pf1 for control of rice sheath blight, *Biocontrol Science and Technology* 9, 67-74.

Voisard, C., Bull, C.T., Keel, C., Laville, J., Maurhofer, M., Schnider, U., Défago, G., Haas, D., 1994. Biocontrol of root diseases by *Pseudomonas fluorescens* CHA0, Molecular Ecology of Rhizosphere Microorganisms 69-89.

Vollenbroich, D., Özel, M., Vater, J., Kamp, R.M., Pauli, G., 1997. Mechanism of inactivation of enveloped viruses by the biosurfactant surfactin from *Bacillus subtilis*, *Biologicals* 25, 289-297.

Volpon, H., Besson, F., Lancelin, J.M., 2000. NMR Structure of antibiotics plipastatins A and B from Bacillus subtilis inhibitors of phospholipase A_2, *FEBS* 485, 76-80.

Volpon, L., Besson, F., Lancelin, J.M., 1999. NMR structure of active and inactive forms of the sterol dependent antibiotic bacillomycin L, *Eur. J. Bioche.* 264, 200-210.

Von Alten, H., Lindermann, A., Schonbeck, F., 1993. Stimulation of vesicular-arbuscular mycorrhiza by fungicides or rhizosphere bacteria, *Mycorrhiza* 2, 167-173.

Ward, E.R., Uknes, S.J., Williams, S.C., Dincher, S.S., Wiederhold, D.L., Alexander, D.C., Ahl-Goy, P., Métraux, J.P., Ryals, J.A., 1991. Coordinate gene activity in response to agents that induce systemic acquired resistance, *Plant Cell* 3 (1991) 1085-1094.

Weller, D.M., 1988. Biological control of soil borne plant pathogens in rhizosphere with bacteria, *Ann. Rev. Phytopathol.* 26, 379-407.

Whipps, J.M., 2001. Microbial interactions and Biocontrol in the Rhizosphere, *Journal of Experimental Biology* 52, 487-511.

Wisniiewski, M.E., Wilson, C.L., 1992. Biological control of post- harvest diseases of fruits and vegetables, *Recent advances in Horticultural Science* 27, 94-98.

Woese, C.R., Stackebrandt, E., Macke, T.J., Fox, G.E., 1985. A phylogenetic definition of the major eubacterial taxa, *Syst. Appl. Microbiol.* 6, 143–151.

Woese, C.R., 1987. Evolution Towards a natural system of organisms, *Microbiol. Rev.* 51, 221-271.

Ye, L., Zhang, H., Xu, H., Zou, Q., Cheng, C., Dong, D., Xu, Y., Li, R., 2010. Phenazine-1-Carboxylic Acid derivatives: Design, Synthesis and Biological Evaluation against *Rhizoctonia solani* Kuhn., *Bioorganic and Medicinal Chemistry Letters* 20, 7369–7371.

Zhender, G.W., Yao, C., Murphy, J.F., Sikora, E.R., Kloepper, J.W., Schuster, D.J., Polston, J.E., 1999. Microbe induced resistance against pathogens and herbivores: evidence of effectiveness in agriculture. In: Induced Plant Defenses Against Pathogens and Herbivores: Biochemistry, Ecology and Agriculture, APS Press, St Paul, MN. Edited by Agarwal AA, Tuzun S, Bent, E, 33.

Zou, C., Li, Z., Yu, D., 2010. *Bacillus megaterium* Strain XTBG34 promotes plant growth by producing 2-Pentylfuran, *The Journal of Microbiology.* 48(4), 460-466.

2015, Modern Methods in Phytomedicine *Pages* ***221–231***
Editor: **T. Parimelazhagan**
Published by: **DAYA PUBLISHING HOUSE, NEW DELHI**

13

The Study of Antioxidant and Antinociceptive Properties of *Acalypha alnifolia* Klein ex. Willd.

P. Revathi, T. Sajeesh and T. Parimelazhagan*

Department of Botany, Bharathiar University, Coimbatore – 641 046, Tamil Nadu

1.0 INTRODUCTION

Free radicals are constantly produced within the human body in response to both internal and external stimuli (Matés *et al.*, 1999). In small amount these products play an important role as growth regulator, signal transducers, and as part of the immune defense system (Atmani *et al.*, 2009). However, excess generation of free radicals and other oxidants will cause oxidative stress. Nowadays, it is well known that oxidative stress is associated with several diseases such as cancer, arteriosclerosis, neurodegenerative diseases, and ageing processes (De Oliveira *et al.*, 2009). Our cells are well protected against free radicals damage either by endogenous antioxidative enzymes such as superoxide dismutase (SOD) and catalase (CAT) or by exogenous chemicals such as α tocopherol, ascorbic acid, carotenoids, polyphenol compounds and glutathione (Cheesman and Slater, 1993) while in case of acquiring natural foods.

Antioxidants play an important role in inhibiting and scavenging radicals, thus providing protection to humans against infections and degenerative diseases

(Venkatadri Rajkumar *et al.*, 2010). However, during recent years people have been more concerned about the safety of their food and the potential effect of synthetic additives on their health. The two most commonly used synthetic antioxidants; butylated hydroxyanisole (BHA) and butylated hydroxytoluene (BHT) have begun to be restricted because of their toxicity and DNA damage induction (Sasaki *et al.*, 2002). Hence potential antioxidants from natural source are in demand.

Oxidative stress and resulting lipid peroxidation is involved in various and numerous pathological states. The detection and measurement of lipid peroxidation is the evidence most frequently cited to support the involvement of free-radical reactions in toxicology and disease. Nitric oxide synthase (NOS) is an example of a family of heme-containing monooxygenases that, under the restricted control of a specific substrate, can generate free radicals. While the generation of nitric oxide (NO^o) depends solely on the binding of L-arginine, NOS produces superoxide and hydrogen peroxide when the concentration of the substrate is low. NOS generates secondary free radicals that may initiate pathological events, along with the cell signaling properties of NO^o, O^o-, and H_2O_2 (Supatra Porasuphatana *et al.*, 2003). Reducing power of a compound is also a supporting feature for its antioxidant activity. The reducing properties are generally associated with the presence of reductones, which have been shown to exhibit antioxidant action by breaking the chain reactions by donating a hydrogen atom (Matsushige *et al.*, 1996). Metal chelating activity is significant as it reduces the concentration of the catalyzing transition metal in lipid peroxidation through the Fenton reaction (Hseu *et al.*, 2008). Inflammatory pain results from the increased excitability of peripheral nociceptive sensory fibres produced by the action of inflammatory mediators. This excitatory effect, in turn, is a result of the altered activity of ion channels within affected sensory fibres (Linley *et al.*, 2010). The effective antioxidant potential extracts are tested for its antinociceptive analysis with animal model.

The Plant *Acalypha alnifolia* comes under the family Euphorbiaceae. This family Euphorbiaceae containing 60 genera and the Genus Acalypha having 23 species in it. The Irula tribes of Marudhamalai hills have been used this plant to treat dysentery (Sethilkumar *et al.*, 2006). Villagers of Dharapuram taluk use the *A. alnifolia* leaf juice mixed with 150 mL boiled cow milk twice a day for five months against diabetes (Balakrishnan *et al.*, 2009). Hooralis of Vilangombai using this plant leave to treat skin problems of infants (Revathi *et al.*, 2013a). As *A. alnifolia* is one of the endemic plants of Nilgiris, its leaf is commonly used as leafy vegetable by the local people (Sasi and Rajendran, 2012). These reports and necessity makes to prospect new plant with good pharmacological effect. Generally, any part of the plant can be used for antioxidant studies but most commonly used part is leaf followed by fruit (Chanda and Dave, 2009). Since some of the antioxidant works have completed, the leaf has taken for disclose its remaining important antioxidant and antinociception analysis.

2.0 MATERIALS AND METHODS

2.1 Plant Collection

The plant *A. alnifolia* leaves were collected from Bharathiar University campus and was authenticated by Botanical Survey of India (BSI/SRC/5/23/2011-12/Tech.-

1137), Southern circle, Coimbatore. The voucher specimen was deposited in Department of Botany, Bharathiar University and BSI Coimbatore. The collected leaves were washed in running tap water and shade dried. The dried leaves are powdered and stored in an air tight container for further use.

2.2 Extraction of Plant Material

The powdered plant leaf was packed in small thimbles separately and extracted successively with different solvents such as petroleum ether, chloroform, acetone, methanol and water in the increasing order of polarity using soxhlet apparatus. The extract obtained was used for the assessment of various antioxidant assays and for further analysis.

2.3 Antioxidant Analysis

2.3.1 Nitric Oxide Scavenging Activity (Sreejayan and Rao, 1997)

Sodium nitroprusside in aqueous solution at physiological pH, spontaneously generates nitric oxide, which interacts with oxygen to produce nitrite ions that can be estimated using Greiss reagent. Scavengers of nitric oxide compete with oxygen leading to reduced production of nitric ions. 2 mL of sodium nitropruside (10mM) in phosphate buffered saline (0.2M, pH-7.4) was mixed with 100 µl sample solution of various extracts and standards and incubated at room temperature for 150 minutes. After the incubation period, 2 mL of Greiss reagent (1 per cent sulfanilamide, 2 per cent H_3PO_4 and 0.1 per cent N- (1-naphthyl) ethylene diamine dihydrochloride) was added to all the test tubes. The same reaction mixture without the sample was used as the negative control. The absorbance of the chromophore formed was read at 546 nm against the blank (phosphate buffer). The scavenging activity (per cent) was calculated as:

Per cent = [Control OD-Sample OD]/Control OD x 100

2.3.2 Metal Chelating Activity (Dinis *et al.*, 1994)

About 100 µl the extract samples were added to 50 µl of 2mM $FeCl_2$ solution. Then the reaction was initiated by the addition of 200 µl of 5 mM ferrozine and the test tubes were vortexed well and left standing at room temperature for 10 min. The reaction mixture containing deionized water in place of sample was considered as the negative control. Absorbance of the solution was then measured spectrophotometrically at 562 nm against the blank (deionized water). EDTA was used as the standard metal chelating agent and the results were expressed as mg EDTA eq./g extract.

2.3.3 Lipid Peroxidation Assay (Afanasev *et al.*, 1989)

A modified thiobarbituric acid-reactive species (TBARS) assay was used to measure the lipid peroxide formed, using egg yolk homogenates as lipid-rich media (Ruberto *et al.,* 2000). Malondialdehyde (MDA), a secondary end product of the oxidation of polyunsaturated fatty acids, reacts with two molecules of TBA yielding a pinkish red chromogen with an absorbance maximum at 532 nm. Egg homogenate (500 µl of 10 per cent, v/v in phosphate-buffered saline pH-7.4) and 300 µl of sample were added to a test tube and made up to 1.0 mL with distilled water. Then, 50 µl of

$FeSO_4$ (0.075 M) and 20 µl of L-ascorbic acid (0.1 M) were added and incubated for 1 h at 37°C to induce lipid peroxidation. Thereafter, 0.2 mL of EDTA (0.1 M) and 1.5 mL of TBA reagent (3 g TBA, 120 g TCA and 10.4 mL 70 per cent $HClO_4$ in 800 mL of distilled water) were added in each sample and heated for 15 min at 100°C. After cooling, samples were centrifuged for 10 min at 3000 rpm and absorbance of supernatant was measured at 532 nm (Afanasev *et al.,* 1989). Inhibition (per cent) of lipid peroxidation was calculated using the equation:

Per cent inhibition = $[(A_0 - A_1) / A_0] \times 100$

Where, A_0 is the absorbance of the control, and A_1 is the absorbance of the tested sample.

2.4 *In vivo* Studies

2.4.1 Animals Used

Adult Wister albino rats weighing between 150 – 200 gm and adult Swiss albino mice weighing between 25 – 30 gm (for analgesic experiment only) of either sex were used for the studies. The study was carried out with consent from Committee for the Purpose of Control, Supervision on Experimental Animals (CPCSEA) and Institutional Animal Ethics Committee (IAEC), Proposal number NCP/IAEC/No: 03/2013-14.

2.4.2 Acute Toxicity

The animals were divided into control and test groups containing six animals each. The test groups got marked doses of acetone and methanol extracts orally upto 2000 mg/kg and observed for mortality till 48 h as per OECD guideline. There was no mortality and hence the doses were fixed as 1/10th and 1/5th of the higher dose were selected and each extract were taken as 1:1 ratio with 10 per cent carboxymethyl cellulose (CMC) for experimental study.

2.4.3 Analgesic Experiment

Formalin Induced Paw Licking Experiment

Wistar albino rats were pretreated with two extract of *Acalypha alnifolia* (acetone and methanol 200 and 400 mg/kg) 60 min before subcutaneous injection of 50 µl of 2.5 per cent formalin into the dorsal surface of the right hind paw was applied. Animals were observed in the chambers. Animals were observed from 0 to 5 min (neurogenic phase) and from 15 to 30 min (inflammatory phase) and the time that they spent licking the injected paw was recorded and considered as indicative of nociception (Hunskar and Hole, 1997).

2.4.4 Statistics

Results were statistically analyzed and values expressed as mean ± standard deviations and compared with the corresponding control group by applying one way analysis of variance (ANOVA). Significance of *in-vivo* experiment and comparison with control group has been assessed by Dunnet's t-test.

3.0 RESULTS AND DISCUSSION

Reactive Oxygen Species are continuously produced in aerobic organisms. In addition, production of ROS by leukocytes and macrophages is an important response in the immune system and a potent microbicidal mechanism of host defence. However, over production of these metabolites can be harmful not only to immune cells but also to surrounding tissue (Joyce, 1987).

3.1 Nitric Oxide Scavenging Activity

Nitric oxide is a potent pleiotropic inhibitor of physiological processes such as smooth muscle relaxation, neuronal signaling, inhibition of platelet aggregation and regulation of cell mediated toxicity. It is a diffusible free radical that plays many roles as an effector molecule in diverse biological systems including neuronal messenger, vasodilation and antimicrobial and antitumour activities (Hagerman *et al.,* 1998). The role of nitric oxide (NO) in numerous disease states has generated a considerable discussion over the past several years. During infections and inflammations, formation of NO is elevated and may bring about some undesired deleterious effects. Hence, it has generated a considerable discussion over the past several years. (Marcocci *et al.,* 1994 a, b). In this present study the NO radical scavenging effect of different extracts (Figure 13.1) has been observed with standard protocol. The percentage of radical inhibition is splendid in acetone extract (51.55 per cent) and followed by chloroform (46.27 per cent) instead of methanol which is always next to acetone in previous study. The effects are compared with standard quercetin (88.23 per cent).

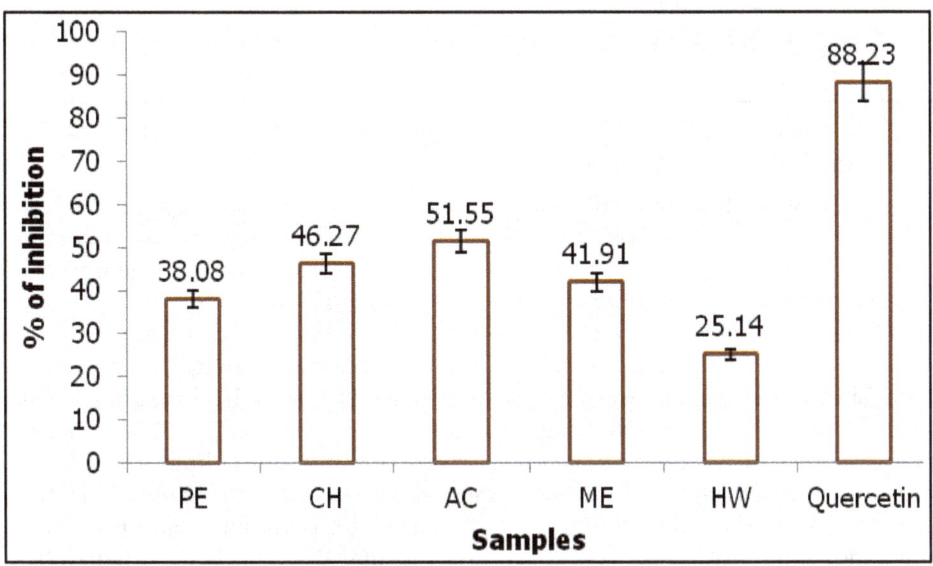

Figure 13.1: Nitric Oxide Radical Scavenging Activity of *Acalypha alnifolia* Leaf Extracts.

PE: Petroleum ether extract; CH: Chloroform extract; AC: Acetone extract; ME: Methanol extract; HW: Hotwater extract.

3.1 Metal Chelating Activity

Chelation of ions is also a route of antioxidation, many of the antioxidants are chelating the metal ions and impeding the molecular damages which cause by the ion interactions. The metal chelators in the extracts might have formed complex ions or coordination compounds with metals, thereby preventing the interaction between metals and lipid intermediates. In the present study the plant *Acalypha alnifolia* leave extracts were evaluated and the results were shown in µg EDTA eq./100µg (Table 13.1). Result shows acetone extract chelating the Fe 2+ ions better (0.543 µg EDTA eq./100µg) followed by methanol extract. The metal chelators are also reported to decrease oil oxidation in an indirect way (Choe and Min, 2006). They can convert iron or copper ions into insoluble complexes and hinder the formation of the complex between metals and lipid hydroperoxides preventing the oil oxidation (Gomathi *et al.,* 2011). Besides, Iron may catalyse free radical production in the joints, leading to lipid peroxidation and membrane disruption (Blake *et al.,* 1981). By the way the lipid peroxidation ability of the extracts has also monitored in this study.

Table 13.1: Metal Chelating Activity of *Acalypha alnifolia* Leaf Extracts

Samples	*Metal Chelating Activity (µg EDTA eq./100µg)*
Petroleum ether extract	0.041±0.07
Chloroform extract	0.209±1.10
Acetone extract	0.543±0.26
Methanol extract	0.328±0.25
Hot water extract	0.158±0.539

3.3 Lipid Peroxidation Activity

Reaction of free radicals with unsaturated lipids may trigger the lipid peroxidation chain reaction, resulting in the oxidative breakdown of cellular membranes (Halliwell and Gulteridge, 1990). Lipid peroxidation has been implicated in the pathogenesis of number of diseases including diabetes. It is well established that bioenzymes are very much susceptible to LPO which is considered to be the starting point of many toxic as well as degenerative process. In fact, many studies have shown the association between the levels of free radical mediated lipid peroxidation products and the progress of diseases (Sottero *et al.,* 2009). Therefore, it is an essential task for radical scavenging antioxidants to suppress lipid peroxidation. The present study indicates the level of inhibition as percentage (Figure 13.2) which was compared with control and standard. The better percentage has observed in methanol extract of *A. alnifolia.* In addition to that the reports depicted radical scavenging antioxidants are effective primarily against the free radical mediated lipid peroxidation and hence radical scavenging ability has also been supportive with this plant extracts previously and present study.

3.4 Antinociception Experiment

Formalin Induced Paw Licking Model

Most of the drugs used at present for analgesic effect are synthetic in nature, prolong use of which cause several side and toxic effects like respiratory depression, constipation, kidney damage, physical dependence as well as gastrointestinal

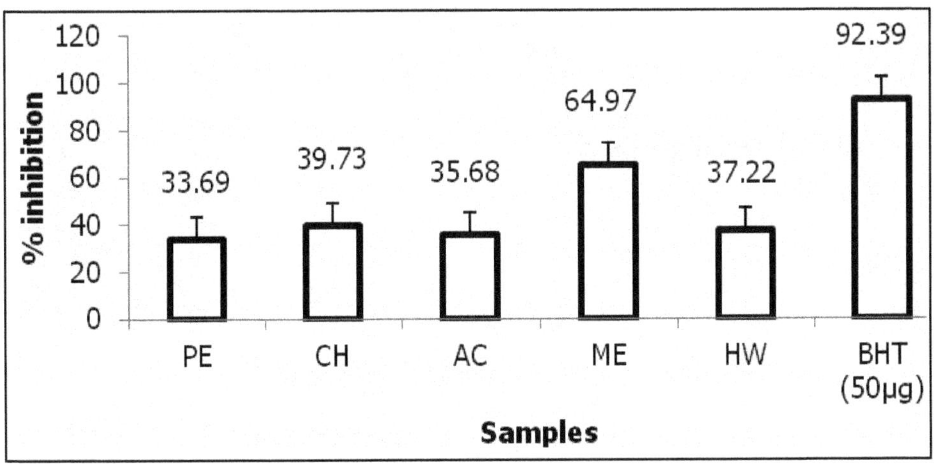

Figure 13.2: Lipid Peroxidation Activity of *Acalypha alnifolia* leaf Extracts.
PE: Petroleum ether extract; CH: Chloroform extract; AC: Acetone extract; ME: Methanol extract; HW: Hotwater extract.

irritation. Screening and scientific evaluation of plant extracts for their analgesic activity may provide new drug molecule that can combat various side effects of the commercially available synthetic drugs, moreover reducing the cost of medication (Barua *et al.,* 2010). Analgesia due to the first phase of formalin 5/10 -15 min is indicative of central effects while the second phase 10 - 60 min is due to inflammation as a result of direct stimulation of pain sensation (Tjolsen *et al.,* 1992). Scores of the first 5-10 min after formalin were recorded as the first phase of analgesia, while the period between 15 and 60 min was recorded as the late phase of pain. Extracts exhibiting considerable analgesic effect on their early phase. Similarly, the extract

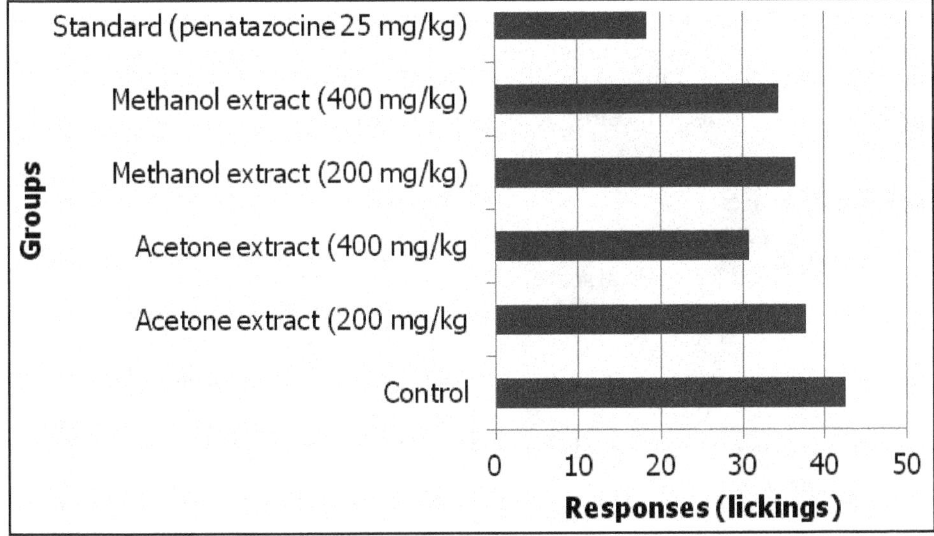

Figure 13.3: Responses of the Mice at Early Phase (5 min).

exhibited both phases of formalin-induced noxious stimuli. The late phase effect is very good which is observed in acetone 400 mg/kg extract group. The observations are conveyed in Figure 13.3 and 13.4.

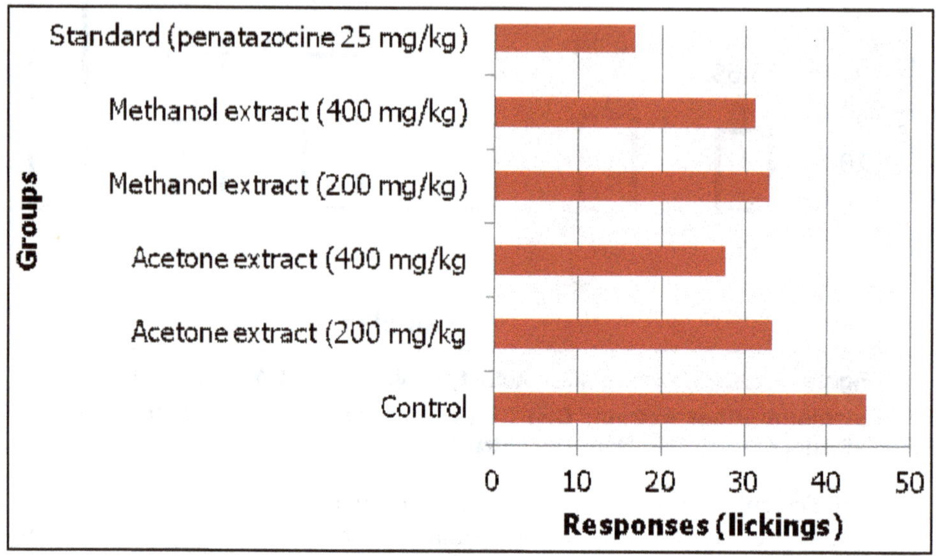

Figure 13.4: Responses of the Mice at Late Phase (20 min).

4.0 CONCLUSION

Results indicated that the selected leafy vegetable *A. alnifolia* is besides acting as good sources of antioxidants, may serve as substitute for synthetic antioxidants with anti-nociception property. In addition to that the plant is generally reported to contain substantial amounts of polyphenols with pharmacological properties which includes flavonoids and tannins (Revathi *et al.*, 2013b, Revathi *et al.*, 2014). In conclusion, the findings in the present work showing that consumption of these leafy vegetable will enhance the health benefits by adsorbing and neutralizing free radicals, quenching singlet oxygen, or decomposing peroxides and also contributes pharmacological benefits.

REFERENCES

Afanas'ev, I.B., Dorozhko, A.I., Brodskii, A.V., Kostyuk, V.A., Potapovitch, A.I., 1989. Chelating and free radical scavenging mechanisms of inhibitory action of rutin and quercetin in lipid peroxidation. *Biochemistry and Pharmacology* 38(11):1763-1769.

Atmani, D., Chaher, N., Berboucha, M., Ayouni, K., Lounis, H., Boudaoud, H., Debbache, N., Atmani, D., 2009. Antioxidant capacity and phenol content of selected Algerian medicinal plants. *Food Chemistry* 112:303–309.

Balakrishnan, V., Prema, P., Ravindran, K.C., Philip Robinson, J., 2009. Ethnobotanical Studies among Villagers from Dharapuram Taluk, Tamil Nadu, India. *Global Journal of Pharmacology* 3 (1): 08-14.

Blake, D.R., Hall, N.D., Bacon, P.A., Dieppe, P.A., Halliwell, B., Gutteridge, J.M.C., 1981. The importance of iron in rheumatoid disease. *The Lancet* 2:1142-1144.

Chanda, S., Dave, R., 2009. Review *In vitro* models for antioxidant activity evaluation and some medicinal plants possessing antioxidant properties: An overview, *African Journal of Microbiology Research* 3(13): 981-996.

Chandana C Barua, Archana Talukdar, Shameem A Begum, Lalit C Lahon, Dilip K Sarma, Debesh C Pathak, Probodh Borah, 2010. Antinociceptive activity of methanolic extracts of leaves of *Achyranthes aspera* Linn. (Amaranthaceae) in animal models of nociception. *Indian Journal of experimental biology* 48; 817-821.

Cheesman, K., Slater, H., 1993. An introduction to free radicals biochemistry. *British Medical Bulletin* 49: 481–493.

Choe, E., Min, D.B., 2006. Mechanisms and factors for edible oil oxidation. *Comprehensive Reviews in Food Science and Food Safety* 5: 169-186.

De Oliveira, A.C., Valentim, L.B., Silva, C.A., Bechara, E.J.H., Barros, M.P., Mano, C.M., Goulart, M.O.F., 2009. Total phenolic content and free radical scavenging activities of methanolic extract powders of tropical fruit residues. *Food Chemistry* 115: 469–475.

Dinis T.C.P., Madeira V.M.C., Almeida M.L.M., 1994. Action of phenolic derivates (acetoaminophen, salycilate and 5-aminosalycilate) as inhibitors of membrane lipid peroxidation and as peroxyl radical scavengers. *Archives of Biochemistry and Biophysics* 315:161-169

Hagerman, A.E., Riedl, K.M., Jones, G.A., Sovik, K.N., Ritchard, N.T., Hartzfeld, P.W., 1998. High molecular weight plant polyphenolics (tannins) as biological antioxidants. *Journal of Agricultural and Food Chemistry* 46, 1887-1892.

Halliwell, B., Gulteridge, J.M.C., 1990. Role of free radicals and catalyticmetal ions in human diseases: An overview. *Methods in Enzymology* 105:105-114.

Hseu, Y., Chang, W., Chen, C., Liao, L., Huang, C., Lu, F., Chia, Y., Hsu, H., Wu, J., Yang, H., 2008. Antioxidant activities of *Toona Sinensis* leaves extracts using different antioxidant models. *Food and Chemical Toxicology* 46: 105–114.

Hunskar, S., Hole, K., 1997. The formalin test in mice: dissociation between inflammatory and non-inflammaotry and pain. *Pain* 30:103-114.

Joyce, D.A., 1987. Oxygen radicals in disease. *Adverse Drug Reactions Bulletin* 127: 476-479.

Linley, J.E., Rose, K., Ooi, L., Gamper, N., 2010, Understanding inflammatory pain: ion channels contributing to acute and chronic nociception. *Pflugers Archiv-European Journal of physiology* 459(5):657-669. doi: 10.1007/s00424-010-0784-6.

Marcocci, L., Maguire, J.J., Droy-Lefaix, M.T., Packer, L., 1994a. The nitric oxide-scavenging properties of *Ginkgo biloba* extract EGb 761. *Biochemical and Biophysical Research Communications* 15: 748–755.

Marcocci, L., Packer, L., Droy-Lefaix, M.T.,Sekaki, A., Gardes-Albert M., 1994b. Antioxidant action of *Ginkgo biloba* extract EGB 761. *Methods in Enzymology* 234: 462–475

Mates, M.J., Gomez, C.P., De Castro, I.N., 1999. Antioxidant Enzymes and Human Diseases. *Clinical Biochemistry* 32: 595-603.

Matsushige, K., Basnet, P., Kadota, S., Namba, T., 1996. Potent free radical scavenging activity of dicaffeoyl quinic acid derivatives from propolis. *Journal of Traditional Medicine* 13: 217-228.

Rajkumar Gomathi, Nagarajan Anusuya, Chinnasamy Chitravadivu, Sellamuthu Manian, 2011. Antioxidant Activity of Lettuce Tree (*Pisonia morindifolia* R.Br.) and Tamarind Tree (*Tamarindus indica* L.) and Their Efficacy in Peanut Oil Stability. *Food Science and Biotechnology* 20(6): 1669-1677.

Revathi, P., Parimelazhagan, T., Manian, S., 2013a. Ethnomedicinal plants and novel formulations used by Hooralis tribe in Sathyamangalam forests, Western Ghats of Tamil Nadu, India. *Journal of medicinal plant research* 7(28):2083-2097.

Revathi, P., Parimelazhagan, T., Manian, S., 2013b. Quantification of phenolic compounds, in vitro antioxidant Analysis and screening of chemical compounds using GC-MS in *Acalypha alnifolia* klein ex willd. - a leafy vegetable. *International Journal of Pharma and Bio Science* 4(2) (B): 973-986.

Revathi, P., Parimelazhagan, T., Manian, S., 2014. Total nutritional capacity and inflammation inhibition effect of *Acalypha alnifolia* klein ex wild. - An unexplored wild leafy vegetable. *Journal of food and drug analysis.* (Status: accepted; DOI: 10.1016/j.jfda.2014.04.004)

Ruberto G., Baratta M.T., Deans S.G., Dorman H.J., 2000. Antioxidant and antimicrobial activity of *Foeniculum vulgare* and *Crithmum maritimum* essential oils. *Planta Medica* 66:687-93.

Sasaki, Y.F., Kawaguchi, S., Kamaya, A., Ohshita, M., Kabasawa, K., Iwama, K., Taniguchi, K., Tsuda, S., 2002. The comet assay with 8 mouse organs: results with 39 currently used food additives. *Mutation Research* - Genetic *Toxicology and Environmental Mutagenesis* 519, 103-119.

Sasi, R., Rajendran, A., 2012. Ethnobotany of some endemic plants of the Nilgiris, Southern western Ghats, India (NCPM/OP/090). *National conference on phytomedicine*: 4th and 5th October, Department of Botany, Bharathiar University.

Senthilkumar, M., Gurumoorthi, P., Janardhanan, K., 2006. Some medicinal plants used by Irular, the tribal people of Marudhamalai hills, Coimbatore, Tamil Nadu. Explorer: Research article, *Natural Product Radiance* 5(5):382-388.

Sottero, B., Gamba, P., Gargiulo, S., Leonarduzzi, G., Poli, G., 2009. Cholesterol oxidation products and disease: an emerging topic of interest in medicinal chemistry. *Current Medicinal Chemistry* 16:685–705.

Sreejayan, Rao, M.N.A., 1997. Nitric oxide scavenging by curcuminoids. *Journal of Pharmacy and Pharmacology* 49:105–107.

Supatra Porasuphatana, Pei Tsai, Gerald M. Rosen, 2003. Review-The generation of free radicals by nitric oxide synthase. *Comparative Biochemistry and Physiology* Part C 134: 281–289.

Tjolsen, A., Berge, O.G., Hunskaar, S., Rosland, J.H., Hole, K., 1992. The formalin test: an evaluation of the method. Pain 51:5.

Venkatadri Rajkumar, Gunjan Guha, R., Ashok Kumar, Lazar Mathew, 2010. Evaluation of Antioxidant Activities of *Bergenia ciliata* Rhizome. *Records of Natural Products* 4:1, 38-48.

2015, Modern Methods in Phytomedicine *Pages* ***233–240***
Editor: **T. Parimelazhagan**
Published by: **DAYA PUBLISHING HOUSE, NEW DELHI**

14

Evaluation of Stem Bark and Leaf of *Scolopia crenata* (Wight and Arn.) Clos. (Flacourtiaceae) for Anticholinesterase Activity

***R. Gomathi*[1] *and S. Manian*[2]**

[1]*Department of Botany, PSG College of Arts and Science (Autonomous), Peelamedu, Coimbatore – 641 004, Tamil Nadu*
[2]*Department of Botany, Bharathiar University, Coimbatore – 641 046, Tamil Nadu*

1.0 INTRODUCTION

Scolopia crenata (Wight and Arn.) Clos. (Family: Flacourtiaceae) is an Indian endemic medicinal tree species distributed in Andhra Pradesh, Karnataka, Tamil Nadu and Kerala. It is an evergreen tree whose branches and branchlets are often spiny whereas the older branches are unarmed. The Chellipale community of Kolli hills, Tamil Nadu uses the bark of the tree to treat cut wounds and the leaves grounded with jaggery are applied over inflammations.The leaves are also used to treat musco-skeletal pain and are taken internally against leucorrhoea (Udayan *et al.*, 2005; Kadavul and Dixit, 2009) while the bark is also used as a diuretic (Smitha *et al.*, 2012).Various parts of the tree are also employed as sedative in herbal formulations with several other herbal ingredients (Oudhia, 2012). Besides these reports, the functional property of *S. crenata* has not been fully explored.

Alzheimer's disease (AD) is the most common form of age related dementia and is characterized by neurodegeneration of the central nervous system that eventually leads to gradual decline of cognitive function and dementia. Cholinesterase inhibitors are the only approved drugs for treating patients with mild to moderately severe AD. Acetylcholinesterase (AChE) inhibitors have been shown to function by increasing acetylcholine within the synaptic region, thereby restoring deficient cholinergic neurotransmission (Krall *et al.*, 1999). Currently, the effective chemicals to treat AD, such as tacrine, donepezil and rivastigmine have been reported to cause gastrointestinal disturbances and problems associated with bioavailability (Schulz, 2003). Therefore, the search for new AChE inhibitors, particularly from natural products with higher efficacy still continues. Recently increasing number of studies have evidenced that plants/plant extracts used to treat cognitive disorders were found to act as AChE inhibitors (Stafford *et al.*, 2008). Several natural polyphenols such as epigallocatechingallate, resveratrol, huperzine A, quercetin, kuwanon U, E and C, kaempferol, and tri- and tetrahydroxy flavone have shown cholinesterase inhibitory effect (Zhang *et al.*, 2009; Min *et al.*, 2010; Kim *et al.*, 2011; Wang *et al.*, 2011). Therefore, the present study attempted to investigate the AChE inhibitory effects of *S. crenata* which might serve as a candidature for pharmaceutical applications.

2.0 MATERIALS AND METHODS

2.1 Chemicals

Drug and chemicals were purchased from Sigma-Aldrich (St. Louis, MO, USA) and Himedia (Mumbai, India). All other reagents used were of the highest purity and analytical grade made in India.

2.2 Plant Collection and Preparation of Polyphenol Extract

Fresh leaves and stem bark of *Scolopia crenata* were collected from Kolli hills, Tamil Nadu, India. The specimen was authenticated by the Botanical Survey of India, Southern Regional Centre, Coimbatore (No. BSI/SRC/5/23/2012-13/Tech./637) and the voucher specimen (No.006173) deposited at the herbarium in the Department of Botany, Bharathiar University, Coimbatore. Fresh plant materials were washed in tap water and shade dried at 25°C.

Polyphenolic contents from the stem bark and leaf of *S. crenata* were extracted, according to the modified method of Sun *et al.* (2009). Twenty gram of plant sample was extracted with 100 mL of extracting solvent mixture (methanol/acetone/water, 3.5: 3.5: 3, v/v/v) containing 1 per cent formic acid for 30 min in a shaking incubator (Orbitek, 4656Z, India) at 20°C. The extraction was performed twice and the pooled extract was centrifuged at 7000 g for 15 min. The supernatant was collected and methanol and acetone were evaporated at 35°C under reduced pressure (Rotary evaporator, Yamato B0410, Japan). The lipophilic pigments in the aqueous phase were then removed with a two-fold volume of petroleum ether by two successive extractions and was further extracted thrice in ethyl acetate (ethylacetate: aqueous phase = 1:1, v/v) in a separatory funnel. The solvent was evaporated under vacuum at 35°C. The resulting residue was re-dissolved in ethanol and stored at -20°C for further investigation.

2.3 HPLC Analysis

The amounts of phenolic acids and flavonoids in the methanol extracts of stem bark and leaf of *S. crenata* were determined using HPLC employing the method described by Amarowicz *et al.* (2010) with minor modifications. The HPLC system (Shimadzu LC-6AD, Shimadzu Corporation, Japan) comprised a FCV-7AL reciprocating double-plunger type pump, a degasser, an autosampler and a SPD-20A UV-vis detector. Luna-5u C18 (Phenomenex 250 × 4.60 mm, i.d., 5 μm) column was used for all separations.The binary mobile phase comprised of deionized water: acetic acid (98:2, v/v) (solvent A) and deionized water: acetonitrile: acetic acid (78:20:2, v/v/v) (solvent B) that was pumped at a flow rate of 1 mL/min for a total run time of 120 min, with the column temperature set at 25°C, and the sample injection volume of 20 μl (1mg/mL). The elution gradient established was 0-10 min, 50 per cent B; 11-20 min, 60 per cent B; 21-30 min, 70 per cent B; 31-40 min, 80 per cent B; 41-50 min, 90 per cent B; 51-110 min, 100 per cent B and 111-120 min, 50 per cent B. Detection was employed at 280 nm as preferred wavelengths for the simultaneous monitoring of phenolic acids and flavonoids. All the samples were filtered in the sample filtration apparatus through a 0.45 μm PVDF membrane filter prior to HPLC injection. LC solution 2.1.software was used for data acquisition, processing and reporting. Four different standard compounds (gallic acid,p-coumaric acid, apigenin and quercetin) were run using the same chromatographic conditions described above for their detection in the sample extracts throughcomparison. The calibration curves defined for each compound were in the range of 0.5-4 μg. Compounds were identified by comparing their retention time and UV-vis spectra with those of standards and the quantities were calculated by comparing peakarea with those of standards.

2.4 AChE Inhibitory Activity

The AChE activity was measured according to the method developed by Ellman *et al.* (1961) with slight modifications. To various concentrations of sample extracts (25–250 μg) in 250 μl, 25 μl of 15 mM acetylthiocholine iodide (ATCI) in water, 125 μl of 3 mM DTNB (5-5′-thiobis-2-nitrobenzoic acid) in Buffer C (50 mMTris-HCl, pH 8, containing 0.1M NaCl and 0.02 M $MgCl_2.6H_20$) and 50 μl of Buffer B (50 mMTris-HCl, pH 8, containing 0.1 per cent bovine serum albumin) were added. Absorbance was measured spectrophotometrically at 405 nm every 45 sec, five times consecutively. Thereafter, 25 μl ofAChE (0.2 U/mL) was added and the final mixture was again read at 405 nm five times consecutively for every 45 sec. The drug, Galanthamine was used as standard AChE inhibitor. The sample concentration providing 50 per cent inhibition (IC_{50}) under the assay condition was calculated from the graph of inhibition percentage against sample concentration.

2.5 Statistical Analysis

The values are expressed as means of three independent samples with triplicate determinations of each (n=9) ± standard deviation. Analysis of variance (ANOVA) and significant difference between means, determined by Duncan's multiple range test ($p<0.05$), were done using SPSS, version 17.0 (Stat Soft Inc., Tulsa, OK).

3.0 RESULTS AND DISCUSSION

3.1 HPLC Analysis

The polyphenols extracts ofstem bark and leaf of *S. crenata* were subjected to HPLC analysis for the identification and quantification of certain polyphenolic compounds present in them. Calibration curves were obtained for the standard phenolic compounds and flavonoids used in the present study by injecting known concentrations of the compounds (Table 14.1). The concentration of the phenolic acids and flavonoids present in the polyphenolicextracts of stem bark and leaf of *S. crenata* are shown in Figures 14.1a and b and Table 14.2. The HPLC chromatograms of the stem bark and leaf, obtained under optimum conditions, displayed the presence of higher concentration of p-coumaric acid (2.30 µg/mg extract) in the extract of stem bark while it was not detected in the leaves. However, gallic acid recorded its maximum content in the extract of leaves than stem bark (1.12 and 0.32 µg/mg extract, respectively). With regard to flavonoids, stem bark contained high amounts of quercetin (3.18 µg/mg extract). Both the stem bark and leaf recorded similar levels of apigenin (respectively, 2.38 and 2.89 µg/mg extract).

Table 14.1: Calibration Curves and Correlation Coefficients (R^2) of Standard Compounds Obtained in HPLC Analysis

Sl.No.	*Compound*	*Linearity curve*	*R^2 value*
1.	Apigenin	y = 322.05x + 16	0.9995
2.	p-Coumaric acid	y = 1050.2x – 5	1
3.	Gallic acid	y = 138.95x + 0.025	0.9971
4.	Quercetin	y = 1899.4x – 111.85	0.9978

Table 14.2: Retention Time (Rt) and Quantification of the Phenolic Compounds Present in the Polyphenol Extracts of Leaf and Stem Bark of *S. crenata*

Peak	*Rt*	*Compound*	*Contentin Stem Bark (µg/mg extract)*	*Contentin leaf (µg/mg extract)*
1.	63.42±0.02	Apigenin	2.38±0.02	2.89±0.17
2.	23.52±0.17	p-Coumaric acid	2.30±0.02	ND
3.	4.21±0.12	Gallic aid	0.32±0.09	1.12±0.11
4.	86.61±0.11	Quercetin	2.06±0.01	3.18±0.35

ND: Not detected.

3.2 AChE Inhibitory Activity

Results of AChE enzyme inhibitory activity of polyphenol extracts of stem bark and leaf of *S. crenata*, expressed as percent inhibition activityis presented in Table 14.3 and Figure 14.2. Both the polyphenol extracts of stem bark and leaf of *S. crenata*

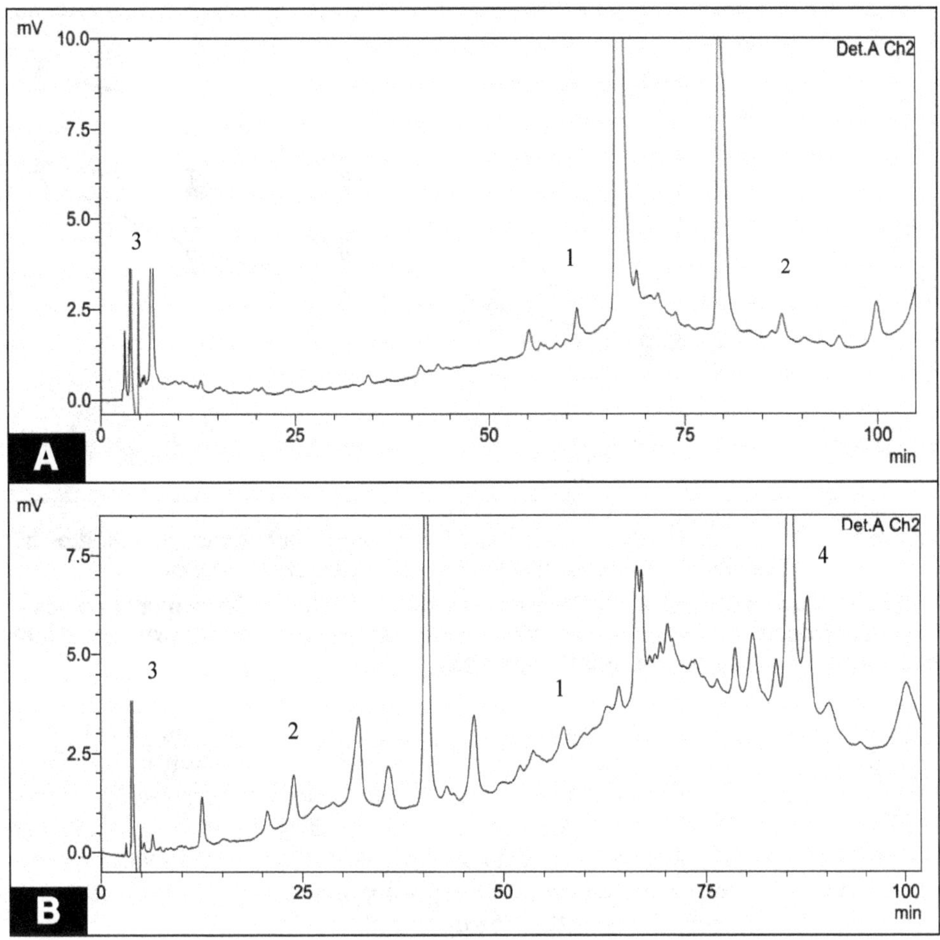

Figure 14.1: HPLC Chromatogram of Polyphenol Extracts of (a) Leaf and (b) Stem Bark of *S. crenata* at 280 nm.

Identified compounds from the HPLC chromatograms: 1. Apigenin; 2.p-Coumaric acid; 3. Gallic acid; 4. Quercetin.

Table 14.3: Inhibitory Activity of Acetylcholinesterase (AChE) of Stem Bark and Leaf Extracts of *S. crenata*

Sample	*AChE Inhibition (IC_{50} µg/mL)*
PL	98.35 ± 1.22^{b}
PSB	105.58 ± 0.19^{c}
Galanthamine	0.58 ± 0.07^{a}

Values are means of three independent samples with triplicate determination of each ± standard deviation (n=9). Mean values followed by different superscript indicates statistical significance ($p<0.05$).PL and PSB are respectively polyphenol extracts of leaf and stem bark of *S. crenata*.

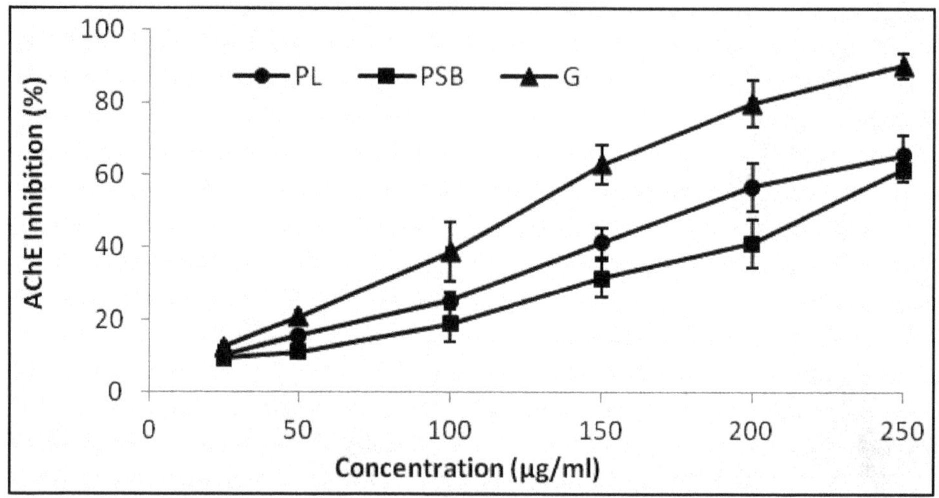

Figure 14.2: Concentration Dependent Inhibition of Acetylcholinesterase Inhibition by Polyphenol Extracts of Stem Bark and Leaf of *S. crenata*.

Values are means of three independent samples with triplicate determination of each ± standard deviation (n=9). PL and PSB are respectively polyphenol extracts of leaf and stem bark of *S. crenata*; G-Galanthamine.

registered inhibitory activity towards AChE. The percentage inhibition exhibited was found to be concentration dependent which increasedwith increasing concentrations of the extracts. Pronounced inhibitory activity was seen at higher concentration of 250 µg in both leaf and stem bark extracts (78 per cent and 72 per cent, respectively). The IC_{50} value in leaf and stem bark were found to be 98.35 µg/mL and 105.58µg/mL, respectively. However, the positive control galanthamine recorded the strongest inhibition of AChE (IC_{50} 0.58 µg/mL).

Inhibition of the enzyme AChE serves as a strategy in several therapeutic applications such as Alzheimer's disease, senile dementia, ataxia, myasthenia gravis and Parkinson's disease. The polyphenol extracts from stem bark and leaves of *S. crenata* conferred an effective inhibition when assessed for AChE inhibitory activity. This may possibly related to the presence of natural polyphenols such as gallic acid, p-coumaric acid, apigenin and quercetin recorded in them. Several natural polyphenols such as quercetin, kaempferol and tannic acid have been reported to play a pivotal role in preventing the pathophysiology of several neurodegenerative disorder (Zhang *et al.,* 2009). However, the exact mechanism of interaction of polyphenols with the cholinergic system is still not clear (Ebrahimi and Schluesener, 2012).

4.0 CONCLUSION

Therefore, taking all into account, it may be proposed that the polyphenol extracts of *S. crenata* may exhibit AChE inhibition due to the following reasons: (i) the presence of phenolic acids and flavonoids in the plant extracts may exert antiradical activity which might prevent lipid peroxidation in the neuronal cell membranes, and/or (ii) traditional analgesic usage of the plant may be anticipated to modulate the intrinsic cholinergic inhibitory pathway through the inhibition of AChE. However, further studies are warranted to elucidate the actual mechanism of action of stem bark and leaf of *S. crenata* on AChE inhibition.

ACKNOWLEDGEMENT

R.Gomathi acknowledges the financial support provided by Department of Science and Technology (DST), Government of India through the INSPIRE Programme.

REFERENCES

Amarowicz, R., Estrella, I., Hernández, T., Robredo, S., Troszyñska, A., Kosiñska, A., Pegg, R.B., 2010.Free radical-scavenging capacity, antioxidant activity, and phenolic composition of green lentil (*Lens culinaris*). *Food Chemistry* 121, 705–711.

Ebrahimi, A., Schluesener, H., 2012. Natural polyphenols against neurodegenerative disorders: Potential and pitfalls. *Ageing Research Reviews* 11, 329-345.

Ellman, G. L., Courtney, K. D., Andres, V., Featherstone, R. M., 1961. A new and rapid colorimetric determination of acetylcholinesterase activity. *Biochemical Pharmacology* 7, 88–95.

Kadavul, K., Dixit, A.K., 2009. Ethnomedicinal studies of the woody species of Kalrayan and Shervarayan Hills, Eastern Ghats, Tamil Nadu. *Indian Journal of Traditional Knowledge* 8, 592-597.

Kim, J.Y., Lee, W.S., Kim, Y.S., Curtis-Long, M.J., Lee, B.W., Ryu, Y.B., Park, K.H., 2011. Isolation of cholinesterase-inhibiting flavonoids from *Moruslhou*. *Journal of Agricultural and Food Chemistry* 59, 4589-4596.

Krall, W.J., Sramek, J.J., Cutler, N.R., 1999. Cholinesterase inhibitors: a therapeutic strategy for Alzheimer's disease. *The Annals of Pharmacotherapy* 33, 441–450.

Min, B.S., Cuong, T.D., Lee, J.S., Shin, B.S., Woo, M.H., Hung, T.M., 2010. Cholinesterase inhibitors from *Cleistocalyxoperculatus*buds. *Archives of Pharmacal Research* 33, 1665–1670.

Oudhia, P., 2012. Part 13 of Punarnava (*Boerhavia diffusa*L.) based Herbal Formulations in Indian Traditional Healing: Pankaj Oudhia's Ethnobotanical Surveys from year 1990 to 2012. http://www.pankajoudhia.com/home-page/boer13.pdf

Schulz, V., 2003. Gingko extract or cholinesterase inhibitors in patients with dementia: what clinical trial and guidelines fail to consider. *Phytomedicine* 10, 74–79.

Smitha, V.P., Raju, K., Mohan, M.C.H., Sreeramulu, H.S., Praneeth, D., 2012. Antioxidantactivity and phytochemical analysis of *Scolopia crenata* (Flacourtiaceae) stem bark. *International Journal of Pharmacy and Pharmaceutical Sciences* 4, 315-319.

Stafford, G.I., Pedersen, M.E., van Staden, J., Jäger, A.K., 2008. Review on plants with CNS-effects used in traditional South African medicine against mental diseases. *Journal of Ethnopharmacology* 119, 513-537.

Sun, J., Yao, J., Huang, S., Long, X., Wang, J.,Garcßa, *E.G.*, 2009. Antioxidant activity of polyphenols and anthocyanin extracts from fruits of *Kadsura coccinea* (Lem.) A.C. Smith. *Food Chemistry* 117, 276-281.

Udayan, P.S., George, S., Thushar, K.V., Balachandran, I., 2005. Ethnomedicine of the Chellipale community of Namakkal district, Tamil Nadu. Indian *Journal of Traditional Knowledge* 4, 437-442.

Wang, C., Tian, J., Wang, Q., 2011. ACE inhibitory and antihypertensive properties of apricot almond meal hydrolysate. *European Journal of Food Research and Technology* 232, 549–556.

Zhang, L., Cao, H., Wen, J., Xu, M., 2009. Green tea polyphenol (")epigallocatechin- 3-gallate enhances the inhibitory effect of huperzineA on acetylcholinesterase by increasing the affinity with serum albumin. *Nutritional Neuroscience* 12, 142–148.

2015, Modern Methods in Phytomedicine *Pages* ***241–250***
Editor: **T. Parimelazhagan**
Published by: **DAYA PUBLISHING HOUSE, NEW DELHI**

15

Evaluation of the Pharmacognostical, Phytochemical and Antioxidant Characteristics of Leaves of *Carica papaya* L.

T.V. Binu, Kerna Book Leena Margret and B. Vijayakumari

Department of Botany,
Avinashilingam Institute for Home Science and Higher Education for Women,
Coimbatore – 641 043, Tamil Nadu

1.0 INTRODUCTION

Search for eternal health and longevity and to seek remedy to relieve discomfort prompted man to develop diverse ways and means of health care. The early man explored his immediate natural surroundings, tried many things like plants, animals and minerals and developed a variety of therapeutic agents (CSIR, 1992). Steadily, a sizeable section of scientists in biological, biochemical and bio medicinal disciplines have embarked on research onmedicinal plants, which are the stable sources of many indigenous drugs (Agarwal *et al.*, 2007 and Mukherjee, 2002). Standardization of natural products is a complex task due to their heterogenous composition, which is in the form of whole plant, plant parts or extracts obtained (Mukherjee, 2002 and Charindy *et al.*, 1999).The phytochemicals of leaves of medicinal plants improve the health status of people and so it is used in pharmaceutical and nutraceutical products of commercial importance. The trend of using natural products has increased in recent and the active plant extracts are frequently screened for new drug (Ncube *et al.*, 2008).

These antioxidants may be constantly replenished since they are used up in theprocess of neutralizing free radicals.Many studies have been carried out on plants which resulted in development of natural antioxidant formulations for applications.However,scientific information on antioxidant properties of various plants, particularly those that are less widely used in medicine is still scarce. So it is an interesting and useful task to find new sources for natural antioxidant function (Miliauskas *et al.*, 2004).

Hence the present investigation was undertaken with the main objective of screening the plant, *Carica papaya* L. (Caricaceae) for its phytochemical, antioxidant and free radical scavenging activities.

2.0 MATERIALS AND METHODS

2.1. Plant Material Collection and Identification

Healthy and disease free leaves of *Carica papaya* L.were collected from C.S.IAnn's Womens Hostel during the months of August – September (2013). The plant material was identified and authenticated by the Department of Botany, Avinashilingam University, Coimbatore-43, Tamil Nadu, India.

2.2 Processing and Extraction

The leaves were cleaned, dried in the shade and pulverized in a mechanical grinder, passed through a 40 mesh sieve and stored in an air tight container. Powdered dried samples (30g) were successively extracted with chloroform (300 mL), methanol (300 mL) and aqueous (300 mL) by using soxhlet apparatus until the decolourisation of the solvents.

2.3 Pharmacognostic Study

The pharmacognostic studies of leaves were evaluated using the following qualitative analysis.

2.3.1 Organoleptic Study

The plant powder characterstics such as colour, odour, taste and nature were evaluated.

2.3.2 Fluorescence Analysis

The fluorescence characters of *Carica papaya* L. leaf powders were observed under ordinary light and long ultra violet light at 360 nm using different chemical reagents.

2.4 Preliminary Phytochemical Analysis

The different leaf extracts of chloroform, methanol and water of *Carica papaya* L. were subjected to qualitative analysis for the identification of various primary and secondary metabolites.

2.4.1 Test for Alkaloids

Mayer's Test

1 mL of Mayer's reagent was added with 1 mL of the extract. The formation of white precipitate was taken as a positive result for the presence of alkaloids.

2.4.2 Test for Flavonoids

Ferric Chloride Test

1 mL of neutral ferric chloride was added with 1 mL of extract. Appearance of brown colour indicates the presence of flavonoids.

2.4.3 Test for Anthocyanin

NaOH Test

1 mL of sodium hydroxide solution was added with 1 mL of the extract. A blue green colouration of the interface was formed which indicates positive result for the presence of anthocyanin.

2.4.4 Test for Starch

1 mL of iodine solution was added with 1 mL of the extract. A brown colour is observed, which showed the presence of starch.

2.4.5 Test for Quinone

0.5 mL of extract was treated with a few drops of concentrated hydrochloric acid and observed the formation of yellowish brown colour, which indicates the presence of quinone.

2.4.6 Test for Cellulose

1 mL of iodine solution was added to 1 mL of the extract. After that a few drops of sulphuric acid was added. Dark brown (or) red colour solution indicates the presence of cellulose.

2.4.7 Test for Phenols

About 1 mL of lead acetate solution was added with 1 mL of the extract. A brown colour precipitate is observed, which showed the presence of phenolic compounds.

2.4.8 Test for Proteins

1 mL of ninhydrin was dissolved in 1 mL of acetone and then small amount of extract was added with ninhydrin. The formation of purple colour revealed the presence of protein.

2.4.9 Test for Carbohydrates

Molisch's Test

Three to five drops of Molisch reagent was added with 1 mL of the extract and then1 mL of concentrated sulphuric acid was added carefully through the side of the test tube. The mixture was then allowed to stand for two minutes and diluted with 5 mL of distilled water. Development of red or dull violet ring at the junction of the liquids, which showed the presence of carbohydrates.

2.4.10 Test for Terpenoids (Salkowski Test)

1 mL of extract was treated with 1 mL of chloroform and 1 mL of concentrated sulphuric acid was added to form a layer. A reddish brown colour indicates the presence of terpenoids.

2.4.11 Test for Steroids (Lieberman-Burchard's Test)

1 mL of chloroform was mixed with 1 mL of extract and then ten drops of acetic anhydride and five drops of concentrated sulphuric acid were added and mixed. The formation of dark red colour or dark pink colour indicates the presence of steroids.

2.4.12 Test for Fixed Oil and Fat

A few drops of Sudan III solution was treated with 1 mL of extract. A shining orange colour showed the presence of fixed oil and fat.

2.5 Non-Enzymatic Antioxidants

2.5.1 Estimation of Reduced Glutathione

The method of Moron *et al.* (1979) was followed to determine the amount of reduced glutathione.1 mL of sample extract was treated with 4 mL of metaphosphoric acid, 0.2g EDTA and 30g NaCl dissolved in 100 mL distilled water after centrifugation 0.2 mL of the protein free supernatant was mixed with 0.2 mL of 0.4 mL Na2HPO4 and 1 mL of DTNB reagent (40mg DTNB in 100 mL of aqueous 1 per cent tri sodium citrate. Absorbance was read at 412nm with in 2minutes GSH concentration was expressed as mol/mg protein.The concentration of Reduced Glutathione was calculated and expressed as mg of protein.

2.5.2 Estimation of Tannin

The principle behind this method is reduction of Phosphotungstomolybdic acid in alkaline solution to produce coloured complex. The Tannins were determined by Folin and Ciocalteu method. 0.1 mL of the sample extract was added with 7.5 mL of distilled water and 0.5 mL of folin- ciocaltue reagent, 1 mL of 35 per cent sodium carbonate solution and diluted to 10 mL with distilled water. The mixture was shaken well, kept at room temperature for 30min. An absorbance was measured at 725nm. Blank was prepared with water instead of the sample. A set of standard solutions of Tannic acid is treated in the same manner as described earlier and read against a blank. The results of Tannin are expressed in terms of Tannic acid in mg/g TAE of extract.

2.5.3 Estimation of α-Tocopherol

α-Tocopherol can be estimated using Emmeric-Engel reaction. 2.5 g of the homogenized sample was weighed accurately into a conical flask. Added 50 mL 0f 0.1 N sulphuric acid slowly without shaking. Stoppered and allowed to stand overnight. The next day contents of the flask were shaken vigorously and filtered through Whatman No.1 filter paper discarding the initial 10-15 mL of filtrate. Aliquots of the filtrate were used for the estimation. In to three stoppered centrifuged tubes (three leaf samples, standard and blank) pippetted out 1.5 mL of each leaf extract, 1.5 mL of standard and 1.5 mL of distilled water respectively. To the three leaf samples and blank was added 1.5 mL of ethanol, 1.5 mL of water was added 1.5 mL of xylene to all the test tubes mixed well, and centrifuged. Transferred 1.0 mL of xylene layer into another stoppered tube, taking care not to include any ethanol or protein. Pipetted out 1.5 mL of the mixture into the spectrophotometer cuvette and read extinction co-efficient of the test and standard against the blank at 460 nm. This in turn, beginning with the blank, added 0.33 mL of ferric chloride solution. Mixed well and after exactly 15 minutes read test and standard against the blank at 520nm.

2.6 Free Radical Scavenging Activity

2.6.1 DPPH Radical Scavenging Activity

The free radical scavenging activity of methanol leaf extracts against *in vitro* 1,1 – diphenyl – 2-picryl-hydrazyl (DPPH) assay was evaluated spectrophotometrically by(Mensor *et al.,* 2001) method. About 3 mL of grated concentration (20 – 100 µg/mL) of each mL leaf extracts and standard were taken in different test tubes, then 1 mL of 0.1 mM DPPH methanol solution was added to these test tubes and shaken vigorously. After 30 min incubation of samples at 25C in the dark, the absorption was measured at 517 nm. Ascorbic acid (Vitamin C) was used as reference and prepared using methanol solvent instead of extract.

2.6.2 Estimation of IC_{50} Value of the Methanol Leaf Extract

IC_{50} (Concentration providing 50 per cent inhibition) of each methanol leaf extract and standard (ascorbic acid) was determined using regression curves in the liner range of concentration.

Where, IC_{50} expressed as µg/mL.

2.7 Statistical Analysis

All the results were expressed as mean± SEM. Statistical significance was determined by using the two, way ANOVA $p < 0.05$ was considered statistically significant

3.0 RESULTS

3.1 Organoleptic Study

The pharmacognostic characters of the leaves of *Carica papaya* L. were analysed and the results were obtained.

The organoleptic study using powders of leaf indicated the external characters like colour, odour and taste. The results of the present study are indicated (Table 15.1). The colour of the leaf was green. The odour was pleasant and the taste was bitter in the case of leaf powder.

Table 15.1: Organoleptic Study of *Carica papaya* L. Leaf Powder

Sl.No.	*Sample*	*Colour*	*Odour*	*Taste*
1.	Leaf	Green	Pleasant	Bitter

3.2 Fluorescence Analysis

The results of sample treatment with chemical reagents are expressed in Table 15.2.

Table 15.2: Fluorescence Analysis of *Carica papaya* L. Leaf Powder

Sl.No.	*Sample*	*Treatment with Chemical Reagent*	*Underordinary Light*	*Under UV Light 360nm*
1.	Leaf	Powder as such	Green	Green
2.	Leaf	Powder with 1N HCl	Green	Light green
3.	Leaf	Powder with Acetic acid	Light green	Dark green
4.	Leaf	Powder with Picric acid	Yellow green	Yellow green
5.	Leaf	Leaf Powder with FeCl3	Brown	Brown
6.	Leaf	Leaf Powder with 50 percent H2SO4	Brown	Light green
7.	Leaf	Leaf Powder with H2O	Light green	Light yellow
8.	Leaf	Leaf Powder with 1N NaOH	Green	Dark Green
9.	Leaf	Leaf Powder withHNO3+Ammonium solution	Brown	Brown
10.	Leaf	Powder with 1N NaOH inMethanol	Dark green	Light green

3.3 Preliminary Phytochemical Screening

For the preliminary phytochemical analysis leaf extracts (chloroform methanol and water) of *Carica papaya* L.were taken. In the present study, a phytochemical screening was carried out to detect the active constituents such as steroids, flavonoids, alkaloids, anthocyanins, phenols, proteins, carbohydrates, quinones, terpenoids, cellulose, starch, fixed oil and fat.

Table 15.3: Preliminary Phytochemical Analysis of Leaf Extracts of *Carica papaya* L.

Sl.No.	*Constituents*	*Chloroform Extract*	*Methanol Extract*	*Water Extract*
1.	Alkaloids	+	+	+
2.	Flavonoids	+	–	+
3.	Steroids	+	+	+
4.	Anthocyanin	–	+	+
5.	Protein	+	+	+
6.	Phenols	+	+	+
7.	Quinones	+	–	–
8.	Carbohydrates	+	+	+
9.	Terpenoids	+	–	–
10.	Starch	+	+	+
11.	Cellulose	+	+	+
12.	Fixed oil and fat	–	–	–

The results indicated that the chloroform extract showed maximum results compared to the other two solvents.

3.4 Non-Enzymatic Antioxidants

As shown in Table 15.4, it is evident that the Reduced glutathione content was maximum in methanol extract (0.87±0.003 mg-g tissue) but minimum in water extract (0.68±0.00 mg-g tissue). In the case of Tannin the result was higher in methanol extract (0.44±0.003mggTAE) but lower in water extract (0.32±0.002mg–gTAE). The maximum α-tocopherol content was present in methanol extract (44.65 ± 0.03mg-gtissue) but lower in water extract (32.41 ± 0.03 mg-g tissue).

Table 15.4: Estimation of Non-enzymatic Antioxidants of Leaf Extracts of *Carica papaya* L.

Solvents	*Reduced Glutathione (mg–g tissue)*	*Tannin (mg–gTAE)*	*α-tocopherol (mg-g tissue)*
Chloroform	0.79 ± 0.02	0.37 ± 0.031	39.63 ± 0.02
Methanol	0.87 ± 0.003	0.44 ± 0.003	44.65 ± 0.03
Water	0.68 ± 0.002	0.32 ± 0.002	32.41 ± 0.03
Standard	0.88 ± 0.006	0.63 ± 0.003	65.07 ± 0.02
SEd	0.0035	0.0074	0.0023
CD (p<0.05)	0.0048	0.0186	0.0415

Values are expressed by mean ± SD of three samples in each group.

TAE: Tannic Acid Equivalent per gm.

3.5 Free Radical Scavenging Activity

3.5.1 *In vitro* Antioxidant Activity of the Samples using DPPH Assay

The antioxidant activity of the samples was evaluated according to their ability for scavenging free radicals using DPPH assay is indicated in Table 15.5.

Table 15.5: Estimation of Antioxidant Activity Using *In vitro* DPPH Assay

Solvents	*DPPH Assay (Per cent of inhibition)*
Chloroform	51.54
Methanol	63.66
Water	61.13
Standard (Ascorbic acid)	70.83

It is evident that the DPPH free radical scavenging activity was more in methanolic extracts (63.66 per cent) but less in chloroform extract (51.54 per cent).

3.5.2 Comparision of IC_{50} Values and Antioxidant Activity of the Methanolic Leaf Extracts

The IC_{50} values of the DPPH assay were depicted in Table 15. 6. Among the samples in the case of DPPH assay methanol leaf extract of graded concentration 20 µl highest activity with low IC_{50} value (11.20 µg/mL) lowest antioxidant was observed in leaf extract with high IC_{50} value (26.94 µg/mL).

Table 15.6: IC_{50} Value of the Methanolic Leaf Extract of *Carica papaya* L.

Graded Concentration	*Methanol (µg/mL)*
20	11.20
40	17.04
60	18.71
80	23.82
100	26.94
IC_{50} value	15.56

4.0 DISCUSSION

Organoleptic study of aerial powder of *Tridax procumbens* indicated the characters like colour, odour and taste. The aerial powder showed deep green colour. The taste was slightly bitter (Saha and Paul, 2012).Many substances, both of plant and animal origin, exhibit fluorescence characteristics both qualitatively and quantitatively when exposed to UV radiation. Since the solvents and pH are capable of modifying the fluorescence of many substances, the powder is treatedwithdifferent chemicals and then observed under UV light (Kokoshi*et al.*, 1958 and Chase and Pratt, 1949).

Prathyusha *et al.* (2010) observed the seeds of *Abrus precatorious* in ordinary light and UV light at 254 nm. The colour of the seeds was brown under ordinary light, but it was yellowish brown under UV light. Phytochemical analysis intend to serve as a major resource for information on analytical and instrumental methodology in plant science as was reported in"phytochemical analysis" Houghton *et al.*(2004). Phytochemicals are non-nutritive plant chemicals that have protective or disease preventive properties (Shanthi *et al.*, 2011).The phytochemical screening of the poly herbal powder showed the presence of alkaloids, carbohydrates, phytosterol and flavonoids and saponin were absent in all the cases. Phytochemicals like alkaloids, carbohydrates, phytosterols, sterols, tannins, protein,amino acids, saponins, fixed oil,fat and flavonoids were analysed in *Solanum xanthocarpum* by Udaya kumar *et al.* (2003).

Qualitative phytochemical analysis of different extracts of *Cardiospermum helicacabum* showed that the leaf and stem contain a broad spectrum of secondary metabolites (Viji and Murugesan, 2010). It is an important antioxidantthat is found to detoxify toxic substance by conjugation (Peklak – Scott *et al.*, 2005).There are some supporting findings of Jayaprakash *et al.* (2002) that the reduction in the number of DPPH molecules can be correlated with the number of available hydroxide groups when the fraction showed significantly higher inhibition percentage, they have stronger hydrogen donating ability and positively correlated with enzymatic and non-enzymatic antioxidant activity.

IC_{50} values (concentration of sample required to scavenge 50 per cent free radical or to prevent lipid peroxidation by 50 percent) were calculated from the regression equations. IC_{50} value is inversely proportional to the antioxidant activity.Higher the IC_{50} lower the ability of the leaf extract to scavenge hydroxyl radicals (Dasgupta and De, 2005).

DPPH assay is one of the widely used methods of screening antioxidant activity of plant extracts (Nanjo *et al.,* 1996).It is evident from the study that DPPH scavenging is widely used to test the free radical scavenging activity of several natural products (Ahn *et al.,* 2007). DPPH is a stable free radical and any molecule that can donate an electron or hydrogen atom to DPPH and can react with it which bleach the DPPH absorption at 517 nm (Huang *et al.,* 2005).

5.0 CONCLUSION

According to the data derived from the present study, methanolic extract of *Carica papaya* L. found to be an effective antioxidant. The biochemical tests and antioxidant levels of the leaves of *Carica papaya* L. proved the plant to be a potent source for Ayurvedic drug preparation.

REFERENCES

Agarwal, S.S., Paridhari, M., 2007. Herbal drug technology, Universities Press Private Limited, Hyderabad, 83: 625.

Ahn, R., Kumazawa, S., Clsui, Y., Nakaura, J., Matsuka, M., Zhu, F., Nakayama, T., 2007. Antioxidant activity and constituents of propolis collected in various areas of China, *Food Chem.,* 101: 1383-1392.

Charindy, C.M., Seaforth, C.E., Phelps R.H., Pollara G.V., Khambay, B.P., 1999. Screening of medicinal plants form trinid and and tobago for antimicrobial and insecticidal properties, *J.Ethno.Pharm.*64:265-270.

Chase, C.R., R.J. Pratt, 1949. Fluorescence of powered vegetable drugs with particular reference to development system of identification, *J Am Assoc* 38:324-331.

Dasgupta, N., De Bratati, 2005. Antioxidant activity of some leafy vegetables of india: A comparative study, *J Food Chem* 10**:** 417-474.

Folin, C., Ciocalteu, V., 1927. Tyrosine and tryptophan determination in protein. *J Biol Chem* 73: 627-650.

Garratt, D.C., 1964 Quantitative analysis of drugs, Chapmand Hall Ltd, Japan 3:456-458.

Houghton, Oh, M.H., Whang, P.J., Cho, W.K., 2004. Screening of Korean herbal medicines used to improve cognitive function for anti cholinesterase activity,*Phytomedicine*,11 (6): 544-548.

Jackson, B.P., Snowdown, D.W., 1968.Powdered vegetable drugs, Cheer Chill Ltd., London, pp. 25-28.

Jayaprakash, G.K., Rao, L.J., 2000. Phenolic constituents from lichen *Parmontrema stuppeum. Food Control* 56**:** 1018-1022.

Kokoshi, C.J., Kokoshi, J.R., Sharma, F.J., 1958. Fluorescence of powdered vegetable drugs under ultra violet radiation, *J Amer Pharmaceu Assn* 38(10):715-717.

Mensor, L., Menzes, L.F.S., Leitao, A.S., Rels, A.S., Santos, T.C., 2001. Screening of Brazaile plant extracts, *Food Chem* 112 (6): 595-596.

Miliauskas, G., Venuskutonis, P.R., Beek, T.A., 2004. Screening of radical scavenging activity of some medicinal and aromatic plants extracts, *Food Chem* 85:231-274.

Moron, M.S., 1979. Levels of glutathione, glutathione reductase and glutathione S transferase activities in rat lung and liver, *Biochem Biophys Acta* 582:109-117.

Mukherjee, P.K., 2002. Quality control of herbal drugs,Business Horizon's Pharmaceutical Publishers, New Delhi,138-141.

Nanjo, F., Goto, K., Seto, R., Suzuki, M., Sekai. M., Hara, Y., 1996.Scavenging effects of tea catechins and their derivatives on 1,1-diphenyl-2-picryl-hydrozyl radical, *Free Radic Biol Med* 21:895-902.

Ncube, N.S., Afolayan, A.J., Okob, A.I., 2008.Assessement techniques of antimicrobial properties of natural compounds of plant origin current methods and future trends, *African Journal of Biotechnology* 7(12):1797-1806.

Peklak, S.C., Townsena A.J., Morrow, C.S., 2005. Dynamics of glutathione conjugation and conjugate effux in detoxification of the carcinogen, 4 – nitroquinoline l – oxide, contributions odf glutathione, glutathione stransferase, and MRPI, *Biochem* 44: 4426 – 4433.

Prathyusha, P., Modupalayam, S., Subramanian, Sivakumar, R., 2010. Pharmacognostical studies on white and red forms of *Abrus precatorius* Linn., *Indian Journal of Natural Products and Resources* 1(4): 476-480.

Rosenberg, H.R., 1992. Chemistry and physiology of the vitamins, Interscience PublishersInc., 5th Edition, New York, pp. 452-543.

Ruch, R.J., Chang, S.J., klaunig, J.F., 1989. Prevention of cytoxicity and inhibition of Intracellular communication by antioxidant catechins isolated from Chinese green tea Carcinogensis 10:1003-1008.

Saha, D., Paul, S., 2012. Pharmacognostic studies of aerial part of methanolic extract of *Tridax* procumbens, *Asian J Pharm Tech* 2(3): 107-109.

Santhi, R., Lakshmi, G., Priyadharshini, A.M., Anandaraj, L., 2011. Phytochemical screening of *Nerium oleander* leaves and *Momordica chavantia* leaves, *International Research Journal of Pharmacy* 2 (1): 131-135.

Udayakumar, R., Velmurugan, K., Sivanesan, K.D., Raghuram, K., 2003. Phytochemcial and antimicrobial studies of extracts of *Solanum xanthocarpum*, *Ancient Science of Life*, 23 (2): 90.

Viji, M., Murugesan, S., 2010. Phytochemical analysis and antibacterial activity of medicinal plant *Cardiospermum halicacabum* Linn., *J Phytol* 2(1): 68-77.

2015, Modern Methods in Phytomedicine *Pages* ***251–259***
Editor: **T. Parimelazhagan**
Published by: **DAYA PUBLISHING HOUSE, NEW DELHI**

16

Evaluation of Antioxidant Activity of the Root Tuber of Medicinal Climber *Pueraria tuberosa* (Roxb. ex Willd.) DC

Viji Zereena[1] and P. Paulsamy[2]

[1]*Department of Botany, NSS College, Nemmara, Palakkad – 678 508*

[2]*Department of Botany, Kongunadu Arts and Science College, Coimbatore – 641 029*

1.0 INTRODUCTION

Pueraria tuberosa (Roxb.ex willd.) DC of Fabaceae family is an important plant used in Indian medicine, commonly called as Vidarikand or Indian Kudzu. The plant is described as rasayana and tonic in Ayuvedic Pharmacopoeia of India. The tubers are sweet to taste and it is used in indigenous system of Indian medicine as tonic, aphrodisiac, anti rheumatic, diuretic and galactogue. Now a days it is used in preparing sexual potency enhancement pills. Kudzu is facing extinction in the wild because of herb hunters who trade the tubers illegally to agents of pharmaceutical or Ayurvedic companies. In the black market, the red variety kudzu tuber of about 10 kg is believed to be very expensive ranging up to lakhs of rupees. In order to know the current status of the plant in study area phytosociological work has to be done. Therefore, before going for the development of management and conservation strategies for any wild species, the data on their current availability status with respect to its population size is the most essential. Hence phytosociological work has been carried out.

Free radical production occurs continuously in all cells as part of normal cellular function. However, excess free radical production originating from endogenous or exogenous sources might play a role in many diseases. Antioxidants prevent the free radical induced tissue damage by preventing the formation of radicals, scavenging them, or by promoting their decomposition. Substantial body of evidence has developed supporting a key role for free radicals in many fundamental cellular reactions and suggesting that oxidative stress might be important in the path of physiology of common diseases including atherosclerosis, chronic renal failure, diabetes mellitus and cancer (Dhiman *et al.,* 2009).

The efficacy of a plant extract as an antioxidant is best evaluated based on the results obtained by commonly accepted assays (Frankel and Meyer 2000, Prior *et al.,* 2005) Even though several synthetic antioxidants like butylated hydroxyl anisole (BHA) and butylated hydroxy tolune (BHT) are available, but are unsafe and their toxicity is a problem of concern. Therefore, in recent years, considerable attention has been directed in identifying and developing natural antioxidants (plant based) that can be used for human consumption without any side effects. The present investigation is carried out to evaluate the antioxidant potential of *Pueraria tuberosa* tubers.

2.0 MATERIALS AND METHODS

2.1 Description of the Study Plant

Pueraria tuberosa commonly known as kudzu, Indian kudzu, or Nepalese kudzu, is a climber with woody tuberculated stem. The tubers are globose or pot-like, about 25 centimeters (9.8 in) across and the insides are white, starchy and mildly sweet. Flowers are bisexual, around 1.5 cm (0.59 in) across and blue or purplish-blue in color. The fruit pods are linear, about 2–5 cm (0.79–1.97 in) long and constricted densely between the seeds. Seeds vary from 3 to 6 in number.

2.2 Preparation of Extracts

P. tuberosa root tubers were collected from Nelliampathy and identified. The collected tubers were cut into small pieces, shade dried, powdered and extracted with organic solvents like petroleum ether, chloroform, acetone, methanol and hot water in the increasing order of polarity using a soxhlet apparatus. The different solvent extracts were concentrated by rotary vaccum evaporator.

2.3 Quantification Assay

2.3.1 Measurement of Total Phenolics

Total phenolics concentration was measured by Folin-ciocalteu assay (siddhuraju and Becker, 2003) fifty microlitre aliquots of the extracts were taken in test tubes and made up to 1 mL with distilled water. To this 0.5 mL of folin-ciocalteu phenol regent (1:1 with water) was added followed by the addition of 2.5Ml of sodium carbonate solution (20 per cent).the reaction mixture is vortexed and the test tubes are incubated in dark for 40 minutes and the absorbance was measured at 725 nm against the reagent blank.analysis was performed in triplicates and the results were expressed as Tannic acid equivalents.

2.3.2 Determination of Tannins

The same extracts were used in tannin estimation using polyvinylpyrrolidine (PVPP) (siddhuraju and Manian, 2007). 100 mg of PVPP was weighed into eppendorf tubes and to this 1mLof distilled water and then 1mL of sample extracts were added. The content was vortexed and kept in the freezer at 4° C for 15 minutes. Then the sample was centrifuged at 4000 rpm for 10 minutes and the supernatant was collected. The supernatants have simple phenolics, while tannins would have been precipitated along with PVPP. The phenolic content of the supernatant was measured and expressed as the content of non tannin phenolics on a dry matter basis.

The tannin content of the sample was calculated as

Tannins per cent = Total phenolics per cent - Nontannin phenolics per cent

2.3.3 Determination of Flavonoids

Flavonoid contents of the extracts were quantified according to Zhishen *et al.,* 1999. About 500 µL of the plant extract was taken in different test tubes, to this 2 mL of distilled water was added.Blank was prepared by taking 2.5 mL of distilled water. Then, 150 µL of $NaNO_2$ as added to all the test tubes and incubated at room temperature for 6 minutes. After incubation 150 µL of $AlCl_3$was added to all the test tubes including blank. All the test tubes were incubated again for 6 minutes at room temperature. Then 2 mL of 4 per cent NaOH was added to all the test tubes which were then made up to 5 mL using distilled water. The contents in all the test tubes were vortexed well and allowed to stand for 15 minutes. Pink color was developed due to the presence of Flavonoids. The absorbancy was measured at 510 nm. Rutin was used as standard. The experiments were done in triplicates and the results were expressed as Rutin equivalents (RE).

2.4 *In Vitro* Antioxidant Assays

2.4.1 Ferric Reducing Oxidant Assay (FRAP)

The antioxidant capacities of different extracts of sample were estimated according to the procedure described by Pulido *et al.* (2000). Frap reagent is freshly prepared and incubated at 37° C and was mixed with 90 mL of distilled water and 30 mL of test sample or methanol (for blank) The final dilution of the test sample in the reaction mixture was 1/34. The FRAP reagent contained 2.5 mL of 20 mM TPTZ solution in 40mM HCl plus 2.5 mL of 20 mM $FeCl_3.6H_2O$ and 25 mL of 0.3 M acetate buffer, pH 3.6. After incubation the absorbance was measured at 593nm against reagent blank. Methanolic solutions of known Fe (II) concentrations in the range of 100-2000 mM ($FeSO_4.7H_2O$) were used for calibration curve preparation. The parameter Equivalent Concentration or EC1 was defined as the concentration of antioxidant having a ferric-TPTZ reducing ability equivalent to that of 1 mM $FeSO_4.7H_2O$. EC1 was calculated as the concentration of antioxidant giving an absorbance increase in the FRAP assay equivalent to the theoretical absorbance value of a 1 mM concentration of Fe (II) solution.

2.4.2 Phosphomolybdenum Assay

The antioxidant capacities of different extracts of sample were estimated according to the procedure described by Prieto *et al.* (1999).An aliquot of 10-40 µL of sample or ascorbic acid in 1mM dimethyl sulphoxide (standard) and distilled water (blank) was added with 1mL of reagent solution (prepared by 0.6 M H_2SO_4, 28 mM sodium phosphate and 4 mM ammonium molybdate) in a test tube. The test tubes are covered with foil and incubated in water bath at 95°C for 90 minutes. The samples are then cooled at room temperature and the absorbance was measured at 695 nm against reagent blank. The results are reported as mean values expressed as milligrams of ascorbic acid equivalents per gram extract.

2.4.3 Metal Chelating Activity

The chelation of ferrous ions by various extracts of *P. tuberosa* was estimated by the method described by Dinis *et al.* (1994) 400 µL of sample and BHT (standard) were added to 50µL solution of 2 mM $FeCl_2$ followed by the addition of 200 µL of 5 mM ferrozine and the mixture was shaken vigorously and left to stand at room temperature for 10 minutes. Absorbance of the solution was the measured at 562 nm against deionised water as blank. The metal chelating capacity of the extracts was evaluated using the equation:

Metal chelating capacity per cent = $[(A_0 - A_1)/A_0]$ x100

Where A_0 =Absorbance of control and A_1 is the absorbance of the sample extract/ standard

2.4.4 DPPH Radical Scavenging Activity

The antioxidant activity of different extracts of *P. tuberosa* was determined in terms of hydrogen donating and radical scavenging ability using the stable radical DPPH according to the method described by Blois (1958). Sample extracts of various concentrations was taken and the volume was adjusted to100 µL with methanol. Aliquots of samples were taken, to this 5 mL of 0.1 mM methanolic DPPH was added. Same quantity of DPPH was added to standard Rutin and shaken vigorously. Negative control was prepared by adding 100 µL of methanol in 5 mL of 0.1 mM methanolic solution DPPH. The tubes were allowed to stand for 20 minutes at 27° C. The absorbance of the samples was measure at 517 nm against the blank (methanol). Radical scavenging activity of the samples was expressed as IC_{50}, which is the concentration of the sample required to inhibit 50 per cent DPPH concentration.

3.0 RESULTS AND DISCUSSION

3.1 Determination of Total Phenolics

The amount of total phenolics was analyzed and shown (Figure 16.1). The total phenolic content was found to be higher in acetone extracts of tuber(443.87mg GAE/ g extract), followed by methanolic extract (149.07mg GAE/g extract) from the results we can conclude that the antioxidant activity of *P. tuberosa* tubers is due to these phenolic compounds. Phenols are a class of antioxidants which can scavenge free radicals (Kessler *et al.,* 2003).

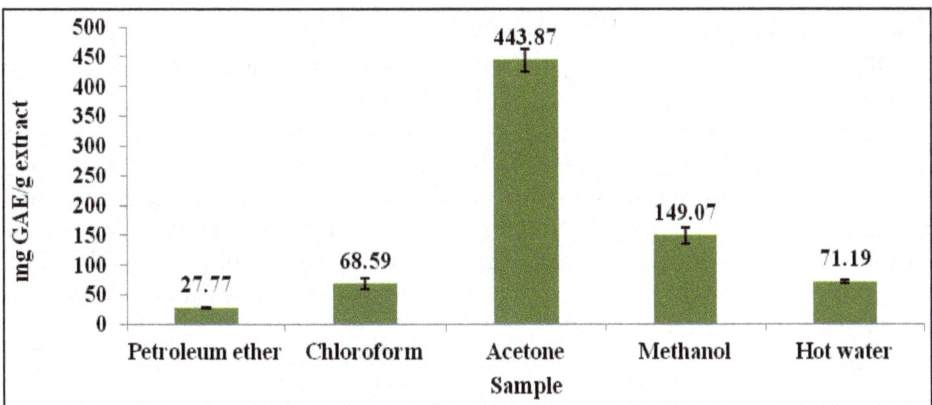

Figure 16.1: Total Phenolic Content of *P. tuberosa* Root Extract.
Values are expressed as mean (n=3) ± Standard Deviation (SD).
GAE: Gallic Acid Equivalents.

3.2 Determination of Tannins

The root extracts of P.tuberosa was analysed for its tannin content. The total tannins were found to be higher in acetone extract 290.41mgGAE/g extract (Figure 16.2). Higher the content of tannins greater will be the capacity to quench free radical. Tannins, poly phenols including flavonoids have been reported to exhibit a wide range of activity and prevent the attack of free radicals in human body (Narasimhan *et al.,* 2006).Tannins inhibit the absorption of Iron which may lead to anaemia (Brune *et al.,* 1989). Tannins are metal chelators and the chelated metal ions are not bioavailable (Karamac, 2009).Therefore higher amount of tannins of *P. tuberosa* have greater capacity to chelate metal ions.

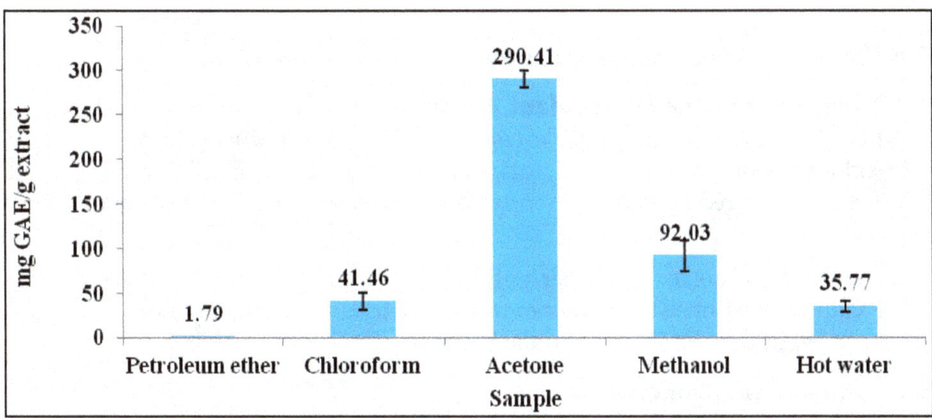

Figure 16.2: Tannin Content of *P. tuberosa* Root Extracts.
Values are expressed as mean (n=3) ± Standard Deviation (SD).
GAE: Gallic Acid Equivalents.

3.3 Determination of Flavonoids

Flavonoid contents were analyzed in the root tubers of *P. tuberose* and were shown in (Figure 16.3). The methanolic extract of tubers posses high (673.76 mg RE/ g extract) followed by chloroform (355.19 mg RE/g extract). Flavonoids are group of natural phenolics which generate H_2O_2 that can scavenge free radicals. Flavonoids not only scavenge free radicals but also posses the capability of chelating metal ions and inhibition of enzymes like NADPH (Miura and Nakatani, 1989).The higher contents of isoflavonoids in Pueraria lobata were inferred to be responsible for its more potent antioxidant activity as compared with that of Pueraria thomsonii.(Jing *et al.,* 2005).

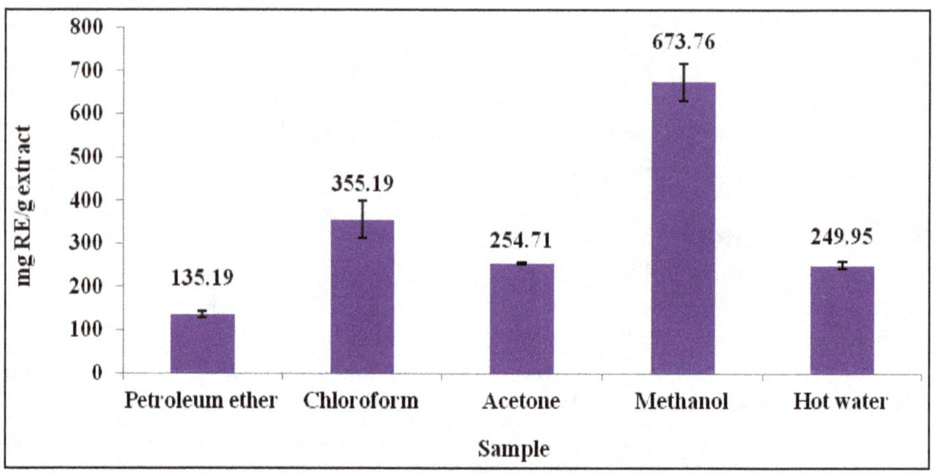

Figure 16.3: Flavonoid Content of *P. tuberosa* Root Extracts.
Values are expressed as mean (n=3) ± Standard Deviation (SD).
RE: Rutin Equivalents.

3.4 *In vitro* Antioxidant Assays

3.4.1 Ferric Reducing Antioxidant Assay

FRAP is a simple and reliable test to measure the reducing potential of an antioxidant reacting with ferric 2, 4, 6 tripyridyl S triazine (Fe (III) TPTZ complex and producing a coloured ferrous 2, 4, 6 tripyridyl S triazine (Fe (II) TPTZ complex at low pH.

Higher the absorbance higher will be the reducing power.It was found that FRAP values were consistently higher in acetone fraction (Table 16.1) when referred to the corresponding Fe (II) methanolic fraction.

3.4.2 Phosphomolybdenum Assay

The phosphomolybdenum assay is based on the reduction of Mo (VI) to Mo (V) by the antioxidant compound which turns green Mo (V) complex with absorption maximum at 695nm.

Total antioxidant capacity of different extracts were analyzed and shown in Table 16.1. The better antioxidant capacity was shown by acetone fraction (162.81

mgAAE/g extract).This can be correlated with the free radical scavenging activity of natural antioxidant Ascorbic acid.

Table 16.1: Phosphomolybdenum, FRAP, Metal Ion Radical Scavenging Activity of *P. tuberosa* Root Extract

Sample Extract	*Phosphomolybdenum (mgAAE/g)*	*FRAP (mmolesFeII/g)*	*Metal ion Chelating (mgEDTA/g Extract)*
Petroleum Ether	14.04±6.02	16.71 ± 1.01	92.71 ± 5.11
Chloroform	53.16±4.25	48.51 ± 1.20	23.59 ± 7.16
Acetone	162.81±6.64	91.32±2.61	112.78 ± 3.75
Methanol	84.54±2.45	65.56±2.15	102.53 ± 5.65
Hot water	21.57±2.21	31.94±1.16	109.92 ± 4.29

Values are mean of triplicate replication n=3 ± standard deviation.

3.4.3 Metal Chelating Activity

Metal chelating action is based on chelation of Fe^{2+} ions by the reagent Ferrozine, which results in the formation of a complex with Fe^{2+} ions (Dinis *et al.,* 1994).When other chelating agents are present it would decrease the formation of red coloured complex. Measurement of the rate of reduction of colour, therefore allows estimation of chelating activity.

In the present study *P.tuberosa* root extracts interact with ferrous and ferrozine complex suggesting that they have chelating properties which may capture Fe^{2+} ions before ferrozine. Metal chelating capacity of acetone extract of tuber was found to be 112.78mgEDTA/g (Table 16.1) from the result we can say that the extracts have protective role against oxidative damage by sequestering iron ions that may catalyse hydroperoxide decomposition reaction.

3.4.4 DPPH Radical Scavenging Activity

The principle of DPPH method is based on the reduction of DPPH in the presence of a hydrogen donating antioxidant.Extracts reduce the colour of DPPH due to the power of hydrogen donating ability (Blois, 1958) DPPH is one of the compounds that possess a proton free radical with a characteristic absorption which decreases significantly on exposure to proton radical scavengers(Yamaguchi *et al.,* 1998). Antioxidants may guard against reactive oxygen species (ROS) toxicities by scavenging reactive metabolites and converting them to less reactive molecules.

Importantly IC_{50} value of the extracts was also calculated to determine the amount of extract needed to quench 50 per cent of radicals. The results of DPPH were expressed inIC_{50} value. Lower the IC_{50} value, higher will be the antioxidant activity. The free radical scavenging activity of the extracts were estimated by comparing with standards such as BHT, BHA, quercetin and rutin and the result were shown in (Table 16.2) In the present study the chloroform fraction was found to be more potent hydroxyl radical scavenger, with an IC_{50} value of 119.59μg/mL compared to other extracts.

Table 16.2: DPPH Radical Scavenging Activity of *P. tuberosa* Root Extracts

Extract	*IC_{50} (µg/mL)*
Petroleum ether	676.88 ± 12.36
Chloroform	119.59 ± 9.18
Acetone	227.06 ± 15.34
Methanol	498.17 ± 21.84
Hot water	882.22 ± 26.71
BHA	5.10 ± 1.21
BHT	9.10 ± 2.14
Quercetin	3.81 ± 1.81
Rutin	4.94 ± 1.39

Values are expressed as mean (n=3) ± Standard Deviation (SD).

4.0 CONCLUSION

In the present study it was found that the root tubers of *Pueraria tuberosa* showed higher amount of phenolics and flavonoids which may be responsible for higher antioxidant activity. The toxicity of synthetic antioxidants like BHA, BHT are a problem of concern and are found to be unsafe,investigations must be directed towards the identification of natural or plant based antioxidants that may be better used for human consumption.

REFERENCES

Anonym: The Ayurvedic Pharmacopoeia of India, Part I, Vol. V, P 193, Ministry of Health and Family Welfare, NewDelhi 2006.

Blois, M.S., 1958. Antioxidant determinations by the use of a stable free radical. *Nature* 181 1199-1200.

Brune, M., Rossander, L., Hallberg, L., 1989. Iron absorption and phenolic compounds: importance of different phenolic structures. *Eur J Clin Nutr* 43: 547-558.

Dhiman, S.B., Kamat, J.P, Naik, D.B., 2009. Antioxidant activity and free radical scavenging reactions of hydroxybenzyl alcohol, biochemical and pulse radiolysis studies. *Chem Biol Interact* 182(2-3):119-127

Dinis, T.C.P, Madeira, V.M.C, Almeida, L.M., 1994 Action of phenolic derivatives (Acetaminophen salicylate and 5 amino salicylate) as inhibitors of membrane lipd peroxidation and as peroxyl radical scavengers. *Archives of Biochemistry and Biophysics* vol.35.No.1161-169

Frankel, E.N., Meyer, A.S., 2000. The problems of using one dimensional methods to evaluate multifunctional food and biological antioxidants. *J Sci Food Agric* 80:1925-1941

Jiang Ren-Wang, Kit-ManLau, Hung-MingLam, WingSzeYam, Lai—KinLeung,Kam-LinChoi, MarymyWaye, Thomas, C.W., Mak, Kam-SangWoo, Kwok-Pui Fung,

2005. A comparative study on aqueous root extracts of Pueraria thomsonii and Pueraria lobata by antioxidant assay and HPLC fingerprint analysis. *J Ethnopharmacol* 96(1-2): 133-138.

Karamac, M., 2009. In vitro study on efficacy of tannin fractions of edible nuts as antioxidants. *Eur J Lipids Sci Technol* 111(11), 1063-1071,

Kessler, M., Ubeaud, G., Jung, L., 2003. Anti and prooxidantactivity of rutin and quercetin derivatives. *J Pharm Pharmacol* 55:131-142

Miura, K., Nakatani, N., 1989. Antioxidative activity of flavonoids from thyme (*Thymus vulgaris* L.). *Agric Biol Chem* 53, 3043-3045.

Narasimhan, S., Shobana, R., Sathy, T.N., 2006 Antioxidants –Natural rejuvenators that heal, detoxify and provide nourishment,In: Rakesh K S, Rajesh A (eds), Herbal drugs A twenty first century prospective, NewDelhi, JP Brothers Medical Publishers,548-557

Prieto, P., Pineda, M., Aguilar M., 1999. Spectrophotometric quantification of antioxidant capacity through the formation of a phosphomolybdenum complex: specific application of vitamin E. *Anal Biochem* 269: 337-341

Prior, R.L., Wu, X.L., *et al.,* 2005 Standardisedmethods for the determination of antioxidant capacity and phenolics in foods and dietary suppliments. *J Sci Food Agric* 53:4290-4302

Pulido, R., Bravo, L., Sauro-Calixo, F., 2000. Antioxidant activity of dietary polyphenols as determined by a modified ferric reducing/antioxidant power assay, *J. Agri. Food chem.,* 48: 3396-3402.

Seong Eun Jin, You Kyung Son, Byung Sun Min, Hyun Ah Jung, Jae sue Choi 2012.Antiinflamatory and antioxidant activities of constituents isolated from *P. lobata* roots. *Arch. Pharm Res* 35, 823-827

Siddhuraju P., Becker K., 2003. Antioxidant properties of various extracts of total phenolic constituents from three different agroclimatic origins of drumstick tree (*Moringa oleifera* L.) leaves. *J. Agric. Food Chem.* 51, 2144-2155.

Siddhuraju, P., Manian, S. 2007. The antioxidant activity and free radical scavenging capacity of dietary phenolic extracts from horse gram (*Macrotyloma uniflorum* (Lam.) Verdc.) seeds. *Food Chemistry* 105, 950-958.

Yamaguchi, T., Takamura, H., Matoba, T., Terao, J., 1998. HPLC method for evaluation of the free radical-scavenging activity of foods by using 1,1,-diphenyl-2-picrylhydrazyl. *Biosci Biotech Biochem* 62:1201–1204.

Zhishen, J., Mengcheng T., Jianming, W., 1999. The determination of flavonoid contents in mulberry and their scavenging effects on superoxide radicals. *Food Chem* 64: 555-559.

2015, Modern Methods in Phytomedicine *Pages* ***261–267***
Editor: **T. Parimelazhagan**
Published by: **DAYA PUBLISHING HOUSE, NEW DELHI**

17

Evaluation of Enzymatic and Non-enzymatic Antioxidant Activity of Leaf Extract of *Ageratina adenophora* (Sprengel) King and H. Rob

S.R. Radha, R. Vanitha and B. Vijayakumari

Department of Botany, Avinashilingam University, Coimbatore – 641 043, Tamil Nadu

1.0 INTRODUCTION

Medicinal Plants have been identified and used throughout human history. India has one of the oldest, richest and most diverse cultural traditions associated with the use of medicinal plants. Plants have the ability to synthesize a wide variety of chemical compounds, that are used to perform important biological functions and to defend against attack from predators such as insects, fungi and herbivorous mammals. The history of medicinal plants can be traced back to Vedic periods about 4500- 1600 BC. Ayurveda has an extensive medical formulation, based on product of plant kingdoms. Medicinal plants constitute the dominant part of drug substances. There are some 1250 Ayurvedic Medicinal plants (Dev and Sukh, 2012) which go into formulating therapeutic preparation as per Ayurvedic and other traditional systems. If folklore medicinal herbs are added to these the total number of plants with medicinal applications used throughout India will exceed 2000. India has often been referred to as the Medicinal Garden of the world. WHO (2001)estimated that 80 per cent of world population rely on medicinal plants for their primary health care needs,out of that 3,50,000 plant species are known so far, about 35,000 are used worldwide for medicinal purposes and less than about 0.5 per cent of these have been investigated for their phytochemical and pharmacological potential (Hostettmann and Marston, 2002). Medicinal plants would be the best source to obtain a variety of drugs and therefore such plants should be investigated to understand better about their properties safety and efficacy Nature will still serve as the man's primary source for the cure of his ailments. However, the potential of higher plants as sources for new drugs is still largely explored. It is widely accepted that antioxidants are radical scavengers, which protect the human body against free radicals that may cause pathological conditions such as ischemia, anaemia, asthma, arthritis, inflammantion, neuro-degeneration, Parkinson's disease, ageing process and perhaps dementias (Ames *et al.,* 1993; Tripathy *et al.,* 2010). Antioxidants are the agents which scavenge free radicals and prevent damage caused by Reactive oxygen sps (Ros). Antioxidants can attack the cells to prevent damage to lipids, proteins, enzymes, carbohydrates and DNA. Antioxidants can be classified into two major classes *i.e.,* enzymatic and non-enzymatic. The enzymatic antioxidants are produced endogenously which include superoxide dismutase, caroteins and non-enzymatic includes ascorbic acid, flavonoids and tannins which are obtained from natural plant sources. A wide range of antioxidants from both natural and synthetic origin has been proposed for use in the treatment of various human diseases. There are some synthetic antioxidant compounds such as butylated hydroquinone which are commonly used in processed foods (Badami and Channabasavaraj, 2007). However it has been suggested that these compounds have shown toxic effects like liver damage mutagenesis. Flavonoids and other phenolic compounds of plant origin have been reported as scavenger's of free radicals. Hence now a days search for natural antioxidant source is gaining much importance. Plants may contain a wide variety of free radical scavenging molecules, such as phenolic compounds (e.g, phenolic acid, flavonoids, quinones, coumarins, lignins, stilbenes, tannins), nitrogen compounds (alkaloids, amines, betalains), vitamins, terpenoids (including carotenoids) and some other endogenous metabolites, which are also rich in antioxidant activity (Velioglu *et al.,* 1998; Zheng and wang, 2001; Cai *et al.,* 2003). The majority of the antioxidant activity is due to the flavones, isoflavones, anthocyanins, coumarins, lignins, catechins and isocatechins (Ahmed *et al.,* 2006). Antioxidant based drug formulations are used for the prevention and treatment of complex disease and cancer (Tilak *et al.,* 2004). Recently there is an increasing interest in finding natural antioxidant from plants because they can protect the human body from the attack of free radicals and retard the progress of many chronic diseases (Chung *et al.,* 2002). In the past few years, there has been considerable interest in natural products endowed with properties. Plants contain wide variety of antioxidant phytochemicals (or) bioactive molecules which can neutralize the free radicals. Now a days, the use of herbal medicine with antioxidant properties in both food and health science is an interesting area (Kitts *et al.,* 2000). This intake of antioxidant compounds present in food is now considered as important for health promotion and protection against damage due to oxidative stress (Fogliano *et al.,* 1999; Handleman *et al.,* 1999).The present study was carried out in order to orient the future investigation towards the finding of new, potent and safe antioxidant and antimicrobial compounds of *Ageratina adenophora.*

Plant Description

Ageratina adenophora, a native of Mexico has been naturalized in many countries (Hashorsky and Lichti, 2007). *Ageratina adenophora* belongs to the family Asteraceae. It is a perennial herb, nearly one meter height and erect, found in open and deforested areas of the central and eastern Nepal at altitude of 500-2000m. The leaf juice of *Ageratina adenophora* is used to stop bleeding of cut and wounds, forming clots (Bhattarai, 1997). Root juice is prescribed to treat fever. Pure juice of the leaf is poured in the eye to treat insomnia. A decoction of the plant has been recommended to treat jaundice and ulcers (Sharma *et al.,* 1998). The plant has many common names including eupatory, sticky snakeroot, croftenweed and mexican devil.*Ageratina adenophora* is otherwise called *Eupatorium glandulosum.* So the present study was undertaken to assess the Enzymatic and non enzymatic antioxidant activities of the leaves of *Ageratina adenophora* (Sprengel) King and H.Rob.

2.0 MATERIALS AND METHODS

An investigation was carried out to analyze the enzymatic and non enzymatic antioxidant activity of leaves of *Ageratina adenophora* (Sprengel) King and H.Rob.

2.1 Collection of Plant Sample

Healthy and disease free leaves of *Ageratina adenophora* (Sprengel) King and H.Rob.were collected from hill areas of the Nilgiris during the months of August – September (2013). The plant material was identified and authenticated by the Department of Botany, Avinashilingam University, Coimbatore-43, Tamilnadu, India. The leaves were cleaned, dried in the shade and pulverized in a mechanical grinder, passed through a 40 mesh sieve and stored in an air tight container.

2.2 Extraction of Plant Sample

Powdered dried samples (30g) were successively extracted with chloroform (300 mL), methanol (300 mL) and aqueous (300 mL) by using soxhlet apparatus until the decolourisation of the solvents.

2.3 Enzymatic Antioxidants

2.3.1 Catalase (Chance, 1995)

Catalase activity was determined by the titrimetric method. To 1 mL of plant extract, 5 mL of 300 µM phosphate buffer (pH 6.8) containing 100 µM hydrogen peroxide (H_2O_2) was added and left at 25°C for 1 min. The reaction was arrested by adding 10 mL of 2 per cent H_2SO_4, and residual H_2O_2 was titrated with potassium permanganate (0.01 N) till pink colour was obtained. Enzyme activity was estimated by calculating the decomposition of µM H_2O_2/min/mg protein.

2.3.2 Peroxidase (Addy and Goodman, 1972)

The reaction mixture consisted of 3 mL of buffered pyrogallol (0.05 M pyrogallol in 0.1 M phosphate buffer (pH 7.0)) and 0.5 mL of 1 per cent H_2O_2. To this 0.1 mL enzyme extract was added and O.D. change was measured at 430 nm for every 30 seconds for 2 min. The peroxidase was calculated using an extinction coefficient of oxidized pyrogallol (4.5 litres/mol).

2.3.3 Polyphenol Oxidase (Mahadevan and Sridhar, 1982)

The reaction mixture contained 3.0 mL of phosphate buffer, 1.0 mL of 0.01 M catechol in phos-phate buffer and 2.0 mL of the plant extract. Changes in absorbance were recorded in a spectrophotom-eter at 495 nm. The concentration of polyphenol oxidase was calculated and expressed as µg/g of protein.

2.4 Estimation of Non-enzymatic Antioxidant

2.4.1 Estimation of Reduced Glutathione (Boyne and Ellman, 1972)

1.0 mL of the plant extract was treated with 4.0 mL of precipitating solution containing 1.67g of glacial metaphosphoric acid, 0.2 g of EDTA and 30 g of NaCl in 100 mL water. After centrifugation, 2.0 mL of the protein free supernatant was mixed with 0.2 mL of 0.4M disodium hydrogen phosphate and 1.0 mL of DTNB reagent. Absorbance was read at 412 nm within 2 min. GSH concentration was expressed as n mol per mg protein.

2.4.2 Estimation of Total Flavonoid (Marinova *et al.*, 2005)

1 mL of extract was added to 10 mL volumetric flask containing 4 mL of distilled water. To above mixture, 0.3 mL of 5 per cent NaNO2 was added. After 5 min, 0.3 mL of 10 per cent $AlCl_3$ was added. At 6th min, 2 mL of 1M NaOH was added and the total volume was made up to 10 mL with distilled water. The solution was mixed well and the absorbance was measured against prepared reagent blank at 510 nm.

2.4.3 α-Tocopherol (Backer *et al.*, 1980)

500mg of fresh tissue was homogenized with 10 mL of a mixture of petroleum ether and ethanol (2:1.6 v/v) and the extract was centrifuged at 10,000 rpm for 20 min. To one mL of extract, 0.2 mL of 2 per cent 2, 2-dipyridyl in ethanol was added and mixed thoroughly and kept in the dark for 5 min. The resulting red colour was diluted with 4 mL of distilled water and mixed well. The resulting colour in the aqueous layer was measured at 520 nm. The α-tocopherol content was calculated using a standard graph made with a known amount of α- tocopherol. The results were expressed in milligrams/gram fresh weight.

3.0 RESULTS

3.1 Estimation of Enzymatic Antioxidant

In the present study, chloroform extract showed maximum result for catalase (36.16± 0.04mg-g) but minimum result was result was recorded in water extract (31.84 ± 0.03 mg^{-g}) (Table 17.1).

In the case of peroxidase the result was higher in methanol extract (396.50 ± 0.30 mg-^{g}) but minimum in chloroform extract (343.67 ± 0.15mg^{-g}). High amount of Polyphenol oxidase was present in chloroform extract (0.62 ± 0.03 mg^{-g}) and minimum amount was noticed in water extract (0.57±0.01 mg^{-g}) (Table 17.1).

Table 17.1: Estimation of Enzymatic Antioxidant Activity of Leaf Extracts of *Ageratina adenophora* (Sprengel) King and H.Rob.

Sl.No.	*Sample*	*Solvent*	*Catalse (mg⁻ᵍ)*	*Peroxidase (mg⁻ᵍ)*	*Estimtaion of Ployphenol Oxidase*
1.	Leaf	Chloroform	36.16±0.04	343.67 ±0.15	0.62±0.04
		Methanol	31.95±0.04	396.50 ±0.30	0.59 ±0.03
		Water	31.84 ±0.03	393.63 ±0.15	0.57 ±0.02
	SEd		0.231	0.135	0.021
	CD(P<0.05)		0.051	0.3008	0.047

Values are expressed by mean ± SD of three samples in each group.

In the present study, chloroform extract showed maximum result for reduced glutathione 0.62 ± 0.03mg^{-g} but minimum result was recorded in water extract (0.57 ± 0.01 mg^{-g}) Flavonoid content was more in chloroform extract (0.62±0.03 mg^{-g}) than the other extracts. The highest amount of α-tocopherol was recorded in water extract (3.15±0.03 mg^{-g}) TAE whereas minimum amount recorded was in methanol extract (2.35 ± 0.05 mg^{-g}) TAE (Table 17.II).

Table 17.2: Estimation of Non-enzymatic Antioxidant Activity of Leaf Extracts of *Ageratina adenophora* (Sprengel) King and H.Rob.

Sl.No.	*Sample*	*Solvent*	*Reduced Glutathione (mg⁻ᵍ)*	*Flavonoid (mg⁻ᵍ)*	*∞-Tocopherol (mg⁻ᵍ)*
1.	Leaf	Chloroform	0.62 ±0.03	0.62 ±0.03	2.83 ±0.04
		Methanol	0.59 ±0.02	0.59 ±0.02	2.35 ±0.03
		Water	0.57 ±0.01	0.57 ±0.01	3.15 ±0.02
	SEd		0.231	0.135	0.021
	CD(P<0.05)		0.051	0.3008	0.047

Values are expressed by mean ± SD of three samples in each group.

4.0 CONCLUSION

From the present study it can be concluded that the leaf of *Ageratina adenophora* (Sprengel) King and H. Rob is a good source of antioxidants. This results point out the significance of *Ageratina adenophora* (Sprengel) King and H.Rob as an important medicinal plant with numerous medicinal properties.

REFERENCES

Addy, S.K., Goodman, R.N., 1972. Polyphenol oxdiase and peroxidase in apple leaves inoculated with a virulent or an a virulent strain for *Ervinia amylovora. Indian phyto pathology*, 25; 575-579.

Ahmad, J., Aquil F., Mehmood, Z., 2009. Antioxidant and free radical scavenging properties of twelve traditionally used Indian medicinal plants, *Turk J Biol* 30(2): 177-183.

Ahmed, J., Aquil F., Mehmood, Z., 2006. Antioxidant and free radical scavenging properties of twelve traditionally used Indian medicinal plants, *Turk J Biol* 30 (2): 177-183.

Ames, B.N., Shigenaga, M.K., Hagen, T.M., 1993.Oxidants, Antioxidants and the degenerative diseases of aging, *Proc Natl Acad Sci* USA, 90:7915-7922.

Badami, S., Channabasavaraj, K.P., 2007. *In vitro* antioxidant activity of 13 medicinal plants of western Ghats, India, *Pharmaceutical Biology* 45(5): 392-396.

Becker, H., Frank, B.D., Angells., Feingold, 1980. Plasm tocopherol in man at various time after ingesting free or ocetylaned tocopherol. *Nutrition reports international*, 21:531-536.

Bhattararai, N.K., 1997. Traditional herbal medicines used to treat wounds and injuries in Nepal, Tropical Doctoo, Suppl, 1: 43-47.

Cai, Y.Z., Sun M., Corke, H., 2003. Antioxidant activity of battalions from plants of the Amaranthaceae, *J. of Agriculture Food Chemistry*, 51(8): 2288-2294.

Chance, C.M., 1995. Assay of Peroxidase and Catalase. *Methods in enzymology* 11(7): 64-75.

Chung,Y.G.,C.T.Chang W.W.Chao, C.F.Lin and S.T.Chou, 2002. Antioxidant activity of 50 per cent ethanolic extracts from red been fermented by *Bacillus subtilis* IMR-NLK, *J.of Agricultural and Food Chemistry*, 50: 2454-2458.

Dev, Sukh, 2012. Prime Ayurvedic Drugs- a modern Scientific Appraisal. Edn 2, Ane Books Pvt. Ltd. New Delhi, India.

Fogliano, V., Verde, V., Randazzo G., Ritieni, A., 1999. Method of measuring the antioxidant activity and its application to monitoring the antioxidant capacity of wines. *J. of Agriculture and Food Chemistry*, 47(3): 1035-1040.

Handleman, G.J., Cao, G., Watter, M.F., Nightingale, Z.D., Paul, G.L., Prior R.L., Blumberg, J.B., 1999. Antioxidant capacity of Oat (*Avena sativa* L.) extracts. Inhibition of low density lipoprotein oxidation and oxygen radical absorbance capacity. *J Agriculture and Food Chemistry* 47(12): 4888-4893.

Hoshovsky, M.C., Lichti, R., 2007. Ageratina adenophora httpill ucce.Ucdavis. Edu/ datastore/detail report. S.no-1829.

Hostettmann, K., Marston, A., 2000. Twenty years of research into medicinal; Results and perspectives, *Phytochem Rev* 1:275-285.

Kitts, D.D., Wjewickerence A.H., Hu, C., 2000. Antioxidant properties of a North American ginreng extract. *Molecular Cell Biology* 203(1): 1-10

Magadevan, A., Sridhar, R., 1982. Methods in physiological plant pathlogy 2nd ed. Sivakami publicatiosn, Madras, India 316.

Sharma, O.P., Dawra, R.K., Kirade N.P., Sharma, P.D., 1998. A review of the toxicosis and biological properties of the genus Eupatorium, *Nature Toxins*, 6: 1-14.

Tilak, J.C., Devasagayam T.P.A., Boloor, K.K., 2004. Review: Free radical and antioxidants in human health. *Curr. Stat. Fut Pros* JAPI 53(3): 794-804.

Tripathy, S., Pradhan D., Anjana, M., 2010. Anti inflammatory and antiarthritic potential of *Ammania bacifera* Linn. *Int J Pharma Biosci* 1(3): 1-7.

Velioglu, Y.S., Mazza, G., Goa L., Oomah, B.D., 1998. Antioxidant activity and total phenolics in related fruits, vegetables and grain products, *J Agric Food Chem* 46(10): 4113-4117.

Zheng, W., Wang, S.Y., 2001. Antioxidant activity and phenolic compounds in selected herbs, *J Agric Food Chem* 49(11): 5165-5170.

2015, Modern Methods in Phytomedicine *Pages* ***269–276***
Editor: **T. Parimelazhagan**
Published by: **DAYA PUBLISHING HOUSE, NEW DELHI**

18

Antimicrobial and Antifungal Activity of Aqueous Extracts of Certain Traditional Plants against Multidrug Resistant Clinical Pathogens

Chanthru Kothandapani[1,2], *Gomathi Rajkumar*[1] *and Sellamuthu Manian*[1,3]*

[1]*Department of Botany;* [2]*Department of Biotechnology, Bharathiar University, Coimbatore – 641 046, Tamil Nadu*
[3]*Department of Life Science, Manian Institute of Science and Technology, Coimbatore – 641 004, Tamil Nadu*

1.0 INTRODUCTION

Popular knowledge of plants used by Indians is an outcome of thousands of years' experience by trial and error methods. From time immemorial, man has been dependent on nature for survival. This dependency led the people living in harmony with nature to evolve a unique system of knowledge about plant wealth. In India, there are about 54 million indigenous people of different ethnic groups inhabiting various terrains (Ramar *et al.,* 2008). There are about 36 tribal communities in Tamil Nadu (Udayan *et al.,* 2006). They depend mostly on medicinal plants for treating diseases.Medicinal plants are being probed as an alternate sourceto get therapeutic compounds based on their medicinalproperties. These Knowledge was given a name

- Ethnobotany in 1895 by a North American botanist John Harshberger to describe studies of "plants used by primitive and aboriginal people" (Balick and Cox, 1996).

During the last two decades, there has been a considerable increase in the study and use of medicinal plants all over the world especially in advanced countries. There was also increase in the international commerce and commercial exploitation of herbal medicines through over the counter labeled products. The renewed interest in the use of medicinal plants may be attributed to cheapness, availability, and accessibility by the local populace, high incidence of side effects of synthetic medicines and environmental friendliness of plant extracts.

In accordance with the traditional knowledge, Antimicrobial screening has been carried out on crude leaf extracts of *O. sanctum, S. robusta, B. pilosa, F. zippelii* and *U. lobata* against commonly seen human pathogens.

2.0 MATERIALS AND METHODS

2.1 Plant Material

The fresh leaves of plants were used for the present study. Samples were collected July 2014 in and around Tamil Nadu. Leaves were dried separately under shade at room temperature for 12 days. The dried leaves were milled to a fine powder and stored in the dark at room temperature in closed containers until required.

2.2 Fresh Leaf Juice and Powder from Fresh Leaf Juice

One Hundred and Fifty grams of fresh leaves of *O. sanctum, S. robusta, B. pilosa, F. zippelii* and *U. lobata* were crushed directly by grinder by adding water, and the resulting extracts were decanted, filtered using Muslin clothand Whatman No. 1 filter paper and stored in sterile bottles (Atata, 2003). The leaf juice was collected in a clean airtight bottle, and stored for antibacterial activity test.15 mL of leaf juice were air-dried and 2.57g fine powder was obtained and stored in airtight bottle for antibacterial activity test.

2.3 Determination of Extraction Yield (Per cent yield)

The yield (per cent, w/w) from all the dried extracts wascalculated as:

Yield (per cent) = (W1 * 100)/W2

where, W1 is the weight of the extract after lyophilizationof solvent, and W2 is the weight of the plant powder.

2.4 Preparation of Various Concentrations of Extract

The extract was reconstituted in distilled water to obtain various concentrations of the extract thus: 2 g of extract was reconstituted in water to obtain 100 mL of a 20 mg/mL solution. A portion of the 20 mg/mL solution was diluted with an equal volume of distilled water to obtain a 10 mg/mL solution. The double dilution procedure was continued to obtain lower concentrations of the extract.

2.5 Test Microorganism

Tested Bacterial Stains

The following 8 bacteria were procured from Department of Biotechnology, Bharathiar University and used for the study: *S. aureus, S. typhi, Shigella* sp, *K. pneumonia, E. coli, P. fluorescence, Citrobacter sp* and *C. albicans.* All these cultures were maintained on nutrient agar plates at 4°C.

Tested Fungal Stains

The following fungi were used for experiments: *Trichophyton rubrum, Trichophytonmentagrophytes, Epidermophytonfloccosum, Scopulariopsis* sp., *Aspergillus flavus, Aspergillus niger, Botrytis cinerea, Candida albicans, Alternaria brassicicola, Alternaria alternata* and *Helminthosporium tetramera.* All these cultures were maintained on nutrient agar plates at 4°C.

2.6 Preparation of the Test Organisms

The isolates were sub-cultured onto selective and differential solid media and re-identified using biochemical tests.

2.7 Antibacterial Activity of Plant Extracts

Disc Diffusion Method

Circular disc of 6 mm diameter were made from the whatman's no 1 filter paper. Discs were impregnated with equal volume (50 mL) of each plant extracts at four different concentrations (0.05 g/mL, 0.1g/mL, 0.2g/mL and 0.4g/mL). The discs were aseptically placed over plates of Muller Hinton agar (MHA,Difco) seeded with each of test pathogens (Koochak *et al.,* 2010). The plates were incubated in an upright position at 37°C for 24 hours and the zone of inhibition was measured (in mm diameter). Inhibition zones with diameter less than 12 mm were considered as having low antibacterial activity. Diameters between 12 and 16 mm were considered moderately active, and these with >16mm were considered highly active and these with >16mm were considered highly active (Indu *et al.,* 2006).

2.8 Minimum Inhibitory Concentration (MIC)

MIC was prepared by serial dilution method. Two fold serial dilution of the test compound was carried out in the L.B broth. To each test tube 10^5 CFU/ mL of actively growing bacterial cultures in the log phase was inoculated. The culture tubes were incubated at 37°C for 24 hours. They were checked for bacterial growth and MIC of all the extracts were determined and expressed in ppm. Tetracyclin was used as positive control.The minimum inhibitory concentration was defined as the lowest concentrationable to inhibit any visible bacterial growth (Prescott *et al.,* 1999; Shahidi Bonjar, 2004).

2.9 Antifungal Activity of Plant Extracts

The antifungal activity was performed according to the standard reference method. The paper disc method was used for testingantifungal activity. The medium (25 mL) inoculated with spore suspension ofexperimental organism was poured intosterilized Petri dishes and left to get at roomtemperature. Whatman's No. 1 filter paperdiscs (6 mm dia). The filter paper discs wereplaced equidistantly on inoculated media. Plateswere incubated at room temperature for 72ho urs. Three plates were employed pertreatment and the average zone of inhibitionwas recorded.MIC was defined as the lowest extract concentration showingno visible fungal growth after incubation time. The antifungal agent fluconazole wasincluded in the assays as positive control.

2.10 Statistical Analysis

The data were statistically analysed by methodsuggested by Panse and Sukhatme. All the experiments were done in three replicates.

3.0 RESULTS AND DISCUSSION

Phytochemical research based on ethnopharmacologicalinformations is generally considered an effective approachto the discovery of antinfective agents from higher plants(Kloucek *et al.,* 2005).The presence of zones of inhibition on the seededagar plates showed that the plant extract possessesantibacterial activity on the tested organisms.

The zone of inhibition was measured for all the crude extracts of plants and the results were displayed in Table 18.1. *O. sanctum* showed higher efficiency against *S. typhi* which is followed by *S. aureus.* The crude extracts of *S. robusta* showed higher efficiency against *K. pneumonia* and *P. fluorescence.B. pilosa* crude extract showed greater efficiency against *P. fluorescence* and *K. pneumonia. F. zippelii* showed good inhibition against *K. pneumonia* and *P. fluorescence. U. lobata* showed higher zone of exhibition against *E. coli* and *Citrobacter sp* as well. However differences were observed between antibacterialactivities of the extracts. These differences could be due to the differences in the chemical composition ofthese extracts as the secondary metabolites of plants have many effects including antibacterial, antifungal and antiviral properties (Noumedem *et al.,* 2013, Jain *et al.,* 2009 and Cowman, 1999). Comparisons with pertinent data from literature indicate that, according to the methodology adopted in studies on antimicrobial activity, the most diverse results can be obtained. Plant extracts have shown inhibitory effect on the growth of the bacteria studied, in distinct forms. It is therefore nature and the number of the active antibacterial principles involved in each plant extract has involved in inhibition.

The MIC tests of all the crude extracts were carried out using micro dilution technique. The MIC results are shown in Table 18.2. All the plant samples showed considerably good zone of inhibition with some exceptions. MIC ranged from 0.380 to 0.642 mg/mL. *O. sanctum* showed best activity against most of the pathogens. This suggests that the extracts of these plants are broad spectrum in their activities. The most sensitive pathogen was *E. coli* (MIC ranging from 0.385 to 0.642 mg/mL). Followed by *S. aureus* (MIC ranging from 0.452 to 0.567 mg/mL).Whereas the other extracts also showed considerable zones of inhibitions. The results represents that all the leaf extracts had variable degree of antibacterial activity and the inhibition of bacterial growth was dose dependent in its action. The activity found increasing when the concentration of the extract increases. All the bacterial pathogens as evidenced by higher zone of inhibitions at higher concentration.

Table 18.1: Antibacterial Activity of different Crude Plant Extract by Disc Diffusion Method 0.00 Represents No Zone of Inhibition

Bacterial Species	*Ocimum sanctum*				*Shorea robusta*				*Bidens pilosa*				*Flacourtia zippelii*				*Urena lobata*			
	50	*100*	*200*	*400*	*50*	*100*	*200*	*400*	*50*	*100*	*200*	*400*	*50*	*100*	*200*	*400*	*50*	*100*	*200*	*400*
Staphylococcus aureus	12.4	14.1	22.0	31.2	19.1	21.0	27.2	33.3	9.4	17.1	24.0	35.2	19.1	27.5	35.1	34.0	4.8	6.7	6.9	7.0
Salmonella typhi	**21.1**	**25.6**	**28.6**	**36.5**	0.0	0.0	5.4	10.1	7.9	8.5	9.9	11.3	9.1	9.9	11.2	11.6	0.0	0.0	5.4	10.1
Shigella sp.	24.5	36.7	42.7	46.9	9.1	10.0	11.2	11.4	9.1	9.3	10.2	10.3	4.1	10.2	11.2	11.3	17.9	18.5	29.9	31.3
Klebsiella pneumonia	0.0	0.0	5.4	9.5	**19.9**	**21.6**	**27.2**	**35.5**	7.1	7.5	8.1	9.0	**20.0**	**26.2**	**28.3**	**37.3**	22.1	28.0	36.2	39.4
E. coli	0.0	3.5	7.3	9.1	9.2	9.3	10.2	10.3	7.0	7.2	8.3	8.5	4.5	6.7	6.7	6.9	**29.4**	**35.8**	**40.2**	**40.6**
Pseudomonas fluorescence	23.7	29.4	33.2	39.2	10.1	13.3	19.6	**11.5**	**9.1**	**9.3**	**11.2**	**11.3**	19.1	21.6	27.2	31.4	9.1	10.3	10.6	11.5
Citrobacter sp.	11.1	16.5	21.1	28.0	8.9	10.0	10.3	11.3	7.1	7.3	8.6	9.5	17.3	22.0	26.4	32.3	9.3	9.5	10.0	10.2
Candida albicans	7.9	8.5	9.9	11.3	18.4	19.1	21.0	26.2	9.1	9.1	10.2	10.3	0.0	0.0	5.4	10.1	17.1	17.5	18.1	19.0

Table 18.2: Antibacterial Activity of different Crude Plant Extract by Minimum Inhibitory Concentration (MIC)

Bacterial Species	*Plants*					*Control*
	Ocimum sanctum (mg/mL)	*Shorea robusta (mg/mL)*	*Bidens pilosa (mg/mL)*	*Flacourtia zippelii (mg/mL)*	*Urena lobata (mg/mL)*	*Tetracyclin*
Staphylococcus aureus	0.567	0.417	0.469	0.427	0.452	0.002
Salmonella typhi	0.489	0.476	0.380	0.602	0.499	0.004
Shigella sp.	0.537	0.532	0.401	0.489	0.542	0.003
Klebsiella pneumonia	0.582	0.531	0.389	0.601	0.582	0.004
E. coli	0.642	0.611	0.531	0.385	0.631	0.006
Pseudomonas fluorescence	0.490	0.459	0.567	0.500	0.489	0.002
Citrobacter sp.	0.511	0.480	0.502	0.469	0.530	0.001
Candida albicans	0.632	0.581	0.409	0.460	0.531	0.002

Antifungal activity of five medicinal plants extract was assayed by agar well diffusion method with different concentrations (via.1.00, 1.20, 1.30 and 1.40) in mg/mL. Average inhibition was taken as result and presented in Table 18.2. Antifungal activity of 5 aqueous plant leaf extracts was macerated and data on effect of aqueous plant extracts on the growth of *T. rubrum, T. mentagrophytes, E. floccosum, Scopulariopsis* sp.,*A. flavus, A. niger, B. cinerea, C. albicans,A. brassicicola, A.alternata* and *H. tetramera* are presented in Table 18.3. The data revealed that significant reduction in growth of *A. niger* was observed.Extracts of 5medicinal plants and the extracts showed significant differences in their mode of action. The result resembles with the result of Raji and Raveendra (2013).

Among the 5 plant aqueous extracts tested for antibacterial and antifungal *O. sanctum* showed over all inhibition in most of the bacterial and fungal pathogens which was followed by *U. lobata* and *S. robusta*.

The implication of the broad spectrum action of some of these extracts is that they can be useful in antiseptic and disinfectant formulation as well as in chemotherapy if the active principle can be isolated (Olukoya *et al.,* 1993). The anti-pseudomonal and anti-staphylococcal activities of some of the effective extracts of these plants can be further explored.

The low number of papers that have appeared to work on screening of antifungal activity as compared to work on antibacterial activity. These results may contribute to are solution of these difficulties. The continuous evaluation of bacterial and fungal resistance to current available antibiotics has necessitated the search for novel and effective antimicrobial compounds. Medicinal plants that have pharmaceutical properties can be natural composite and may act as new anti-infectious agents. So, today foremost demand is in search of effective, cheapest and improved antimicrobial compounds. Moreover, the present work gives additional information of the antibacterial activities of these plant extracts against multi-resistant bacteria.

Table 18.3: Antifungala Ativity of different Crude Plant Extract by Minimum Inhibitory Concentration (MIC)

Bacterial Species	*Plants*				
	Ocimum sanctum (mg/mL)	*Shorea robusta (mg/mL)*	*Bidens pilosa (mg/mL)*	*Flacourtia zippelii (mg/mL)*	*Urena lobata (mg/mL)*
Trichophyton rubrum	1.000	–	0.746	–	0.389
Trichophytonmentagrophytes	0.937	–	0.531	1.000	–
Epidermophytonfloccosum	0.559	0.532	–	0.567	0.499
Scopulariopsis sp	–	0.231	0.381	0.602	–
Aspergillus flavus	1.000	–	0.643	–	0.502
Aspergillus niger	1.000	0.342	0.530	0.582	1.000
Botrytis cinerea	–	0.611	–	–	0.581
Candida albicans	0.847	0.542	0.736	0.427	–
Alternaria brassicicola	–	–	0.476	0.389	0.531
Alternaria.alternata	0.611	0.489	0.826	0.827	–
Helminthosporium tetramera	–	0.635	0.745	0.459	0.531

4.0 CONCLUSION

Aqueous extract of leaves of *O. sanctum, S. robusta, B. pilosa, F. zippelii* and *U. lobata* showed antibacterial as well as antifungal activity on some bacterial and fungal organisms tested. Further, studies are required to identify the active principles of the extract and their mode of extraction. This supports the traditional use of the leaves for treatment of ailments associated with other fungi and bacteria. Research should be carried out on the toxicity of the plant in order to know the safety and toxicity of the plant and establish a safe dosage regimen since the infusion of the leaves is taken orally by local people for various treatments.

REFERENCES

Atata, R., Sani, A., Ajewole, S.M., 2003. Effect of stem back extracts of *Enantia chloranta* on some clinical isolates. *Biokemistri*, 15 (2): 84-92.

Balick, M., Cox, P., 1996. Plants, People and Culture. The Science of Ethnobotany. Scientific American Library, USA, 228 pp.

Cowan MM: Plant products as antimicrobial agents., 1999. Clin Microbiol Rev, 12:564–582.

Elgayyar, M., Drugon, FA., Golden, DA., Mount, JR., 2001. Antimirobial activity of essential oils from plants against selected pathogenic and saprophytic microorganisms. *J Food Prot.* 64(7), 1019 – 1024.

Indu, MN., Hatha, AAM., Abirosh, C., Harsha, U., Vivekanandan, G, 2006. Antimicrobial Activity of Some of The South-Indian Spices Against Serotypes *of Escherichia Coli, Salmonella, Listeria monocytogenes* and *Aeromonas hydrophila. Brazilian Journal of Microbiology.* 37:153-158.

Jain, P., Bansal, D., Bhasin, P. 2009. Antibacterial activity of aqueous plantextracts against *Escherichia coli* and *Bacillus substilis. Drug invension today* 2(4):220-222.

Kloucek, P., Polesny, Z., Svobadova, B., Vloka, E., Kokoska, L., 2005 Antibacterial screening of somePeruvian medicinal plants used in Calleria District.*Journal of Ethnopharmacology*. 99, 309-312.

Koochak, H., Seyyed, MS., Hussein, M., 2010. Preliminary study on the antibacterial activity of some medicinal plants of Khuzestan (Iran). *Asian Pacific Journal of Tropical Medicine*.3(3)180-184.

NCCLS. Reference method for broth dilution antifungalsusceptibility testing of filamentous fungi. 2002. Approved standardM38-A. Wayne, standard.

Noumedem, J., Mihasan, M., Lacmata, S., Stefan, M., Kuiate, J., Kuete, V., 2013. Antibacterial activities of the methanol extracts of ten Cameroonian vegetables against Gram-negative multidrug-resistant bacteria. *BMC Complement Altern Med.* 13:26.

Olukoya, D.K., Ndika, N., Odugbemi, T.O, 1993. Antibacterial activity of some medicinal plants in Nigeria. *Journal ofEthnopharmacology*. 39: 69-72.

Panse, V.G., Sukhatme, P.V., Statistical methods for agricultural workers 1985.ICAR, New Delhi.

Prescott, M.L., Harley, J., Donald, P., Klein A., 1999. In 'Antimicrobial chemotherapy.' Microbiology 2nd edition published by C. Brown Publishers, U.S.A. Pp 325.

Raji, R., Raveendra, K., 2013. Antifungal activity of selected plant extracts against phytopathogenic fungi *Aspergillus niger. A J of Plant Science and Res* 3(1):13-15.

Ramar, P.S., Maung, M.T., Ponnampalam, G., Ignacimuthu, S., 2008. Ethnobotanical survey of folk plants for the treatment of snakebites in Southern part of Tamilnadu, India. *J Ethno pharmacol* 17; 115(2):302-12.

Shahidi Bonjar, G. H., 2004. Evaluation of Antibacterial properties of Iranian Medicinal plants against *Micrococcusaureus, Serratia marcescens, Klebsiella pneunomiae* and *Bordella bronchoseptica. A J Sci* 3(1):82-86.

Thippeswamy, T., Lokesh, S., 1997. Effect ofleaf extracts on seed mycoflora,germination and seedling vigor of sunflower variety APSH 11. *International journal oftropical plant diseases.* 15 (1): 53-58.

Udayan, P.S., George, S., Tushar, K.V., Balachandran, I., 2006. Medicinal Plants Used by the Malayali Tribes of Servarayan Hills, Yerkad, Salem District, Tamil Nadu, India. *Zoo's Print Journal* 21(4): 2223-2224.

2015, Modern Methods in Phytomedicine *Pages* ***277–291***
Editor: **T. Parimelazhagan**
Published by: **DAYA PUBLISHING HOUSE, NEW DELHI**

19

Flavonoids and their Health Effects

K. Preethi

Department of Microbial Biotechnology, Bharathiar University, Coimbatore, Tamil Nadu

1.0 INTRODUCTION

Phytochemicals are defined as bioactive non-nutrient plant compounds. They are non-essential nutrients, and are not essentially required by the human body for sustaining life. These are chemicals derived from plants. In a narrower sense the terms are often used to describe the large number of secondary metabolites found in plants. Many of these are known to provide protection against insect attacks and plant diseases. They also exhibit a number of protective functions for humans. It is well-known that plant produces these chemicals to protect themselves but recent research demonstrate that they can also protect humans against diseases. There are more than thousand known phytochemicals. The term is generally used to refer to those chemicals that may have biological significance, for example carotenoids or flavonoids, other well-known phytochemicals are lycopene in tomatoes, isoflavones in soy and flavanoids in fruits. Among the phytochemicals, phenols, have received a great deal of attention because of their antioxidant activity. Various plants have been shown to possess significant antioxidant property (Okamura *et al.*, 1993; Chen *et al.*, 1999 and Kweon *et al.*, 2001).

2.0 SOURCES OF PHYTOCHEMICALS

Foods containing phytochemicals are already part of our daily diet. In fact, most foods contain phytochemicals except for some refined foods such as sugar or alcohol.

Some foods, such as whole grains, vegetables, beans, fruits and herbs, contain many phytochemicals. The easiest way to get more phytochemicals is to eat more fruit (blueberries, cranberries, cherries, apple,) and vegetables (cauliflower, cabbage, carrots, broccoli.). It is recommended to take daily at least 5 to 9 servings of fruits or vegetable. Fruits and vegetables are also rich in minerals, vitamins and fibre and low in saturated fat (Dillard and German, 2000)

3.0 MECHANISM OF ACTION OF PHYTOCHEMICALS

There are many phytochemicals and each works differently. Some possible modes of actions of phytochemicals are outlined below:

- **Antioxidant** - Most phytochemicals have antioxidant activity and protect our cells against oxidative damage and reduce the risk of developing certain types of cancer. Phytochemicals with antioxidant activity: allyl sulfides (onions, leeks, garlic), carotenoids (fruits, carrots), flavonoids (fruits, vegetables) and polyphenols (tea, grapes).
- **Hormonal action** - Isoflavones found in soy, imitate human estrogens and help to reduce menopausal symptoms and osteoporosis.
- **Stimulation of enzymes** - Indoles, which are found in cabbages, stimulate enzymes that make the estrogen less effective and could reduce the risk for breast cancer. Other phytochemicals, which interfere with enzymes, are protease inhibitors (soy and beans), terpenes (citrus fruits and cherries).
- **Interference with DNA replication** - Saponins found in beans interfere with the replication of cell DNA, thereby preventing the multiplication of cancer cells. Capsaicin, found in hot peppers, protects DNA from carcinogens.
- **Anti-bacterial effect** - The phytochemical allicin from garlic has anti-bacterial properties.
- **Physical action** - Some phytochemicals bind physically to cell walls thereby preventing the adhesion of pathogens to human cell walls. Proanthocyanidins are responsible for the anti-adhesion properties of cranberry. Consumption of cranberries will reduce the risk of urinary tract infections and will improve dental health. (Mark *et al.*, 2004)

4.0 TYPES OF PHYTOCHEMICALS

- There are hundreds of phytochemicals found in foods of plant origin. Based on their chemical structure, phytochemicals can be broken into the following groups (Arts and Hollman, 2005). Phytochemicals- Phenolic acids, Flavonoids, Stilbenes/Lignans.

Some of the best known phytochemicals are listed below along with their benefits.

- **Bioflavonoids:** These are helpful in the absorption of vitamin C and protect it from oxidation (damage). Citrus fruits, such as lemons, limes, grapefruit, and oranges, are particularly good sources of bioflavonoids.

- **Carotenoids:** These may protect against cardiovascular disease. Carotenoids are found in orange-fleshed melon, carrots, sweet potatoes, and butternut squash.
- **Glucosinolates:** Found in vegetables, these help the liver in its detoxification function. They help regulate certain white blood cells involved in immunity. They may also help reduce tumour growth, particularly in the breast, liver, colon, lung, stomach, and oesophagus.
- **Organosulphides:** These give onions and leeks their pungent odour. They stimulate anti-cancer enzymes, slow the formation of blood clots, and are known to boost the immune system.
- **Flavonoids:** These may protect the body from inflammation, allergic reactions and viral infections.
- **Indoles:** These phytochemicals are thought to help prevent breast cancer.
- **Isoflavones:** These may inhibit oestrogen-promoted cancers and lower high levels of blood cholesterol.
- **Limonoids:** Found in the peel of citrus fruits, these phytochemicals appear to protect lung tissue.
- **Phytoestrogens:** These protect the body against cardiovascular disease and osteoporosis. Phytoestrogens may also slow the progression of cancer. They are found in soya products and linseeds.
- **Lycopene:** Found in tomatoes, this may protect against cancers of the cervix, stomach, bladder, colon, and prostate, as well as cardiovascular disease.
- **Para-coumaric acid:** This phytochemical helps prevent cancer by interfering with the development of cancer-causing nitrosamines in the stomach.
- **Phenols and polyphenol:** These protect plants from chemical damage and perform the same function in humans. Found in tea, polyphenol is thought to protect against stomach cancer.
- **Phytosterols:** These include stanols, which can reduce the absorption of cholesterol from the diet and therefore lower cholesterol levels in the blood. Stanols are found in soya products and fortified margarines.
- **Terpenes:** These may block action of cancer-causing factors (carcinogens) and may inhibit hormone-related cancers such as ovarian cancer.

Let us have a detailed look on phenol compounds and its classifications. Phenolic compounds are classified as simple phenols or polyphenols based on the number of phenol units in the molecule. Simple phenols (C6) seldom occur naturally, so plant phenolics are divided into the following main groups.

- Phenolic acid that are hydroxylated derivatives of benzoic acid (C6-C1) which are quite common in the free state as well as combined as esters or glycosides (gallic acid)
- Phenolic acid derived from cinnamic acid (C6-C3) (coumaric, caffeci, ferulic acid) which are widely distributed and occurs rarely in the free state and are very offeuesterfied, and
- Glycosidic phenylpropanoid esters.

Depending on the degree of oxidation of the central pyram ring, they can be subdivided into several classes of flavonoids and flavonoid-related compounds: flavones, flavonols, flavanones, isoflavones, flavans, flavanols and anthocyanins, proanthocyanidins.

The largest and best studied natural phenols are the flavonoids, which include several thousand compounds, among them the flavonols, flavones, flavan-3ol (*catechins*), flavanones, anthocyanidins and isoflavonoids are important. We shall see about Flavonoids in detail in this chapter.

4.1 Flavonoids

Flavonoids, an amazing array of over 6,000 different substances found in virtually all plants, are responsible for many of the plant colors that dazzle us with their brilliant shades of yellow, orange, and red.

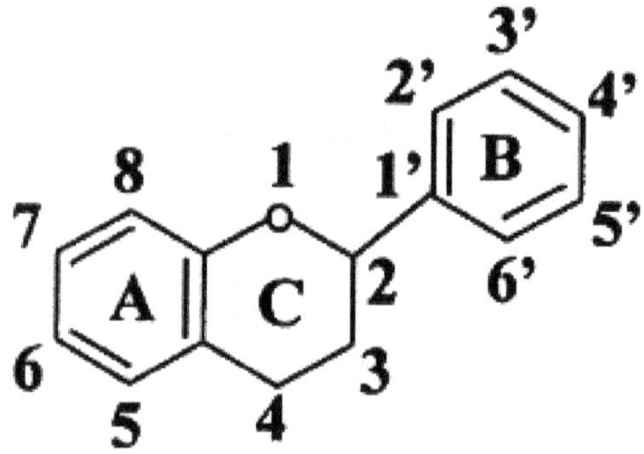

Figure 19.1: Nuclear Structure of Flavanoids-Dietary Flavonoids are Diverse and Vary According to Hydroxylation Pattern, Conjugation between the Aromatic Rings, Glycosidic Moieties, and Methoxy Groups.

Flavonoids are the most diverse group of phytochemicals. Research suggests that flavonoids, in particular, may be an important phytochemical group that contributes to the reduced mortality rates observed in people consuming high levels of plant-based foods (Hertog *et al.*, 1993). Flavonoids are water soluble polyphenolic molecules containing 15 carbon atoms. Flavonoids belong to the polyphenol family. Flavanoids can be visualized as two benzene rings which are joined together with a short three carbon chain. One of the carbons of the short chain is always connected to a carbon of one of the benzene rings, either directly or through an oxygen bridge, thereby forming a third middle ring, which can be five or six-membered.

4.2 Distribution and Dietary Sources of Flavonoids

Flavonoids (specifically flavanoids such as the catechins) are "the most common group of polyphenolic compounds in the human diet and are found ubiquitously in plants. Flavonols, the original bioflavonoids such as quercetin, are also found ubiquitously, but in lesser quantities. The widespread distribution of flavonoids, their variety and their relatively low toxicity compared to other active plant compounds (for instance alkaloids) mean that many animals, including humans, ingest significant quantities in their diet. Foods with a high flavonoid content includes, onions, blueberries and other berries, black tea, green tea and oolong tea, bananas, all citrus fruits, *Ginkgo biloba*, red wine, sea-buckthorns, and dark chocolate (with a cocoa content of 70 per cent or greater). Further information on dietary sources of flavonoids can be obtained from the US Department of Agriculture flavonoid database. Flavonoids exist naturally in cocoa, but because they can be bitter, they are often removed from chocolate, even dark chocolate although flavonoids are present in milk chocolate, milk may interfere with their absorption.

Flavonoids are found in most plant material. The most important dietary sources are fruits, tea and soybean. Green and black tea contains about 25 per cent percent flavonoids. Other important sources of flavonoids are apple (quercetin), citrus fruits (rutin and hesperidin).

Table 19.1: Some Examples of Flavonoids and their Food Sources

Class	*Example*	*Common Food Source*
Flavonol	Quercetin	Apples, Onions
Flavanol	Catechin	Tea, coffee, chocolate
Isoflavone	Genistein	soy
Flavonone	Hesperitin	Grapefruit
Anthocyanidin	Cyanidin	Berries

4.3 Flavonoids in Plants

Flavonoids are widely distributed in plants, fulfilling many functions. Flavonoids are the most important plant pigments for flower coloration, producing yellow or red/blue pigmentation in petals designed to attract pollinator animals. In higher plants, flavonoids are involved in UV filtration, symbiotic nitrogen fixation and floral pigmentation. They may also act as chemical messengers, physiological regulators, and cell cycle inhibitors. Flavonoids secreted by the root of their host plant help *Rhizobia* in the infection stage of their symbiotic relationship with legumes like peas, beans, clover, and soy. Rhizobia living in soil are able to sense the flavonoids and this triggers the secretion of Nod factors, which in turn are recognized by the host plant and can lead to root hair deformation and several cellular responses such as ion fluxes and the formation of a root nodule. In addition, some flavonoids have inhibitory activity against organisms that cause plant diseases, *e.g. Fusarium oxysporum* (Galeotti *et al.,* 2008).

4.4 Physical and Chemical Properties of Flavonoids

Flavonoids are crystallic substances with certain melting point. Cathechins, leucoanthocyanidins, flavanes, isoflavanes, flavanones, flavanonoles are colorless crystalls, flavones, flavonoles, chalcones and aurones are yellow or vividly yellow. Anthocyanes are sap pigments and the actual colour of the plant organ is determined by the pH of the sap. For example, the blue colour of the cornflower and the red of roses are due to the same glycosides and both of these plants on hydrolysis with hydrochloric acid yield cyanidin hydrochloride. Changes of anthocyanes colour depend upon pH: in acid medium their colour is red, in alkaline medium – blue. As a general rule, glycosides are water-soluble and soluble in alcohols. Flavonoid glycosides are soluble in diluted alcohols and hot water. Aglycones are, for the most part, soluble in apolar organic solvents: when they have at least one free phenolic group, they dissolve in alkaline hydroxide solutions. Flavonoid aglycones soluble in diethylether, acetone, alcohols, almost are insoluble in water.

Flavanoles (cathechins) are optically active. *i.e.*, among four optical isomers (D-and L-cathechins, D- and L-epicathechins) only L-epicathechin possesses P-vitaminic activity. Flavanones and flavonononones are unstable compounds. Treated with oxidants, they turn into chalcones and leucocyanidins accordingly. Flavonoid O-glycosides may be treated with acid, alkaline or fermentative hydrolysis. Rutin occurs as a yellow crystalline powder, soluble in alkali but only slightly soluble in water. Rutin on hydrolysis yields quercetin, rhamnose and glucose, while hesperidin yields hesperetin (or methyl eriodictyol), rhamnose and glucose. C-linkage between aglycone and sugar is verystrong, therefore hydrolysis of C-glycosides is carried out with Killiani's reagent (mixture of concentrated HCl and acetic acid) (Danylo Halytskyi)

4.5 Biological Properties of Flavanoids

As a dietary component, flavonoids are thought to have health-promoting properties due to their high antioxidant capacity both *in vivo* and *in vitro* systems. Flavonoids have the ability to induce human protective enzyme systems. The number of studies has suggested protective effects of flavonoids against many infectious (bacterial and viral diseases) and degenerative diseases such as cardiovascular diseases, cancers, and other age-related diseases. Flavonoids also act as a secondary antioxidant defense system in plant tissues exposed to different abiotic and biotic stresses. Flavonoids are located in the nucleus of mesophyll cells and within centers of ROS generation. They also regulate growth factors in plants such as auxin. Biosynthetic genes have been assembled in several bacteria and fungi for enhanced production of flavonoids.

4.6 Types of Flavonoids

Over 5000 naturally occurring flavonoids have been characterized from various plants. They have been classified according to their chemical structure, and are usually subdivided into the subgroups showing in Table 19.2.

4.7 Functions of Flavonoids

4.7.1 Protection of Cell Structures

Most flavonoids function in the human body as antioxidants. In this capacity, they help neutralize overly reactive oxygen-containing molecules and prevent these overly reactive molecules from damaging parts of cells. Particularly in oriental medicine, plant flavonoids have been used for centuries in conjunction with their antioxidant, protective properties. Scultellaria root, cornus fruit, licorice, and green tea are examples of flavonoid-containing foods widely used in oriental medicine. While flavonoids may exert their cell structure protection through a variety of mechanisms, one of their potent effects may be through their ability to increase levels of glutathione, a powerful antioxidant, as suggested by various research studies.

Table 19.2: Flavonoids -Types

Flavonoids	☆ Anthoxanthins
	☆ Flavones
	☆ Flavonols
	☆ Flavanones
	☆ Flavanonols
	☆ Flavans
	☆ Flavan-3-ols
	☆ Proanthocyanidins
	☆ Flavan-4-ols
	☆ Flavan-3,4-diols
	☆ Anthocyanidins or Anthocyaninsaglycone
Isoflavonoids	☆ Isoflavones (Pyranoisoflavones)
	☆ Isoflavans
	☆ Pterocarpans
Neoflavonoids	☆ Dalbergichromene
Aurones	☆ Aureusidin
	☆ Leptosidin
Other categories	☆ C-methylated flavonoids
	☆ O-methylated flavonoids
	☆ Flavonolignans
	☆ Furanoflavonoids
	☆ Pyranoflavonoids
	☆ Prenylflavonoids
	☆ Methylenedioxy
	☆ Castavinols

4.7.2 Vitamin C Support

The relationship between flavonoids and vitamin C was actually discovered by mistake. Dr. Albert Szent-Gyorgyi, the Nobel Prize winning researcher who discovered flavonoids, was attempting to make a preparation of vitamin C for one of his patients with blood vessel problems. The preparation he gave the patient was not 100 per cent pure–it contained other substances along with the vitamin C. It worked amazingly well.

Later, when Dr. Szent-Gyorgyi purchased a pure solution of vitamin C, he found it was not nearly so effective with his patient. He suspected flavonoids as the magic addition to vitamin C in his first impure preparation. Present-day research has clearly documented the synergistic (mutually beneficial) relationship between flavonoids and vitamin C. Each substance improves the antioxidant activity of the other, and many of the vitamin-related functions of vitamin C also appear to require the presence of flavonoids.

4.7.3 Inflammation Control

Inflammation–the body's natural response to danger or damage–must always be carefully regulated to prevent overactivation of the immune system and unwanted immune response. Many types of cells involved with the immune system–including T cells, B cells, NK cells, mast cells and neutrophils–have been shown to alter their behavior in the presence of flavonoids. Prevention of excessive inflammation appear to be a key role played by many different chemical categories of flavonoids.

4.7.4 Antibiotic Activity

In some cases, flavonoids can act directly as antibiotics by disrupting the function of microorganisms like viruses or bacteria. The antiviral function of flavonoids has been demonstrated with the HIV virus, and also with HSV-1, a herpes simplex virus.

4.8 Health Benefits of Flavonoids

Flavonoids are becoming very popular because they have many health promoting effects. Some of the activities attributed to flavonoids include: anti-allergic, anti-cancer, antioxidant, anti-inflammatory and anti-viral. The flavonoids quercetin is known for its ability to relieve hay fever, eczema, sinusitis and asthma. Epidemiological studies have illustrated that heart diseases are inversely related to the flavonoid intake (Groff *et al.,* 1995) Apart from this flavanoids are doing many good things for us such as

- ✰ Help protect blood vessels from rupture or leakage
- ✰ Enhance the power of your vitamin C
- ✰ Protect cells from oxygen damage
- ✰ Prevent excessive inflammation throughout your body

The contribution of flavonoids to the total antioxidant activity of components in food can be very high because daily intake can vary between 50 to 500 mg. Red wine contains high levels of flavonoids, mainly quercetin and rutin. The high intake of red wine (and flavonoids) by the French might explain why they suffer less from coronary heart disease then other Europeans, although their consumption of cholesterol rich foods is higher (French paradox). Many studies have confirmed that one or two glasses of red wine daily can protect against heart disease. Tea flavonoids have many health benefits. Tea flavonoids reduce the oxidation of low-density lipoprotein, lowers the blood levels of cholesterol and triglycerides. Soy flavonoids (isoflavones) can also reduce blood cholesterol and can help to prevent osteoporis. Soy flavonoids are also used to ease menopausal symptoms (Yao *et al.,* 2004).

Before any chemical compound can be approved as a pharmaceutical drug or any food can be labeled with a health claim, it must undergo extensive *in vitro, in vivo,* and clinical testing to confirm both safety and efficacy. National and international regulatory authorities like the US Food and Drug Administration (FDA) and European Food Safety Authority (EFSA) are responsible for assessing this evidence and granting such approval. At the current time, neither the FDA nor the EFSA has approved any health claim for flavonoids, or approved any flavonoids as pharmaceutical drugs. Moreover, several companies have been cautioned by the FDA over misleading health claims.

4.9 Therapeutical Potential of Flavonoids

4.9.1 Antioxidant Activity

Flavonoids possess many biochemical properties, but the best described property of almost every group of flavonoids is their capacity to act as antioxidants. The antioxidant activity of flavonoids depends upon the arrangement of functional groups about the nuclear structure. The configuration, substitution and total number of hydroxyl groups substantially influence several mechanisms of antioxidant activity such as radical scavenging and metal ion chelation ability. (Kelly *et al.,* 2002; Pandey *et al.,* 2012). Mechanisms of antioxidant action can include (1) suppression of ROS formation either by inhibition of enzymes or by chelating trace elements involved in free radical generation; (2) scavenging ROS; and (3) upregulation or protection of antioxidant defenses. Flavonoid action involves most of the mechanisms mentioned above. Some of the effects mediated by them may be the combined result of radical scavenging activity and the interaction with enzyme functions. Flavonoids inhibit the enzymes involved in ROS generation, that is, microsomal monooxygenase, glutathione S-transferase, mitochondrial succinoxidase, NADH oxidase, and so forth (Halliwell and Gutteridge, 1998).

Lipid peroxidation is a common consequence of oxidative stress. Flavonoid protects lipids against oxidative damage by various mechanisms. Free metal ions enhance ROS formation by the reduction of hydrogen peroxide with generation of the highly reactive hydroxyl radical. Due to their lower redox potentials flavonoids (Fl-OH) are thermodynamically able to reduce highly oxidizing free radicals (redox potentials in the range 2.13–1.0V) such as superoxide, peroxyl, alkoxyl, and hydroxyl radicals by hydrogen atom donation Because of their capacity to chelate metal ions (iron, copper, etc.), flavonoids also inhibit free radical generation. Quercetin in particular is known for its iron-chelating and iron-stabilizing properties. Trace metals bind at specific positions of different rings of flavonoid structures (Kumar *et al.,* 2013).

4.9.2 Hepatoprotective Activity

Several flavonoids such as catechin, apigenin, quercetin, naringenin, rutin and venoruton are reported for their hapatoprotective activities. Different chronic diseases such as diabetes may lead to development of hepatic clinical manifestations. glutamate-cysteine ligase catalytic subunit (Gclc) expression, glutathione and ROS levels are reported to be decreased in liver of diabetic mice. Anthocyanins have drawn increasing attention because of their preventive effect against various diseases (Tapas, 2008).

Hepatoprotective activities were observed in flavonoids isolated from *Laggera alata* against carbon-tetrachloride (CCl_4^-) induced injury in primary cultured neonatal rat hepatocytes and in rats with hepatic damage. Flavonoids at a concentration range of 1–100μg/mL improved cell viability and inhibited cellular leakage of hepatocyte aspartate aminotransferase (AST) and alanine aminotransferase (ALT) caused by CCl_4. Similarly in an in vivo experiment flavonoids at of 50, 100, and 200mg/kg oral doses significantly reduced the levels of AST, ALT, total protein, and albumin in serum and the hydroxyproline and sialic acid levels in liver. Histopathological examinations also revealed the improvement in damaged liver with the treatment of flavonoid (Wu *et al.,* 2006).

4.9.3 Antibacterial Activity

Flavonoids are known to be synthesized by plants in response to microbial infection; thus it should not be surprising that they have been found *in vitro* to be effective antimicrobial substances against a wide array of microorganisms. Flavonoid rich plant extracts from different species have been reported to possess antibacterial activity. Several flavonoids including apigenin, galangin, flavone and flavonol glycosides, isoflavones, flavanones, and chalcones have been shown to possess potent antibacterial activity (Mishra, 2013).

Antibacterial flavonoids might be having multiple cellular targets, rather than one specific site of action. One of their molecular actions is to form complex with proteins through nonspecific forces such as hydrogen bonding and hydrophobic effects, as well as by covalent bond formation. Thus, their mode of antimicrobial action may be related to their ability to inactivate microbial adhesins, enzymes, cell envelope transport proteins, and so forth. Lipophilic flavonoids may also disrupt microbial membranes (Cowan, 1999).

Catechins, the most reduced form of the C3 unit in flavonoid compounds, have been extensively researched due to their antimicrobial activity. These compounds are reported for their *in vitro* antibacterial activity against *Vibrio cholerae, Streptococcus mutans, Shigella* and other bacteria. The catechins have been shown to inactivate cholera toxin in *Vibrio cholera* and inhibit isolated bacterial glucosyltransferases in *S. mutans,* probably due to complexing activities. Robinetin, myricetin, and (")-epigallocatechin are known to inhibit DNA synthesis in *Proteus vulgaris.* It is suggested that the B ring of the flavonoids may intercalate or form hydrogen bond with the stacking of nucleic acid bases and further lead to inhibition of DNA and RNA synthesis in bacteria. Another study demonstrated inhibitory activity of quercetin, apigenin, and 3,6,7,32,42-pentahydroxyflavone against *Escherichia coli* DNA gyrase (Ohemeng *et al.,* 1993).

Naringenin and sophoraflavanone G have intensive antibacterial activity against methicilline resistant *Staphylococcus aureus* (MRSA) and *Streptococci.* An alteration of membrane fluidity in hydrophilic and hydrophobic regions may be attributed to this effect which suggests that these flavonoids might reduce the fluidity of outer and inner layers of membranes. The correlation between antibacterial activity and membrane interference supports the theory that flavonoids may demonstrate antibacterial activity by reducing membrane fluidity of bacterial cells. The 5,7-dihydroxylation of the A ring and 22, 42-or 22, 62-dihydroxylation of the B ring in the flavanone structure is important for anti-MRSA activity. A hydroxyl group at position 5 in flavanones and flavones is important for their activity against MRSA. Substitution with C8 and C10 chains may also enhance the anti staphylococcal activity of flavonoids belonging to the flavan-3-ol class. It is also shown that 5-hydroxyflavanones and 5-hydroxyisoflavanones with one, two, or three additional hydroxyl groups at the 7, 22 and 42 positions inhibited the growth of *S. mutans* and *Streptococcus sobrinus* (Osawa *et al.*, 1992).

4.9.4 Anti-Inflammatory Activity

Inflammation is a normal biological process in response to tissue injury, microbial pathogen infection, and chemical irritation. Inflammation is initiated by migration of immune cells from blood vessels and release of mediators at the site of damage. This process is followed by recruitment of inflammatory cells, release of ROS, RNS, and proinflammatory cytokines to eliminate foreign pathogens, and repairing injured tissues. In general, normal inflammation is rapid and self-limiting, but aberrant resolution and prolonged inflammation cause various chronic disorders (Pan *et al.*, 2010).

The immune system can be modified by diet, pharmacologic agents, environmental pollutants and naturally occurring food chemicals. Certain members of flavonoids significantly affect the function of the immune system and inflammatory cells. A number of flavonoids such as hesperidin, apigenin, luteolin, and quercetin are reported to possess anti-inflammatory and analgesic effects. Flavonoids may affect specifically the function of enzyme systems critically involved in the generation of inflammatory processes, especially tyrosine and serine-threonine protein kinases. The inhibition of kinases is due to the competitive binding of flavonoids with ATP at catalytic sites on the enzymes. These enzymes are involved in signal transduction and cell activation processes involving cells of the immune system. It has been reported that flavonoids are able to inhibit expression of isoforms of inducible nitric oxide synthase, cyclooxygenase, and lipooxygenase, which are responsible for the production of a great amount of nitric oxide, prostanoids, leukotrienes, and other mediators of the inflammatory process such as cytokines, chemokines, or adhesion molecules. Flavonoids also inhibit phosphodiesterases involved in cell activation. Much of the anti-inflammatory effect of flavonoid is on the biosynthesis of protein cytokines that mediate adhesion of circulating leukocytes to sites of injury. Certain flavonoids are potent inhibitors of the production of prostaglandins, a group of powerful proinflammatory signaling molecules (Manthey, 2000).

Reversal of the carrageenan induced inflammatory changes has been observed with silymarin treatment. It has been found that quercetin inhibit mitogen stimulated immunoglobulin secretion of IgG, IgM, and IgA isotypes *in vitro*. Several flavonoids are reported to inhibit platelet adhesion, aggregation, and secretion significantly at 1–10 mM concentration. The effect of flavonoid on platelets has been related to the inhibition of arachidonic acid metabolism by carbon monoxide. Alternatively, certain flavonoids are potent inhibitors of cyclic AMP phosphodiesterase, and this may in part explain their ability to inhibit platelet function (Cumella *et al.*, 1987).

4.9.5 Anticancer Activity

Dietary factors play an important role in the prevention of cancers. Fruits and vegetables having flavonoids have been reported as cancer chemopreventive agents. Consumption of onions and/or apples, two major sources of the flavonol quercetin, is inversely associated with the incidence of cancer of the prostate, lung, stomach, and breast. In addition, moderate wine drinkers also seem to have a lower risk to develop cancer of the lung, endometrium, esophagus, stomach, and colon. The critical relationship of fruit and vegetable intake and cancer prevention has been thoroughly documented. It has been suggested that major public health benefits could be achieved by substantially increasing consumption of these foods (Mishra *et al.,* 2013).

Several mechanisms have been proposed for the effect of flavonoids on the initiation and promotion stages of the carcinogenicity including influences on development and hormonal activities. Major molecular mechanisms of action of flavonoids are given as follows: (1) downregulation of mutant p53 protein, (2) cell cycle arrest, (3) tyrosine kinase inhibition, (4) inhibition of heat shock proteins, (5) estrogen receptor binding capacity, (6) inhibition of expression of Rase proteins (Duthie *et al.,* 2000).

Recently it has been shown that the flavanol epigallocatechin-3-gallate inhibited fatty acid synthase (FAS) activity and lipogenesis in prostate cancer cells, an effect that is strongly associated with growth arrest and cell death. In contrast to most normal tissues expression of FAS is markedly increased in various human cancers. Upregulation of FAS occurs early in tumor development and is further enhanced in more advanced tumors (Koen *et al.,* 2005).

Quercetin is known to produce cell cycle arrest in proliferating lymphoid cells. In addition to its antineoplastic activity, quercetin exerted growth-inhibitory effects on several malignant tumor cell lines in vitro. These included P-388 leukemia cells, gastric cancer cells (HGC-27, NUGC-2, NKN-7, and MKN-28), colon cancer cells (COLON 320 DM), human breast cancer cells, human squamous and gliosarcoma cells, and ovarian cancer cells. Markaverich *et al.* (1988) proposed that tumor cell growth inhibition by quercetin may be due to its interaction with nuclear type II estrogen binding sites (EBS). It has been experimentally proved that increased signal transduction in human breast cancer cells is markedly reduced by quercetin acting as an antiproliferative agent.

4.9.6 Antiviral Activity

Natural compounds are an important source for the discovery and the development of novel antiviral drugs because of their availability and expected low side effects. Naturally occurring flavonoids with antiviral activity have been recognized since the 1940s and many reports on the antiviral activity of various flavonoids are available. Search of effective drug against human immunodeficiency virus (HIV) is the need of hour. Most of the work related with antiviral compounds revolves around inhibition of various enzymes associated with the life cycle of viruses. Structure function relationship between flavonoids and their enzyme inhibitory activity has been observed. Gerdin and Srensso demonstrated that flavan-3-o1 was more effective than flavones and flavonones in selective inhibition of HIV-1, HIV-2, and similar immunodeficiency virus infections. Baicalin, a flavonoid isolated from *Scutellaria baicalensis* (Lamieaceae), inhibits HIV-1 infection and replication. Baicalein and other flavonoids such as robustaflavone and hinokiflavone have also been shown to inhibit HIV-1 reverse transcriptase. Another study revealed inhibition of HIV-1 entry into cells expressing CD4 and chemokine coreceptors and antagonism of HIV-1 reverse transcriptase by the flavone O-glycoside. Catechins are also known to inhibit DNA polymerases of HIV-1. Flavonoid such as demethylated gardenin A and robinetin are known to inhibit HIV-1 proteinase. It has also been reported that the flavonoids chrysin, acacetin, and apigenin prevent HIV-1 activation via a novel mechanism that probably involves inhibition of viral transcription (Critchfield *et al.,* 1996).

5.0 CONCLUSION

Prevention and cure of diseases using phytochemicals especially flavonoids are well known. Fruits and vegetables are natural sources of flavonoids. Variety of flavonoids found in the nature possesses their own physical, chemical, and physiological properties. Structure function relationship of flavanoids is epitome of major biological activities. Medicinal efficacy of many flavanoids as antibacterial, hepatoprotective, anti-inflammatory, anticancer and antiviral agents is well established. These substances are more commonly used in the developing countries. Therapeutic use of new compounds must be validated using specific biochemical tests. With the use of genetic modifications, it is now possible to produce flavonoids at large scale. Further achievements will provide newer insights and will certainly lead to a new era of flavonoid based pharmaceutical agents for the treatment of many infectious and degenerative diseases.

REFERENCES

Arts, I.C., Hollman, P.C., 2005. Polyphenols and disease risk in epidemiologic studies. *Am J Clin Nutr* 81(1): 317S-325S.

Chen Y, Wong M, Rosen R, *et al.,* 1999. 2,2-Diphenyl-1-picryihydrazyl radical scavenging active componene[ts from Polygonum multifloxem Thumb. *Journal of Agricultural Food chemistry* 47: 2226 - 2228.

Cowan, M.M., 1999. Plant products as antimicrobial agents, *Clinical Microbiology Reviews,* 12(4): 564–582, 1999.

Critchfield, J.W., Butera, S.T., Folks, T.M., 1996. Inhibition of HIV activation in latently infected cells by flavonoid compounds, *AIDS Research and Human Retroviruses* 12(1): 39–46.

Cumella, J.C., Faden, H., Middleton, F., 1987. Selective activity of plant flavonoids on neutrophil chemiluminescence (CL), *Journal of Allergy and Clinical Immunology* 77(131).

Danylo Halytskyi, LVIV National Medical University, Department of Pharmacognosy and Botany.(Lecture notes)

Davis W.L., Matthew, S.B., 2000. Antioxidants and cancer III: quercetin, *Alternative Medicine Review* 5(3): 196–208.

Dillard, C.J., German, J.B., 2000. Phytochemicals: nutraceuticals and human health. *Journal of the Science of Food and Agriculture* 80: 1744-1756.

Duthie, G.G., Duthie, S.J., Kyle, J.A.M., 2000. Plant polyphenols in cancer and heart disease: implications as nutritional antioxidants, *Nutrition Research Reviews* 13(1): 79–106.

Galeotti, F., Barile, E., Curir, P., Dolci, M., Lanzotti, V., 2008. Flavonoids from carnation (Dianthus caryophyllus) and their antifungal activity. *Phytochemistry Letters* **1**: 44.

Groff, J.L., Gropper, S.S., Hunt, S.M., 1995. Advanced Nutrition and Human Metabolism. West Publishing Company, New York.

Halliwell, B., Gutteridge, J.M.C., 1998. Free Radicals in Biology and Medicine, Oxford University Press, Oxford, UK.

Hertog, M.G., *et al.,* 1993. Dietary antioxidant flavonoids and risk of coronary heart disease: the Zutphen Elderly Study. Lancet, 342(8878): 1007-11.

Kelly, E.H., Anthony, R.T., Dennis, J.B., 2002. Flavonoid antioxidants: chemistry, metabolism and structure-activity relationships, *Journal of Nutritional Biochemistry*. 13(10): 572–584, 2002.

Koen, B., Ruth, V., Guido, V., Johannes, V.S., 2005. Induction of cancer cell apoptosis by flavonoids is associated with their ability to inhibit fatty acid synthase activity, *Journal of Biological Chemistry* 280(7): 5636–5645.

Kumar, S., Mishra, A., Pandey, A.K., 2013. Antioxidant mediated protective effect of *Parthenium hysterophorus* against oxidative damage using in vitro models, *BMC Complementary and Alternative Medicine* 13(120).

Kweon, M, H, Hwang, H.J, Sung, H.C., 2001. Identification and Antioxidant activity of Novel chlorogenic acid derivatives from Bamboo (*Phyllostachys edulis*) *Journal of Agricultural Food Chemistry* 49: 4646-4655.

Manthey, J.A., 2000. Biological properties of flavonoids pertaining to inflammation, *Microcirculation* 7(1): S29–S34.

Mark S Meskin, Wayne R Bidlack, Audra J Davies, Douglas S Lewis, R Keith Randoiph, 2004. Phytochemicals: Mechanism of Action. *American Society for Clinical Nutrition.*

Markaverich, B.M., Roberts, R.R., Alejandro, M.A., Johnson, G.A., Middleditch, B.S., Clark, J.H., 1988. Bioflavonoid interaction with rat uterine type II binding sites and cell growth inhibition, *Journal of Steroid Biochemistry* 30(1–6): 71–78.

Mishra, A., Kumar, S., Pandey, A.K., 2013. Scientific validation of the medicinal efficacy of Tinospora cordifolia, *The Scientific World Journal* Article ID 292934.

Ohemeng, K.A., Schwender, C.F., Fu, K.P., Barrett, J.F., 1993. DNA gyrase inhibitory and antibacterial activity of some flavones, Bioorganic and Medicinal Chemistry Letters, 3(2), 225–230.

Okamura, H., Mimura, A., Yakou, Y., Niwano, M., Takahara, Y., 1993. Antioxidant activity of tannins and flavonoids in *Eucalyptus rostrata*. *Phytochemistry* 33: 557-561.

Osawa, K., Yasuda, H., Maruyama, T., Morita, H., Takeya, K., Itokawa, H., 1992. Isoflavanones from the heartwood of *Swartzia polyphylla* and their antibacterial activity against cariogenic bacteria, *Chemical and Pharmaceutical Bulletin* 40(11): 2970–2974.

Pan, M.H., Lai, C.S., Ho, C.T., 2010. Anti-inflammatory activity of natural dietary flavonoids, *Food and Function* 1(1): 15–31, 2010.

Pandey, A.K., Mishra, A.K., Mishra, A., 2012. Antifungal and antioxidative potential of oil and extracts derived from leaves of Indian spice plant *Cinnamomum tamala, Cellular and Molecular Biology* 58: 142–147.

Tapas, A.R., Sakarkar, D.M., Kakde, R.B., 2008. Flavonoids as nutraceuticals: a review, *Tropical Journal of Pharmaceutical Research* 7: 1089–1099.

US FDA, Guidance for Industry: Evidence-Based Review System for the Scientific Evaluation of Health Claims.

Wu, Y., Wang, F., Zheng Q., *et al.,* 2006. Hepatoprotective effect of total flavonoids from Laggera alata against carbon tetrachloride-induced injury in primary cultured neonatal rat hepatocytes and in rats with hepatic damage, *Journal of Biomedical Science* 13(4): 569–578.

Yao, L.H., Jiang, Y.M., Shi, J., Tomas-Barberan, F.A., Datta, N., Singanusong, R., Chen, S.S., 2004. Flavonoids in food and their health benefits. *Plant Foods Human Nutrition* 59(3):113-22.

2015, Modern Methods in Phytomedicine *Pages* ***293–301***
Editor: **T. Parimelazhagan**
Published by: **DAYA PUBLISHING HOUSE, NEW DELHI**

20

Cancer Chemopreventive Effects of Garlic-Derived Second Generation Organosulfides: An Overview

Rajan Ramachandran, Manickam Dakshinamoorthy Balakumaran and Puthupalayam Thangavel Kalaichelvan

Centre for Advanced Studies in Botany, School of Life Sciences, University of Madras, Guindy Campus, Chennai – 600 025, Tamil Nadu

1.0 INTRODUCTION

Natural products have long been used as a source of novel therapeutic agents. Due to their vast structural diversity and amazing biological properties, compounds of biological origin still continue to play a crucial role in the drug discovery process. Indeed, about half of the drugs introduced between 1940 and 2006 were of natural origin or inspired by natural products and have an impressive role in preventing cancer (Newman and Cragg, 2007; Koehn and Carter, 2005). Chemoprevention is regarded as one of the most promising and realistic approaches in the prevention of human cancer. Many classes of cancer chemopreventive agents, both naturally occurring and synthetic compounds, have been thoroughly investigated for their efficiency in preventing cancer (Kelloff *et al.,* 2000). Among them, the most intensively investigated are sulfur-containing compounds, which occur naturally in Allium vegetables (Milner, 2001). Besides its beneficial health effects, garlic *(Allium sativum)* has also been reported to possess tumor inhibitory properties against various types of cancer (Pinto and Rivlin, 2001). Experimental as well as epidemiological studies have provided some evidence to support the assumption that garlic intake could reduce the cancer risk (Khanum *et al.*, 2004), including the reduction of esophageal, mammary, skin, pulmonary, forestomach, colon and lung tumors (Yu *et al.*, 2005; Hosono *et al.*, 2005). Belman (1983) has demonstrated that topical application of garlic and onion oil inhibited the incidence of tumor promoted by phorbol–myristate–acetate. The cancer chemopreventive effects of garlic have been observed against benzo[a]pyrene-induced forestomach and pulmonary cancer in mice (Sparnins *et al.*, 1986), *N*-nitrosomethylbenzylamine-induced esophageal cancer in rats (Wargovich *et al.*, 1988), azoxymethane-induced colon carcinogenesis in rats (Reddy *et al.*, 1993) and 2-amino-1-methyl-6-phenylimidazo[4,5-b]pyridine-induced mammary tumorigenesis in rats (Suzui *et al.*, 1997).

In recent years, numerous studies have shown that sulfur-containing compounds, especially garlic compounds, induced apoptosis in various cell lines and also in experimental models of carcinogenesis (Wu *et al.*, 2002). Preclinical studies have provided convincing evidence that several natural organosulfur compounds (OSCs) are highly effective in affording protection against cancer induced by a variety of chemical carcinogens. Several mechanisms have been proposed to explain the cancer-preventive effects of Allium vegetables and related OSCs. These include inhibition of mutagenesis, modulation of enzyme activities, inhibition of DNA adduct formation, free-radical scavenging and effects on cell proliferation and tumor growth. The compounds present in garlic can effectively modify common metabolic events leading to suppression in cancer development. The constitutive compounds of garlic can selectively inhibit tumor proliferation by a number of factors such as controlling DNA repair mechanisms, chromosomal stability and cell cycle regulation. Modulation of carcinogen activation might be one of the mechanisms by which garlic constituents offer protection against chemically induced cancers. Research to date indicates that garlic constituents may function as a double-edged sword in the prevention of chemically induced cancers by inhibiting carcinogen activation and enhancing detoxification of activated carcinogenic intermediates through the induction of Phase-2 enzymes, including glutathione transferases (GST) and quinone reductase (Sparnins *et al.*, 1988).

The main sulfur compound in intact garlic is γ-glutamyl-S-alk(en)yl-L-cysteine, which is hydrolyzed and oxidized to yield alliin (*S*-allylcysteine sulfoxide) (Block, 1985). Alliin accumulates naturally during storage of the bulbs at cold temperature and is the odorless precursor of the OSCs believed to be responsible for the anticancer property of garlic (Belman, 1983; Sparnins *et al.*, 1986). Processing of garlic bulbs (crushing, cutting or chewing) releases a vacuolar enzyme alliinase that acts on alliin to give rise to extremely unstable and odoriferous compounds, including allicin. Allicin and other thiosulfinates decompose to oil-soluble OSCs, including diallyl sulfide (DAS), diallyl disulfide (DADS), diallyl trisulfide (DATS), dithiins and ajoene (4,5,9-trithiadodeca-1,6,11-triene-9-oxide) (Block, 1985) and their chemical structures were given in Figure 20.1. These second generation organosulfides represent an important source of effective and versatile chemopreventive agents and some of which are strong modifiers of chemical carcinogenesis. The goal of this chapter is to summarize cancer chemopreventive effects of natural organosulfides derived from garlic and to discuss the molecular mechanisms by which these organosulfides afford cancer prevention.

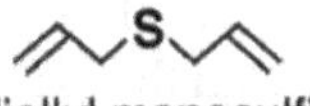

diallyl monosulfide (DAS)

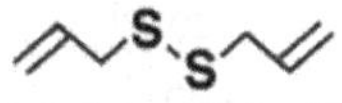

diallyl disulfide (DADS)

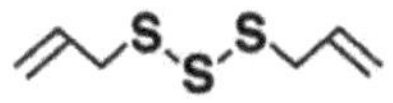

diallyl trisulfide (DATS)

Figure 20.1: Chemical Structures of Allyl Sulfides.

2.0 CANCER-PREVENTIVE EFFECTS OF DIALLYL DISULFIDE

Diallyl disulfide (DADS), an important component of garlic, has been shown to inhibit the tumor growth at colon, lung and breast of human (Iciek *et al.,* 2001). The anti-proliferative effect of DADS was attributed to suppression of the rate of cell division and induction of apoptosis in human cancer cells. Studies on the Allium derivatives have shown that DADS induced apoptosis and nonsteroidal anti-inflammatory drug–activated gene (NAG-1) protein expression via p53-dependent mechanisms in human colorectal HCT 116 cells (Bottone *et al.,* 2002). Similarly, Hong *et al.* (2000) have shown that the apoptotic mechanism induced by organoallyl sulfur compounds such as DAS, DADS or garlic extract is regulated through p53-dependent or p53-independent related Bax/Bcl-2 dual pathway in non–small cell lung cancer cells. Further, DADS induced apoptosis in breast cancer cells by increasing the expression of apoptotic protein Bax and reducing the expression of anti-apoptotic protein Bcl-2 and activating caspase-3, final caspase in the apoptotic pathway (Nakagawa *et al.,* 2001). Caspases are a family of proteases which play a pivotal role in the execution of apoptosis. Caspase-3 has been shown to play an important role in chemotherapy, growth factor withdrawal, Fas and retinoic acid induced apoptosis. Caspase-3 is a most likely candidate to mediate DADS induced apoptosis, as evidenced from the increased protease activity of caspase-3 in DADS treated PC-3 cells. Filomeni *et al.* (2003) have demonstrated that treatment of neuroblastoma cells (SHSY5Y) with DADS resulted in cell cycle arrest at G2/M phase, which eventually led to apoptosis through the activation of mitochondrial pathway (Bcl-2 down regulation, cytochrome C release into the cytosol and activation of caspase-9 and caspase-3). DADS also induced apoptosis in T24 cells and underwent caspase-3 activation. These results are highly corroborated with the findings of Kwon *et al.* (2002), who have shown that DADS induced apoptosis in HL-60 cancer cells with all the features of apoptosis. Thus, it is apparent that multiple mechanisms are being involved in the apoptosis inducing effects of DADS.

Recent data indicate that pleiotropic biological effects of DADS may involve the modulation of gene expression. With regard to drug-metabolizing enzymes, for instance, DADS enhanced the expression of CYP 1A1, 2B1 and 3A1 genes at the mRNA and protein levels (Knowles and Milner, 2003). In addition, a study using the cDNA array technology in HCT-15 human colon cancer cells has provided further evidence that DADS up- or down-regulated the expression of a wide range of genes, suggesting that DADS may modulate the expression of specific genes through a modification of histone acetylation. The acetylation of histones is a key process in activating transcription and has been reported to induce selectively the expression of specific genes such as the p21waf1/cip1 cyclin dependent kinase inhibitor to effect cell cycle arrest (Richon *et al.,* 2000). Recently, it has been reported that in colon tumor cells, DADS modulated the expression of genes involved in the regulation of cell proliferation. As p21waf1/cip1 protein is involved in the cell cycle progression and could induce cell cycle arrest in G1 or G2 phases, increased p21waf1/cip1 expression could explain the G2/M phase cell cycle arrest induced by DADS in PC-3 cells. Other studies have shown that in colon cell lines, the two HDAC inhibitors butyrate and trichostatin A (TSA) induced p21waf1/cip1 expression (Blottiere *et al.,* 2003). DADS has also been shown to inhibit the growth of H-ras oncogene transformed tumors in nude mice by inhibiting membrane association of tumoral p21/ras (Singh, 2001). These experiments have permitted to partially elucidate the DADS induced inhibition of the proliferation of PC-3 cells, suggesting that these cellular effects might be associated with the anticarcinogenic properties of DADS on prostate cancer (Arunkumar *et al.,* 2007).

3.0 CANCER-PREVENTIVE EFFECTS OF DIALLYL SULFIDE

Diallyl sulfide (DAS), a volatile organosulfide from garlic, has received considerable attention as potential chemopreventive agent. DAS is a potent antioxidant with anti-inflammatory and cancer preventive properties. *In vitro* and *in vivo* studies have suggested that DAS may impart chemopreventive effects against many kinds of cancers including skin cancer (Hu *et al.,* 1996; Arora and Shukla, 2002). DAS is also documented to inhibit the cancer of forestomach, colon, esophagus, mammary gland and lung (Herman-Antosiewicz *et al.,* 2004). The anticarcinogenic properties of DAS have been shown in rodents with a variety of chemical carcinogens (Hu *et al.,* 1996; Arora and Shukla, 2002). The ability of DAS, DASO and DASO2 to competitively inhibit a major carcinogen activating enzyme, CYP2E1, is a viable mechanism in systems in which CYP2E1 substrates are used as carcinogens. These compounds may inhibit the activation of other carcinogens at low efficiency. The induction of GST and phase II enzymes may also play a role. Garlic constituent DAS and its metabolites diallyl sulfoxide and diallyl sulfone have competitively inhibited the activity of cytochrome P-450 2E1 in a time-dependent and NADPH-dependent manner using pseudo-first-order kinetics (Brady *et al.,* 1991). Induction of cytochrome P-450 2B1 by treatment with DAS in rat liver microsomes has also been reported (Brady *et al.,* 1991). The chemopreventive activity of DAS has been attributed to its ability to modulate phase I and II detoxifying enzymes, scavenging of free radicals and its antimutagenic potential (Smith and Yang, 2000). DAS also exhibited a protective effect against cancer in a broad range of target organs such as skin, lung, esophagus, colon and liver (Hu *et al.,* 1996; Arora and Shukla, 2002), suggesting that the antitumorigenic properties of DAS are not limited to a single tissue or a carcinogen.

The tumor suppressor gene p53 is regarded as a key factor in maintaining the balance between cell growth and cell death (Mowat, 1998). The importance of p53 gene can be drawn from the fact that this gene is reported to be mutated in approximately 80 per cent of the all human malignancies. Because of its role in regulation of cell cycle, alterations in p53 are critical events in carcinogenesis. The wtp53 in response to toxic insults to DNA triggers a chain of cell cycle regulatory events to check the proliferation of altered cells to repair or minimize the damage (Mowat, 1998). In tumors, loss of wtp53 function prevents the activation of this growth control pathway (Hollstein *et al.*, 1991). The failure to induce transcriptionally active wtp53 plays a role in the unregulated growth of the tumors and also in the failure to respond to chemotherapeutic agents, which normally trigger wtp53 and regulates cell cycle arrest or cell death. Because the balance between wtp53 and mutp53 determines the fate of the cell, many chemopreventive agents are known to exert their anticancer effects by modulating their expression levels.

The up-regulation of wtp53 may induce the expression of several p53-regulated downstream genes including DR5, bax, fas and p21/waf1, which may cause growth arrest or apoptosis. The up-regulation of wtp53 by chemopreventive agents is most likely responsible for the transcriptional induction of p21/waf1 by directly interacting with its regulatory elements. Thus, the up-regulation of p21/waf1 could be responsible for the growth inhibitory effects of DAS because of its role in cell cycle arrest. Moreover, another plausible reason could be the prevention of mutation of p53 to oncogenic forms by promoting DNA repair. This may also be attributed to the antioxidant and antigenotoxic properties of the DAS, as oxidants produced during tumorigenic transformation promote genetic instability and mutational damage to the DNA. Thus, it could be suggested that DAS exerts its anticancer effect either by stimulating a cancer suppressor gene to prevent the action of carcinogenic influences or by preventing the mutation of other proto-oncogenes that may function together with mutp53 (Arora *et al.*, 2004). However, further insights into the effect of DAS on cyclin-dependent kinase complexes operated by cyclin dependent kinase 2 and Cdc2 (cyclin-dependent kinase 1) and cell cycle checkpoints are required (Arora *et al.*, 2004).

4.0 SULFUR CHEMISTRY: IMPORTANT FOR CANCER PREVENTIVE ACTIVITY

The garlic organosulfides are all sulfur-rich containing either a sulfide, disulfide or polysulfide functional group in their backbone which is likely the pharmacophore responsible for their biological activity. In support of this, the disulfide bond in ajoene was found to be important for both antimicrobial and anticancer properties, where the removal of disulfide to the sulfide rendered the analog inactive (Kaschula *et al.*, 2012). Disulfides and polysulfides are considered as thiol oxidising agents, able to thiolate cysteine residues in proteins to form mixed disulfides. Diallyltrisulfide (DATS) and allylmethyltrisulfide (AMTS) inhibited forestomach cancer, DATS, which contains two allyl groups, being more potent than AMTS, the analogue derivative with only one allyl group (Sparnins *et al.*, 1988). This finding underlines the importance of the allyl group for cancer-preventive activity. On the other hand, the trisulfide derivatives, AMTS and DATS, were not effective against pulmonary adenoma formation, suggesting that the number of sulfur atoms in the molecule is also important, possibly determining the organ sites at which protection is achieved against carcinogenesis (Sparnins *et al.,* 1988).

Among many organosulfur compounds tested, only diallylsulfide and allylmethyltrisulfide are effective in significantly reducing levels of hepatic CYP2E1 protein, indicating that the presence of an allylic side chain coupled to a single sulfur atom is necessary-perhaps ideal-for inhibition of the CYP2E1 protein. This finding correlates highly with the chemopreventive effects in models of carcinogenesis wherein CYP2E1 activates the carcinogen. Phase II enzymes involve a general response of an organism to xenobiotic plant exposure. When the lipid-soluble compounds are examined, no significant structure-activity effect or conclusion can be drawn. However, although it is difficult to generalize, research indicates that garlic compounds do induce Phase II enzymes, particularly glutathione-s-transferase, uridine diphosphate-glucuronyl transferase and quinone reductase. Generally, the greater the number of sulfur atoms in a given compound, the better inducer it is of these Phase II enzymes. Trisulfurs are more effective inducers of glutathione-s-transferase than are disulfurs and monosulfurs (Wattenberg *et al.,* 1989).

5.0 CONCLUSION

Many dietary bioactive food components interact with the immune system with the potential to reduce the risk of cancer (Ferguson and Philpott, 2007). The cancer-preventive effects of Allium vegetables, mainly garlic and related organic allyl sulfur components, have been well-demonstrated using a wide range of human cancer cells. Evidence to date indicates that these OSCs can suppress the growth of cancer cells of different anatomical locations in association with cell cycle arrest, mainly in the G2/M phase of the cell cycle. These speculative mechanisms should be verified in human studies to establish a causative link between some molecular properties and the cancer-preventive activity. The pharmacologic role of allyl sulfides in prevention and treatment of cancer has received considerable attention in the recent past, but the mechanism of action of allyl sulfide compounds is poorly defined, especially its role in cancer controlling genes.

REFERENCES

Arora, A., Shukla, Y., 2002. Induction of apoptosis by diallyl sulfide in DMBA induced mouse skin tumors. *Nutr Cancer* 44(1): 89-94.

Arora, A., Siddiqui, I.A., Shukla, Y., 2004. Modulation of p53 in 7,12 dimethylbenz[a]anthracene-induced skin tumors by diallyl sulfide in Swiss albino mice. *Mol Cancer Ther* 3(11): 1459-1466.

Arunkumar, A., Vijayababu, M.R., Gunadharini, N., Krishnamoorthy, G., Arunakaran, J., 2007. Induction of apoptosis and histone hyperacetylation by diallyl disulfide in prostate cancer cell line PC-3. *Cancer Letters* 251(1): 59-67.

Belman, S., 1983. Onion and garlic oils inhibit tumor promotion. *Carcinogenesis* 4(8): 1063-5.

Block, E., 1985. The chemistry of garlic and onions. *Sci Am* 252(3): 114-9.

Blottiere, H.M., Buecher, B., Galmiche, J.P., Cherbut, C., 2003. Molecular analysis of the effect of short-chain fatty acids on intestinal cell proliferation. *Proc Nutr Soc* 62(1): 101-106.

Bottone, F.G. Jr., Baek, S.J., Nixon, J.B., Eling, T.E., 2002. Diallyl disulfide (DADS) induces the antitumorigenic NSAID activated gene (NAG-1) by a p53- dependent mechanism in human colorectal HCT 116 cells. *J Nutr* 132(4): 773-8.

Brady, J.F., Ishizaki, H., Fukuto, J.M., Lin, M.C., Fadel, A., Gapac, J.M., Yang, C.S., 1991. Inhibition of cytochrome P-450 2E1 by diallyl sulfide and its metabolites. *Chem Res Toxicol* 4(6): 642-7.

Ferguson, L.R., Philpott, M., 2007. Cancer prevention by dietary bioactive components that target the immune response. Curr. *Cancer Drug Targets* 7(5): 459-464.

Filomeni, G., Aquilano, K., Rotilio, G., Ciriolo, M.R., 2003. Reactive oxygen species-dependent c-Jun NH2-terminal kinase/c-Jun signaling cascade mediates neuroblastoma cell death induced by diallyl disulfide. *Cancer Res* 63(18): 5940-5949.

Herman-Antosiewicz, A., Singh, S.V., 2004. Signal transduction pathways leading to cell cycle arrest and apoptosis induction in cancer cells by *Allium* vegetable-derived organosulfur compounds: a review. *Mutat Res* 555(1-2): 121-131.

Hollstein, M., Sidransky, D., Vogelstein, B., Harris, C.C., 1991. p53 mutations in human cancers. *Science* 253(5015): 49-53.

Hong, Y.S., Ham, Y.A., Choi, J.H., Kim, J., 2000. Effects of allyl sulfur compounds and garlic extract on the expression of Bcl- 2, Bax, and p53 in non small cell lung cancer cell lines. *Exp Mol Med* 32(3):127-134.

Hosono, T., Fukao, T., Ogihara, J., Ito, Y., Shiba, H., Seki, T., Ariga, T., 2005. Diallyl trisulfide suppresses the proliferation and induces apoptosis of human colon cancer cells through oxidative modification of {beta}-tubulin. *J Biol Chem* 280(50): 41487-41493.

Hu, J.J., Yoo, J.S., Lin, M., Wang, E.J., Yang, C.S., 1996. Protective effects of diallyl sulfide on acetaminophen induced toxicities. *Food Chem Toxicol* 34(10): 963-9.

Iciek, M.B., Rokita, H.B., Wlodek, L.B., 2001. Effects of diallyl disulfide and other donors of sulfane sulfur on the proliferation of human hepatoma cell line (HepG2). *Neoplasma* 48(4): 307-312.

Kaschula, C.H., Hunter, R., Stellenboom, N., Caira, M.R., Winks, S., Oqunleye, T., Richards, P., Cotton, J., Zilbeyaz, K., Wang, Y., Siyo, V., Ngarande, E., Parker, M.I., 2012. Structure-activity studies on the anti-proliferation activity of ajoene analogues in WHCO1 oesophageal cancer cells. *Eur J Med Chem* 50: 236-254.

Kelloff, G.J., Crowell, J.A., Steele, V.E., Lubet, R.A., Malone, W.A., Boone, C.W., Kopelovich, L., Hawk, E.T., Lieberman, R., Lawrence, J.A., Ali, I., Viner, J.L., Sigman, C.C., 2000. Progress in cancer chemoprevention: development of diet derived chemoprevention agents. *J Nutr* 130(2): 467S-471S.

Khanum, F., Anilakumar, K.R., Viswanathan, K.R., 2004. Anticarcinogenic properties of garlic: a review. Crit. Rev. *Food Sci Nutr* 44(6): 479-488.

Knowles, L.M., Milner, J.A., 2003. Diallyl disulfide induces ERK phosphorylation and alters gene expression profiles in human colon tumor cells. *J Nutr* 133(9): 2901–2906.

Koehn, F.E., Carter, G.T., 2005. The evolving role of natural products in drug discovery. Nat. Rev. *Drug Discov* 4(3): 206–220.

Kwon, K.B., Yoo, S.J., Ryu, D.G., Yang, J.Y., Rho, H.W., Kim, J.S., Park, J.W., Kim, H.R., Park, B.H., 2002. Induction of apoptosis by diallyl disulfide through activation of caspase-3 in human leukemia HL-60 cells. *Biochem Pharmacol* 63(1): 41-7.

Milner, J.A., 2001. A historical perspective on garlic and cancer. *J Nutr* 131(3): 1027S–1031S.

Mowat, M.R., 1998. p53 in tumor progression: life, death and everything. *Adv Cancer Res* 74: 25-48.

Nakagawa, H., Tsuta, K., Kiuchi, K., Senzaki, H., Tanaka, K., Hioki, K., Tsubura, A., 2001. Growth inhibitory effects of diallyl disulfide on human breast cancer cell lines. *Carcinogenesis* 22(6): 891-897.

Newman, D.J., Cragg, G.M., 2007. Natural products as sources of new drugs over the last 25 years. *J Nat Prod* 70(3): 461-477.

Pinto, J.T., Rivlin, R.S., 2001. Antiproliferative effects of Allium derivatives from garlic. *J Nutr* 131(3): 1058S-60S.

Reddy, B.S., Rao, C.V., Rivenson, A., Kelloff, G., 1993. Chemoprevention of colon carcinogenesis by organosulfur compounds. *Cancer Res* 53(15): 3493-8.

Richon, V.M., Sandhoff, T.W., Rifkind, R.A., Marks P.A., 2000. Histone deacetylase inhibitor selectively induces p21WAF1 expression and gene-associated histone acetylation. *Proc. Natl. Acad. Sci. USA*, 97(18): 10014-10019.

Singh, S.V., 2001. Impact of garlic organo sulfides on p21H-ras processing. *J Nutr* 131(3): 1046S-1048S.

Smith, T.J., Yang, C.S., 2000. Effect of organosulfur compounds from garlic and cruciferous vegetables on drug metabolizing enzymes. *Drug Metabol Drug Interact* 17(1-4): 23-49.

Sparnins, V.L., Barany, G., Wattenberg, L.W., 1988. Effects of organosulfur compounds from garlic and onions on benzo[a]pyrene-induced neoplasia and glutathione Stransferase activity in the mouse. *Carcinogenesis* 9(1): 131-134.

Sparnins, V.L., Mott, A.W., Barany, G., Wattenberg, L.W., 1986. Effects of allyl methyl trisulfide on glutathione S-transferase activity and BP-induced neoplasia in the mouse. *Nutr Cancer* 8(3): 211-5.

Suzui, N., Sugie, S., Rahman, K.M., Ohnishi, M., Yoshimi, N., Wakabayashi., K, Mori, H., 1997. Inhibitory effects of diallyl disulfide or aspirin on 2-amino-1-methyl-6-phenylimidazo[4,5-b]pyridine-induced mammary carcinogenesis in rats. *Jpn J Cancer Res* 88(8): 705-11.

Wargovich, M.J., Woods, C., Eng, V.W., Stephens, L.C., Gray, K., 1988. Chemoprevention of N-nitrosomethylbenzylamine-induced esophageal cancer in rats by the naturally occurring thioether, diallyl sulfide. *Cancer Res* 48(23): 6872-5.

Wattenberg, L.W., Sparnins, V.L., Barany, G., 1989. Inhibition of N-nitrosodiethylamine carcinogenesis in mice by naturally occurring organosulfur compounds and monoterpenes. *Cancer Res* 49(10): 2689-92.

Wu, C.C., Sheen, L.Y., Chen, H.W., Kuo, W.W., Tsai, S.J., Lii, C.K., 2002. Differential effects of garlic oil and its three major organosulfur components on the hepatic detoxification system in rats. *J Agric Food Chem* 50(2): 378-383.

Yu, F.S., Yu, C.S., Lin, J.P., Chen, S.C., Lai, W.W., Chung, J.G., 2005. Diallyl disulfide inhibits N-acetyltransferase activity and gene expression in human esophagus epidermoid carcinoma CE 81T/VGH cells. *Food Chem Toxicol* 43(7): 1029-1036.

2015, Modern Methods in Phytomedicine *Pages* ***303–319***
Editor: **T. Parimelazhagan**
Published by: **DAYA PUBLISHING HOUSE, NEW DELHI**

21

Comparative Evaluation on Antioxidant Activity and Phytoconstituents of *Thunbergia fragrans* Roxb. from Two different Altitudes in India

Rajan Murugan, Rahul Chandran, Shanmugam Saravanan and Thangaraj Parimelazhagan*

Bioprospecting Laboratory, Department of Botany, Bharathiar University, Coimbatore – 641 046, Tamil Nadu

1.0 INTRODUCTION

Human skin is exposed to environmental factors such as radiation, smoking, pollutants, organic solvents, pesticides (Buyukokuroglu *et al.*, 2001) and internal factors including reactive oxygen species (ROS) products from normal cell metabolism, aerobic respiration, stimulated polymorphonuclear leukocytes or macrophages that increase the level of oxidative stress (Pinnell, 2003; Svobodova *et al.*, 2006). Oxidation provides energy to living organisms for various metabolic processes. ROS are generated in the normal metabolism of living organisms and besides of their useful role in signal transduction; they are also involved in the distribution of several

* *Corresponding Author*

degenerative diseases like rheumatism, cough, inflammation and wounds (Halliwell and Gutteridge, 1984) Synthetic and natural antioxidants are of particularly importance in maintaining the oxidative stress level under the critical point in human organism. Phytochemicals such as phenolics, tannins and flavonoids and also non volatile compounds also act as an antioxidant agent was well documented.

Traditionally, *Thunbergia fragrans* Roxb. (Acanthaceae) leaf paste is used for treating wounds (Karunyal and Andrews, 2010). The roots and leaves are used as antidote to snakebite and in the treatment of rheumatism and cough (Dinesh, 2005). *T. fragrans* have been reported to have anti-bacterial (Achara, 2012), anti-inflammatory and anti-diarrheal property (Suresh *et al.,* 1994). Phytochemical constituents and antioxidant property of plant depend on a number of factors such as variety, location and environmental conditions (Middleton *et al.,* 2000; Alpinar *et al.,* 2009). Based on the traditional claims surrounding *T. fragrans* and the lack of scientific studies of its potential antioxidant property and its phytoconstituents, the objective of this study was to elucidate the effect of altitude and climatic conditions on total phenolic and flavonoid constituents and to evaluate the antioxidant activity of *T. fragrans* whole plant from two different regions in Tamil Nadu, India. The results may be helpful in understanding of geographical and climatic conditions on biological activity of this useful plant. Further, the acetone extract was selected for gas chromatography-mass spectroscopy (GC/MS) study to validate its prominent antioxidant capacity.

2.0 MATERIALS AND METHODS

2.1 Collection and Identification of Plant Material

The fresh whole plant of *Thunbergia fragrans* Roxb. were collected from higher altitude (TFHA) in Kotagiri, Nilgiris and another plant was collected from lower altitude (TFLA) in Coimbatore, Tamil Nadu, India. The taxonomic identity of the plant was confirmed by Dr. A. Rajendran and voucher specimen (No. 006163) was deposited at Botany Department Herbarium, Bharathiar University, Coimbatore, Tamil Nadu, India. The plant materials were washed under running tap water to remove the surface pollutants and the whole plant was separated mechanically. The separated materials were air dried under shade. The dried sample were powdered and used for further studies.

2.2 Chemicals

DPPH (2,2-diphenyl-1-picrylhydrazyl), ferrous chloride, ferric chloride, EDTA disodium salt, ferrous sulphate, ascorbic acid, butylated hydroxyanisole (BHA) and butylated hydroxytoluene (BHT) were obtained from Himedia, Merch, and Sigma (Mumbai, India). All other chemicals and solvents used were of analytical grade.

2.3 Extraction of Plant Material

The powdered materials of whole plant were packed in small thimbles separately and extracted successively with organic solvents such as petroleum ether, chloroform, acetone and methanol in the increasing order of polarity using Soxhlet apparatus. Each time before extracting with the next solvent, the thimble was air dried. Finally, the material was macerated using hot water with constant stirring for 24 h and the water extract was filtered. Whole plant of *T. fragrans* from two different altitudes was extracted under the same environmental conditions (Room temperature 30 – 40 °C). The different solvent extracts were concentrated by rotary vacuum evaporator (RE300; Yamato Scientific America Inc., Santa Clara, CA, USA) and then air dried. The dried extract obtained with each solvent was weighed. The percentage of yield was calculated in terms of the air dried weight plant material (1 mg/mL) the extract obtained was used for the assessment of various antioxidant assays and for further analysis.

2.4 Quantification Assays

2.4.1 Determination of Total Phenolic and Tannin Contents

The total phenolic content was determined according to method described by Siddhuraju and Becker (2003). 50 µL aliquots of the extracts (1 mg/mL) were taken in the test tubes and made up to the volume of 1 mL with distilled water. Then 0.5 mL of Folin - Ciocalteu phenol reagent (1:1 with water) and 2.5 mL sodium carbonate solution (20 per cent) were added sequentially in each tube. Soon after vortexing the reaction mixture, the test tubes were placed in dark for 40 min and the absorbance was recorded at 725 nm against the reagent blank. Gallic acid at different concentrations ranging from 20 to 80 µg/mL was used for preparation of calibration curve. The analysis was performed in triplicates and the results were expressed as gallic acid equivalents (GAE).

Using the same extract the tannins were estimated after treatment with polyvinyl polypyrrolidine (PVPP) (Siddhuraju and Manian, 2007). One hundred mg of PVPP was weighed into a 100×12 mm eppendrof tube and to this 1 mL distilled water and then 1 mL of the sample extracts were added. The content was vortexed and kept in the freezer at 4 °C for 15 min. Then the sample was centrifuged at 1,681× *g* for 10 min at room temperature and the supernatant was collected. This supernatant has only simple phenolics other then the tannins (the tannins would have been precipitated along with the PVPP). The phenolic content of the supernatant was measured and expressed as the content of non-tannin phenolics on a dry matter basis. From the above results, the tannin content of the sample was calculated as follows:

Tannins (per cent) = Total phenolics (per cent) - Non-tannin phenolics (per cent)

2.4.2 Determination of Flavonoid Contents

The flavonoid contents of all the extracts were quantified according to the method described by Zhishen *et al.* (1999). About 500 µL of all the plant extracts were taken in different test tubes and 2 mL of distilled water was added to each of the test tube. A test tube containing 2.5 mL of distilled water served as blank. Then, 150 µL of 5 per cent $NaNO_2$ was added to all the test tubes followed by incubation at room temperature for 6 min. After incubation, 150 µL of 10 per cent $AlCl_3$ was added to all the test tubes including the blank. All the test tubes were incubated for 6 min at room temperature. Then 2 mL of 4 per cent NaOH was added to all the test tubes which were made up to 5 mL using distilled water. The contents in all the test tubes were vortexed well and they were allowed to stand for 15 min at room temperature. The pink color developed due to the presence of flavonoids was read spectrophotometrically at 510 nm. Rutin was used as the standard for the quantification of flavonoids. Rutin at different concentrations ranging from 20 to 100 µg/mL was used for preparation of calibration curve. All the experiments were done in triplicates and the results were expressed in rutin equivalents (RE).

2.5 *In vitro* Antioxidant Assays

2.5.1 DPPH Radical Scavenging Activity

The antioxidant activity of the extract was determined in terms of hydrogen donating or radical scavenging ability using the stable radical DPPH, according to the method of Blois (1958). Sample extracts at various concentrations were taken and the volume was adjusted to 100 µL with methanol. About 5 mL of a 0.1 mM methanolic solution of DPPH was added to the aliquots of samples and standards (BHT and quercetin) and shaken vigorously. Negative control was prepared by adding 100 µL of methanol in 5 mL of 0.1 mM methanolic solution DPPH. The tubes were allowed to stand for 20 min at 27 °C. The absorbance of the samples was measured at 517 nm against the blank (methanol). Radical scavenging activity of the samples was expressed as IC_{50} which is the concentration of the sample required to Inhibit 50 per cent of $DPPH^{\bullet}$ concentration.

2.5.2 ABTS Radical Cation Scavenging Activity

The total antioxidant activity of the samples was measured by ABTS radical cation decolorization assay according to the method of Re *et al.* (1999). $ABTS^{\bullet}$ + was produced by reacting 7 mM ABTS aqueous solution with 2.4 mM potassium persulfate in the dark for 12-16 h at room temperature. Prior to assay, this solution was diluted in ethanol (about 1:89 v/v) and equilibrated at 30 °C to give an absorbance of 0.700±0.02 at 734 nm. The stock solution of the sample extracts were diluted such that after introduction of 10µL aliquots into the assay, they produced between 20 and 80 per cent inhibition of the blank absorbance. After the addition of 1 mL of diluted ABTS solution to 10µL of sample or Trolox (final concentration 0-15 µM) in ethanol, absorbance was measured at 30 °C exactly 30 min after the initial mixing. Triplicate determinations were made at each dilution of the standard, and the percentage inhibition was calculated against the blank (ethanol) absorbance at 734 nm and then was plotted as a function of Trolox concentration. The unit of total antioxidant activity (TAA) is defined as the concentration of Trolox having equivalent antioxidant activity expressed as µMol/g sample extracts.

2.5.3 Ferric Reducing Antioxidant Power (FRAP) Assay

The antioxidant capacities of different extracts of samples were estimated according to the procedure described by Pulido *et al.* (2000). FRAP reagent (900 µL), prepared freshly and incubated at 37 °C, was mixed with 90 µL of distilled water and 30 µL of test sample or methanol (for the reagent blank). The test samples and reagent blank were incubated at 37 °C for 30 min in a water bath. The final dilution of the test sample in the reaction mixture was 1/34. The FRAP reagent was prepared by mixing 2.5 mL of 20 mM TPTZ in 40 mM HCl, 2.5 mL of 20 mM $FeCl_3.6H_2O$, and 25 mL of 0.3 M acetate buffer (pH 3.6). At the end of incubation, the absorbance readings were taken immediately at 593 nm against the reagent blank, using a spectrophotometer.

Methanolic solutions of known Fe (II) concentration, ranging from 100 to 2000 μM, ($FeSO_4.7H_2O$) were used for the preparation of the calibration curve. The parameter equivalent concentration was defined as the concentration of antioxidant having a ferric-TPTZ reducing ability equivalent to that of 1 mM $FeSO_4.7H_2O$. Equivalent concentration was calculated as the concentration of antioxidant giving an absorbance increase in the FRAP assay equivalent to the theoretical absorbance value of a 1 mM concentration of Fe (II) solution.

2.5.4 Phosphomolybdenum Assay

The antioxidant activity of samples was evaluated by the green phosphomolybdenum complex formation according to the method of Prieto *et al.* (1999). A triplicate of 100 μL of sample and different concentrations of standard (ascorbic acid in 1 mM dimethyl sulphoxide) was added with 3 mL of reagent solution (0.6 M sulphuric acid, 28 mM sodium phosphate and 4 mM ammonium molybdate) in a test tube. The test tubes were covered with foil and incubated in a water bath at 95°C for 90 min. After the samples had cooled to room temperature, the absorbance of the mixture was measured at 695 nm against the reagent blank. A typical blank solution contained 3 mL of reagent solution and the appropriate volume of the same solvent used for the sample, and it was incubated under the same conditions as the rest of the samples. The results reported (Total antioxidant capacity) are mean values expressed as milligrams of ascorbic acid equivalents per gram extract.

2.5.5 Metal Chelating Activity

The chelating of ferrous ions by various extracts of whole plant was estimated by the method of Dinis *et al.* (1999). Briefly, 400 μL of samples and EDTA (standard) were added to 50 μL solution of 2 mM $FeCl_2$ in separate test tubes. The reaction was initiated by the addition of 200 μL of 5 mM ferrozine and the mixture was shaken vigorously and left standing at room temperature for 10 min. Absorbance of the solution was then measured spectrophotometrically at 562 nm against the blank (deionized water). The metal chelating capacities of the extracts were evaluated using the following equation. The results were expressed as mg EDTA equivalents/100 g extract.

2.5.6 Nitric Oxide Radical Scavenging Activity

The procedure is based on the method where sodium nitroprusside in aqueous solution at physiological pH spontaneously generates nitric oxide, which interacts with the oxygen to produce nitrite ions that can be estimated using Griess reagent. Scavengers of nitric oxide compete with oxygen, leading to reduced production of nitrite ions. The nitric oxide scavenging activity of different solvent extracts of *T. fragrans* on nitric oxide radical was measured according to the method of Sreejayan and Rao (1997). Sodium nitroprusside (10 mM) in phosphate buffered saline, was mixed with Whole plant extracts and incubated at room temperature for 150 min. Griess reagent (0.5 mL), containing 1 per cent sulphanilamide, 2 per cent H_3PO_4 and 0.1 per cent N-(1-naphthyl) ethylene diamine dihydrochloride, was added to the mixture after incubation time. The absorbance of the chromophore formed was read at 546 nm. BHA and rutin were used as a reference compounds. Radical scavenging activity was expressed as the inhibition percentage of free radical by the sample and was calculated.

Per cent of Inhibition = [(Control OD - Sample OD)/Control OD] ×100

2.5.7 Superoxide Radical Scavenging Activity

The assay was based on the capacity of the plant extract to inhibit formazan formation by scavenging superoxide radicals generated in riboflavin–light–NBT system (Beauchamp and Fridovich, 1971). Each 3-mL reaction mixture contained 50 mM sodium phosphate buffer (pH 6), 20 mg riboflavin and 12 mM EDTA, and 0.1 mg NBT. Reaction was started by illuminating the reaction mixture with 300 µg of concentrations of sample extracts for 90 s. Immediately, after illumination, the absorbance was measured at 590 nm. The percentage inhibition of superoxide anion generation was calculated.

Per cent of inhibition = [(Control OD- Sample OD)/Control OD] ×100

2.6 GC-MS Analysis

2.6.1 Preparation of Extract

One µL of the acetone extract of whole plant (TFHA) was employed for GC/MS analysis.

2.6.2 Instruments and Chromatographic Conditions

GC-MS analysis was carried out using Clarus 600 gas chromatograph system equipped with Clarus 600 C mass spectrometer (Perkin Elmer precisely, USA). An Elite-5MS fused silica capillary column coated with a 5 per cent diphenyl/95 per cent dimethyl polysiloxane stationary phase (60m X 0.25 mm, film thickness 0.10 µm; Perkin Elmer precisely, USA) was used for GC-MS. The injector temperature was kept at 200 °C whereas the oven temperature was programmed from70 °C to 300 °C for a total run time of 30 min. Helium was used as carrier gas at a flow rate of 1.0 mL/min. An appropriate blank was run from which the solvent delay was fixed to 4 min. The electron ionization mode with ionization energy of 70 eV; ion source temperature of 200 °C; GC interface temperature of 240 °C; scan interval of 0.2 s and fragments range from 50 to 600 m/z were set for the MS analysis. About 1 µL of the acetone extract was injected manually in a splitless mode. The mass spectra of the respective peaks obtained in the GC/MS were compared with the mass fragmentation patterns of standards in the NIST (National Institute of Standard Technology) library.

2.7 Statistical Analyses

All the experiments were done in triplicates and the results were expressed as mean ± SD. The data were statistically analyzed using one way ANOVA followed by Duncan's test. Mean values were considered statistically significant when $p > 0.05$.

3. RESULTS

3.1 Total Phenolics, Tannin and Flavonoid Contents

The amount of total phenolics, tannins and flavonoids of *T. fragrans* extracts were analyzed and shown in Table 21.1. The results were expressed as gallic acid

Table 21.1: Determination of Total Phenolic, Tannin and Flavonoid Contents of *T. fragrans*

Samples	*Total Phenolics (g GAE/100 g extract)*		*Tannins (g GAE/100 g extract)*		*Flavonoids (g RE/100g extract)*	
	TFHA	*TFLA*	*TFHA*	*TFLA*	*TFHA*	*TFLA*
Petroleum ether	17.00 ± 0.62^{c}	1.62±0.39^{d}	13.22±0.42^{c}	1.18±0.16^{c,d}	7.10±2.11^{d}	11.90±0.75^{d}
Chloroform	9.70±1.27^{d}	4.32±1.00^{c}	8.28±0.24^{d}	2.06±0.34^{c}	38.60±0.32^{c}	16.50±0.42^{c}
Acetone	32.62±0.38^{a}	18.58±0.17^{a}	27.92±0.34^{a}	13.94±1.99^{a}	46.40±0.85^{b}	61.50±0.62^{a}
Methanol	25.28±1.26^{b}	17.00±0.56^{b}	20.86±2.45^{b}	4.90±1.39^{b}	52.10±0.88^{a}	29.40±0.78^{b}
Water	3.14±0.16^{e}	4.46±0.28^{c}	2.24±1.32^{e}	0.10±0.06^{d}	7.20±2.54^{d}	7.80±1.20^{e}

Values are mean of triplicate determination (n=3) ± standard deviation.

Statistically significant at $p < 0.05$ where a > b > c > d> e

TFHA: *T. fragrans* higher altitude; TFLA: *T.fragrans* lower altitude; GAE: Gallic acid equivalents; RE: Rutin Equivalents.

equivalents (y = 0.025, R^2 = 0.989). Acetone extract of whole plant (TFHA) showed higher total phenolic contents (32.62 g GAE/100 g extract, $p > 0.05$) while tannin contents were found to be 27.92 g GAE/100 g of acetone extract (TFHA). Moreover, flavonoid contents showed 52.10 g RE/100 g of methanol extract (TFLA).

3.2 Scavenging Activity of DPPH Radical

The effect of free radical scavenging activity was determined using stable DPPH radical against four different concentrations [20, 40, 60, 80 µg/mL (Figure 21.1)]. The lower IC_{50} values showed higher free radical scavenging activity. The IC_{50} values of acetone extract TFHA and TFLA were 22.4 and 38.1 µg/mL respectively compared with standards (Quercetin and BHT).

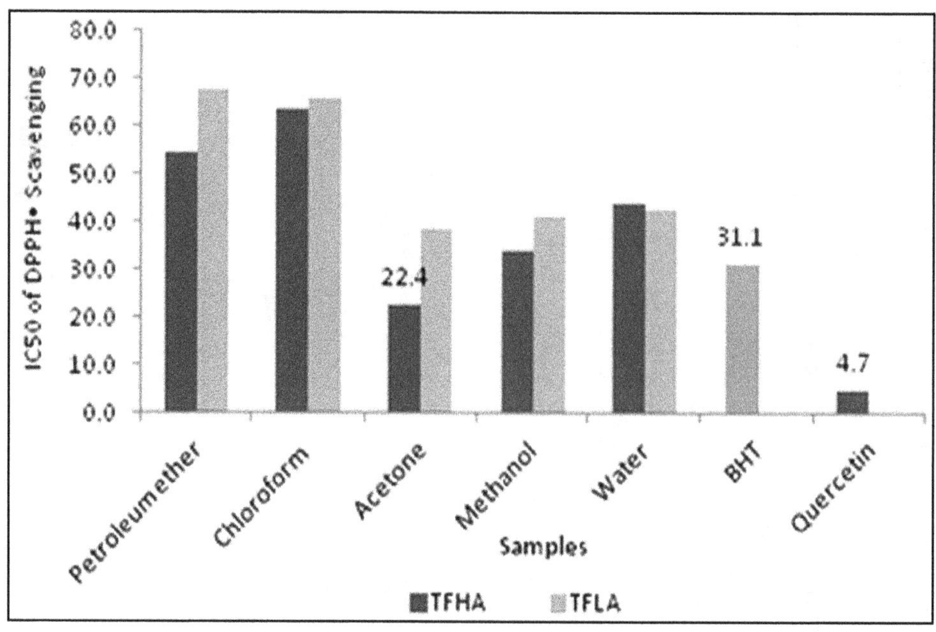

Figure 21.1: DPPH Radical Scavenging Activity.

3.3 Scavenging of ABTS Radical Cation

Antioxidant activity as determined by decolorization of the ABTS cation radical method. ABTS model can be used to assess the scavenging activity for both the polar and non-polar samples (Re *et al.,* 1999). Among various solvent extracts, TFHA acetone extract possessed the highest ABTS radical scavenging activity (5487.72 µMol TEAC/ g extract, $p > 0.05$) than other solvent extracts. These data (Table 21.2) are consistent with reports that the radical scavenging activity may be due to the presence of phenolics, flavonoids and tannins.

3.4 Ferric Reducing Antioxidant Power Assay

The FRAP activity was determined in terms of Fe (II) equivalents (y=0.027, R^2=0.108) and shown in Table 21.2. Among the different solvent extracts acetone

Table 21.2: ABTS$^{\cdot+}$, FRAP, Phosphomolybdenum, Metal Ion Chelating Assay of *T. fragrans*

Samples	*ABTS^{+} Assay (µMol TEAC/g extract)*		*FRAP Assay (µMol Fe (II)/g extract)*		*Phosphomolybdenum (mg AAE/g extract)*		*Metal Chelating Activity (mg EDTA/100 g extract)*	
	TFHA	*TFLA*	*TFHA*	*TFLA*	*TFHA*	*TFLA*	*TFHA*	*TFLA*
Petroleum ether	870.74±193.17^{d}	1511.99±455.66^{c}	43.5±14.35^{e}	35.7±4.79^{e}	2.8±0.51^{d}	3.5±0.45^{c}	1484.2±61.66^{d}	1626.3±88.19^{b}
Chloroform	1019.24±356.73^{c}	897.74±183.74^{b}	335.5±68.83^{d}	657.9±56.28^{c}	14.2±1.44^{c}	11.4±3.00^{b}	1494.7±7.89^{d}	1335.1± 28.12^{d}
Acetone	5487.72±73.01^{a}	5399.97±143.66^{a}	1738.4±4.21^{a}	967.1±15.53^{b}	20.3±2.22^{a}	13.3±1.74^{b}	1858.8±21.86^{c}	1538.6±55.47^{c}
Methanol	1464.74±61.86^{b}	1424.24±81.84^{b}	781.5±14.6^{b}	1305.3±76.26^{a}	16.8±0.80^{b}	19.2±0.53^{a}	2277.2±49.97^{a}	1743.9±68.71^{a}
Water	1214.99±193.1^{b,c}	796.50±405.17^{c}	391.2±24.98^{c}	185.8±17.49^{d}	2.2±0.12^{d}	2.8±0.14^{c}	2021.1±33.54^{b}	604.4±49.20^{e}

Values are mean of triplicate determination (n=3) ± SD;Statistically significant at $p < 0.05$ where a > b > c > d> e.

AAE: Ascorbic acid equivalents; Fe(II): Ferrous sulphate Equivalents; EDTA: Ethylene diamine tetra acetic acid equivalents; TEAC: Trolox equivalents antioxidant capacity.

extract (TFHA) showed higher ferric reducing ability (1738.4 µMol Fe (II) equivalents/g extract, $p > 0.05$) compared to TFLA methanol extract (1305.3 µMol Fe (II) equivalents/g extract).

3.5 Phosphomolybdenum Assay

The total antioxidant capacity of different solvent extracts of *T. fragrans* were analyzed and shown in Table 21.2. The results were expressed as mg ascorbic acid equivalents (AAE)/g extract (y=0.05, R^2=0.989) respectively. The good antioxidant capacity was shown by TFHA acetone extract (20.3 mg AAE/g extract, $p > 0.05$) and the moderate antioxidant activity was observed in TFLA methanol extract (19.2 mg AAE/g extract) respectively.

3.6 Scavenging of Metal Ions

The metal chelating activity of different extracts was shown in Table 21.2 and was determined by ferrozine method. The results were expressed as mg EDTA equivalents/100 g extract. Methanol extract of TFHA showed higher ferric ions chelating ability (2277.2 mg EDTA/100 g extract, $p > 0.05$) compared to TFLA methanol extract (1743.9 mg EDTA equivalents/100 g extract).

3.7 Scavenging of Nitric Oxide Radical

The plant extracts and reference compounds that possess NO scavenging activity inhibited nitrite formation by competing with oxygen to react with NO. The results were expressed as IC_{50} values. Rutin an extreme Natural antioxidant, exhibited potent NO radical scavenging activity with a 64.58 per cent whereas BHA showed 53.39 per cent activity. TFHA extracts showed greater NO radical scavenging activity compare to TFLA extracts (Figure 21.2).

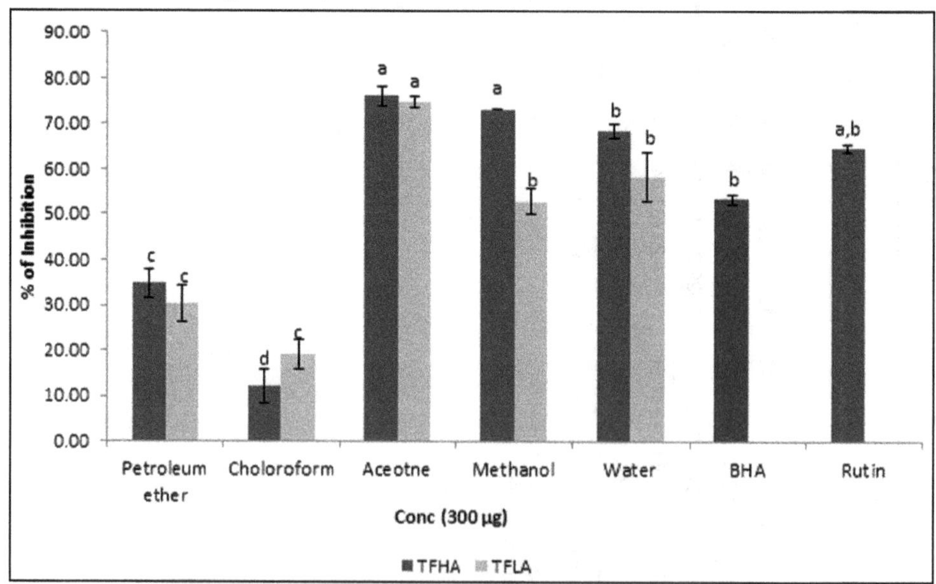

Figure 21.2: Nitric Oxide Radical Scavenging Activity. Statistically significant at $p < 0.05$ where a > b > c > d> e

3.8 Scavenging of Super Oxide Radical

The *T. fragrans* of different solvent extracts against superoxide radical are presented in Figure 21.3. The TFHA acetone extract (66.55 per cent) was highly effective against superoxide radical, whereas TFLA acetone extract inhibit the superoxide at 51.28 per cent ($p > 0.05$). The percentage inhibitions against superoxide radical scavenging of Ascorbic acid obtained were 55.97 per cent ($p > 0.05$).

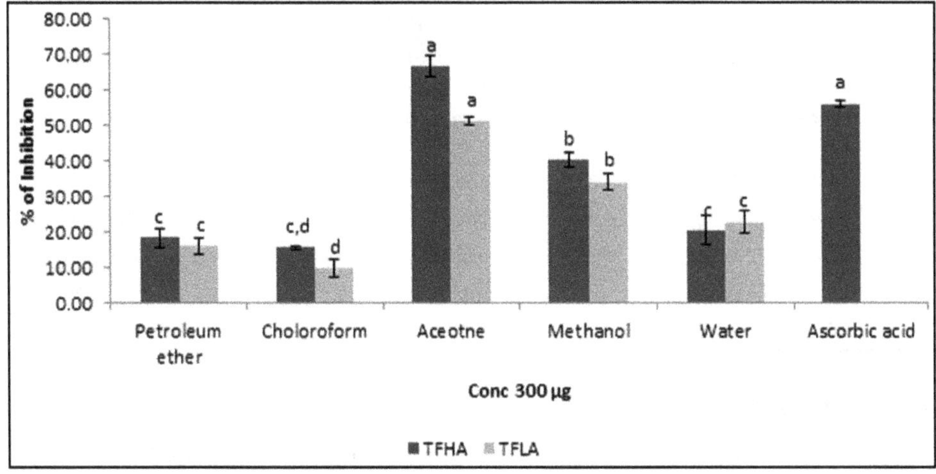

Figure 21.3: Superoxide Radical Scavenging Activity. Statistically significant at $p < 0.05$ where a > b > c > d> e

3.9 Identification of Bioactive Compounds

The chromatogram of GC/MS analysis of acetone extract of whole plant (TFHA), in accordance with the above discussed experimental conditions with a total run time of 30 min and the comparison of mass fragmentation pattern of compounds to that of in NIST library revealed the presence of 10 phytocomponents of different groups (Figure 21.4, Table 21.3). Among the 10 phytocomponents, methyl 2, 4-di-O-methyl-.beta.-D-xylopyranoside (37.96 per cent) and methyl 3,4-di-O methyl.beta.-L-arabinopyranoside (30.74 per cent), were found to be present in major amount.

On the other hand, aziridine (2.52 per cent), acetamide (3.83 per cent), permethylspermine (4.23 per cent), 1, 3-propanediamine (3.00 per cent) were minor components present in whole plant extract.

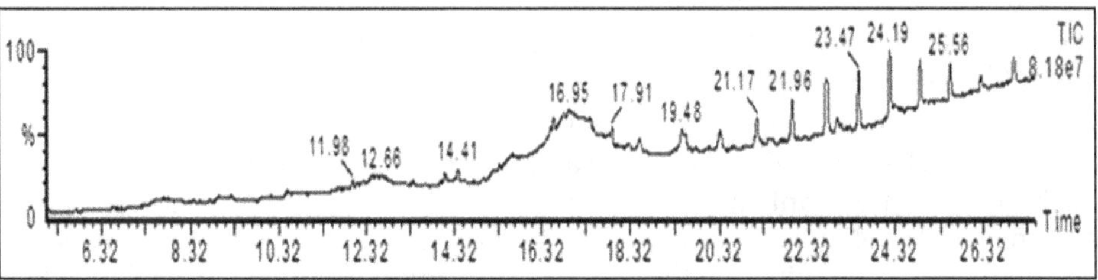

Figure 21.4: Chromatogram of Whole Plant of TFHA Acetone Extract.

Table 21.3: Chemical Composition of Whole Plant of TFHA Acetone Extract

Sl.No.	Ret. Time	Peak Area	Area (Per cent)	Compound Name	Formula
1.	11.98	864043	2.52	Aziridine	$C_{12}H_{25}N$
2.	14.41	869614	2.53	Chloroacetic Acid	$C_{10}H_{19}O_2Cl$
3.	16.95	13034236	37.96	Methyl 2,4-Di-O-Methyl-. Beta.-D-Xylopyranoside	$C_8H_{16}O_5$
4.	17.91	10555905	30.74	Methyl 3,4-Di-O-Methyl. Beta.-L-Arabinopyranoside	$C_8H_{16}O_5$
5.	19.48	1451684	4.23	3,6-Diazahomoadamantan-9-One Hydrazone	$C_9H_{16}N_4$
6.	21.17	1313861	3.83	Acetamide	$C_{16}H_{26}O_2N_4$
7.	21.96	1453020	4.23	3-Piperidinamine	$C_7H_{16}N_2$
8.	23.47	1967157	5.73	1,1-(Diethylcarbamoyl)Succinimide	$C_9H_{14}O_3N_2$
9.	24.19	1799262	5.24	Permethylspermine	$C_{16}H_{38}N_4$
10.	25.56	1031824	3.00	1,3-Propanediamine	$C_7H_{18}N_2$

4.0 DISCUSSION

It is well known that phenolic, tannin and flavonoid compounds exist in many plants, which have attracted a great deal of scientific interest because of the health promoting effects as antioxidants. The most important classes of natural antioxidants include tocopherols, flavonoids, vitamins and phenolic acids, which are derived from plant sources (Hassas *et al.*, 2009). Tannins are generally defined as naturally occurring polyphenolic compounds of high molecular weight to form complexes with the proteins (Yang and Russell, 1992). Flavonoids are naturally occurring in plants and are thought to have positive effects on human health. Studies on flavonoid derivatives have shown a wide range of anti-inflammatory activity (Di *et al.*, 1999; Montoro *et al.*, 2005). In the present study the TFHA extracts have high concentration of phenolics, tannins and flavonoids concentration. Therefore, *T. fragrans* have shown greater potential to reduce or scavenge free radicals or produces more beneficial effects.

Relative stable DPPH radical has been widely used to test the ability of compounds to act as free radical scavengers and thus to evaluate the antioxidant activity. The percentage inhibitory activity of free radicals by 50 per cent has been used widely as a parameter to measure antioxidant activity. Among the different extracts, acetone extract (TFHA) showed higher free radical scavenging activity than other solvent extracts due to the presence of higher amount of phenolic contents.

The Trolox equivalents antioxidant capacity (TEAC) was measured using the improved ABTS radical decolorization assay, one of the most commonly employed methods for antioxidant capacity, which measures the ability of a compound to scavenge ABTS cation radical. Trolox, a water soluble vitamin E analog, serves as a positive control inhibiting the formation of the radical cation in a dose dependent manner. Hagerman *et al.* (1998) have reported that the high molecular weight phenolics (tannins) have more ability to quench free radicals ($ABTS^{\bullet+}$) and their effectiveness depends on the molecular weight, the number of aromatic rings and nature of hydroxyl groups substitution than the specific functional groups. Hence, TFHA acetone extract showed high total antioxidant activity due to the presence of tannin contents.

The FRAP assay is a simple and inexpensive procedure that measures the total antioxidant level in a sample. It utilizes the reducing potential of the antioxidants to react with a ferric tripyridyltriazine (Fe III -TPTZ) complex and produce a colored ferrous tripyridyltriazine (Fe (II) -TPTZ) form (Sharififar *et al.,* 2009). Generally, the reducing properties are associated with the presence of phenolic compounds, which exert their action by breaking the free radical chain through donating a hydrogen atom. From this, a good reduction of Fe^{3+}-TPTZ complex to blue colored Fe^{2+}-TPTZ occurs at low pH by the presence of phenolic constituents in the TFHA acetone extract than TFLA extracts.

The total antioxidant activity of the whole plant extracts was measured spectrophotometrically through phosphomolybdenum method, based on the reduction of Mo(VI) to Mo(V) by the plant sample and the subsequent formation of green phosphate/Mo(V) compounds with a maximum absorption at 695 nm. Recent studies have shown that many flavonoid and related polyphenols contribute significantly to the phosphomolybdate scavenging activity of medicinal plants (Khan *et al.,* 2012; Benzie and Strain, 1996). The present study showed that TFHA has the highest antioxidant capacity for phosphomolybdate reduction. Hence this could also be related to the presence of polyphenolic content present in the sample.

The method of metal chelating activity is based on chelating of Fe^{2+} ions by the reagent ferrozine, which is quantitative formation of a complex with Fe^{2+} ions (Dinnis *et al.,* 1994). The formation of a complex is probably disturbed by the other chelating reagents, which would result in the reduction of the formation of red-colored complex. In this assay, *T. fragrans* whole plant extracts interfered with the formation of ferrous and ferrozine complex, suggesting that they have chelating activity and capture ferrous ion before ferrozine. Chelating agents may serve as secondary antioxidants because they reduce the redox potential thereby stabilizing the oxidized form of the metal ions (Rajesh *et al.,* 2008). Among the different solvent extracts, methanol extract (TFHA) showed better scavenging ability compared to other extracts. The present study reveals that the methanol extract of TFHA has a marked capacity for iron binding, suggesting that their action as peroxidation protector.

From the results, TFHA is a potent scavenger of nitric oxide. Nitric oxide radical generated from sodium nitroprusside reacts with oxygen to form nitrite. Nitric oxide is also implicated in inflammation, cancer and other pathological conditions (Moncada *et al.,* 1991). From this, we conclude that the therapeutic properties of *T. fragrans* may be useful for treating radical related inflammation and cancer.

Superoxide is known to involve in the accumulation of ROS/RNS in cells leading to redox imbalance associated physiological consequences (Pervaiz and Clemen, 2007). Percentage inhibition of superoxide radical scavenging activity was found higher TFHA extracts. Superoxide ($O_2^{\bullet-}$) radical is known to be very harmful to cellular components as a precursor of the more reactive oxygen species, contributing to the tissue damage and various diseases. The results revealed *T. fragrans* may inhibit the tissue damage and oxidative stress by quench the free radicals.

Among the different antioxidant assays, acetone extract of TFHA showed higher antioxidant activity in DPPH, phosphomolybdenum and FRAP assays compared to TFLA extracts. Therefore, acetone extract (TFHA) selected for further GC-MS analysis.

Gas chromatography and Mass spectroscopy is a commonly used method for identify the non-volatile compounds, which is the major analytical technique in quality control of herbal medicine. The present study identied several compounds, which were not described for *T. fragrans* earlier. From the GC-MS results, aziridine has antimicrobial activity (Priya *et al.,* 2011), Acetamide has antioxidant and anti-inflammatory property (Autore *et al.,* 2010), permethylspermine have antioxidant property (Shanab and Emad, 2012), 1, 3-propanediamine used as an antioxidant and hypolipidimic agent (Sanjay *et al.,* 2001). Hence, we assume that these components could also be the contributing factor for antioxidant capacity of whole plant (TFHA).

5.0 CONCLUSION

The present study revealed that geographical and climatic conditions of different region could lead to significant differences in the antioxidant activity due to the differences of phenolic and flavonoid contents. *T. fragrans* whole plant can be a good source of plant antioxidants, with a potential use in pharmaceutical fields. Many bioactive compounds were identified in *T. fragrans* by Gas Chromatography Mass Spectrometry. The presence of these bioactive compounds may emerge as most effective therapeutic agent to counter the problem of multidrug resistance against oxidative stress. Further investigation of individual compounds, their *in vivo* antioxidant activities and mechanistic studies are needed.

ACKNOWLEDGMENT

The authors are thankful to Dr. P.V.L. Rao, Director and Dr. Kathirvelu, Joint Director, Defence Research and Development Organization, Bharathiar University, Coimbatore, Tamil Nadu, India for providing laboratory facilities to perform GC-MS analysis.

REFERENCES

Achara Dholvitayakhun, Nathanon Trachoo, 2012. Antibacterial Activity of Ethanol Extract from Some Thai Medicinal Plants against *Campylobacter jejuni. Int. J. Biol. Sci.* 6, 235-238.

Alpinar, K., Ozyurek, M., Kolak, U., Guclu, K., Aras, C., Altun, M. and *et al.,* 2009. Antioxidant Capacities of Some Food Plants Wildly Grown in Ayvalik of Turkey. *Food Sci. Tech. Res.* 15, 59-64.

Autore, G., Caruso, A., Marzocco, S., Nicolaus, B., Palladino, C., Pinto, A., *et al.,* 2010. Acetamide derivatives with antioxidant activity and potential anti-inflammatory activity. *Molecules.* 15(3), 2028-2038.

Beauchamp, C., Fridovich, I., 1971. Superoxide dismutase: Improved assays and an assay applicable to Acrylamide gels. *Anal. Biochem.* 44, 276–277.

Benzie, I.F.F., Strain, J.J., 1996. The ferric reducing ability of plasma (FRAP) as a measure of Antioxidant Power. The FRAP Assay. *Anal. Biochem.* 239, 70–76.

Blois, M.S., 1958. Antioxidants determination by the use of a stable free radical. *Nature.* 4617, 1199-1200.

Buyukokuroglu, M.E., Gulcin, I., Oktay, M., Kufrevioglu, O.I., 2001. *In vitro* antioxidant properties of dantrolene sodium. *Pharmacol. Res.* 44, 491–494.

Di Carlo, G., Mascolo, N., Izzo, A.A, Capasso, F., 1999. Flavonoids: old and new aspects of a class of natural therapeutic drugs. *Life Sci.* 65, 337–353.

Dinesh Kumar Tyagi, 2005. Pharma forestry- field to medicinal plants. Atlantic Publisher Ltd., India. 100.

Dinis, P., Pineda, M., Aguilar, M., 1999. Spectrophotometric quantity of antioxidant capacity through the formation of a phosphomolybdenum complex: specific application to the determination of Vitamin E. *Anal. Biochem.* 269, 337-341.

Dinnis, T.C.P., Madeira, V.M.C., Almeida, L.M., 1994. Action of phenolic derivatives (Acetoaminophen, Salycilate and 5-aminosalycilate) as inhibitors of membrane lipid peroxidation and as peroxyl radical scavengers. *Arch. Biochem Biophys.* 315, 161–169.

Hagerman, A.E., Riedl, K.M., Jones, G.A., Sovik, K.N., Ritchar, N.T., Hartzfeld, P.W., 1998. High molecular weight plant polyphenolics (tannins) as biological antioxidants. *J. Agric. Food Chem.* 46, 1887-1892.

Halliwell, B., Gutteridge, J.M.C., 1984. Oxygen toxicity, oxygen radicals, transition metals and disease. *Biochem.* 219,1–14.

Hassas Roudsari, M., Chang, P., Pegg, R., Tyler, R., 2009. Antioxidant capacity of bioactive extracted from canola meal by subcritical water, ethanolic and hot water extraction. *Food Chem.* 114, 717-727.

Karunyal Samuel, J., Andrews B., 2010. Traditional medicinal plant wealth of Pachalur and periyur hamles dindigul district, tamil nadu. *Indian. J. Tradit. Knowl.* 9(2), 264-270.

Khan, R.A., Khan, M.R., Sahreen, S., 2012. Assessment of flavonoids contents and *in vitro* antioxidant activity of *Launaea procumbens. Chem. Central.* 6, 43.

Middleton, E., Kandaswami, C., Theoharides, T.C., 2000. The effects of plant flavonoids on mammalian cells: Implications for inflammation, heart disease and cancer. *Pharmacol Rev.* 52, 673-751.

Moncada, S., Palmer, R.M., Higgs, E.A., 1991. Nitric oxide: physiology, pathophysiology, and pharmacology. *Pharmacol Rev.* 43, 109–42.

Montoro, P., Braca, A., Pizza, C., De Tommasi, N., 2005. Structure-antioxidant activity relationships of flavonoids isolated from different plant species. *Food Chem.* 92, 349–355.

Pervaiz, S., Clemen, M., 2007. Superoxide anion oncogenic reactive oxygen species? The *Int. J. Biochem. Cell. Biol.* 39, 1297-1304.

Pinnell, S.R., 2003. Cutaneous photodamage, oxidative stress, and topical antioxidant protection. *J. Am. Acad. Dermatol.* 48, 1–19.

Prieto, P., Pineda, M., Aguilar, M. 1999. Spectrophotometric quantity of antioxidant capacity through the formation of a phosphomolybdenum complex: specific application to the determination of vitamin E. *Anal. Biochem.* 269, 337-341.

Priya, V., Jananie, R.K., Vijayalakshmi, K., 2011. GC/MS determination of bioactive components of *Trigonella foenum-grecum. J. Chem. Pharm. Res.* 3(5), 35-40.

Pulido, R., Bravo, L., Sauro-Calixto, F., 2000. Antioxidant activity of dietary polyphenols as determined by a modified ferric reducing/antioxidant power assay. *J.Agri. Food Chem.* 48, 3396-3402.

Rajesh Manian, Nagarajan Anusuya, Perumal Siddhuraj, Sellamuthu Manian, 2008. The antioxidant activity and free radical scavenging potential of two different solvent extracts of *Cammelia sinensis* (L.) O. Kuntz, *F. bengalensis* L. and *F. racemosa* L. *Food Chem.* 107, 1000-1007.

Re, R., Pellegrini, N., Proteggente, A., Pannala, A., Yang, M., Rice-evans C., 1999. Antioxidant activity applying an improved ABTS radical cation decolorization assay. *Free Radic. Biol. Med.* 26, 1231–1237.

Sanjay Batra, Amiya, P., Bhaduri, Bhawani, S., Joshi, Raja Roy, Ashok K., Khanna, Ramesh Chander, 2001. Syntheses and biological evaluation of alkanediamines as antioxidant and hypolipidemic agents. *Bioorg. Med Chem.* 9, 3093–3099.

Shanab, M.M., Emad Shalaby, A., 2012. The First record of biological activities of the egyptian red algal species *Compsopogon helwanii* Sanaa. *Int. J. Biosci. Biochem. Bioinforma.* 2, 4.

Sharififar, F., Dehghn, Nudeh, G., Mirtajaldini, M., 2009. Major flavonoids with antioxidant activity from *Teucrium polium* L. *Food Chem.* 112, 885–888.

Siddhuraju, P., Becker, K., 2003. Studies on antioxidant activities of *Mucuna* seed (*Mucuna pruriens* var. *utilis*) extracts and certain non-protein amino/imino acids through *in vitro* models. *J. Agric. Food Chem.* 51, 2144-2155.

Siddhuraju, P., Manian, S., 2007. The antioxidant and free radical scavenging capacity of dierary phenolic extracts from horse gram (*Macrotyloma uniflorum* (Lam.) Verdc.) seeds. *Food Chem.* 105, 950-958.

Sreejayan, N., Rao, M.N.A., 1997. Nitric oxide scavenging by curcuminoids. *J. Pharm Pharmacol.* 49, 105–107.

Suresh, B., Dhanasekaran, S., Kumar, R.V., Balasubramanian, S., 1994. Ethnopharmacological studies on the medicinal plants of Nilgiris. *Indian Drugs.* 32, 340-52.

Svobodova, A., Walterova, D., Psotova, J., 2006. Influence of silymarin and its flavonolignans on H_2O_2-induced oxidative stress in human keratinocytes and mouse fibroblasts. *Burns.* 32, 973–979.

Yang, C.M.J., Russell, J.B. 1992. Resistance of proline-containing peptides to ruminal degradation *in vitro*. *Appl. Environ. Microbiol.* 5:3954–3958.

Zhishen, J., Mengecheng, T., Jianming, W., 1999. The determination of flavonoid contents on mulberry and their scavenging effects on superoxide radical. *Food Chem.* 64, 555-559.

2015, Modern Methods in Phytomedicine *Pages* ***321–343***
Editor: **T. Parimelazhagan**
Published by: **DAYA PUBLISHING HOUSE, NEW DELHI**

22

Chasmophytes: A Potential Source of Medicine

Binu Thomas and Rajendran Arumugam

Department of Botany, School of Life Sciences, Bharathiar University, Coimbatore – 641 046, Tamil Nadu

1.0 INTRODUCTION

India, with its great biodiversity, has a tremendous potential and advantages in the emerging field of herbal medicines. Medicinal plants as a group comprise approximately 7500 and include representatives of about 17,000 species of higher plants (Shiva, 1996). Around 70 per cent of Indian medicinal plants are found in the tropical zone and more plants are seen in the forest of Western and Eastern Ghats (Ignacimuthu *et al.,* 1998.). The Western Ghats region is considered as one of the most important biogeographical zones of India, as it is one of the richest centers of endemism.

India is represented by rich culture, traditions and natural biodiversity and offer unique opportunity for the drug discovery researchers. It is one of diverse countries in the world, rich in medicinal plants. Ethnomedicine is the mother of all modern drugs and recently the importance of the traditional knowledge based medicines are being utilized throughout the world (Singh, 2002). Studies by national and international organizations have shown that for 75-90 per cent of the rural population of the world, the local herbalists alone attend to their medical problems (Pullaiah, 2006).

The Western Ghats region is a great emporium with treasure of ethnobotanical wealth. Most of the tribes as well as local inhabitants in the Western Ghats regions are utilize the medicinal herbs for various ailments after centuries of trials (Silja *et al.,* 2008). The tribes have developed their own traditional ways of diagnosis and treatment of diseases by trial and error to fulfill their basic requirements (Rajith *et al.,* 2010). There are numerous drugs have entered the international pharmacopoeia via the study of ethno pharmacology and traditional medicine (Binu Thomas *et al.,* 2012).

The present study is an attempt has made to explore the medico-potentiality of chasmophytic plants [Plants in the rocky cliffs and crevices (Warming, 1895.)] which are distributed in the rocky habitats of Coimbatore district, Southern Western Ghats region of Tamil Nadu. The rock crevices plays a key role in forming a major habitat for many plants. They represent a good indicator of rich biodiversity within small area (Binu Thomas *et al.,* 2013).

2.0 MATERIALS AND METHODS

2.1 Study Area: Coimbatore District, Southern Western Ghats of Tamil Nadu

Coimbatore district (between 10°-10′and 12°-00' of Northern latitude and 76° – 40' and of 8°-00' of Eastern longitude) is situated in the state of Tamilnadu in Southern India. The average rainfall received in Coimbatore district is 670 - 699 mm for the past twenty years out of the total rainfall 25 percent is received during South West monsoon, 49 percent during Oct. - Nov. and remaining 20 percent during Mar. - May. The temperature begins to increase after March. April is the hottest month with the daily mean maximum temperature of 38.2°C and minimum of 25.6°C. The maximum temperature may go up to 41°C some days. The maximum and minimum temperature is 41.5°C and 16°C respectively (Figure 22.1). The present study has been carried out in four different localities of Southern Western Ghats of Coimbatore district such as Velliangiri, Madukkarai, Maruthamalai and Kanuvai hills (Figure 22.1).

2.2 Documentation

The present investigation was undertaken to study the chasmophytic medicinal plants and its diversity from selected hillocks of Southern Western Ghats of Coimbatore, Tamil Nadu.

During present study, an attempt was made to find out the medicinal uses of potential chasmophytic species from the selected hillocks of Coimbatore district, Tamilnadu, during December 2013 – July 2014. Several intensive and extensive field trips were conducted. In this field visits, personal interviews and direct observation was done to document the traditional knowledge of tribal (tribe *Malasars* in Velliangiri and *Irulars* in Maruthamalai hills) local people (Maruthamalai and Kanuvai hills) about the therapeutic value of the chasmophytes in different parts of the study areas in all seasons (Figure 22.2). The plant specimens were collected and identified taxonomically by using available Floras and Literature (Gamble, 1915 – 1936; Matthew, 1983). The voucher specimens were deposited in the Herbaria of Department of Botany, Bharathiar University, Coimbatore, Tamil Nadu for future reference.

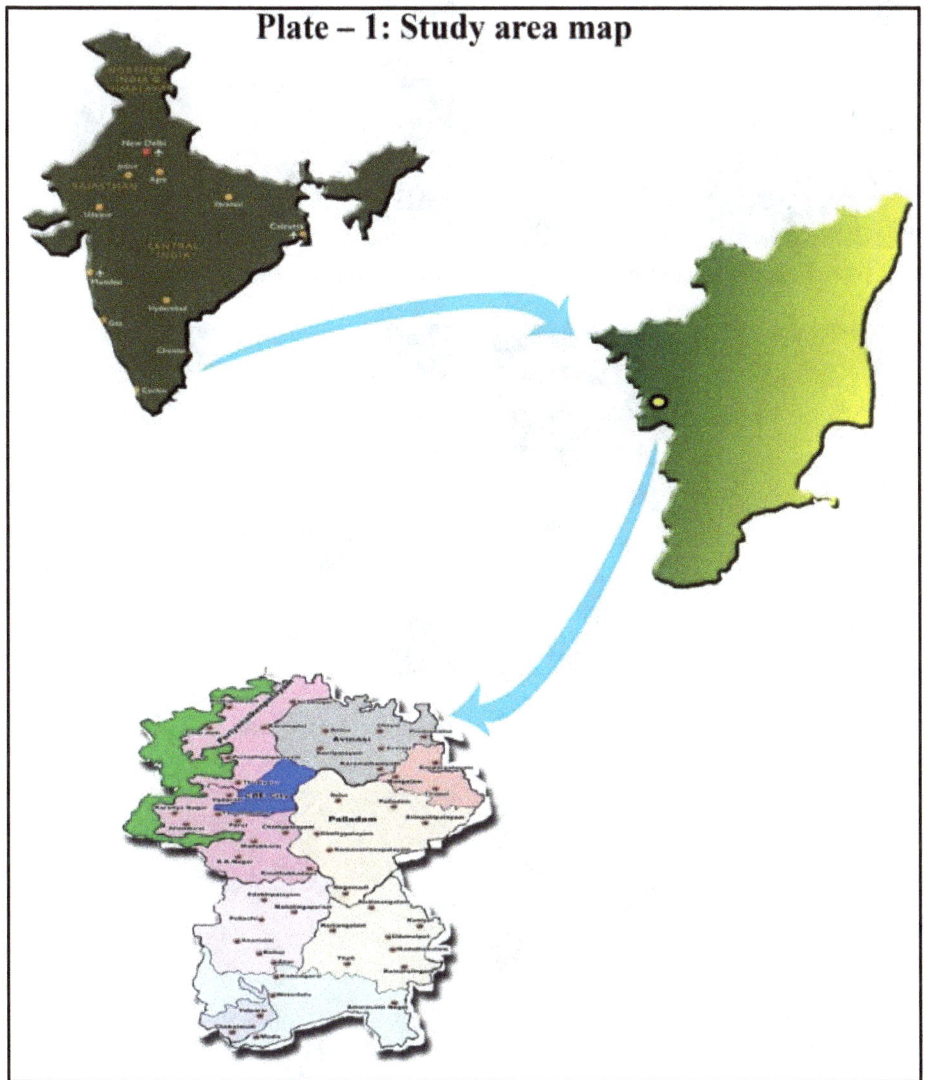

Figure 22.1: Map of India with Tamil Nadu State and Coimbatore District.

Plate-2: Ethno-botanical observations

A) Discussing *Irular* of Maruthamalai

B). Interviewing with *Irular* informant

C). *Malasars* of Velliangiri Hills

D). Interviewing with *Malasar* informant

E). Tribal settlements in Velliangiri Hills

F). Collection of data from local inhabitant in Madukkarai

Figure 22.2: Documentation of Traditional Knowledge from the Study Area.

3.0 RESULTS AND DISCUSSION

The present study on Medico-potential chasmophytes of selected hillocks of Coimbatore district, Southern Western Ghats of Tamil Nadu, revealed that, a total of 186 species distributed in 128 and 65 families (Table 22.1). Among these 186 species, 160 species from Angiosperms and 26 species from pteridophytes. Similarly, out of 65 total families, 50 families consisting angiosperms and remaining 15 families are including pteridophytes.

In order to infer the dominant Angiosperm families reveals that, Asteraceae is the first dominant one with 18 species, followed by Lamiaceae with 13 species, Commelinaceae with 11 species, Poaceae with 10 species, Acanthaceae and Fabaceae with 9 species, Rubiaceae with 5 species, Oxalidaceae with 4 species and the remaining families includes 2 and 1 species respectively (Figure 22.3). Similarly the dominant Angiosperm genera includes *Cyanotis* is the first dominant genus with 5 species followed by *Euphorbia* and *Plectranthus* with 4 species. While the genera like *Cleome, Indigofera, Justicia, Kalanchoe, Persicaria* and *Portulaca* with 3 species each. All others includes 2 and 1 species respectively.

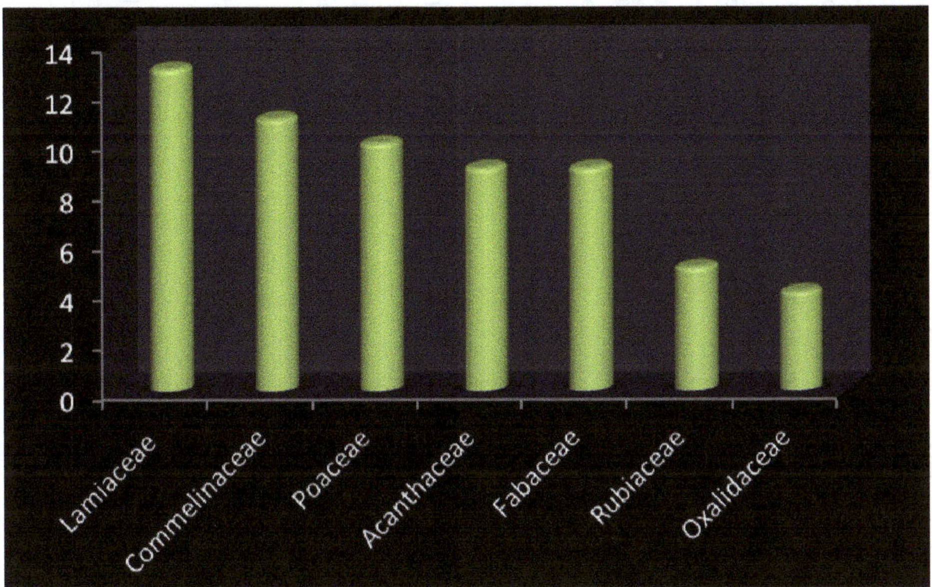

Figure 22.3: Analysis of Dominant Angiosperm Families.

The analysis of dominant pteridophytic families includes Adiantaceae is the first dominant one with 4 species followed by Dryopteridaceae, Pteridaceae and Selaginellaceae with 3 species each. While the families like Aspleniaceae, Cheilathaceae and Hemionitidaceae with 2 species each. All other families includes single each (Figure 22.4). Similarly the dominant pteridophytic genera consist of *Adiantum* is the first dominant genus with 4 species followed by Pteris and Selaginella with 3 species and Dryopteris with 2 species respectively and all other genera having single species each.

Table 22.1: List of Medico-Potential Chasmophytes

Sl No.	*Botanical Name*	*Family*	*Part(s) Used*	*Mode of Drug Oreparation*
1.	*Actiniopteris radiata* (Sw.) Link	Actiniopteridaceae	Whole plant	Plant paste is applied over the cuts and wounds.
2.	*Adiantum capillus - veneris* L.	Adiantaceae	Whole plant	Whole plant decoction is used to cure cough and cold.
3.	*Adiantum incisum* Forssk.	Adiantaceae	Leaves	The leaf powder is mixed with butter and used for controlling the internal burning of the body.
4.	*Adiantum lunulatum* Burm.	Adiantaceae	Rhizome	The rhizome powder is used as antidote against snake bite.
5.	*Adiantum raddianum* C. Presl.	Adiantaceae	Leaves	Leaf juice orally taken for indigestion.
6.	*Agave cantula* Roxb.	Agavaceae	Leaves	The slice of fleshy leaves constitutes a good poultice.
7.	*Ageratum conyzoides* L.	Asteraceae	Leaves	Leaf juice is taken internally for dyspepsia and anaemia.
8.	*Ageratum houstonianum* Mill.	Asteraceae	Leaves	Leaves are crushed along with lime and applied on cuts caused by iron knives; leaf paste applied over regions affected by alopacia (abnormal hair loss).
9.	*Aloe vera* (L.) Burm.	Aloaceae	Leaves	Laef sap is applied as hair tonic and it is act as an anti dandruff.
10.	*Alternanthera pungens* Kunth.	Amaranthaceae	Whole plant	The decoction of plant is reportedly taken to treat gonorrhoea.
11.	*Alysicarpus glumaceus* (Vahl) DC.	Fabaceae	Leaves	Hot decoction of leaves and roots given for cough, cold and fever.
12.	*Alysicarpus monilifer* (L.) DC.	Fabaceae	Leaves	Leaf paste is applied on freshly wounded cuts.
13.	*Ammannia baccifera* L.	Lythraceae	Leaves	The ash of leaves with coconut oil is applied to herpatic eruptions.The leaf paste is applied externally to cure muscular pains.
14.	*Anaphalis lawii* (Hook.f.) Gamble.	Asteraceae	Whole plant	Whole plant is air-dried, powdered and consumed as health powder.
15.	*Anemia wightiana* Gardner	Schizaeaceae	Whole plant	The whole plant is used for the treatment of rheumatism.
16.	*Anisochilus carnosus* (L.f.) Wall.	Lamiaceae	Leaves	Leaf paste is applied over for various skin problems.
17.	*Anisochilus dysophylloides* Benth.	Lamiaceae	Whole plant	The entire plant is boiled with water and it is used for the bathing purpose of children to cure fever.
18.	*Anisomeles indica* (L.) O. Ktze.	Lamiaceae	Leaves	Crushed leaf paste is applied on wounds. Leaf extract is boiled with water and cooked as curry to cure dysentery and stomach ailments.

Contd...

Table 22.1–*Contd...*

Sl No.	*Botanical Name*	*Family*	*Part(s) Used*	*Mode of Drug Oreparation*
19.	*Anisomeles malabarica* (L.) R. Br.	Lamiaceae	Leaves	Leaf paste is applied against insect bites.Leaves crushed with 10 gm of cumin and the extract mixed with either honey or breast milk. After that 2 teaspoon of it orally administered by new born babies to control vomiting and untime motion.
20.	*Apluda mutica* L.	Poaceae	Whole plant	The whole plant Juice: diuretic and also for gonorrhoea.
21.	*Arisaema leschenaultii* Blume	Araceae	Tuber	Tuber ground with water and the paste applied externally on the spot of insect bite.Tubers are dried in sunlight, fried in ghee and made into paste along with cumin; coriander and pepper and it is taken orally for bleeding piles.
22.	*Aristida setacea* Retz.	Poaceae	Whole plant	The paste is made from whole plant parts and stem bark of *Pongamia pinnata.* After that, it is applied topically on affected places to heal wounds.
23.	*Asplenium decrescens* Kunze.	Aspleniaceae	Leaves	The leaves show anti-bacterial properties. The crushed leaves applied over the affected parts of the body to heal microbial infection.
24.	*Asplenium inaequilaterale* Willd.	Aspleniaceae	Leaves	The crushed leaves are applied over the affected parts of the body to cure body pain.
25.	*Axonopus compressus* (Sw.) P. Beauv.	Poaceae	Whole plant	The paste of whole plant is applied over forehead to get relief from headache.
26.	*Barleria buxifolia* L.	Acanthaceae	Root	Root paste is applied on boils, burns and swellings.
27.	*Begonia malabarica* Lam.	Begoniaceae	Stem	The juice of succulent stem is medicinally used as blood purifier.Stem of the plant is soaked in milk overnight and it is administered orally on empty stomach as healthy tonic such as ' *Kayakalpa*'.
28.	*Bidens pilosa* L.	Asteraceae	Whole plant	The warm juice of the fresh plant is used to treat earache and conjunctivitis. It also used for cuts and wounds.
29.	*Biophytum reinwardtii* (Zucc.) Klotzsch	Oxalidaceae	Leaves	Leaf paste rubbed over head and then bath is taken to cure dandruff. It also to prevent falling of hair.
30.	*Biophytum sensitivum* (L.) DC.	Oxalidaceae	Whole plant	Crushed leaves are inserted in to the mouth of a patient at the time of senselessness for epilepsy. Italso used for inflammation by tying for 3 days over the affected area.Plant paste is applied on forehead for migraine.

Contd...

Table 22.1–*Contd...*

Sl No.	Botanical Name	Family	Part(s) Used	Mode of Drug Oreparation
31.	*Blechnum orientale* L.	Blechnaceae	Leaves	Consumption leaf juice is used to cure intestinal wounds.
32.	*Blumea mollis* (D. Don) Merr.	Asteraceae	Leaves	Leaf paste is applied for various fungal allergies on skin.
33.	*Blumea membranaceae* Wall. *ex* DC.	Asteraceae	Leaves	Leaf paste is applied for various fungal allergies.
34.	*Boerhavia diffusa* L.	Nyctaginaceae	Leaves	Leaf paste is applied on the cuts and wounds to stop bleeding.
35.	*Boerhavia erecta* L.	Nyctaginaceae	Leaves	Leaf decoction is used for diarrhoea and dysentery.
36.	*Bulbostylis barbata* (Rottb.) Clarke	Cyperaceae	Whole plant	The decoction of whole plant is given for dysentery.
37.	*Canscora diffusa* (Vahl.) R. Br.	Gentianaceae	Leaves	Leaves mixed with a pinch of calcium and it is powdered. The paste is made from this powder and it is applied on cuts for one time a day for 4 days.
38.	*Caralluma adscendens* (Roxb.) Haw. var. *attenuata*	Asclepiadaceae	Tender shoots	Tender shoots are ground along with 4 – 5 pieces of *Allium sativum* and *Tamirandus indica*. This paste is administered orally to cure digestive disorders.
39.	*Caralluma umbellata* Haw.	Asclepiadaceae	Tender shoots	Young shoot paste with gingili oil and applied for foot sores. Stem burn in direct fire and regularly eaten for five days in empty stomach for ulcer.
40.	*Catharanthus pusillus* (Murr.) G. Don.	Apocynaceae	Leaves and flowers	Crushed leaves and flowers are used for ulcers.
41.	*Catharanthus roseus* (L.) G. Don.	Apocynaceae	Leaves and flowers	Vincristin is the alkaloid obtained from the plant is used for cancer therapeutics.
42.	*Centella asiatica* (L.) Urban.	Apiaceae	Leaves	Leaf juice is mixed with honey and it is taken in empty stomach to enhance the memory power; crushed plants are also used against boils and skin diseases.
43.	*Chamaecrista absus* (L.) Irwin and Barneby	Caesalpiniaceae	Leaves	Crushed leaf paste is applied over wounds.
44.	*Chamaecrista kleinii* (Wight and Arn.) Matthew	Caesalpiniaceae	Leaves	Handful of leaf is made in to paste and applied two times in a day for eczema.

Contd...

Table 22.1–*Contd...*

Sl No.	*Botanical Name*	*Family*	*Part(s) Used*	*Mode of Drug Oreparation*
45.	*Cheilanthes tenuifolia* (Burm.f) Sw.	Cheilanthaceae	Leaves	The paste is made from fronds and it is applied on abscess.
46.	*Cheilanthes mysurensis* Wall. *ex* Bedd.	Cheilanthaceae	Leaves	The juice obtained from the leaves mixed with hot water and taken orally along with honey to treat throat pain.
47.	*Chenopodium ambrosioides* L.	Chenopodiaceae	Aerial parts	Decoction of aerial parts used as a blood purifier.Leaf decoction mixed with ash of banana leaf and is given to goats and sheep to expel worms.
48.	*Chlorophytum laxum* R.Br.	Anthericaceae	Tubers	The tuber powder mixed with water and this infusion is given orally to cure piles.
49.	*Chlorophytum tuberosum* (Roxb.) Baker	Anthericaceae	Tubers	The drug obtained from tubers is used as rejuvenator and health tonic.The tuber powder with sugar is used as sex tonic.
50.	*Christella dentata* (Forssk.) Brown.	Thelypteridaceae	Leaf	Leaf paste is applied for skin diseases.
51.	*Cissus quadrangularis* L.	Vitaceae	Tender shoots	The juice prepared from tender shoot and leaf is mixed with butter milk. It is taken orally for treating menstrual disorders.Whole plant is made in to paste and plastered over the fractured part of the body till cure.
52.	*Cleome gynandra* L.	Cleomaceae	Leaves	Leaf paste is used to heal boils.
53.	*Cleome monophylla* L.	Cleomaceae	Leaves and seeds	Leaves and seeds are boiled in coconut oil and cooled. 4-5 drops of this solution is pour in to ear, twice a day for 3 – 4 days to cure ear pain.The plant ground with 1 or 2 pepper and the paste is applied on forehead for migraine.
54.	*Cleome viscosa* L.	Cleomaceae	Leaves	The leaf juice is mixed with fruit powder of *Pedalium murex*. The administration of this solution during menstrual problems.
55.	*Commelina benghalensis* L.	Commelinaceae	Whole plant	Whole plant is used to treat leprosy.Juice extracted from the stem is mixed with the stem juice of *Canna indica* and fruits of *Areca catechu*. This mixture is applied topically on affected places to heal wounds.
56.	*Commelina ensifolia* R. Br.	Commelinaceae	Aerial parts	Water extract of fresh aerial part is applied externally to heal burn injuries, itches and boils.
57.	*Crassocephalum crepidioides* (Benth.) S. Moore	Asteraceae	Leaves	Crushed leaves are applied on skin for various skin allergies.

Contd...

Table 22.1–*Contd...*

Sl No.	Botanical Name	Family	Part(s) Used	Mode of Drug Oreparation
58.	*Crotalaria albida* Heyne *ex* Roth.	Fabaceae	Whole plant	Whole plant is used as purgative.
59.	*Curculigo orchiodes* Gaertn.	Hypoxidaceae	Rhizome	Rhizome paste is taken internally with milk for the treatment of leucorrhoea, diabetes and tonsillitis.
60.	*Cyanotis arachnoidea* C.B. Clarke	Commelinaceae	Whole plant	Given as a laxative to cattle.
61.	*Cyanotis fasciculata* (Heyne *ex* Roth) Schult.	Commelinaceae	Leaves	The juice obtained from succulent leaf is used to cure fungal infections and mouth sores.
62.	*Cyanotis pilosa* Schult. and Schult.f.	Commelinaceae	Whole plant	Given as a laxative to cattle.
63.	*Cyanotis tuberosa* (Roxb.) Schult	Commelinaceae	Tubers	Dried tubers are made in to a paste with curd and it is given orally to induce motion of blood.
64.	*Cyanotis villosa* (Spreng.) Schult.f	Commelinaceae	Stem	Crushed stem of this plant with stem juice of *Canna indica* are ground into a paste and it is applied externally to heal wounds.
65.	*Cyathula prostrata* (L.) Blume.	Amaranthaceae	Whole plant	Entire plant is made in to a paste and applied externally against muscle ailments.
66.	*Cymbopogon flexuosus* (Nees *ex* Steud.) Will.	Poaceae	Leaves	Juice obtained from the leaf is used against infestations of lice and other insects. The leaves are also pressed and placed where insects attack occurs.
67.	*Cyperus difformis* L.	Cyperaceae	Roots	Grounded roots are applied against various skin allergies.
68.	*Desmodium repandum* (Vahl.) DC.	Fabaceae	Whole plant	The paste is prepared from entire plant is applied to the body followed by a bath after thirty minutes as an antidote against poisonous bites.
69.	*Desmodium triflorum* (L.) DC.	Fabaceae	Whole plant	Whole plant paste with aerial roots of *Ficus benghalensis* is used as a plaster for bone fracture.Fresh leaf paste is applied over wounds and abscess.
70.	*Didymocarpus gambleanus* Fischer	Gesneriaceae	Leaves	Leaf paste is applied on pimples and external cracks.

Contd...

Table 22.1–*Contd...*

Sl No.	Botanical Name	Family	Part(s) Used	Mode of Drug Oreparation
71.	*Didymocarpus humboldtiana* Gard.	Gesneriaceae	Whole plant	Whole plant parts with stem bark of *Pongamia pinnata* and leaf of *Abutilon indicum* are ground into a paste and applied topically on affected places to heal wounds.
72.	*Drymaria cordata* Edgew. and Hook.	Caryophyllaceae	Whole plant	The paste prepared from whole plant is applied on the fractured bone. It is bandaged with the help of cotton cloth. After 20-25 days the fractured bone gets jointed. Juice obtained from the leaves also used for fever.
73.	*Dryopteris atrata* (Kunze) Ching.	Dryopteridaceae	Rhizome	Rhizome powder is taken with water daily twice for rheumatism
74.	*Dryopteris cochleata* (Buch. Ham. *ex* D. Don) C. Chr.	Dryopteridaceae	Whole plant	Plant paste is applied on wounds to prevent infection.
75.	*Ecbolium viride* (Forssk.) Alston.	Acanthaceae	Leaves	Leaf juice is used to cure ear boils.
76.	*Elephantopus scaber* L.	Asteraceae	Whole plant	Crushed leaves are applied on wounds to abate bleeding.The root juice is taken internally for stomach problems.The mixture of whole plant with gingelly and coconut oil (100 mL each) is heated and applied externally for sprain.
77.	*Eleusine indica* (L.) Gaertn.	Poaceae	Whole plant	Whole plant decoction is given to children for convulsion.
78.	*Elytraria acaulis* (L. f.) Lindau	Acanthaceae	Roots	Root paste is given once in a day for 21 days for leucorrhoea. It also with black pepper is applied on snake bite.
79.	*Emilia sonchifolia* (L.) DC.	Asteraceae	Leaves	Leaf paste is used against sprains and muscle spasm. The extract of crushed leaves applied externally on cuts and wounds.
80.	*Eulophia epidendraea* (Koen.) Schltr.	Orchidaceae	Bulb	Crushed bulb is fried with mustard oil and this residue is applied on rheumatism, thrice a day till cure.
81.	*Euphorbia antiquorum* L.	Euphorbiaceae	Latex	Latex obtained from the stem is applied for burn injury.
82.	*Euphorbia hirta* L.	Euphorbiaceae	Latex	The latex is applied on skin for various skin problems and lip cracks.
83.	*Euphorbia rothiana* Spreng.	Euphorbiaceae	Latex	The latex applied on the body to cure sores.
84.	*Euphorbia thymifolia* L.	Euphorbiaceae	Whole plant	Dried leaves and seeds are aromatic, astringent and stimulant. It also used as laxative.The paste of whole plant along with black pepper is given for stomach pain and dysentery.

Contd...

Table 22.1–*Contd...*

Sl No.	Botanical Name	Family	Part(s) Used	Mode of Drug Oreparation
85.	*Evolvulus alsinoides* L.	Convolvulaceae	Whole plant	The plant paste is given orally with milk or hot water to enhance the memory power. 10 – 30 gms of fresh leaves along with whole plant of *Mollugo nudicalis* Lam. is boiled in water. This boiled extract is taken orally two times per day for the period of three days to cure fever.
86.	*Evolvulus nummularis* L.	Convolvulaceae	Whole plant	Whole plant paste is applied over the throat to reduce tonsillitis pain.
87.	*Exacum pedunculatum* L.	Gentianaceae	Whole Plant	Whole plant decoction taken three days to cure fever with dysentery.
88.	*Galinsoga parviflora* Cav.	Asteraceae	Leaves	The crushed leaves are applied on skin for various skin problems.
89.	*Glinus oppositifolius* (L.) A. DC.	Molluginaceae	Whole plant	Leaf juice along with *Allium sativum* is used as purgative and also for eczema.Whole plant paste is applied externally against various types of skin diseases such as scabies, itches etc.
90.	*Glossocardia bosvallea* (L.f.) DC.	Asteraceae	Whole plant	A spoonful of plant juice is given along with cow milk twice a day to cure typhoid.
91.	*Gynura nitida* DC.	Asteraceae	Roots	Crushed roots are boiled with milk and administered orally for snake bite. The healer keeps a small piece of roots in his mouth and suck out impure blood from the spot of bite after widening it with sharp knife.
92.	*Hedyotis corymbosa* (L.) Lam.	Rubiaceae	Leaves	Leaf Juice is applied externally to reduce burning sensation. It also administered orally for 5 days to cure jaundice.
93.	*Hedyotis herbacea* L.	Rubiaceae	Whole plant	Whole plant paste is applied over cuts and wounds.
94.	*Heteropogon contortus* (L.) P. Beauv. *ex* Roem.	Poaceae	Inflorescence and roots	Decoction of inflorescence is given for asthma as bronchodilator. One teaspoonful of air-dried root powder is mixed with water and taken orally for the treatment of snakebite.
95.	*Hibiscus ovalifolius* (Forssk.) Vahl	Malvaceae	Flowers and fruits	The fruits are nutritive, flowers are anti diabetic
96.	*Hybanthus enneaspermus* (L.) F. Muell.	Violaceae	Whole Plant	The plant extract is mixed with goat milk and pinch of sugar and it is administered orally for leucoderma and also as health tonic: 20 mL of the whole plant juice is taken with cow's milk for a period of four to five months to treat diabetes.The consumption of plant juice can improve sexual vigour in male.

Contd...

Table 22.1–*Contd...*

Sl No.	*Botanical Name*	*Family*	*Part(s) Used*	*Mode of Drug Oreparation*
97.	*Hypericum mysurense* Heyne.	Hypericaceae	Leaves	Leaf paste used to remove body hairs.
98.	*Hyptis suaveolens* (L.) Poit.	Lamiaceae	Roots and seeds	Roots chewed with betel leaf for stomach pains. The root decoction used as an appetiser.
99.	*Impatiens balsamina* L.	Balsaminaceae	Leaves and flowers	Both leaves and flowers are used as substitute for henna. The plant is used as emetic and diuretic.An alcoholic extracts of flowers are reported to posses marked antibiotic activity against some pathogenic fungi and bacteria.
100.	*Impatiens chinensis* L.	Balsaminaceae	Whole plant	Whole plant extract is applied externally for burns.
100a.	*Indigofera linnaei*Ali.	Fabaceae	Leaves	Consumption of leaf juice in early morning with milk to cure spermatorrea.
101.	*Indigofera nummularifolia* (L.) Livera.	Fabaceae	Whole plant	Plant juice is diuretic and also for epilepsy.
102.	*Indigofera uniflora* Buch.-Ham. *ex* Roxb.	Fabaceae	Leaves	External application of leaf paste for various skin allergies.
103.	*Indoneesiella echioides* (L.) Sreem.	Acanthaceae	Roots and leaves	The root paste is applied for tooth-ache.Leaf paste is mixed with coconut oil and applied on grey hair for blackening.
104.	*Isodon wightii* (Benth) Hara	Lamiaceae	Leaves	The leaf juice can enhance the resistant power in cancer patients.
105.	*Justicia nagpurensis* Graham.	Acanthaceae	Leaves	Crushed leaves are boiled with coconut oil and it is applied externally over rheumatic swellings.
106.	*Justicia prostrata* (Roxb. *ex* Clarke) Gamble	Acanthaceae	Leaves	Leaf decoction is god for cough.
107.	*Justicia tranquebariensis* L.f.	Acanthaceae	Roots and leaves	Root paste is applied for treating tooth ache.The leaves ground with a piece of gingili oil and applied on knees for joint pains.
108.	*Kalanchoe lanceolata* (Forssk.) Pers.	Crassulaceae	Leaves	Leaf paste is applied over the skin for various skin allergies. It also applied externally for joint pain.
109.	*Kalanchoe pinnata* (Lam.) Pers.	Crassulaceae	Leaves	Leaf juice is an effective medicine for the treatment of dysentery and cholera. Crushed leaves are rubbed on whole body during fever.

Contd...

Table 22.1–*Contd...*

Sl No.	*Botanical Name*	*Family*	*Part(s) Used*	*Mode of Drug Oreparation*
110.	*Kalanchoe schweinfurthii* Penzig.	Crassulaceae	Stem and leaves	The paste made from the succulent stem and leaves of the plant is applied over the cuts and wounds. Leaf extract applied externally for joint pain.
111.	*Knoxia mollis* Wight and Arn.	Rubiaceae	Leaves	Leaf is ground into a paste and applied topically on affected places to heal wounds.
112.	*Leucas aspera* (Willd.) Spreng.	Lamiaceae	Leaves and flowers	The decoction obtained from the leaves and flowers is used for cough and cold.
113.	*Leucas biflora* (Vahl.) R.Br.	Lamiaceae	Leaves	Leaf decoction is used as eye drop twice a day in case of conjunctivitis. Four to five leaves are also prescribed to chew with the leaf of *Piper betel* L. for women who suffering from white discharge.
114.	*Lindernia ciliata* (Colsm.) Pennell	Scropulariaceae	Whole plant	Whole plant paste along with black pepper is given for gonorrhoea.
115.	*Medinilla beddomei* Clarke	Melastomataceae	Leaves	Fresh leaves are eaten for reducing body heat.
116.	*Mitracarpus villosus* (Sw.) DC.	Rubiaceae	Leaves	Leaves and stem bark of *Syzygium cumini* are ground into a paste and then it is heated with gingelly oil. The mixture thus obtained is applied topically on affected places to heal wounds.
117.	*Mollugo nudicaulis* Lam.	Molluginaceae	Whole plant	Whole plant decoction administered orally for treating fever.
118.	*Mollugo pentaphylla* L.	Molluginaceae	Whole plant	The whole plant is powdered along with pepper (*Piper nigrum*) and the oral administration of this extract is used for infantile convulsions by gonads, 1-2 spoonful when symptoms occur till cure.
119.	*Murdannia dimorpha* (Dalz.) Brueck	Commelinaceae	Roots	Root paste with goat milk is prescribed orally to cure asthma.
120.	*Murdannia nudiflora* (L.) Brenan	Commelinaceae	Whole plant	Whole plant paste with common salt is applied on the affected area to cure leprosy.
121.	*Ocimum americanum* L.	Lamiaceae	Shoots and leaves	Fresh leaves and shoots are crushed and applied over forehead to get relief from headache.

Table 22.1–*Contd...*

Sl No.	Botanical Name	Family	Part(s) Used	Mode of Drug Oreparation
123.	*Opuntia stricta* Haw. var. *dillenii* (Ker-Gawl.) L.	Cactaceae	Fruit and Stem	1–½ teaspoon stem juice with sugar and it is given internally for constipation. The juice of the fruit is mixed with sugar, boiled and given internally three times daily for cough and asthma. Flowers are crushed and applied as anti-dandruff agent.
124.	*Opuntia vulgaris* Mill.	Cactaceae	Succulent parts	Crushed succulent part of the plant is applied over inflammation.
125.	*Oxalis corniculata* L.	Oxalidaceae	Aerial vegetative portion	About 20 mL. of extract of aerial vegetative portion is mixed with a glass of water and teaspoonful of sugar. This mixture is given to take 5 times in a day until cure stomach pain; Chewed as a mouth freshener.The fine paste of whole plant with black pepper is applied to boils and wounds.10-20 gms of young leaves along with flower of banana (*Musa paradisica* L.) boiled with water. This boiled extract is filtered and taken orally with honey to arrest dysentery in children.
126.	*Parahemionitis cordata* Roxb. *ex* Hook.	Hemionitidaceae	Aerial parts	Aerial parts with raw rice is cooked and eaten to regulate menstrual cycle.
127.	*Pedalium murex* L.	Pedaliaceae	Fruits	Two teaspoonfuls of fruit powder is mixed with leaf juice of *Cleome viscosa* and is administered from the 5th day of menses before going to bed daily once for 7 days for menstrual problems.
128.	*Pentanema indicum* (L.) Ling.	Asteraceae	Leaves and roots	Leaf juice used as a lotion for insect sting.Root juice is used to cure cough and jaundice.
130.	*Peperomia pellucida* (L.) Kunth.	Piperaceae	Whole plant	The whole plant extract is applied on skin infections caused by bacteria.The consumption of crude extract of the plant is good for diabetic patiens. Roots are ground with water and taken orally to check miscarriage during preganency.
131.	*Peperomia tetraphylla* (G.Forst.) Hook. and Arn.	Piperaceae	Leaves	Leaves are boiled with sesamum oil and applied all over the body to cure body pain.
132.	*Peristrophe paniculata* (Forssk.) Brummitt.	Acanthaceae	Whole plant	The plant extract is given three times in a day to cure fever and abdominal pain.

Contd...

Table 22.1–*Contd...*

Sl No.	*Botanical Name*	*Family*	*Part(s) Used*	*Mode of Drug Oreparation*
133.	*Persicaria barbata* (L.) Hara.	Polygonaceae	Roots	Roots are used as astringent and cooling agent. It also for the treatment of ulcers.
134.	*Persicaria chinensis* (L.) Gross.	Polygonaceae	Roots and leaves	Roots ground along with milk and taken orally for the treatment of fever and diarrhoea; crushed leaves are also used as shampoo.
135.	*Persicaria glabra* (Willd.) Gomez.	Polygonaceae	Leaves	Leaf paste with black pepper is taken with honey to cure fever and colic pain.
136.	*Phyllanthus maderaspatensis* L.	Euphorbiaceae	Leaves	Leaf paste is diluted with 200 mL of goat's milk and oral administration of it for 2 times in a day for 7 days to cure jaundice.
137.	*Phyllanthus virgatus* G. Forst.	Euphorbiaceae	Whole plant	All parts of a plant is used to cure jaundice.Crushed leaves mixed with butter milk and it is used as wash for itch in children.
138.	*Pilea microphylla* (L.) Liebm.	Urticaceae	Whole plant	The crushed plants are applied to sores and bruises.
139.	*Pityrogramma calamelanos* (L.) Link.	Hemionitidaceae	Whole plant	Whole plant parts are boiled with water and the decoction thus obtained is taken orally in early morning to treat kidney stone.
140.	*Plantago erosa* Wall.	Plantaginaceae	Leaves	Leaf paste used against varicose veins; also used as an antiseptic for wounds.
141.	*Plectranthus amboinicus* (Lour.) Spreng.	Lamiaceae	Leaves	Leaf juice is mixed with honey and it is taken orally as carminative.The few drops of leaf juice is administered 2-3 times daily for fever.
142.	*Plectranthus barbatus* Andr.	Lamiaceae	Leaves	Oral administration of 50 mL of the filtered leaf extract by children for 2 times in a day to reduce body heat and fever.
143.	*Plectranthus caninus* Roth.	Lamiaceae	Leaves	Inhaling of strongly aromatic crushed leaf paste is good for reducing headache.
144.	*Plectranthus mollis* (Ait.) Spreng.	Lamiaceae	Leaves	During evening, the chopped plant material was spread near the bed for repelling mosquitos.
145.	*Plumbago zeylanica* L.	Plumbaginaceae	Roots	Root bark paste is mixed with milk, boiled and oral administration of it as febrifuge. Root bark paste mixed with *Allium sativum* (Garlic) and it is applied for muscle pain.

Contd...

Table 22.1–*Contd...*

Sl No.	*Botanical Name*	*Family*	*Part(s) Used*	*Mode of Drug Oreparation*
146.	*Polycarpaea corymbosa* (L.) Lam.	Caryophyllaceae	Whole plant	Freshly collected plant is made into paste and applied externally as an antidote for scorpion sting and insect bite. Poultice made from the fresh leaves and it is applied over boils and inflammatory swellings.Two teaspoonful of powder prepared from whole plant is given orally with honey thrice a day till recovery to treat jaundice.
147.	*Portulaca oleracea* L.	Portulacaceae	Whole plant	The consumption of an entire plant juice is good for healthy heart.
148.	*Portulaca pilosa* L.	Portulacaceae	Tuberous root stock	Tuberous root stock is eaten as raw or after it has been roasted on hot coals, the outer skin can be peeled before eating, it is considered for health.
149.	*Portulaca quadrifida* L.	Portulacaceae	Whole plant	The decoction of whole herb is used for asthma and cough.
150.	*Pouzolzia wightii* Bennett.	Urticaceae	Whole plant	Whole plant paste with *Curcuma longa,* plastered over the fractured bone.The paste is made from the plant powder along with stem bark of *Melia azedarch* L. and *Andrographis paniculata* Wall *ex* Nees. is applied externally for scorpion bite.
151.	*Pouzolzia zeylanica* (L.) Bennett.	Urticaceae	Leaves	A crushed leaf along with eggs and it is used for cuts and fractures.Leaf juice is rubbed to the right eye with little finger of the left hand in three times per day to cure eye inflammation.
152.	*Protasparagus racemosus* (Willd.) Oberm.	Asparagaceae	Leaves and rhizome	The paste of mature leaves is applied externally to cure heel cracks. Rhizome juice is used internally for leucorrhoea and epilepsy.
153.	*Pteris biaurita* L.	Pteridaceae	Rhizome	The rhizome is ground into paste and applied over the affected parts of the body to get relief from body pain.
154.	*Pteris confusa* T.G.	Pteridaceae	Rhizome	Rhizome paste is applied over the boils.
155.	*Pteris vittata* L.	Pteridaceae	Whole plant	The whole plant parts are ground into paste and it is applied over the affected places for wound healing.
156.	*Pteridium aquilinum* (L.) Kuhn.	Dennstaedtiaceae	Whole plant	The whole plant parts are ground into paste and applied over cuts and wounds.
157.	*Pyrrosia porosa* (C. Presl) Hoven.	Polypodiaceae	Whole plant	The whole plant paste is applied over cuts made through knives.

Contd...

Table 22.1–*Contd...*

Sl No.	Botanical Name	Family	Part(s) Used	Mode of Drug Oreparation
158.	*Rubia cordifolia* L.	Rubiaceae	Stem and leaves	Crushed stem and leaves are made in to paste and given to cattle daily once for 3-4 days for curing of dislocation of bones in cattle.Leaf paste is applied over swellings in the gum.
159.	*Rungia repens* (L.) Nees.	Acathaceae	Whole plant	Whole plant is dried and pulverized and it is given in doses of one table spoon daily for jaundice.The paste of the plant also applied topically on affected places to heal wounds.
160.	*Sansevieria roxburghiana* Schult. and Schult.	Dracaenaceae	Leaves	Leaf juice (150 mL) is given internally as an antidote for snake bite.Roots ground with turmeric powder and this paste is used as antidote for snake venom.
161.	*Scilla hyacinthina* (Roth.) Macbr.	Hyacinthaceae	Bulbs	Bulbs are boiled with water and tied over the boils and wounds till cure.
162.	*Scoparia dulcis* L.	Scropulariaceae	Whole plant	Whole plant is powdered and administered orally for 6-7 days with water to reduce blood pressure. It also used for the treatment of kidney stone.
163.	*Sebastiana chamaelea* (L.) Muell.–Arg.	Euphorbiaceae	Whole plant	The juice of the plant is astringent and it is used for diarrhoea.
164.	*Selaginella delicatula* (Desv.) Alston	Selaginellaceae	Whole plant	Plant juice is antibacterial and it is used for the healing of wounds.
165.	*Selaginella involvens* (Sw.) Spring.	Selaginellaceae	Roots and leaves	The root decoction is used to cure cough.The fronds are ground with fresh rhizome of *Curcuma longa* L. (Manjal) and it is applied for poisonous bites.
166.	*Selaginella intermedia* (Bl.) Spring.	Selaginellaceae	Whole plant	The whole plant paste applied over the forehead to get relief from headache.
167.	*Sesamum laciniatum* Klein. *ex* Willd.	Pedaliaceae	Leaves	Leaf juice applied on wounds.
168.	*Sesamum orientale* L.	Pedaliaceae	Seeds	Seed paste is applied on skin for sun burns in every morning and evening until recovery.
169.	*Sida cordata* (Burm.f.) Borss.	Malvaceae	Roots	Root extract mixed with 2-3 gms of powdered ginger and is given thrice in a day as aphrodisiac.
170.	*Smithia racemosa* Heyne *ex* Wight and Arn.	Fabaceae	Whole plant	Whole plant powder is mixed with honey and it is taken internally for body strength.

Contd...

Table 22.1–*Contd...*

Sl No.	Botanical Name	Family	Part(s) Used	Mode of Drug Oreparation
171.	*Sonchus oleraceus* L.	Asteraceae	Roots	The juice obtained from the roots is used to check bleeding from mouth and nose.
172.	*Spermacoce hispida* L.	Rubiaceae	Root	Root paste is applied on the affected part for the treatment of sprains.Root paste along with rhizome paste of *Curcuma longa*, egg white and jiggery, it is applied for bone fracture for 9 days.The juice prepared from 10 gms of root with cow's milk is taken regularly twice a day to induce lactation.
173.	*Spilanthes calva* DC	Asteraceae	Flowers	A crushed flower mixed with the urine of opposite sex of a patient and is used against dog bites and snake bites. Flower heads are chewed in case of tooth ache.
174.	*Synedrella nodiflora* (L.) Gaertn.	Asteraceae	Roots	Juice from the crushed roots taken for swelling of abdomen, especially in cattle.
175.	*Tectaria wightii* (C.B. Clarke) Ching.	Dryopteridaceae	Roots	The decoction of root is mixed with pepper and cumin seeds. Then it is boiled with water. After this boiled solution is taken orally twice in a day to cure asthma.
176.	*Tephrosia pumila* (Lam.) Pers.	Fabaceae	Whole plant	Whole plant powder is used to cure fever.
177.	*Themeda triandra* Forssk.	Poaceae	Whole plant	Powder of whole plant parts is ground with the leaves of *Toddalia asiatica* and *Pongamia pinnata*. The mixture thus obtained is mixed with coconut oil and applied topically on affected places to treat wounds.
178.	*Torena bicolor* Dalz.	Scropulariaceae	Whole plant	Plants boiled with gingili oil and rubbed on forehead 3 to 4 times. After that, hot water bath taken to control migraine.
179.	*Trianthema deccandra* L.	Aizoaceae	Leaves and roots	Dried leaves and roots are powdered and it is mixed with hot water. It is taken orally for 5-10 days for the treatment of kidney troubles.
180.	*Trianthema portulacastrum* L.	Aizoaceae	Stem and roots	Stem and roots are crushed; this extract is used for the treatment of rheumatism. Root decoction is taken internally to treat constipation and asthma.
181.	*Tribulus terrestris* L.	Zygophyllaceae	Fruits and leaves	Whole plant decoction is used to cure urinary disorders.Fruits are cooling agent diuretic. It is used for impotency and urinary disorders. The paste prepared from leaves used for the treatment of stones in bladder.

Contd...

Table 22.1–*Contd...*

Sl No.	*Botanical Name*	*Family*	*Part(s) Used*	*Mode of Drug Oreparation*
182.	*Tridax procumbens* L.	Asteraceae	Leaves	Fresh leaf juice is applied for cuts and wounds.
183.	*Urginea indica* (Roxb.) Kunth.	Hyacinthaceae	Bulbs	Bulbs (2 nos.) are made into a paste and applied externally for treating various skin allergies, daily twice in morning and evening for five days or until symptoms disappear.The paste obtained from the bulbs is used for piles.
184.	*Vernonia cinerea* (L.) Less.	Asteraceae	Whole plant	Plant crushed with lime and it is applied on wounds.Flowers are grind with water and applied over the head for hair growth.The paste of this plant with *E. sonchifolia* applied on throat to clear tonsils.5-10 mL of leaf extract is taken orally once in a day regularly to cure dysentery in children.
185.	*Viola pilosa* Blume	Violaceae	Whole plant	Whole plant is crushed and applied on scabies.The leaf paste is plastered over the fractured bones.
186.	*Xenostegia tridentata* (L.) Austin and Staples	Convolvulaceae	Roots	Root decoction taken internally to get rid of rheumatism and urinary disorders.

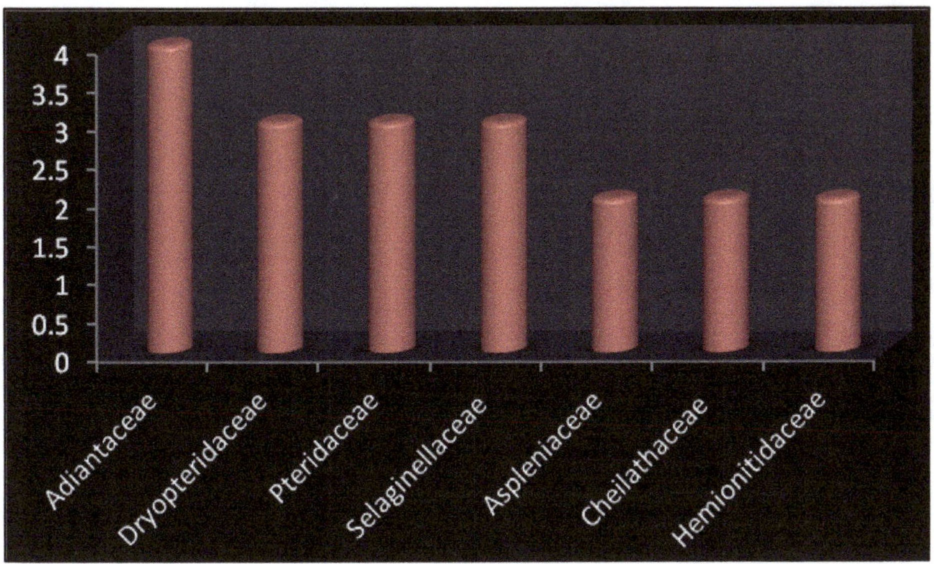

Figure 22.4: Analysis of Dominant Angiosperm Families.

The various plant parts used for the preparation different medico-formulations such as Leaves (67 Nos.), Whole plant (60 Nos.), Root (19 Nos.), Shoot (14 Nos.), Rhizome (7 Nos.), Flower (5 Nos.), Tuber (4 Nos.), Fruit (3 Nos.), Latex (3 Nos.), Bulb (2 Nos.) and Seeds (2 Nos.) respectively (Figure 22.5). These plant parts are either used singly or with the combination of other for the treatments of various ailments like fever, head ache, cuts and wounds, skin problems, inflammatory swellings, stomach problems, diabetes, menstrual problems, rheumatism, jaundice, diarrhea

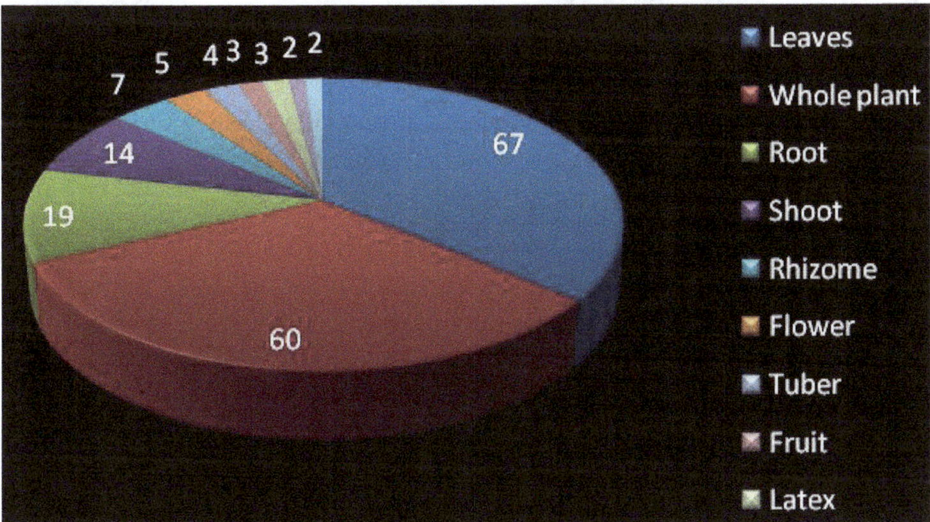

Figure 22.5: Analysis of Plant Parts Used for the Preparation different Medicinal Formulations.

and dysentery, hair tonic, asthma, urinary problems, kidney stone, poisonous bites etc.

The present investigation was the result of traditional or indigenous knowledge of tribe *Malasars* of Velliangiri and *Irulars* of Maruthamalai hills and local and native people of Madukkarai and Kanuvai hills of Coimbatore district. This result also reveals that, both tribal as well as local inhabitants are having strong believes in the efficacy of herbal medicine for the treatments of various ailments which occur in their day to day life. More over the present study also highlights the medico-potentiality of chasmophytes which are distributed in the rocky cliffs and crevices of the study area.

4.0 CONCLUSION

The present study revealed that herbal medicines are still in common use by both tribal and native communities of the study area. Thus the study ascertains the value of a great number of chasmophytic plants which are used for the preparation various medicinal formulations in the health care system of both tribal and local inhabitants of the study area. Ethno-medico-botany plays a great role in exploiting the medicinal chemicals from the medicinal plants used by tribal or rural people of indigenous communities. Such ethno-botanical information serve as a base for new compounds with active principles for phytochemical, pharmacogonostical, pharmacological and clinical research.

The results of the study provides the evidence that, the medicinal plants continue to play an important role in the health care system of both tribal and local inhabitants of the study area. One of the major threat to the medicinal plant resource of the study area is over exploitation and lack of suitable conservation management of the existing ecosystem. More over, the younger generation is ignorant about the vast medicinal resources available in their surroundings and is more inclined towards market resources. The conservation of such valuable medicinal plants for future generation is an urgently needed.

REFERENCES

Binu Thomas, Rajendran, A., Aravindhan, V., 2013. Wound healing Chasmophytes of Velliangiri hills, Southern Western Ghats of Coimbatore district, Tamil Nadu, India. In: Parimelazhagan, T. (Ed.), *Turning plants into medicines-Novel approaches*. New India. Publishers, New Delhi, India. pp. 141 – 148 (ISBN: 978-93-81450-46-8).

Binu Thomas, Rajendran, A., Chandrashekara, U.M., Sivalingam, R., 2012. Ethnomedicinal plant knowledge of Tribe *Muthuvas* of Mannavan Shola Forest of Southern Western Ghats, Kerala, India. *Bot. Report* 1 (1):5-9.

Gamble J.S., Fischer, C.E.C., 1915-1936. *The Flora of Presidency of Madras.* Part 1- 11 (Part 1-7 by Gamble and 8- 11 by Fischer) Adlard and Sons Ltd., London. (Repr. ed. Vols. 1-3, 1957).

Ignacimuthu, S., Sankaranarayanan, K., Kesavan, L., 1998. Medico-ethnobotanical survey among *Kanikar* tribals of Mundanthurai Sanctuary. *Fitoterop*. 69: 409-414.

Matthew, K.M., 1983. *The Flora of Tamil Nadu Carnatic.* Vols. 3 (1-3). Rapinet Herbarium, St. Joseph's College, Tiruchirapalli.

Pullaiah, T., 2006. *Encyclopedia of World Medicinal Plants* vol. 1-5. Regency Publications New Delhi, India.

Rajith, N.P., Navas, M., MuhamadThaha, A., Manju, M.J., Anish, N., Rajasekharan, S., George, V.A., 2010. Study on traditional mothercare plants of rural communities of South Kerala. *Ind. J. Trad. Knowl.* 9: 203-208.

Shiva, M.P., 1996. *Inventory of forestry resourses for sustainable management and biodiversity conservation.* Indus publishing company, New Delhi.

Silja, V.P., Varma, K.S., Mohanan, K.V., 2008. Ethnomedicinal plant knowledge of the *Mullu Kuruma* tribe of Wayanad district, Kerala. *Ind. J. Trad. Knowl.* 7(4): 604-612.

Singh, G.S., 2002. Minor forest products of Sariska National Park: An ethnobotanical profile. In: Trivedi, P.C. (Ed.) *Ethnobotany.* Aavishkar Publications and Distributors, Jaipur. pp. 289-310.

Warming, C., 1895. *Ecology of plants: An introduction to the study of plant-communities.* Clarendon Press, Oxford.

2015, Modern Methods in Phytomedicine *Pages* ***345–357***
Editor: **T. Parimelazhagan**
Published by: **DAYA PUBLISHING HOUSE, NEW DELHI**

23

Nutraceutical Properties of *Measa indica* (Roxb.) A.DC.

*Shanmugam Saravanan, Baby John Prakash, Muniyandi Kasipandi and Thangaraj Parimelazhagan**

Bioprospecting Laboratory, Department of Botany, Bharathiar University, Coimbatore – 641 046, Tamil Nadu

1.0 INTRODUCTION

Now a day's fruits and vegetables are very essential for improving health care products, growing population epidemiological investigations reported that consumption of fruits and vegetables are reduced the adverse effect of chronic diseases like cancer, cardiovascular disease and diabetes (WHO 2003). In human nutrition and health system the traditional herbal medicines plays a vital role because of their nutraceutical properties and bioactive principles (Tulipani *et al.*, 2008). Minerals and other trace elements are significantly play a vital role in the biochemical activities and associated with maintenance of intracellular oxidative balance (Shoham and Youdim, 2000). Mezquitaand Vigoa, (2000) reported that consumption of fruits and vegetables satisfies a recommended amount of dietary supplements, natural antioxidant and also nutraceutical balances. The association between a diet rich in fruits and vegetables anda decrease in the risk of cardiovascular diseases and certain typesof cancer is based on epidemiological evidence and, by hypothesis, on their

* *Corresponding Author.* E-mail: drparimel@gmail.com

antioxidant contents (Alonso *et al.,* 2004). The action of these antioxidant compounds is related to the attenuation of oxidative events that could contribute to the pathophysiology of these diseases (Pietta, 2000), and some vitamins, phenolic compounds and carotenoids stand out among them.

Maesa indica commonly known as wild berry, belongs to the Family *Myrsinaceae*. The plant is confined to forested areas, sporadic in distribution and comes under the endangered plant List. Whole plant of *M. indica* excluding roots has spermicidal property (Pokharkar *et al.,* 2010). *M. indica* have been reported to exhibit direct of virucidal activity against Newcastle Disease Virus, Vaccinia Virus and Herpes Simplex Virus (Jassim and Naji, 2003). North Indian peoples used the leaves of *M. indica* as a curries and also used in treatment of blood purification and anthelminthic ailment. Locally the plant is called as "Vavding", the quercetin is the major phytochemical constituent of this plant (Gaitonde and Naik, 1989).Even though, the plant leaves has been reported to have several medicinal properties but still remain the nutritional, phytochemical properties are unexplored. Hence, the present investigation is aimed to evaluate the nutritional analysis and *in vitro* antioxidant properties of *M. indica*

2.0 MATERIALS AND METHODS

2.1 Collection of Plant Materials

The fresh leaves of *M. indica* were collected from Kotagiri hills during the month of November. The collected plant material was identified and their authenticity was confirmed by comparing the voucher specimen at the herbarium of Botanical survey of India, Southern circle, Coimbatore, Tamil Nadu. Freshly collected plant material was cleaned to remove adhering dust and then dried under shade. The dried sample were powdered and used for further studies.

2.2 Chemicals

All the nutritional and antioxidant chemicals and standards were obtained from Himedia (Mumbai, India) and Sigma–Aldrich (St. Louis, MO, USA). Petroleum ether, chloroform, acetone, methanol and all the culture media were purchase from Himedia. All other reagents used were of analytical grade.

2.3 Successive Solvent Extraction

The powdered plant material was successively extracted by Soxhlet extractor with petroleum ether, chloroform, acetone and methanol. Finally, the material was macerated using hot water with occasional stirring for 24 hr and the water extract was filtered. The extracts were concentrated by rotary vacuum evaporator (Yamato BO410, Japan) and the percentage yield was expressed in terms of dry weight of plant powder material. The extracts thus obtained were used directly for the estimation of total phenolics and also for the assessment of antioxidant potential through various biochemical assays.

2.4 Nutritional Analysis

2.4.1 Proximate Composition

The content of moisture and ash fiber was determined according to the methods defined in Association of Official Analytical Chemists (AOAC, 1995). The protein was estimated by Lowry *et al.* (1951) method. The carbohydrate and starch content was estimated by Sadasivam and Manickam (1992) using Glucose as a standard. The samples were analyzed and the results expressed on dry weight basis.

2.4.2 Estimation of Amino Acids

Amino acids content of leaves were determined with the procedure of Ishida *et al.* (1981). Extracted samples were filtered through a 0.45 µm membrane filter and 20 µL of the filtrate was injected in to a HPLC (model LC 10 AS, Shimadzu, Mount holly, New Jersey) equipped with a cation exchange column packed with a strongly acidic cation exchange resin, *i.e.*, styrene divinyl benzene co-polymer with sulphonic group. The amino acid analysis was with the non-switching flow method and fluorescence detection after post-column derivatization with o-phthaldehyde. Amino acid standards were used to calculate amino acid concentrations in samples.

2.4.3 Mineral Quantification

Amount of total nitrogen (N) content was estimated through micro Kjeldahl method; phosphorus (P) by treating the samples with ammonium molybdate and freshly prepared ascorbic acid and analyzed by spectrophotometer (Hitachi U-2001 Japan); Potassium (K), Sodium (Na), and Calcium (Ca) were determined by Flame Photometer by the method of Allen [18]. The microelements (Fe, CO, Cu, Mg, Mn and Zn) were determined through Atomic Absorption Spectrophotometer (Tee *et al.*, 1996).

2.5 *In vitro* Antioxidant Studies

2.5.1 Quantification of Total Phenolics, Tannins and Flavonoid Content

The total phenolic content of different solvent extracts of *M. Indica* leaves was determined by Folin ciocalteu method. The same extract was used to estimate the tannins, after treatment with polyvinyl polypyrrolidone (PVPP). The amount of total phenolics and tannins were calculated as the Gallic acid equivalents (GAE) (Siddhuraju and Becker, 2003). The total flavonoid content of leaves sample was assessed by the method described by Zhishen *et al.* (1999). The estimation was performed in triplicate analysis and the results were expressed as rutin equivalent (RE).

2.5.2 FRAP Assay

The antioxidant capacities of leaves samples were estimated according to the procedure described by Pulido *et al.* (2000). Freshly prepared FRAP reagent was mixed with distilled water and 50µL of test sample or methanol (for the reagent blank). The test samples and reagent blank were incubated at 37°C for 30 min in a water bath. The absorbance readings were taken at 593 nm for the test samples and the results were calculated in ascorbic acid equivalents.

2.5.3 Metal Chelating Activity

The ferrous ion chelation of leaf extracts was estimated by the method of Dinnis *et al.* (1994). Briefly, 50 µL of $FeCl_2$ was added to 1mL of different concentration of the extract. The reaction was initiated by the addition of 0.2 mL of ferrozine solution. The mixture was vortex and left for 10 min at room temperature. The absorbance of the solution was measured at 562 nm against deionized water which was used as blank. All the reagents without addition of sample extract were used as negative control. BHT was taken as standard and the results of metal chelating activity of leaf extract was expressed as EDTA equivalence.

2.5.4 Phosphomolybdenum Assay

The antioxidant activity of samples was evaluated by the phosphomolybdenum method (Prieto *et al.*, 1999). An aliquot of 50 µL of sample solution (1 mM in dimethyl sulphoxide) was combined in a 4 mL vial with 1 mL of reagent solution (0.6 M sulphuric acid, 28 mM sodium phosphate and 4 mM ammonium molybdate). The reaction mixer was incubated in a water bath at 95 °C for 90 min. After the samples had cooled to room temperature, the absorbance of the mixture was measured at 765 nm against a blank. The results were expressed as grams of ascorbic acid equivalents per gram extract (AAE).

2.5.5 ABTS Cation Radical Scavenging Activity

The total antioxidant activity of the plant extract was measured by ABTS$^{\bullet+}$ radical cation decolorization assay according to the method of Re *et al.* (1999). ABTS radical was generated by reacting 7 mM ABTS aqueous solution with 2.4 mM potassium persulfate in the dark for 12-16 h at room temperature. Proceeding to assay, this ABTS was diluted with ethanol (about 1: 89 v/v) and equilibrated at 30°C to give an absorbance at 734 nm of 0.70 ± 0.02. The concentration of extracts that produced between 20-80 per cent inhibitions of the blank absorbance was determined and adapted. After the addition of 1 mL of diluted ABTS$^{\bullet+}$ solution to 20 µL of seed or Trolox standards (final concentration 0-15 µM) in ethanol, absorbance was measured at 30 °C exactly 30 min after the initial mixing. The unit of total antioxidant activity (TAA) is defined as the concentration of Trolox having equivalent antioxidant activity expressed as µM/g sample extract as dry matter.

2.5.6 DPPH Radical Scavenging Activity

The antioxidant capacity of leaf extracts was examined in terms of contribution of hydrogen atoms or free radical scavenging ability, using the stable radical 2,2-diphenyle-2-picrylhydrazyl(DPPH), according to the method of Blois, (1958). IC_{50} values of the extract *i.e.*, concentration of extract necessary to decrease the initial concentration of DPPH by 50 per cent was calculated.

2.5.7 Superoxide Radical Scavenging Activity

The assay was based on the capacity of the plant extract to inhibit formazan formation by scavenging superoxide radicals generated in riboflavin–light–NBT system, Beauchamp and Fridovich (1971). The inhibition percentage of superoxide radical generation was calculated as following formula.

Per cent inhibition = [(Control OD – Sample OD)/Control OD] X 100

2.6 Statistical Analysis

The results were statistically analyzed and expressed as mean (n=3) ± standard deviation. Values are analyzed by Duncan's multiple test range (One Way ANOVA by statistical software SPSS 20 version).

3.0 RESULTS AND DISCUSSION

3.1 Nutritional Evaluation

Proximate compositions of *M. indica* leaf were shown in Table 23.1. The moisture content of the leaf was determined by calculating their initial and final weights before and after drying the samples respectively. After hot air oven treatment the moisture content of leaf was found to be 53 per cent. The ash content of the leaf was determined as 0.125 g/2 g. The crude protein was found to be 48.45 per cent in leaf. Carbohydrate content was19.32 per cent in the leaf. According to Pearson *et al.* (1976) plant food that provide more than 12 per cent of its calorific value from protein are considered good source of protein. Furthermore, adults, pregnant and lactating mothers require 34-56 g, 13-19 g and 71 g of protein respectively. In accordance with these, *M. indica* leaf might serve as an efficient source of protein rich food.

Table 23.1: Proximate Composition of *M. indica* Leaf

Parameters	*Leaf*
Moisture	53 per cent
Ash	0.125 g/2 g
Carbohydrates	19.32 (mg/g)
Starch	0.87(mg/g)
Proteins	48.45(mg/g)
Free amino acids	10.10(mg/g)

3.2 Amino Acids

The results of all the amino acid analysis were shown in Table 23.2. The results shows that the leaf exhibit higher amount of cysteine (76.77 mg/g protein) and histidine (1.44 mg/g protein), when compared to other amino acids. Amino acids play a vital role as intermediates in plant and animal metabolism, and join together to form proteins. Proteins provide structural material for the human body and function as enzymes, hormones, and antibodies. Dietary proteins are the major source of amino acids. Most proteins are broken down by enzymes into amino acids and absorbed from the small intestine. Nine amino acids, called essential, must come from the diet, including arginine, histidine, isoleucine, leucine, lysine, methionine, phenylalanine, tryptophan, and valine. The amino acids arginine, methionine, and phenylalanine are considered essential for reasons not directly related to lack of metabolic pathway, but because the rate of their synthesis is insufficient to meet the needs of the body (Spallholz *et al.,* 1999). Histidine is considered an essential amino acid in children. Vegetables contain all essential amino acids, but some may be in lower proportions than are required for humans (Young and Pellett, 1994). The presence of one or more essential amino acids in adequate amounts would increase the nutritive value of the protein. Hence, the leaf as a source of amino acids can usually be assessed by comparison with FAO/WHO (1985) suggested pattern of essential amino acids which will serve as a better pathway for nutritional supplement.

Table 23.2: Amino Acids Quantification of *Maesa indica* Leaf

Lysine	*Arginine*	*Histidine*	*Glutamine*	*Glycine*	*Cystine*	*Serine*	*Alanine*	*Glutamic Acid*
8.87	22.70	1.44	28.27	22.29	76.77	8.87	13.21	12.38
Valine	*Threonine*	*Proline*	*Isoleucine*	*Phenyl Alanine*	*Tyrosine*	*Leucine*	*Aspartic Acid*	*Methio-nine*
11.35	22.29	12.38	13.21	37.56	44.99	10.94	ND	ND

3.3 Mineral Quantification

The mineral contents of *M. indica* leaf are shown in Table 23.3. It was found that leaf are rich in N, P, K, Ca, Mn, Fe, Cu, Zn, Mo and Mg. All the mineral contents varied widely in the leaves of *M. indica.* Calcium has been reported to be effective in building of skeletal structures and muscle functioning while magnesium is important in the ionic balance and enzyme co-factors. It has been shown that *M. indica* leaf also had higher content of Mg (1200 ppm), which was beneficial to human body so that they were thought to be used as food materials useful in health.

Table 23.3: Mineral Quantification of *M. indica*

Parameters (PPM)													
N	*P*	*K*	*NH_3*	*Na*	*Mg*	*Mn*	*Ca*	*Fe*	*Zn*	*Cu*	*Si*	*B*	*Mo*
2500	412	327.08	30.03	100	1200	ND	1482	1200	10.83	ND	14102	ND	300

3.4 *In vitro* Antioxidant Assays

3.4.1 Quantification of Total Phenolics, Tannins and Flavonoids

The amount of total phenolics, tannins and Flavonoids of different extracts of leaf of *M. indica* were analyzed and shown in Table 23.4. The total phenolics were found to be higher in acetone extract of leaf (18.74 g GAE/100 g extract). Among the different extracts, leaf acetone extracts showed better phenolic content when compared to other extract. The tannins were found to be higher in acetone extract (14.64 g GAE/100 g extract) of leaf. Polyphenols (phenolics, tannins and flavonoids), widely distributed in plants, have gained much attention, due to their antimutagenic, antitumor, antioxidant, and free radical scavenging abilities, which potentially have beneficial implications for human health (Govindarajan *et al.,* 2007). The chloroform (59.5 g RE/100 g extract) extract of leaf was found to have higher flavonoid content compare to the other solvents. The acetone extract of leaf (27.42 g RE/100 g extract) possess moderate flavonoid content. Flavonoids are one of the most diverse and widespread group of natural compounds and are probably the most important natural phenolics. These compounds possess a broad spectrum of chemical and biological activities including radical scavenging properties (Saravanan and Parimelazhagan, 2014; Arunachalam *et al.*, 2012). Similarly, *Myrsinaceae* family, members of *Aegiceras corniculatum* (Banerjee *et al.*, 2008) and *Labisia pumila* Benth, (Karimi *et al.*, 2011) fruit contain higher amount of phenolics and flavonoids.

Table 23.4: Quantification of Total Phenolics, Tannins and Flavonoid of *M. Indica* Leaf

Samples	*Total Phenolics (g GAE/100g extract)*	*Tannins (g GAE/100g extract)*	*Flavonoids (g RE/100g extract)*
Petroleum ether	3.70±0.10	1.42±0.10	6.5±2.38
Chloroform	2.87±0.09	0.81±0.04	59.5±1.95[a]
Acetone	18.74±0.08[a]	14.64±0.07[a]	27.42±1.26[b]
Methanol	8.95±0.07[b]	6.87±0.09[b]	25.83±3.51[b]
Hot water	4.45±0.09	1.57±0.10	13.33±0.58[c]

Values are mean of triplicate determination (n=3) standard deviation

GAE: Gallic acid equivalents; RE: Rutin equivalents.

Values followed by superscript indicates statistical significance $p<0.05$

3.4.2 Ferric Reducing Antioxidant Power (FRAP) Assay

The FRAP is a simple assay and may offer putative index of antioxidant activity. The reducing potential of an antioxidant reacting with a ferric 2, 4, 6- tripyridyl-S-triazine (Fe (III)-TPTZ) complex and producing a colored ferrous 2, 4, 6- tripyridyl-S-triazine (Fe (II))-TPTZ) complex by a reductant at low pH, was adopted. The result shows (Table 23.5) that the ferric reducing capacity of acetone leaf extract was much higher (1635.56mM/g) and least in petroleum ether (286.44 mM/g). Hence from the results it can be concluded that acetone extract of *M. indica* leaf show good ferric reducing power. In the same way, *Aegiceras corniculatum* (Banerjee *et al.*, 2008) and *Labisia pumila* Benth, leaf (Karimi *et al.*, 2011) showed maximum reducing ability.

Table 23.5: FRAP, Metal Chelating, Phosphomolybdenum and ABTS$^{•+}$ Assays

Sample Extracts	*FRAP (mMoles Fe (II)/g)*	*Metal Chelating (g EDTA equivalents/ 100 g extract)*	*Phosphomalybdenum (mg AA equivalents/ g extract)*	*ABTS$^{•+}$ (μM TE/ g extract)*
Petroleum ether	286.44±1.02[d]	1.74±0.21	51.6±2.5[d]	3570.73±5.85[c]
Chloroform	536.67±2.00[c]	0.65±0.16	148.3±4.6[c]	4151.23±10.1[b]
Acetone	1635.56±2.78[a]	4.11±0.05[a]	332.0±21.5[a]	6621.71±20.2[a]
Methanol	974.671.33[b]	2.48±0.06[b]	202.6±4.7[b]	5521.47±11.6[a]
Hot water	976.89±1.39[b]	2.01±0.09[b]	144.57±2.55[c]	2797.86±15.4[d]

Values are mean of triplicate determination (n=3) ± standard deviation.

EDTA: Ethylenediamine tetra acetic acid; AAE : Ascorbic acid equivalent; TE: Trolox Equivalents.

Values followed by superscript indicates statistical significance $p<0.05$.

3.4.3 Metal Chelating Activity

Iron is essential for life because it is required for oxygen transport, respiration, and activity of many enzymes. In complex systems, such as food and food preparations, various different mechanisms may contribute to oxidative processes, such as Fenton reaction, where transition metal ions play a vital role. Different reactive oxygen species might be generated and various target structures such as lipids, proteins, and carbohydrates, can be affected. Therefore, it is important to characterize the extracts by a variety of antioxidant assays (Halliwell, 1997). The method of metal chelating activity is based on chelating of Fe^{2+} ions by the reagent ferrozine, which is quantitative formation of a complex with Fe^{2+} ions (Dinnis *et al.,* 1994). The metal chelating activity of different solvent extracts of *M. indica* leaf were shown in Table 23.5. Among the different extracts, acetone and methanol extracts showed better chelating ability in compared to other solvent extracts. The metal chelating capacity of acetone extract of leaf was found to be 4.11g EDTA equivalents/100 g extract whereas, methanol extract was found to be 2.48g EDTA equivalents/100 g extract. These results showed that the extracts could chelate irons and the values are substantial. Chelating agents may serve as secondary antioxidants because they reduce the redox potential thereby stabilizing the oxidized form of the metal ions (Rajesh *et al.,* 2008).

3.4.4 Phosphomolybdenumassay

The phosphomolybdenum method is based on the reduction of Mo (VI) to Mo (V) by the antioxidant compounds and the formation of green phosphate/Mo(V) complex. The total antioxidant capacity of different solvent extracts of leaf of *M. indica* were analyzed and shown in Table 23.5. Among different solvent extracts, leaf extracts showed higher activity in most of its solvents. This assay is successfully used to quantify vitamin E, and being simple and independent of other antioxidant assays commonly employed, it was decided to extend its application to plant extract (Prieto *et al.,* 1999). Hydrogen/electron transfer from antioxidants to DPPH radical and Mo (VI) complex occur in the DPPH radical and phosphomolybdenum assays, respectively (Hyder *et al.,* 2001). Previously species of *Phyllanthus* were evaluated by this method by Kumaran and Karunakaran (2007).

3.4.5 ABTS Radical Scavenging Activity

$ABTS^{\bullet+}$ assay is an excellent tool to determine the antioxidant activity of hydrogen-donating antioxidants. The results of ABTS cation radical scavenging activity of different extracts leaf of *M. indica* was shown in Table 23.5. The acetone and methanol extracts of leaf showed higher radical scavenging activity (6621.71and 5521.47 μmoles TE/g extract respectively) as compared to that of other solvent extracts. However, the leaf acetone extract showed to have more total antioxidant activity in most of its solvent extracts. Hagerman *et al.* (1998) have reported that the high molecular weight phenolics (tannins) have more ability to quench free radicals ($ABTS^{\bullet+}$). As in, *Evolvulus alsinoides, Ipomoea digitata, Ipomoea turpethum* and *Ipomoea batatas* have already been reported to exhibit strong $ABTS^{\bullet+}$ scavenging activity (Huang *et al.,* 2006; Surveswaran *et al.,* 2007).

3.4.6 DPPH Radical Scavenging Activity

The model of stable DPPH free radicals can be used to evaluate the anti-oxidative activity in a relatively short time. DPPH is a commercial oxidizing radical which can be reduced by antioxidants. The disappearance of the DPPH radical can be monitored by decreased optical density at 517 nm (Blois 1958).Leaf extracts of *M. indica* were analyzed and shown in Figure 23.1. The IC_{50} of chloroform extract of leaf was found to be 71.74µg/mL whereas that of petroleum ether extract was 73.58 µg/mL. Two natural antioxidants rutin and quercetin; together with two synthetic antioxidants butylated hydroxyanisole (BHA) and butylated hydroxytoluene (BHT) were used as reference compounds. Likewise, *Labisia pumila* Benth, leaf (Karimi *et al.*, 2011) and *Ardisia compressa* (Ramírez-Mares *et al.*, 2010) showed higher DPPH free radical scavenging activity then standards. These results revealed that all the extracts of *M. indica* contain impressive inhibitors, which may act as primary antioxidants that react with DPPH radicals.

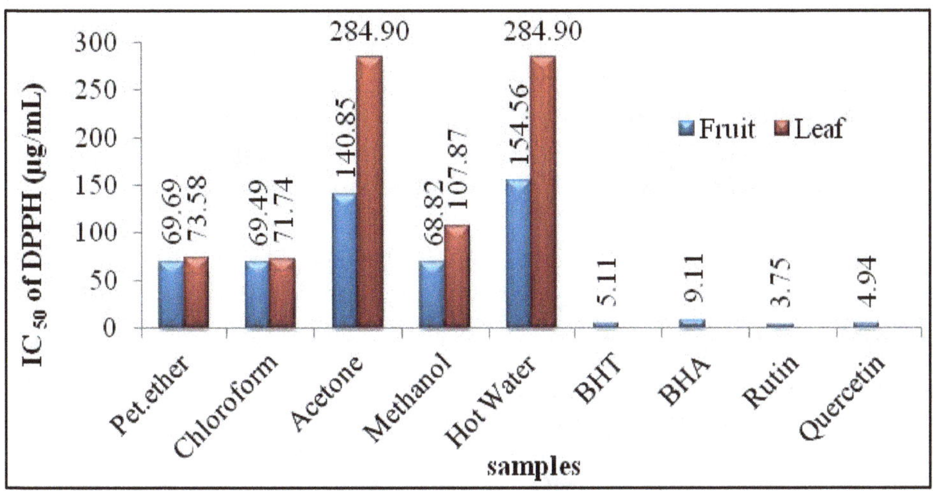

Figure 23.1: DPPH Free Radical Scavenging Activity of *M. indica*. Values are mean of aliquots determination (n=4).

3.4.7 Superoxide Radical Scavenging Activity

The superoxide radical is known to be produced *in vivo* and can result in the formation of H_2O_2 via dismutation reaction. Moreover, the conversion of superoxide and H_2O_2 into more reactive species, *e.g.*, the hydroxyl radical, has been thought to be one of the unfavorable effects caused by superoxide radicals (Halliwell, 1991).The results of superoxide anion scavenging activity of leaf of *M. indica* are shown in Figure 23.2. Free radical scavenging activity was found to be higher in acetone extract of leaf 32.94 per cent. All the extracts of *M. indica* showed higher free radical scavenging activity comparable to that of BHT and BHA. This may be explained by the interaction of different phenolics and flavonoids in these extracts (Zhishen *et al.*, 1999). It was reported that the superoxide anion scavenging activity could be due to the action of a free hydroxyl group of phenolic compounds. Furthermore, flavonoid molecule with

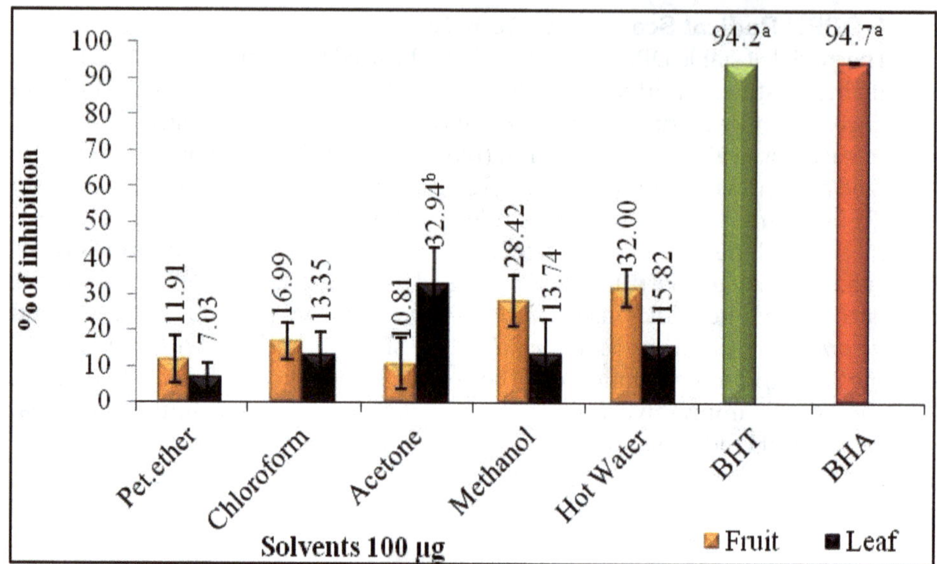

Figure 23.2: Superoxide Radical Scavenging Activity of *M. indica*. Values are mean of triplicate determination (n=3) ± standard deviation, followed by superscript indicates statistical significance p<0.05.

poly hydroxylated substitution on ring A or B and free 3-hydroxyl substitution could possess high superoxide scavenging activity (Siddhuraju *et al.,* 2002).

4.0 CONCLUSION

In the present study, the nutritional composition of *M. indica* leaf exhibited a good source of carbohydrate, protein, and fiber but limited in fat. In the micronutrient composition, the calcium, magnesium, and iron are valuable sources. In addition to the nutritional composition, the leaf also exhibited higher antioxidant potential. From this the plant leaves can serve as an effective source for the development of nutraceutical products.

REFERENCES

Alonso, M.G., de Teresa, S.P., Buelga, C.S., Gonzalo, J.C.R., 2004. Evaluation of the antioxidant properties of fruits. *Food Chem* 84, 13–18.

AOAC. Official Methods of Analysis.16thed. Method 1094.Association of Official Analytical Chemists, Arlington, VA, USA(1995).

Arunachalam, K., Parimelazhagan, T., 2012. Evaluation of nutritional composition and antioxidant properties of underutilized *Ficus talboti* King fruit for nutraceuticals and food supplements. *J Food Sci Technol.*

Banerjee,D., Shrabana, C., Alok, K., Hazra, Shivaji, B., Jharna, R., Biswapati, M., 2008. Antioxidant activity and total phenolics of some mangroves in Sundarbans. *Afric J Biotechnol.* 7 (6), 805-810.

Beauchamp, C., Fridovich, I., 1971. Superoxide dismutase: Improved assays and an assay applicable to acrylamide gels. *Anal Biochem.* 44: 276-277.

Blois, M.S., 1958. Antioxidants determination by the use of a stable free radical. *Nature.* 4617, 1199-1200.

Dinnis, T.C.P., Madeira, V.M.C., Almeida, L.M., 1994. Action of phenolic derivatives (acetoaminophen, salycilate and 5-aminosalycilate) as inhibitors of membrane lipid peroxidation and as peroxyl radical scavengers. *Arch. Biochem. Biophysics.* 315,161–169.

FAO/WHO/UNU, 1985. Energy and protein requirements. WHO Technical Report Series No. 724. World Health Organization,Geneva, Switzerland.

Gaitonde, R.V., Naik, P.L., 1989. Estimation of Quercetin in the Leaves of Maesa-Indica Wall by a New Colorimetric Method. *Current Sci* (*Bangalore*).58, 982-3.

Govindarajan, R., Singh, D.P, Rawat, A.K.S., 2007. High-performance liquid chromatographic method for the quantification of phenolics in'Chyavanprash' a potent Ayurvedic drug. *J. Pharm. Biomed. Anal.* 43, 527-532.

Hagerman, A.E., Reidl, K.M., Jones, G.A., Sovic, K.N., Ritcard, N.T., Hartzfield, P.W. Riechel, T.L., 1998. High molecular weight polyphenolics (tannins) as biological antioxidants. *J. Agric. Food Chem.* 46, 1887–1892.

Halliwell, B., 1991. Reactive Oxygen Species in living system: Source, biochemistry, and role in human disease. *Americ J Med*, 91, 14-22.

Halliwell, B., 1997. Antioxidants; the basics-what they are and how to evaluate them. *Adv. Pharmacol.* 38: 3-20.

Huang, D.J, Chen, H.J., Wen-chi Houc, W., Lin, C.D., Lin, Y.H., 2006. Sweet potato (*Ipomoea batatas* [L.] Lam 'Tainong 57') storage root mucilage with antioxidant activities *in vitro. Food Chem.* 98, 774-781.

Hyder, R.C., Lio, Z.D., Khodr, H.H., 2001. Metal chelation of polyphenols. *Method Enzymol.* 335: 192-203.

Ishida, Y., Fugita, T., Asai, K., 1981. New detection and separation method for aminoacid by high performance liquid chromatography. *J. Chromato publishers* 204: 143-148.

Jassim, S.A., Naji, M.A., 2003. Novel antiviral agents: A medicinal plant perspective. *J ApplMicrobiol.* 95, 412-427.

Karimi, K., Hawa, Z. Jaafar, E., Sahida, A., 2011. Phenolics and flavonoids profiling and antioxidant activity of three varieties of Malaysian indigenous medicinal herb *Labisia pumila* Benth. *J Med Plants Resear.* 5(7), pp. 1200-1206.

Kumaran, A., Karunakaran, J., 2007. Antioxidant activities of methanolic extracts of five *Phyllanthus* species from India. *Lebens-Wiss. Technol.* 40, 344–352.

Lowry, O.H., Rosebrough, N.J., Farr, A.L., Randall, R.J., 1951. Protein measurement with the folin-phenol reagent. *J. Biol. Chem.* 193: 265–275.

Mezquita, P.C., Vigoa, Y.G., 2000. La acerola. Fruta marginada de America con altocontenido en acido ascorbico. *Alimentaria* 1, 113–125.

Pearson, 1976. The Chemical Analysis of Foods. 7th Ed, Churchill Livingstone, P. 493.

Pietta, P.G., 2000. Flavonoids as antioxidants. *J Nat Prod* 63,1035–1042.

Pokharkar, R.D., Saraswat, R.K., Sheetal Kotkar, 2010. Survey of Plants having antifertility activity from western ghat area of Maharastra state. *J Herb med toxicol.* 4(2), 71-75.

Prieto, P., Pineda, M., Aguilar, M., 1999. Spectophotometric quantitative of antioxidant capacity through the formation of a phosphomolybdenum complex: Speci.c application to the determination of vitamin *E. Analyt. Biochem.*269: 337–341.

Pulido, R., Bravo, L., Sauro-Calixto, F., 2000. Antioxidant activity of dietary polyphenols as determined by a modified ferric reducing/antioxidant power assay. *J. Agric. Food Chem.* 48: 3396-3402.

Rajesh, M., Anusuya, N., Siddhuraj, P., Manian, S., 2008. The antioxidant activity and free radical scavenging potential of two different solvent extracts of *Cammelia sinensis F. bengalensis* L., and *F. racemosa* L. *Food Chem*, 107, 1000-1007.

Ramírez-Mares, M.V., Sánchez-Burgos, J. A., Hernández-Carlos, B., 2010. Antioxidant, Antimicrobial and Antitopoisomerase Screening of the Stem Bark Extracts of *Ardisia compressa Pak J Nutrition* 9 (4): 307-313.

Re, R., Pellegrini, N., Proteggente, A., Pannala, A., Yang, M., Rice- Evans, C., 1999. Antioxidant activity applying an improved ABTS radical cation decolorization assay. *Free Radic. Biol. Med.*26: 1231–1237.

Sadasivam, S., Manikam, A., 2008. *Biochemical Methods*. New Age international (P) Limited Publishers, Chennai. Edition 3.

Saravanan, S., Parimelazhagan, T., 2014. In vitro antioxidant, antimicrobial and anti-diabetic properties of polyphenols of *Passiflora ligularis* Juss. fruit pulp. *Food Sci Hum Well,* 3 56-64.

Shoham, S., Youdim, M.B, 2000. Iron involvement in neural damageand microgliosis in models of neurodegenerative diseases. *CellMol Biol* 46:743–760

Siddhuraju, P., Becker, K., 2003. Studies on antioxidant activities of *Mucuna* seed (*Mucuna pruriens* var. *utilis*) extracts and certain non-protein amino/imino acids through *in vitro* models. *J. Sci. Food Agric.*83: 1517-1524.

Siddhuraju, P., Mohan, P.S., Becker, K., 2002. Studies on the antioxidant activity of Indian Laburnum (*Cassia fistula* L.): a preliminary assessment of crude extracts from stem bark, leaves, flowers and fruit pulp. *Food Chem.* 79, 61–67.

Spallholz, J.E., Boylan, L.M., Driskel, J.A., 1999. Nutrition: Chemistry and biology, 2nd ed. Boca Raton, FL: CRC Press, 345pp.

Surveswaran, S., Cai, Y., Corke, H., Sun, M., 2007. Systematic evaluation of natural phenolic antioxidants from 133 Indian medicinal plants. *Food Chem.* 102: 938-953.

Tee, E.S., Rajam, K., Young, S.I., Khor, S.C., Zakiyah, H.O, 1996. Laboratory procedures in nutrient analysis of foods (pp. 1-82). Kuala Lumpur, Malaysia: Division of Human Nutrition, Institute for Medical Research.2338 K.W. Kong *et al.,/Food Res Inter*44 (2011) 2332–2338.

Tulipani, S., Mezzetti, B., Capocasa, F., Bompadre, S., Beekwilder, J., de Vos, C.H.R.,Capanoglu, E., Bovy, A., Battino, M., 2008. Antioxidants, phenolic compounds and nutritional quality of different strawberry genotypes. *J AgricFood Chem* 56, 696–704.

WHO, 2003. Guidelines for the Regulation of Herbal Medicines in the South-East Asia Region: Developed at the Regional Workshop on the Regulation of Herbal Medicines, Bangkok.

Young, V., Pellett, P., 1994. Plant proteins in relation to human protein and amino acid nutrition. *American Journal of Clinical Nutrition,* 59, 1203S-1212S.

Zhishen, J., Mengecheng, T., Jianming, W., 1999. The determination of flavonoid contents on mulberry and their scavenging effects on superoxide radical. *Food Chem.* 64**:** 555-559.

2015, Modern Methods in Phytomedicine *Pages* ***359–367***
Editor: **T. Parimelazhagan**
Published by: **DAYA PUBLISHING HOUSE, NEW DELHI**

24

Lithophytic Diversity of Yercaud Hills of the Eastern Ghats in Tamil Nadu

*M. Parthipan and A. Rajendran**

Floristics Laboratory, Department of Botany, School of Life Sciences, Bharathiar University, Coimbatore – 641 046, Tamil Nadu

1.0 INTRODUCTION

The natural science of faunas and floras on intertidal rocky shores has been a topic of interest for decades in many parts of the world. An interesting phenomenon of ecology of rocky intertidal habitats is the great and interesting difficulty of dealings among similar sorts of plants. Yercaud hills ridges lithophytic are clearly distinct and unique habitats, their vegetation communities differ evidently from those of surrounding areas.

The isolated rock habitats in India are well known as the "inselbergs" and the vegetation on the so called inselbergs have been considered by many botanists and environmentalists. Exposures of rock such as support flowering plant communities found in these situations. Based on the soil factor, the plants were classified in to different ecological groups such as oxylophytes (plants usually found on acid soil), halophytes (plants usually found on saline soil), psammophytes (plants usually found on sand) and lithophytes (Plants usually found on rocks) (Warming, 1895).

* *Corresponding Author.* E-mail: arajendran222@yahoo.com.

Lithophytes are a type of flowering plants that grows in or on rocks. It feed off moss, nutrients in rain water, litter and even their own dead tissue (Tozer, 2005). The term lithophytes come from Greek words, 'Litho'-stone, and 'Phyte' – a plant that lives on stones. There are also many orchids which do not grow on trees at all, but only on specific kinds of rocks (Kolbek *et al.,* 1997). The term "lithophyte" is used to describe any plant that grows attached to a rock; a type of Epiphytic life.

Lithophytic habitat occurs throughout the hills, and is particularly characteristic of high altitudes, but is also found at low altitudes notably in Yercaud hills of Eastern Ghats. Lithophytes belong to growth forms that are capable over the rock surface and it absorbing water derived from rain, dew, melting snow or water running down the rocks (Alves and Kolbek, 1993). Lithophytes require haptera by which they can attach themselves to rock, unless the flowering plant body itself adheres closely to these rocks. Many rock habitats, especially cliff faces, rock ledges, gorges and boulder fields are inaccessible to grazing animals and are unmanaged.The lithophytic vegetation consists of plant communities that colonise on the rock faces. The type of plant community develops on rock is largely determined by the base status of the rock face.

The group of vegetation on rock walls and rocky slopes represent specific habitat with extreme ecological conditions such as extreme drought, fluctuations of temperature and nutrient deficiency. The lithophytic vegetation in rock surface shows xerophytic characters such as morphological and anatomical modifications which make the flowering plants to survive under adverse conditions imposed by climate, soil and the biotic influence. Some of the notable anatomical changes show by lithophytic communities such as the presence of glandular hairs, water storing cells secretary organs, thick cuticle, presence of thickened cells *etc* (Schimper, 1898).

Now recently lithophytic plants classified in 5 types such as Cryptoendolithic, Chasmoendolithic, Euendolithic, Epilithic and Hypolithic. The Cryptoendolithic are plants habitat within the interior of the rock, Chasmoendolithic plants habitat in fissures and cracks within rock, Euendolithic plants habitat formed by active boring/ penetration by microorganisms, Epilithic plants habitat on surface of rocks and Hypolithic plants habitat on underside of rock.

In Yercaud, several plant species have been found growing in soilless environments and attach with rock. Although rock habitats environments are certainly unfavorable to plant life, the number of organisms living in these habitats is surprising, thanks both for their diversity and to the presence of unexpected systematic groups. Cliffs may host algae, lichens, bryophytes, ferns and various angiosperms. It is surprising to see how plants manage not only to grow but flowering plants also to produce often brightly coloued, vigorously blooming flowers. From the phytogeographic viewpoint, these environments are sometimes more important than other high-altitude habitats.

2.0 MATERIALS AND METHODS

The present investigation was undertaken with a view of lithophytic diversity of The Yercaud hills in Eastern Ghats of Tamil Nadu. Yercaud is located in a part of

Sanyasimalai Reserve Forest of Shevaroy Hills, North-East of Salem, Tamil Nadu. Yercaud got its name from the Tamil words, yeri and kaadu, meaning lake and forest respectively. Yercaud Town is situated above 4800' from mean sea level. Surveying and collecting were conducted in the study sites at bimonthly intervals from 2011to 2013. Most plant taxa were collected and made in the form of herbarium specimens (including both dry and spirit specimens).Were only observed and noted. Ecological data, habitats, habit and some diagnostic characters of each specimen, such as color, smell, etc.The plants collected the study areas were given in field numbers and identified with the help of the Gamble's "*Flora of the Presidency of Madras*" (1915-1935), Matthew's "*Flora of the Tamil Nadu Carnatic*" (1983) and "*Flora of Tamil Nadu* by Henry *et al.* (1981-89). And also comparing authentic sheets available in the Madras herbarium (MH) of Botanical Survey of India, Southern circle Coimbatore, Tamil Nadu. Each specimen was carefully examined in fresh condition. The voucher specimens are deposited in the Department of Botany, Bharathiar University Herbarium (BUH), Coimbatore, Tamil Nadu, India for future reference and consultations.

3.0 RESULTS AND DISCUSSION

In the present study, an attempt has been made to enumerate the lithophytic plants which are collected from rocky habitats of Yercaud Hills of Salem, Tamil Nadu. A total of 152 lithophytes belonging to 61 families distributed in 119 genera were collected from the study area. The life forms of lithophytic flowering plants were analyzed and found that, there are about tree-35, shrub- 31, herb- 48, climber-16, erect- 15 and grass- 7 (Table 24.1).

Common distribution plants 107-species of colected and 45-species uncommon or rare plants also collected. The pattern of distribution of each species and genus is also given. The dominant families are represented in the form of chart. The total life forms of lithophytic flowering plants are depicted in the form of Pie Chart.

An attempt was also made to find out the dominant families and found that the families such as Fabaceae is the first dominant family with 10-species followed by Euphorbiaceae, Poaceae, Caselpinaceae and Asteraceae (with 7-species each) along with Asclepiadaceae (with 6-species). Solanaceae, Acanthaceae, Lamiaceae and Moraceae are the third dominant families with 5-species and other dominant families such as Mimosaceae, Myrtaceae, Convolvulaceae, Verbinaceae, Agavaceae, and Commelinaceae with 4-species each. Oxalidaceae, Meliaceae, Bignonaceae, and Liliaceae 3-species each.

Out of 152- species, a total of 41- species are Chasmoendolithic, 7-species are Cryptoendolithic, 32- species are Euendolithic, 33- species are Epilithic, 39- species are Hypolithic (Table 24.2). Lithophytic plants are morphologicaly find out to abnormal growth characters and documented. Totally 3-species leaf size variation, 35- species stem variation, 4- species tuber variation, 130-species root variation and 22- species whole plant size variation also recorded.

Lithophytic flowering plants criticaly discussed and separate various uses based. Totaly 55- species of wild ornamental, 10- species of edible leaves, 4-speies of edible stem, 24-species of edible fruit, 2 -species of edible tubers and 39-species economicaly cultivated Amoung 152-species variousouly usesd.

Table 24.1: List of Diversity of Lithophyic Yercaud Hills in The Eastern Ghats of Tamil Nadu

Sl.No.	*Botanical Name*	*Family*	*Habit*	*Distribution*
1.	*Michelia champaca* L.	Magnoliaceae	Tree	Common
2.	*Annona reticulatea* L.	Annonaceae	Shrub	Common
3.	*Annona squamosa* L.	Annonaceae	Shrub	Common
4.	*Argemone mexicana* L.	Papaveraceae	Herb	Uncommon
5.	*Brassica juncea* (L.) Czern. and. Coss.	Brassicaceae	Herb	Common
6.	*Cleome viscosa* L.	Cleomeceae	Herb	Common
7.	*Cleome gynandra* L.	Cleomeceae	Herb	Common
8.	*Polycarpaea corymosa* (L.)Lam.	Caryophyllaceae	Herb	Uncommon
9.	*Talinum portulacifolium* (Forssk.) Asch. and Schw.	Portulacaceae	Herb	Uncommon
10.	*Sida acuta* Burm. f.	Malvaceae	Erect	Common
11.	*Sida cordifolia* L.	Malvaceae	Erect	Common
12.	*Waltheria indica* L.	Sterculiaceae	Erect	Common
13.	*Muntingia calabura* L.	Elaeocarppaceae	Tree	Common
14.	*Tribulus terresteis* L.,	Zygophyllaceae	Herb	Common
15.	*Oxalis corniculata* L.	Oxalidaceae	Herb	Common
16.	*Oxalis latifolia* H.B.K.	Oxalidaceae	Herb	Uncommon
17.	*Biophytum sensitivum* (L.) DC.	Oxalidaceae	Herb	Common
18.	*Toddalia asiatica* (L.) Lam.	Rutaceae	Shrub	Common
19.	*Limonia acidissima* L.	Rutaceae	Tree	Common
20.	*Melia azadarach* L.	Meliaceae	Tree	Common
21.	*Cipadessa baccifera* (Roth.) Miq.	Meliaceae	Shrub	Common
22.	*Azadirachta indica* Juss.	Meliaceae	Tree	Common
23.	*Ziziphus mauritiana* Lam.	Rhamnaceae	Shrub	Common
24.	*Zizyphus oenoplia* (L.) Mill.	Rhamnaceae	Shrub	Uncommon
25.	*Cissus quadrangularis* L.	Vitaceae	Shrub	Common
26.	*Cardiospermum halicarabum* L.	Sapindaceae	Climber	Common
27.	*Dodonaea viscosa* (L.) Jacq.	Sapindaceae	Shrub	Common
28.	*Mangifera indica* L.	Anacardiaceae	Tree	Common
29.	*Moringa concanensis* Nimmo *ex* Gibs.	Moringaceae	Tree	Common
30.	*Alysicarpus monilifer* (L.) DC.	Fabaceae	Herb	Uncommon
31.	*Desmodium laxiflorum* DC.	Fabaceae	Shrub	Common

Contd...

Table 24.1–*Contd...*

Sl.No.	*Botanical Name*	*Family*	*Habit*	*Distribution*
32.	*Desmodium triflorum* (L.) DC.	Fabaceae	Herb	Common
33.	*Abrus precatorius* L.	Fabaceae	Shrub	Common
34.	*Mucuna pruriens* (L.) DC.	Fabaceae	Climber	Uncommon
35.	*Pongamia pinnata* (L.) Pierre	Fabaceae	Tree	Common
36.	*Clitoria ternatea* L.	Fabaceae	Climber	Common
37.	*Indigofera linifolea* (L. f.) Retz.	Fabaceae	Herb	Uncommon
38.	*Indigofera nummularifolia* (L.) Livera	Fabaceae	Herb	Common
39.	*Tephrosia purpurea* (L.) Pers.	Fabaceae	Shrub	Common
40.	*Cassia fistula* L.	Caesalpinaceae	Tree	Common
41.	*Cassia auriculata* L.	Caesalpinaceae	Shrub	Common
42.	*Cassia siamea* Lam.	Caesalpinaceae	Tree	Common
43.	*Cassia occidentalis* L.	Caesalpinaceae	Shrub	Common
44.	*Caesalpinia pulcherrima* (L.) Sw.	Caesalpinaceae	Shrub	Common
45.	*Delonix regia* (Boj. *ex* Hook.) Raf.	Caesalpinaceae	Tree	Common
46.	*Tamarindus indica* L.	Caesalpinaceae	Tree	Common
47.	*Mimosa pudica* L.	Mimosaceae	Herb	Common
48.	*Acacia nilotica* (L.) Willd. *ex.* Del.	Mimosaceae	Tree	Common
49.	*Albizia lebbeck* (L.) Willd.	Mimosaceae	Tree	Common
50.	*Pithacellobium dulce* (Roxb.) Benth.	Mimosaceae	Tree	Common
51.	*Terminalia catappa* L.	Combretaceae	Tree	Common
52.	*Callistemon citrinus* (Curtis) Stapf.	Myrtaceae	Tree	Uncommon
53.	*Eucalyptus tereticornis* Sm.	Myrtaceae	Tree	Common
54.	*Psidium guajava* L.	Myrtaceae	Tree	Common
55.	*Syzygium cumini* (L.) Skeels	Myrtaceae	Tree	Common
56.	*Lawsonia inermis* L.	Lythraceae	Shrub	Common
57.	*Passiflora foetida* L.	Passifloraceae	Climber	Common
58.	*Passiflora subpeltata* Orteg.	Passifloraceae	Climber	Uncommon
59.	*Coccina grandis* (L.) Voigt	Cucurbitaceae	Climber	Common
60.	*Opuntia dillenii* (Ker. Gawl.) Haw.	Cactaceae	Erect	Common
61.	*Cereus pterogonus* Lem.	Cactaceae	Shrub	Uncommon
62.	*Mollugo nudicaulis* Lam.	Aizoaceae	Herb	Common
63.	*Alangium salvifolium* (L.f.) Wanger.	Alangiaceae	Tree	Common
64.	*Canthium parviflorum* Lam.	Rubiaceae	Tree	Uncommon
65.	*Richardia scabra* L.	Rubiaceae	Herb	Uncommon
66.	*Emilia sonchifolia* (L.) DC.	Asteraceae	Erect	Uncommon
67.	*Ageratum conyzoides* L.	Asteraceae	Herb	Uncommon

Contd...

Table 24.1–*Contd...*

Sl.No.	*Botanical Name*	*Family*	*Habit*	*Distribution*
68.	*Bidens biternata* (Lour). Merr. and Sherff.	Asteraceae	Herb	Common
69.	*Chromolaena odorata* (L.) King and Robinson	Asteraceae	Erect	Common
70.	*Tithonia diversifolia* (Hemsley) A. Gray.	Asteraceae	Shrub	Common
71.	*Tridax procumbens* L.	Asteraceae	Herb	Common
72.	*Vernonia cinerea* (L.) Less.	Asteraceae	Herb	Common
73.	*Plumbago zeylanica* L.	Plumbaginaceae	Erect	Uncommon
74.	*Jasminum angustifolium* (L.) Willd.	Oleaceae	Climber	Common
75.	*Catharanthus roseus* (L.) G. Don	Apocyanaceae	Herb	Common
76.	*Caralluma adscendens* (Roxb.) Haw.	Asclepiadaceae	Herb	Uncommon
77.	*Caralluma diffusa* (Wight) N.E. Br.	Asclepiadaceae	Herb	Uncommon
78.	*Hemidesmus indicus* (L.) R. Br.	Asclepiadaceae	Climber	Common
79.	*Calotropis gigantea* (L.) R. Br. Dryand.	Asclepiadaceae	Shrub	Common
80.	*Calotropis procera* (Ait.) R. Br.	Asclepiadaceae	Shrub	Uncommon
81.	*Sarcostemma brunoianum* Wight and Arn.	Asclepiadaceae	Climber	Uncommon
82.	*Trichodesma indicum* (L.) R. Br.	Boranginaceae	Herb	Common
83.	*Evolvulus alsinoides* (L.) L.	Convolvulaceae	Herb	Common
84.	*Ipomoea indica* (Burm.f.) Merr.	Convolvulaceae	Climber	Common
85.	*Ipomoea obscura* (L.) Ker. -Gawl.	Convolvulaceae	Climber	Common
86.	*Ipomaea hederifolia* L.	Convolvulaceae	Climber	Uncommon
87.	*Datura metel* L.	Solanaceae	Erect	Common
88.	*Solanum surattense* Burm.	Solanaceae	Herb	Uncommon
89.	*Physalis minima* L.	Solanaceae	Herb	Common
90.	*Physalis peruviana* L.	Solanaceae	Herb	Uncommon
91.	*Capsicum frutescens* L.	Solanaceae	Herb	Uncommon
92.	*Jacaranda mimosifolia* D. Don	Bignonaceae	Tree	Uncommon
93.	*Millingtonia hortensis* L.	Bignonaceae	Tree	Common
94.	*Tabebuia rosea* (Berol.)DC.	Bignonaceae	Tree	Uncommon
95.	*Pedalium murex* L.	Pedaliaceae	Herb	Common
96.	*Andrographis paniculata* (Burm. f.) Wall. *ex* Ness.	Acanthaceae	Erect	Common
97.	*Barleria cristata* L.	Acanthaceae	Shrub	Common
98.	*Thunbergia fragrans* Roxb.	Acanthaceae	Climbers	Uncommon
99.	*Justicia betonica* L.	Acanthaceae	Shrub	Uncommon
100.	*Justicia prostrata* C.B.Clark.	Acanthaceae	Herb	Common

Contd...

Table 24.1–*Contd...*

Sl.No.	*Botanical Name*	*Family*	*Habit*	*Distribution*
101.	*Lantana camera* L.	Verbenaceae	Shrub	Common
102.	*Vitex negunda* L.	Verbenaceae	Tree	Common
103.	*Stachytarpheta jamaicensis* (L.) Vahl.	Verbenaceae	Herb	Common
104.	*Stachytarpheta mutabillis* (Jacq.) Vahl.	Verbenaceae	Herb	Uncommon
105.	*Leucas aspera* (Willd.) Link.	Lamiaceae	Erect	Common
106.	*Ocimum americanum* L.	Lamiaceae	Erect	Common
107.	*Ocimum canum* Sims.	Lamiaceae	Herb	Uncommon
108.	*Anisomeles malabarica* (L.) R. Br. *ex* Sims.	Lamiaceae	Shrub	Common
109.	*Leonotis neptifolia* (L.) R.	Lamiaceae	Herb	Uncommon
110.	*Bougainvillea spectabilis* Willd.	Nyctaginaceae	Shrub	Common
111.	*Gomphrena celosioides* Mart.	Amaranthaceae	Herb	Common
112.	*Gomphrena globosa* L.	Amaranthaceae	Erect	Common
113.	*Chenopodium ambrosioides* L.	Chenopodiaceae	Erect	Uncommon
114.	*Aristolochia indica* L.	Aristolochiaceae	Climber	Common
115.	*Euphorbia antiquorum* L.	Euphorbiaceae	Shrub	Common
116.	*Euphorbia hirta* L.	Euphorbiaceae	Herb	Common
117.	*Euphorbia tirucalli* L.	Euphorbiaceae	Shrub	Common
118.	*Euphorbia tortilis* Rottler *ex* Ain.	Euphorbiaceae	Shrub	Uncommon
119.	*Jatropha gosspifolia* L.	Euphorbiaceae	Shrub	Common
120.	*Phyllanthus madraspatensis* L.	Euphorbiaceae	Herb	Common
121.	*Phyllanthus emblica* L.	Euphorbiaceae	Tree	Common
122.	*Artocarpus heterophyllus* Lam.	Moraceae	Tree	Common
123.	*Ficus bengalensis* L.	Moraceae	Tree	Common
124.	*Ficus hispida* L.	Moraceae	Tree	Uncommon
125.	*Ficus microcarpa* L.	Moraceae	Tree	Uncommon
126.	*Ficus religiosa* L.	Moraceae	Tree	Common
127.	*Pilea microphylla* (L.) Liebm.	Urticaceae	Herb	Uncommon
128.	*Habenaria plantaginea* Lindl.	Orchidaceae	Erect	Uncommon
129.	*Canna indica* L.	Cannaceae	Herb	Common
130.	*Curculigo orchiodes* Gaertn.	Hypoxidaceae	Herb	Uncommon
131.	*Agave angustifolia* Haw.	Agavaceae	Shrub	Common
132.	*Agave americana* L.	Agavaceae	Shrub	Common
133.	*Agave cantula* Roxb.	Agavaceae	Herb	Uncommon
134.	*Sanseviera roxburgniana* Schult. and Schult.f.	Agavaceae	Shrub	Common
135.	*Tacca leontopetaloides* (L.) Kuntze.	Taccaceae	Erect	Uncommon

Contd...

Table 24.1–*Contd...*

Sl.No.	Botanical Name	Family	Habit	Distribution
136.	*Aloe vera* (L.) Burm.	Liliaceae	Herb	Common
137.	*Asparagus fysoni* Macbr.	Liliaceae	Climber	Uncommon
138.	*Asparagus racemosus* Willd.	Liliaceae	Climber	Common
139.	*Commelina benghalensis* L.	Commelinaceae	Herb	Common
140	*Commelina forskalei* Vahl.	Commelinaceae	Herb	Uncommon
141.	*Commelina ensifolia* R.Br.	Commelinaceae	Herb	Uncommon
142.	*Cyanotis cristata* (L.) D. Don.	Commelinaceae	Herb	Common
143.	*Borassus flabellifer* L.	Palmaceae	Tree	Common
144.	*Phoenix sylvestris* (L.) Roxb.	Palmaceae	Tree	Common
145.	*Andropogon pumilus* Roxb.	Poaceae	Grass	Common
146.	*Aristida setaceae* Retz.	Poaceae	Grass	Uncommon
147.	*Cynodon dactylon* (L.) Pres.	Poaceae	Grass	Common
148.	*Dactyloctenium aegyptium* (L.) Willd.	Poaceae	Grass	Common
149.	*Ergrostis japonica* (Thump.) Trin.	Poaceae	Grass	Uncommon
150.	*Heteropogon contortus* (L.) P. Beauv. *ex* Roem. and Schutt.	Poaceae	Grass	Common
151.	*Setaria verticillata* (L.) P. Beauv.	Poaceae	Grass	Common
152.	*Bambusa arundinaceae* (Retz.) Willd.	Bambusaceae	Tree	Common

Table 24.2: Total Lithophytes with Respect to the Number Various Lithic Type

Sl.No.	Type of Lithophytic	No. Plants	No. Genus	No. Species
1.	Chasmoendolithic	41	37	41
2.	Cryptoendolithic	7	7	7
3.	Euendolithic	32	26	32
4.	Epilithic	33	26	33
5.	Hypolithic	39	35	39
	Total Number of genus and species	152	131	152

4.0 CONCLUSOIN

A more extensive and intensive survey in this area will definitely add some more interesting information and the available data can be utilized for carrying out Lithometric and other environmental studies in the area in future.The presence study also evidenced that some areas of the rock may be bare rock surfaces, crusted by sparse soil layers and some of them may be cleft and gaps between rock and depressions on top or on the slopes are occupied by diverse plants species highly adapted to the specific environment. This type of studies may be useful to assess the impact of climatic change on their growth lithometric analysis and distribution, to use abundance distribution of species within genera, to evaluate phytogeographic patterns and also to evaluate the environmental determination of biodiversity in the nature. It is recommended that, the lithophytic habitats has to be protected for the conservation of lithophytic bioresources which will be benefit to the mankind.

The present study also that dedect that roots of higher plants penetrate deep into rock surfaces and grow to a large size capable of growing in rock crevices. The presence study conclude that the lithic habitats are very suitable sites for comparative studies on mechanism and effectiveness of ecological adaptation in plants.

REFERENCES

Alves, R., Kolbek, J., 1993. Penumbral rock communities in Campo-Rupestre sites in Brazil. *J. Veget. Sci.* 4: 357-366.

Gamble, J.S., 1915- 1936. Flora of Presidency of the Madras. Vols. I-III. Adlard and Sons Ltd., London.

Henry, A. N., Kumar, G.R., Chitra, V., 1987. Flora of Tamil Nadu, India, Analysis I (2), Botanical Survey of India, Coimbatore.

Henry, A. N., Chitra, V., Balakrishnan, N.P., 1989. Flora of Tamil Nadu, India Analysis I (3), Botanical survey of india, Coimbatore.

Kolbek, J., Jarolimek, I., Valachovic, M., 1997. Plant communities of rock habitats in North Korea and communities of semi-dry rocks, *Biol.* 52: 503-522.

Matthew, K.M., 1983. The Flora of Tamilnadu Carnatic. Parts (1-3) Rapinat Herbarium, Tiruchirapalli.

Schimper, A.F., 1898. Plant geography up on a physiological basis. Engl. Ed., Oxf

Tozer, W. C., Hackell, D., Miers, D. B., Silvester, W. B., 2005. Ex tree isotopic depletion of nitrogen in New Zealand lithophytes and epiphytes. *Oecol.* 144: 628-635.

Warming, E., 1895. Ecology of plants: An introduction to the study of plant communities. Ind. Ed. 1997.

2015, Modern Methods in Phytomedicine *Pages* ***369–387***
Editor: **T. Parimelazhagan**
Published by: **DAYA PUBLISHING HOUSE, NEW DELHI**

25

Biological Synthesis of Nanoparticles: Process and Methodological Advancements at Glance

P. Prema Sudha

Assistant Professor, Department of Nanoscience and Technology, Bharathiar University, Coimbatore, Tamil Nadu

1.0 INTRODUCTION

Nanotechnology is a field of science which deals with production, manipulation and use of materials ranging in nanometers. In nanotechnology nanoparticles research is an important aspect due to its innumerable applications. The emergence of nanotechnology has provided an extensive research in recent years by intersecting with various other branches of science especially with physics, chemistry, biology, material science and medicine and forming impact on all forms of life.The prefix nano is derived from Greek word nanos meaning "dwarf" in Greek that refers to things of one billionth (10^{-9} m) in size. The primary concept of nanotechnology was presented by Richard Feynman in a lecture entitled "There's plenty of room at the bottom" at the American Institute of Technology in 1959. Nanoparticles are usually 0.1 to 1000 nm in each spatial dimension and are commonly synthesized using two strategies: top-down and bottom-up (Fendler, 1998). In top-down approach, the bulk materials are gradually broken down to nanosized materials whereas in bottom up approach, atoms or molecules are assembled to molecular structures in nanometer range. Bottom-up approach is commonly used for chemical and biological synthesis of nanoparticles. Unlike bulk materials, nanoparticles have characteristic physical, chemical, electronic, electrical, mechanical, magnetic, thermal, dielectric, optical and biological properties (Schmid, 1992; Daniel, 2004). Nanoparticles are of great scientific interest as they bridge the gap between bulk materials and atomic or molecular structures. A bulk material has constant physical properties regardless of its size, but at the nanoscale this is often not the case. Several well characterized bulk materials have been found to possess most interesting properties when studied in the nanoscale.

2.0 TYPES OF NANOPARTICLES

Nanoparticles are classified into major types *viz.* organic and inorganic nanoparticles. Carbon nanoparticles are called the organic nanoparticles. Magnetic nanoparticles, noble metalnanoparticles (platinum, gold and silver) and semiconductor nanoparticles (titanium dioxide and zinc oxide) are grouped as inorganic nanoparticles. Inorganic nanoparticles are increasingly used in drug delivery due to their distinctive features such as ease of use, good functionality, biocompatibility, ability to targeted specifi cell and controlled release of drugs (Xu *et al.,* 2006).

2.1 Nanoparticles Synthesis

Development of reliable and eco-friendly processes for synthesis of nanoparticles is an important step for introduction of applications of nanotechnology. Synthesis of noble metal nanoparticles for applications such as catalysis, electronics, optics, environmental, and biotechnology is an area of constant interest (Albrecht *et al.,* 2006, Hussain *et al.,* 2003). Different physical and chemical processes are currently widely used to synthesize metal nanoparticles, which allow one to obtain particles with the desired characteristics (Figure 25.1). However, these production methods are usually expensive, labor-intensive, and are potentially hazardous to the environment and living organisms. Thus, there is an obvious need for an alternative, cost-effective and at the same time safe and environmentally sound method of nanoparticle production (Raveendran *et al.,* 2003; Sharma *et al.,* 2009).

2.2 Biological Synthesis of Nanoparticles

Use of chemical and physical method in the synthesis of nanoparticles is very expensive and cumbersome. The chemical and physical methods of nanoparticle synthesis lead to the presence of some toxic chemicals absorbed on the surface that may have adverse effects in applications, so there is a growing need to develop environmentally benign nanoparticles. Researchers have used biological extracts for the synthesis of nanoparticles, by adopting simple protocols, involving in the process of reduction of metal ions by using biological extracts as a source of reductants either extracellularly or intracellularly (Figure 25.2).

Synthesis of nanoparticles may be triggered by several compounds such as carbonyl groups, terpenoids, phenolics, flavonones, amines, amides, proteins, pigments, alkaloids and other reducing agents present in the plant extracts and microbial cells (Chandran *et al.,* 2006, Mohanpuria *et al.,* 2008; Leela *et al.,* 2008;

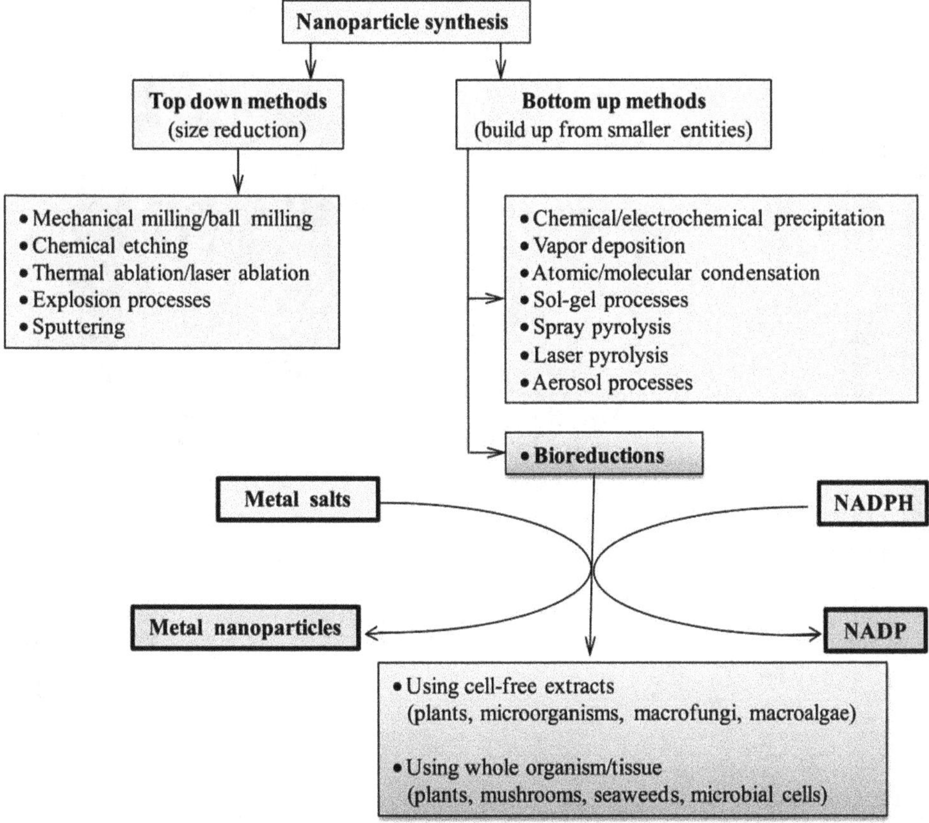

Figure 25.1: Various Approaches for Making Nanoparticles and Cofactor Dependent Bioreduction.

Tripathy *et al.*, 2010; Vineet Kumar *et al.*, 2010). Biogenic synthesis is useful not only because of its reduced environmental impact (Anastas and Zimmerman, 2007; Dahl *et al.*, 2007; Shankar *et al.*, 2004) compared with some of the physico-chemical production methods, but also because it can be used to produce large quantities of nanoparticles that are free of contamination and have a well-defined size and morphology (Hutchison, 2008). Biosynthetic routes can actually provide nanoparticles of a better defined size and morphology than some of the physicochemical methods of production (Raveendran *et al.*, 2003). The exact mechanism of nanoparticles synthesis by biological extracts is yet to be understood.

2.3 Biosynthesis of Nanoparticles by Plant Extracts

Plant extract mediated synthesis is an increasing focus of attention (Ali *et al.*, 2011; Ankamwar, 2010; Babu and Prabu, 2011; Banerjee, 2011). Recent reports of plants towards production of nanoparticles is said to have advantages such as easily available, safe to handle and broad range of biomolecules such as alkaloids, terpenoids, phenols, flavanoids, tannins, quinines etc. are known to mediate synthesis of nanoparticles (Figure 25.3). Plant extracts may act both as reducing agents and

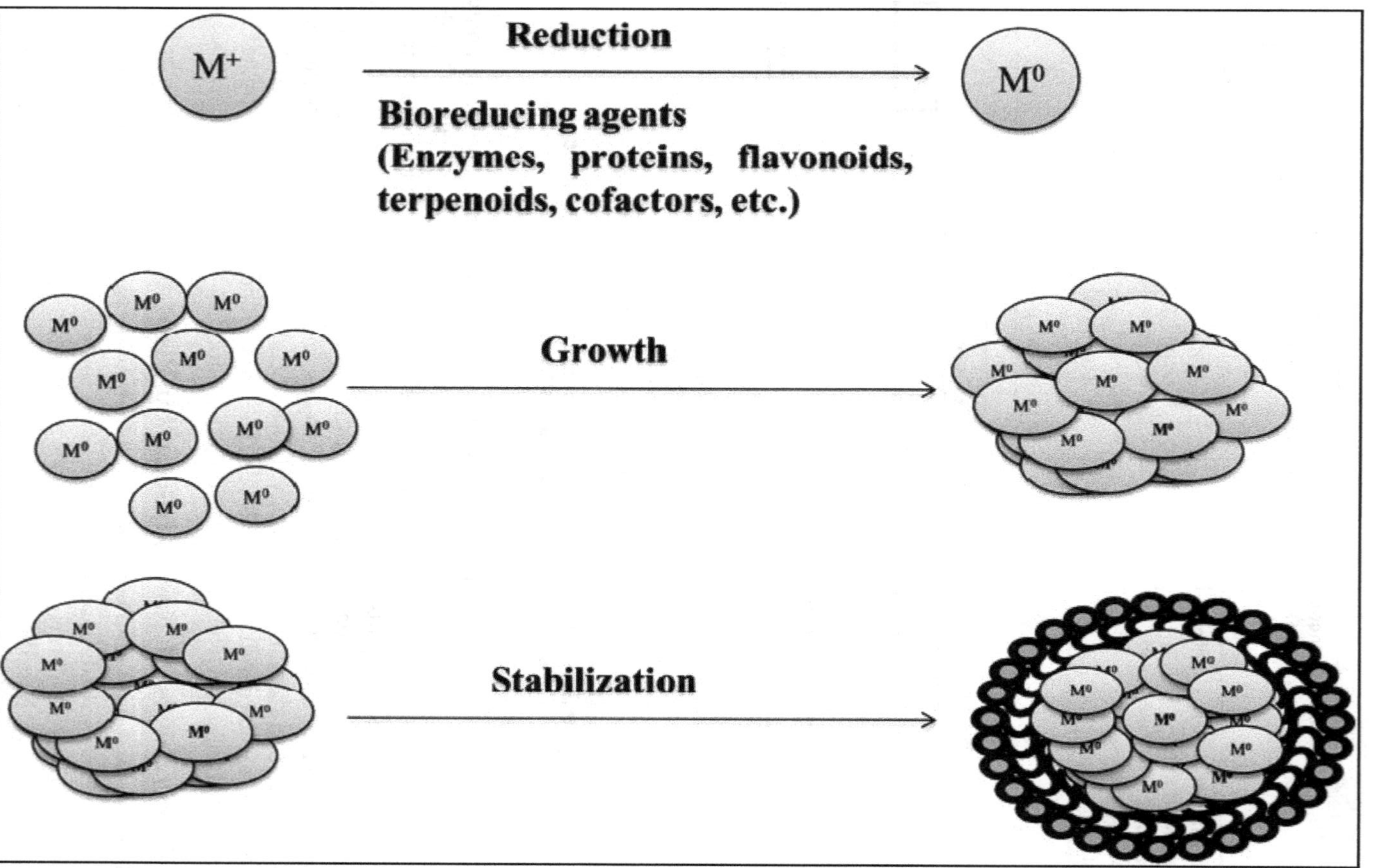

Figure 25.2: Mechanisms of Nanoparticle Synthesis by Bioreducing Agents (M+–metal ion).

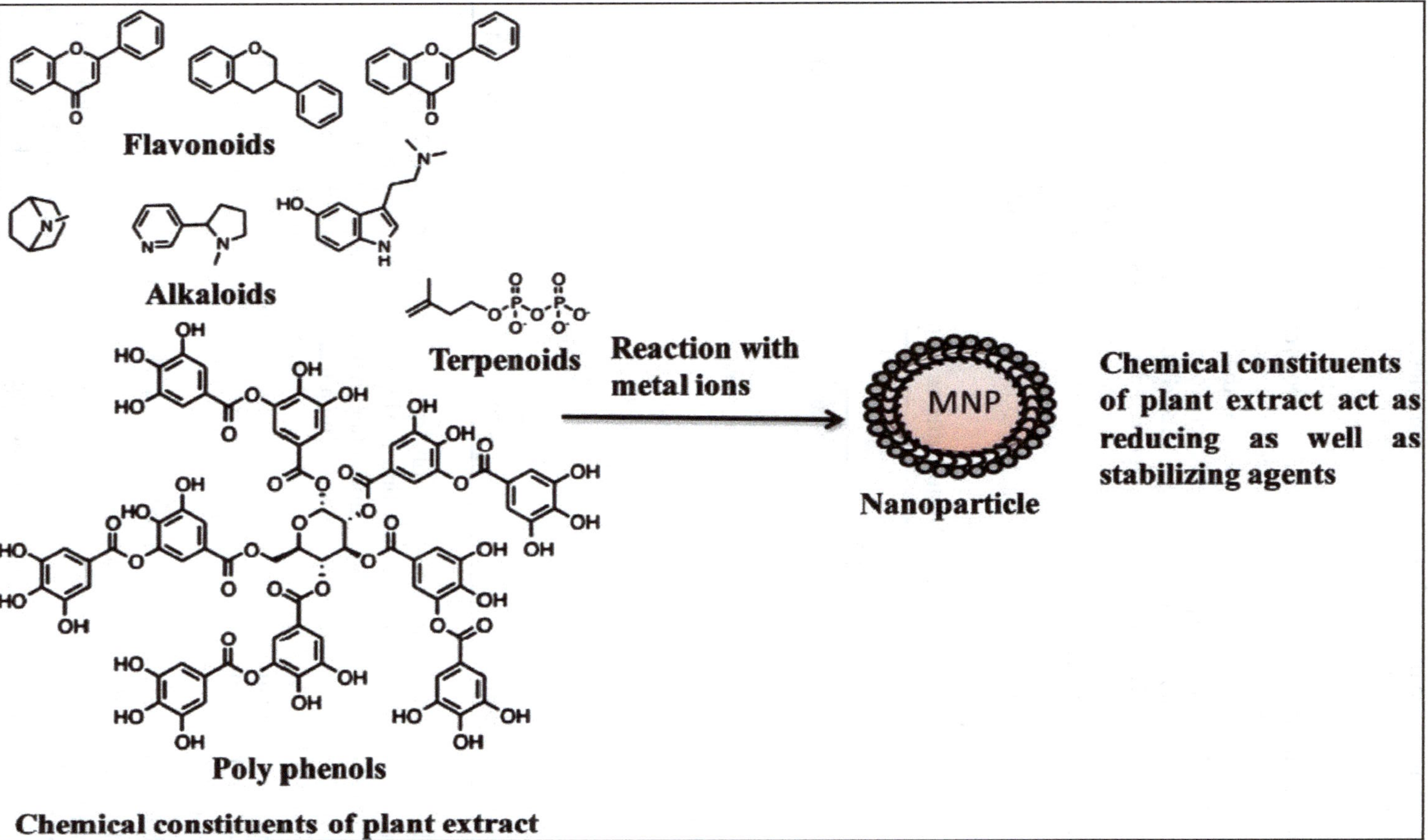

Figure 25.3: Possible Chemical Constituents of Plant Extract Responsible for the Bioreduction of Metal Ions (Dubey *et al.*, 2009; Huang *et al.*, 2007).

stabilizing agents in the synthesis of nanoparticles.The source of the plant extract is known to influence the characteristicsof the nanoparticles (Kumar and Yadav, 2009). This is because different extracts contain different concentrations and combinations of organic reducing agents (Mukunthan and Balaji, 2012). Typically, a plant extract-mediated bioreduction involves mixing the aqueous extract with an aqueous solution of the relevant metal salt. The reaction occurs at room temperature and is generally complete within a few minutes. In view of the number of different chemicals involved,the bioreduction process is relatively complex. In producing nanoparticles using plant extracts, the extract is simply mixed with a solution of the metal salt at room temperature.The reaction is complete within minutes. Nanoparticles of silver, gold and many other metals have been produced this way (Li *et al.,* 2011). This *in vitro* approach has been demonstrated using extracts from a variety of different plant species in combination with a variety of acids and salts of metals, such as copper, gold, silver, platinum, iron, and many others (Ghosh *et al.,* 2012; Khan *et al.,* 2013; Rai Met, 2013).

Table 25.1: Biosynthesis of Nanoparticles using some Plant Extracts

Plant	*Type of Nanoparticle*	*Size and Shape*	*Reference*
Acalypha indica	Ag	20–30 nm; spherical	Krishnaraj *et al.* (2010)
Aloe vera	Au, Ag	50–350 nm; spherical, triangular	Chandran *et al.* (2006)
Boswellia ovalifoliolata	Ag	30–40 nm	Ankanna *et al.* (2010)
Coleus amboinicus Lour	Ag	25.8±0.8 nm	Subramanian (2012)
Coleus aromaticus	Ag	44 nm	Vanaja and Annadurai (2012)
Datura metel	Ag	16–40 nm; quasilinearsuper structures	Kesharwani *et al.* (2009)
Desmodium triflorum	Ag	5–20 nm	Ahmad *et al.* (2010)
Trachyspermum copticum	Ag	6–50 nm	Vijayaraghavan *et al.* (2012)

The nature of the plant extract, its concentration, the concentration of the metal salt, the pH, temperature and contact time are known to affect the rate of production of the nanoparticles, their quantity and other characteristics (Dwivedi and Gopal, 2010). Synthesis of silver nanoparticles using a leaf extract of Polyalthialongifolia was reported by Prasad and Elumalai (2011). Silver and gold ions could be reduced to nanoparticles using a leaf extract of *Cinnamomum camphora*. The reduction was ascribed to the phenolics, terpenoids, polysaccharides and flavones compounds present in the extract (Huang *et al.,* 2011).

2.4 Biosynthesis of Nanoparticles by Bacteria

Microbial synthesis of nanoparticles is a green chemistry approach that inter connects nanotechnology and microbial biotechnology. Interactions between metals and microbes have been exploited for various biological applications in the fields of bioremediation, biomineralization, bioleaching, and biocorrosion (Klaus-Joerger *et al.,* 2001) and the microbial synthesis of nanoparticles has been emerged as a promising field of research as nanobiotechnology interconnecting biotechnology and nanotechnology. The biosynthesis of nanoparticles with the use of microorganisms depends on culture conditions and hence standardizing these conditions for high synthesis of nanoparticles is necessary. Many marine microorganisms are known to produce nanostructured mineral crystals and metallic nanoparticles with properties similar to chemically synthesized materials, while exercising strict control over size, shape and composition of the particles (Asmathunisha *et al.,* 2013).

Several microorganisms have been utilized to grow Ag NPs intracellularly or extracellularly (Kowshik *et al.,* 2003; Ahmad *et al.,* 2003). For instance, Ag containing nanocrystals of different compositions were synthesized by *Pseudomonas stutzeri* AG259 bacterium. In *Fusarium oxysporum* fungus, the reduction of Ag^+ ions was attributed to an enzymatic process involving NADH-dependent reductase (Ahmad Aetal, 2003). The white rot fungus, *Phanerochaete chrysosporium*, also reduced Ag^+ ion to form Ag NPs; aprotein was suggested to cause the reduction. Possible involvement of proteins in synthesizing Ag NPs was observed in filamentous *Cyanobacterium, Plectonema boryanum* UTEX 485 (Lengke *et al.,* 2007). Moreover, Ag^+ reduction by culture supernatants of *Klebsiella pneumonia, Escherichia coli* (*E. coli*), and *Enterobacter cloacae* (Enterobaceteriacae) produced rapid formations of Ag NPs (Shahverdi, 2007).

2.5 Biosynthesis of Nanoparticles by Fungi

Myconanotechnology is the interface between 'Mycology' and 'Nanotechnology' and has considerable potential, partly due to the wide range and diversity of the fungi (Rai *et al.,* 2009). When focusing on the synthesis of nanoparticles using fungi, it was observed that nanoparticles of good monodispersity and well dimensions could be synthesized. As fungi are found to secrete high amount of protein they might result in the significant mass productivity of nanoparticles.

Fungi can accumulate metal ions by physico-chemical and biological mechanisms including extracellular binding by metabolites and polymers, binding to specific polypeptides, and metabolism-dependent accumulation (Holan *et al.,* 1995). The possible use of fungi has gained much importance, as they are easy to culture in bulk. Also, the extracellular secretion of enzymes has anadded advantage in the downstream processing and handling of biomass (Gade *et al.,* 2008) when compared to the bacterial fermentation process which involves use of sophisticated instruments to obtain clear filtrate from the colloidal broth. Moreover, fungi are excellent secretors of protein compared to bacteria and actinomycetes, resulting into higher yield of nanoparticles (Sastry *et al.,* 2003). Thus, using these dissimilatory properties of fungi, it could be extensively used for the rapid and ecofriendly biosynthesis of metal nanoparticles.

Silver nanoparticles are being extensively synthesized using various fungi such as *Fusarium oxysporum* (Ahmad 2003), *Fusarium semitectum* (Basavaraja, 2008), *Aspergillus fumigatus* (Bhainsaet *et al.,* 2006) *Pleuro tussojarcaju* R, Nithya *et al.* (2009), *Penicillium brevicompactum* (Shaligram *et al.,* 2009), *Clostridium versicolor* (Sanghi, 2009). Sastry *et al.* (2003) have synthesized silver nanoparticles intracellularly within the cell walls of *Verticillium* sp. Vigneshwaran *et al.*(2007) employed *Aspergillus flavus* to produce silver nanoparticles.

2.6 Biosynthesis of Nanoparticles by Virus Mediated

Biological approaches to nanocrystal synthesis have been extended to intact biological particles Biomolecules like fatty acids, amino acids, and polyphates are used as template in the growth of semiconductor nanocrystals. Biological materials like DNA (Alivisatos, 1996; Mirkin, 1996; Braun, 1998), protein cages (Wong, 1998), biolipid cylinders (Archibald, 1993), viroid capsules Douglas, 1998), bacterial rapidosomes Pazirandeh (1992), S-layers (Shenton, 1997) and multicellular superstructures Davis (1997) have been used in template-mediated synthesis of inorganic nanoparticles and microstructures. Interestingly, tobacco mosaic virus (TMV) was used as template for the synthesis of iron oxides by oxidative hydrolysis, co-crystallization of CdS and PbS, and the synthesis of SiO_2 by sol–gel condensation. It happened with the help of external groups of glutamate and aspartate on the external surface of the virus Shenton (1999).

3.0 CHARACTERIZATION OF NANOPARTICLES

In the recent years, much advancement is brought to the technology for synthesis and characterization of nanoparticles. Several techniques have been reported worldwide for the characterization of herbal and medicinal plants based nanoparticles.The current review of methodology provides the technological advancement for biological and eco-friendly synthesis and characterization of nanoparticles. Nanoparticles are generally characterized by their size, shape, surface area, and dispersity (Jiang *et al.*, 2009). Homogeneity of these properties is important in many applications. The common techniques of characterizing nanoparticles are as follows: UV–visible spectrophotometry, Dynamic light scattering (DLS), Scanning electron microscopy (SEM), Transmission electron microscopy (TEM), Fourier transform infraredspectroscopy (FTIR), powder X-ray diffraction (XRD) and energy dispersive spectroscopy (EDS) (Feldheimand Foss, 2002; Sepeur, 2008; Shahverdi *et al.*, 2011).

3.1 Ultra Violet-Visible (UV-Vis) Spectrophotometry

A number of studies have reported monitoring of bio-reduction of silver ions inaqueous solution by UV-Vis spectrophotometer.According to these studies, reduction of pure Ag+ ions was usually monitored after 3 to 5 hours of diluting the small aliquot of the sample into distilled water (Verma *et al.*, 2010).

3.2 Fourier Transform-Infra Red (FT-IR) Spectroscopy

An infrared spectrum (IR) represents a fingerprint of a sample with absorption peaks which correspond to the frequencies of vibrations between the bonds of the atoms making up the material. Because each different material is a unique combination of atoms, no two compounds produce the exact same IR. Therefore, IR results in a positive identification (qualitative analysis) of every different kind of material. According to a study carried out by Mallikarjuna *et al.* (2011). The bioreduced silver nitrate solution was centrifuged at 10,000 rpm for 15min and the dried samples were grinded with KBr pellets used for FTIR-measurements.

3.3 Scanning Electron Microscopy (SEM)

The SEM uses a focused beam of high-energy electrons to generate a variety of signals at the surface of solid specimens. The signals that derive from electron sample interactions reveal information about the sample including external morphology (texture), chemical composition, crystalline structure and orientation of materials making up the sample. In most applications, data are collected over a selected area of the surface of the sample, and a two-dimensional image is generated that displays spatial variations in these properties. For this purpose, thin films of the samples were prepared by the investigators on carbon coated copper grids by just dropping a very small amount of the sample on the grid, extra solution was removed using a blotting paper and the films on the SEM grid were allowed to dry under a mercury lamp for 5 mins (Linga Rao *et al.,* 2011; Devi *et al.,* 2012; Priya *et al.,* 2011).

3.4 Energy Dispersive X-Ray Spectra (EDX/EDS)

EDS is an analytical technique which utilizes X-rays that are emitted from the specimen when bombarded by the electron beam to identify the elemental composition of the specimen. When the sample is bombarded by the electron beam of the SEM, electrons are ejected from the atoms on the surface of specimen. A resulting electron vacancy is filled by an electron from a higher shell, and an X-ray is emitted to balance the energy difference between the two electrons. The EDS X-ray detector measures the number of emitted X-rays versus their energy. The energy of the X-ray is characteristic of the element from which the X-ray was emitted. A spectrum of the energy versus relative counts of the detected X-rays is obtained and evaluated for qualitative and quantitative determinations of the elements (Herguth *et al.,* 2004).

3.5 Transmission Electron Microscopy (TEM)

TEM is commonly used for imaging and analytical characterization of the nanoparticles to assess the shape, size, and morphology (Zargar *et al.,* 2011). The outstanding resolutionachieved by TEM is an excellent fit for these extremely challenging studies. According to several studies carried out by different investigators, thin films of the samples were prepared on carbon coated copper grids by dropping a very small amount of the sample on the grid, extra solution was removed using a blotting paper. The films thus prepared on the TEM grid were then allowed to dry undetr a mercury lamp for 5 min (Saxena *et al.,* 2010).

3.6 Atomic Force Microscopy (AFM)

AFM is an important biophysical technique for studying the morphology of nanoparticles and biomolecules.Using the AFM, individual particles and groups of particles can be resolved. Software-based image processing of AFM data can generate quantitative information from individual nanoparticles and between groups of nanoparticles. For individual particles, size information (length, width, and height) and other physical properties (such as morphology and surface texture) can be measured. AFM can be performed both in liquid or gas medium. This capability can be very advantageous for nanoparticle characterization. AFM has several advantages over SEM and TEM for characterizing nanoparticles. Images from an AFM represent data in three dimensions, so that it is possible to measure the height of the nanoparticles quantitatively (Mucalo *et al.,* 2002, Vasenka, 1993).

3.7 X-Ray Diffraction (XRD)

XRD is a versatile, non-destructive analytical method for identification and quantitative determination of various crystalline forms. According to studies, the solution of silver nanoparticles obtained was purified by repeated centrifugation at 10, 000 rpm for 20 min followed by re-dispersion of the pellet of silver nanoparticles into 10 mL of distilled water. After freeze drying of the purified silver particles, the structure and composition of silver nanoparticles were analyzed by XRD (Mallikarjuna, 2011, Priya *et al.,* 2011). The diffraction occurs as waves interact with a regular structure whose repeat distance is about the same as the wavelength. The phenomenon is common in the natural world and occurs across abroad range of scales.

3.8 Dynamic Light Scattering (DLS)

DLS is one of the most popular techniques which is used to determine the size of particles. Shining a monochromatic light beam, such as laser, onto a solution with spherical particles in Brownian motion causes a Doppler shift when the light hits the moving particle, changing the wavelength of incoming light. This change is related to the size of the particle. Using DLS, it is possible to compute the sphere size distribution and give a description of the particle's motion in the medium measuring the diffusion coefficient of the particle by using autocorrelation function (Saxena *et al.,* 2010).

4.0 GENARILIZED PROCESS AND PROCESS OUTLOOK

The use of natural sources like biological systems becomes essential to development of a reliable and ecofriendly process for synthesis of metallic nanoparticles. Whatever the biological system used, in order to exploit the system to its maximum potential, it is very much essential to understand the biochemical and molecular mechanism of nanoparticle synthesis by bioreductants. However, the biological systems have been relatively unexplored for the presence of relatively potential bioreductants, and there are many opportunities for budding nanobiotechnologists to use the biological systems for metallic nanoparticle synthesis. Figure 25.4 emphasize the generalized process from synthesis to the means of characterization of biosynthesized nanoparticles.

5.0 APPLICATIONS OF NANOPARTICLES

The unique properties of nanomaterials encourage belief that they can be applied in a wide range of fields, from medical applications to environmental sciences (Figure 25.5). Studies conducted by nanotechnology experts mapping the risks and opportunities of nanotechnology have revealed enormous prospects for progress in both life sciences and information technology (Donaldson, 2004). Nanoparticles have a greater surface area per weight than larger particles and this property makes them

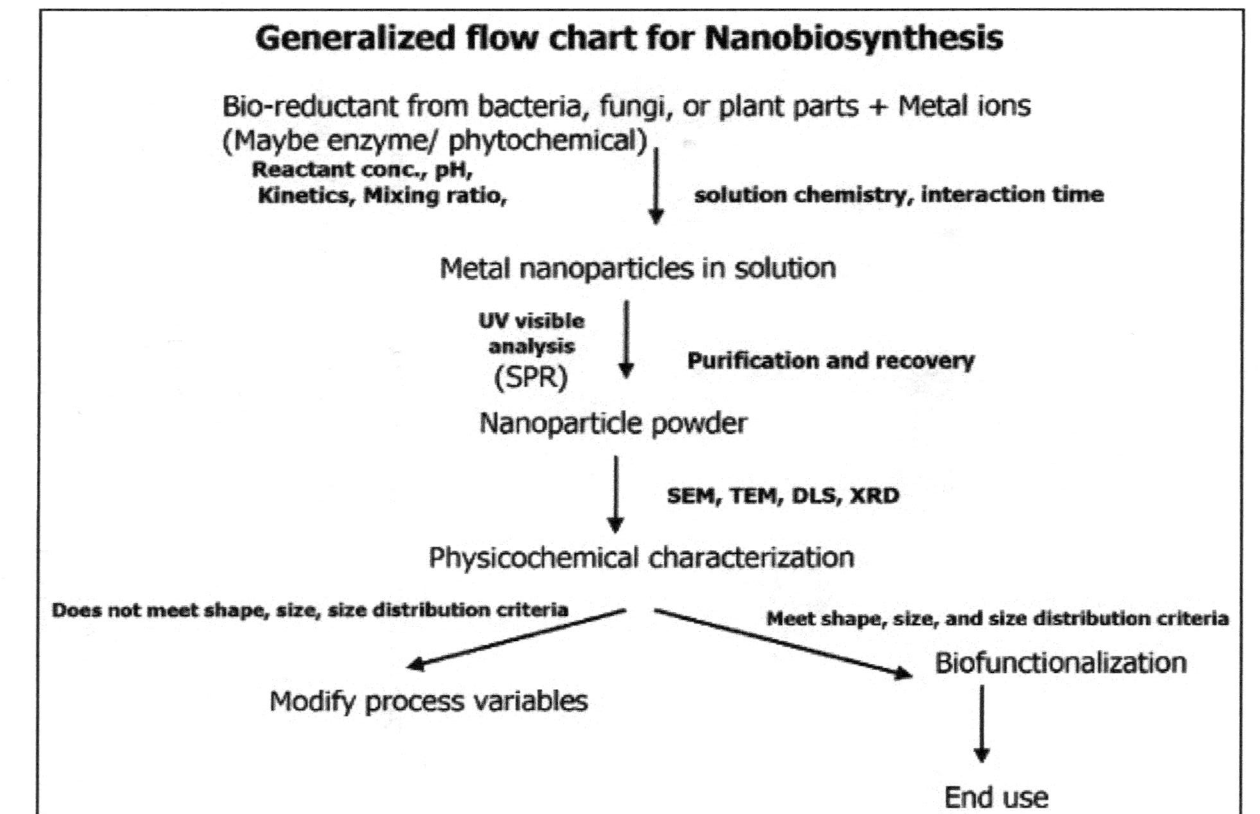

Figure 25.4: Generalized Flowchart for Biosynthesis of Nanoparticles and Further Characterization.

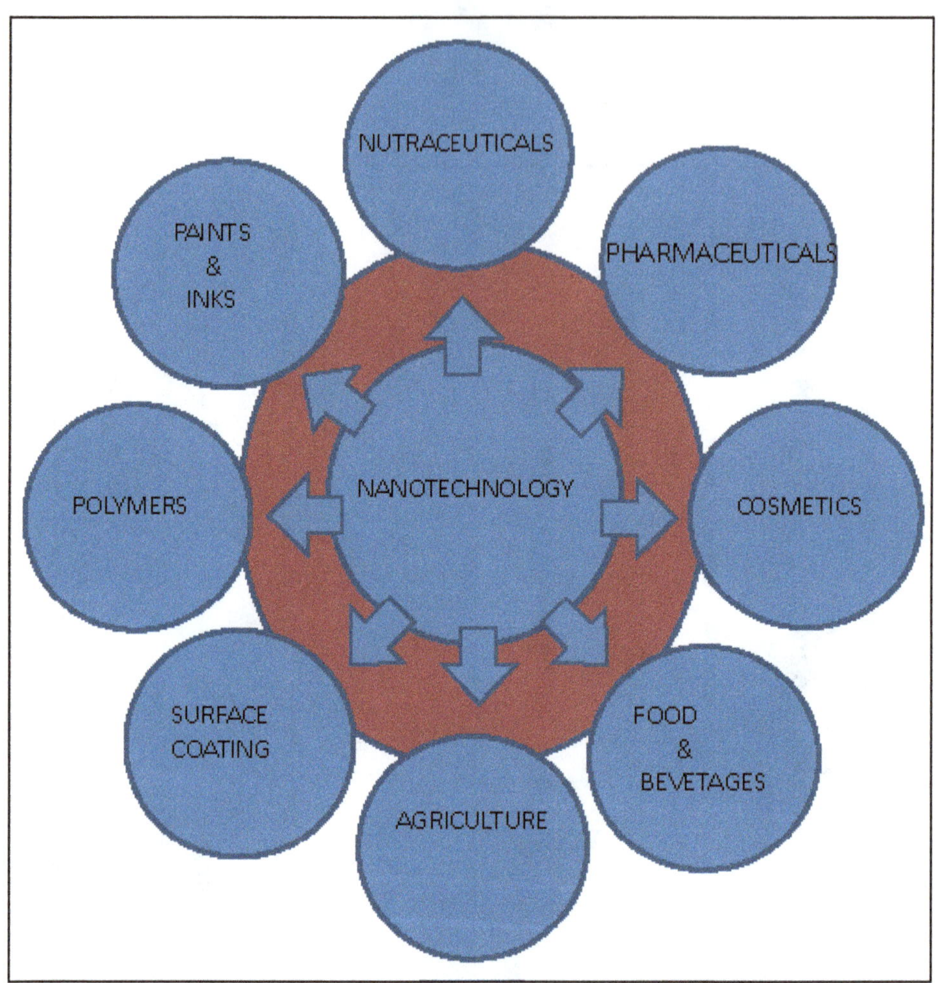

Figure 25.5: Application of Nanoparticle in Range of Fields.

to be more reactive to certain other molecules and they are used or being evaluated for use in many fields (Kathiresan *et al.,* 2009).

The unique properties of NPs have enabled their use as promising tools to study biological processes. Many innovative techniques using NPs are being developed to activate cellsignaling pathways, to induce protein production, and to improve upon current techniques used in molecular and cellular biology research. NPs such as QDs have been extensively studied for many biological applications that use fluorescence (Edina *et al.,* 2014).

Nanoparticles can be used in targeted drug delivery at the site of disease to improve the uptake of poorly soluble drugs the targeting of drugs to a specific site, and drug bioavailability. Nanoparticles may prove effective tools for improving stem cell therapy, new research suggests. Chemical engineers have successfully used

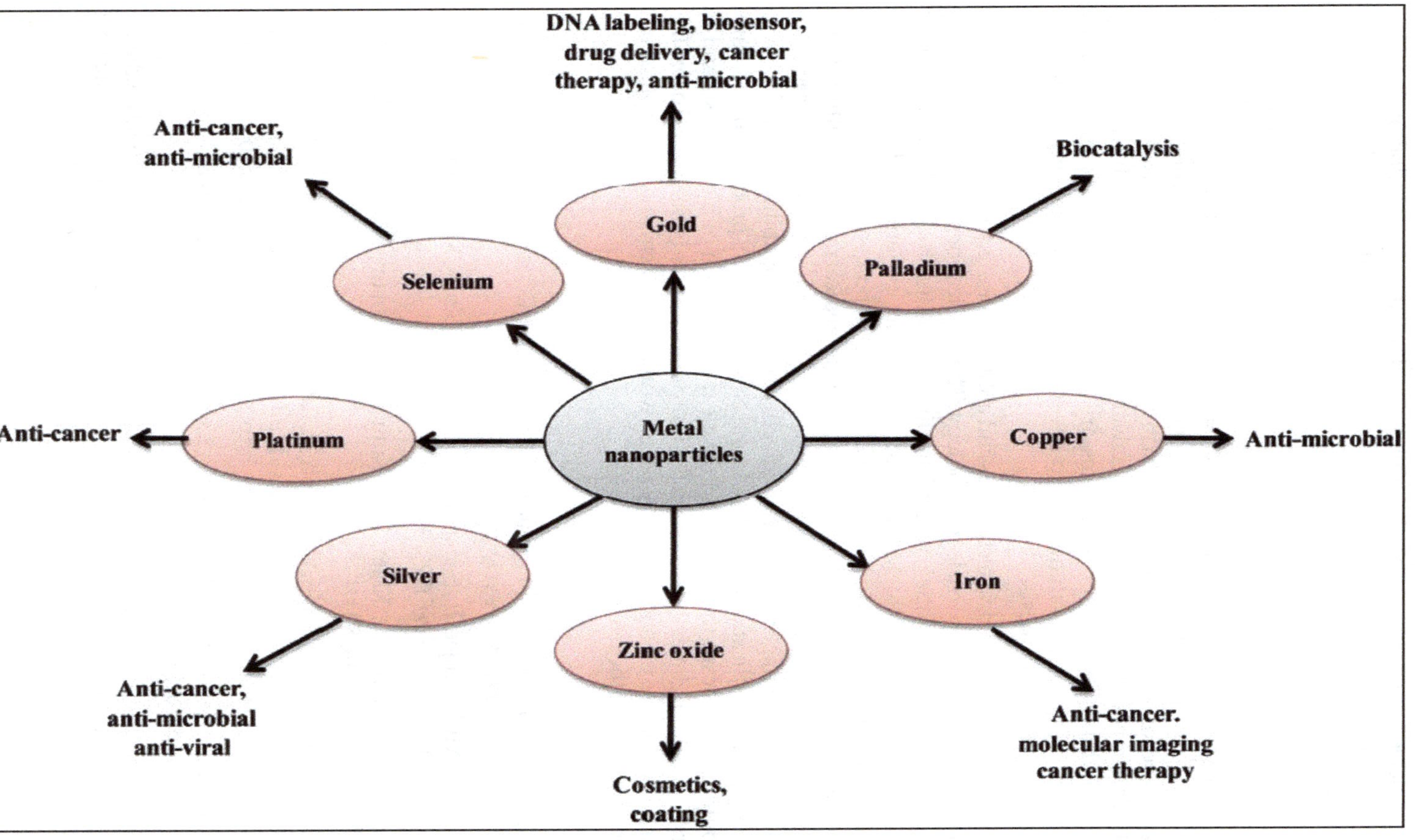

Figure 25.6: Types of Metal Nanoparticles and their Applications in Biotechnology.

nanoparticles to enhance stem cells' ability to stimulate regeneration of damaged vascular tissue and reduce muscle degeneration in mice (Abhilash, 2010).

6.0 CONCLUSION

During the current scenario nanotechnology motivates progress in all sphere of life, hence biosynthetic route of nanoparticles synthesis will emerge as safer and best alternative to conventional methods. Obviously, the synthesis of metal nanoparticles using bioreductive agents, despite obvious limitations, has a significant potential and a number of substantial advantages relative to traditional methods of nanoparticle synthesis. However, to compete cost-effectively with nanoparticles obtained through physical and chemical methods, it is necessary to scale these methods of nanoparticle production using plant material and to develop schemes for keeping expenses in check during their synthesis.

Some modern analytical techniques and antimicrobial assays have been applied to characterize the nanoparticle morphology. The applications of such eco-friendly nanoparticles in bactericidal, wound healing and other medical and electronic applications make the biological approach potentially exciting for the large- scale synthesis of other nanoparticles. From a technological point of view, the silver and gold nanoparticles have potential applications in biomedical fields, agriculture, foods and herbal research among several others. The biological procedures have considerable advantages such as cost effectiveness, compatibility for medical and pharmaceutical applications as well as large scale commercial production. In future, it would be significant to understand the clear mechanism of biosynthesis and to technologically engineer the nanoparticles inorder to achieve better control over size, shape, and absolute monodispersivity which would further enhance the applications ofnanoparticles in the related fields.

REFERENCES

Abhilash, M., 2010. Potential applications of Nanoparticles. *International Journal of Pharma and Bio Sciences* 1(1).

Ahmad, A., Mukherhjee, P., Senapati, S., Mandal, D., Khan, M.I., Kumar, R., Sastry, M., 2003. *Colloids Surf.* B 28, 313.

Ahmad, A., Senapati, S., Khan, M.I., Kumar, R., Ramani, R., Srinivas, V., 2003. *Nanotechnology* 14, 824.

Ahmad, N., Sharma, S., Singh, V., Shamsi, S., Fatma, A., Mehta, B., 2010. Biosynthesis of silver nanoparticles from Desmodium triflorum: a novel approach towards weed utilization. *Biotechnol Res Int.* http://dx.doi.org/10.4061/2011/454090.

Albrecht, M.A., Evans, C.W., Raston, C.L., 2006. Green Chem and Health implication of nannoparticles. 8, 417.

Ali, D.M., Thajuddin, N., Jeganathan, K., Gunasekaran, M., 2011. Plant extract mediated synthesis of silver and gold nanoparticles and its antibacterial activity against clinically isolated pathogens. *Colloids Surf B Biointerfaces* 85, 360–5.

Alivisatos, A., Johnsson, K., Peng, X., Wilson, T., Lowth, C., Bruchez, M., Schultz, P., 1996. *Nature* 382, 609–611.

Ankamwar, B., 2010. Biosynthesis of gold nanoparticles (green-gold) using leaf extract of *Terminalia catappa*. *Eur J Chem* 7, 1334–9.

Ankanna, S., Prasad, T.N.V.K.V., Elumalai, E.K., Savithramma, N., 2010. Production of biogenic silver nanoparticles using *Boswellia ovalifoliolata* stem bark. *Dig J Nanomater Biostruct* 5, 369–72.

Archibald, D.D., Mann, S., 1993. Template mineralization of self-assembled lipid microstructures. *Nature* 364, 430-3.

Asmathunisha, N., Kathiresan, K., 2013. A review on biosynthesis of nanoparticles by marine organisms. *Colloids and Surfaces B: Biointerfaces* 103, 283– 287.

Babu, S.A., Prabu, H.G., 2011. Synthesis of AgNPs using the extract of *Calotropis procera* flower at room temperature. *Mater Lett* 65, 1675–7.

Banerjee, J., Narendhirakannan, R., 2011. Biosynthesis of silver nanoparticles from *Syzygium cumini* (L.) seed extract and evaluation of their *in vitro* antioxidant activities. *Dig J Nanomater Biostruct* 6, 961–8.

Basavaraja, S., Balaji, S.D., Lagashetty, A., Rajasab, A.H., Venkataraman, A. Mater, 2008. Extracellular biosynthesis of silver nanoparticles using the fungus *Fusarium semitectum Materials Research Bulletin* 43(5), 1164-1170.

Bhainsa, K.C., D'Souza, S.F., 2006. Extracellular biosynthesis of silver nanoparticles using the fungus *Aspergillus fumigatus*. *Colloids and Surfaces B: Biointerfaces* 47(2), 160-164.

Braun, E., Eichen, Y., Sivan, U., Ben-Yoseph, G., 1998 DNA-templated assembly and electrode attachment of a conducting silver wire. *Nature* 391, 775–778.

Chandran, S.P., Chaudhary, M., Pasricha, R., Ahmad, A., Sastry, M., 2006. Synthesis of gold nanotriangles and silver nanoparticles using Aloe vera plant extract. *Biotechnol Prog* 22, 577–83.

Fendler, J.H., 1998. Nanoparticles and nanostructured films: preparation, characterization and applications. JohnWiley and Son.

Daniel, M.C., Astruc, D., 2004. Gold nanoparticles: assembly, supramolecular chemistry, quantum-size-related properties, and applications toward biology, catalysis, and nanotechnology. *Chem Soc Rev* 104, 293.

Dahl, J.A., Maddux, B.L.S., Hutchison, J.E., 2007. Toward greener nanosynthesis. *Chem Rev* 107, 2228–69.

Donaldson, K., Stone, V., Tran, C., Kreyling, W., Borm, P.J.A., 2004. *Nanotoxicology Occup. Environ. Med* 61, 727-728.

Devi, N.N., Shankar, P.D., Femina, W., Paramasiva, T., 2012. Antimicrobial efficacy of green synthesized silver nanoparticles from the medicinal plant *Plectranthus amboinicus, International J Pharm Sci Res*, 12, available online.

Davis, S.A., Burkett, S.L., Mendelson, N.H., Mann, S., 1997. Bacterial templating of ordered macrostructures in silica and silica-surfactant mesophases. *Nature* 385, 420–423.

Douglas, T., Young, M., 1998. Host-guest encapsulation of materials by assembled virus protein cages. *Nature* 393, 152-155.

Dwivedi, A.D., Gopal, K., 2010. Biosynthesis of silver and gold nanoparticles using *Chenopodium album* leaf extract. *Colloids Surf A* 369, 27–33.

Edina, C., Wang, Andrew, Z., Wang, 2014. Nanoparticles and their applications in cell and molecular biology. *Integr. Biol* 6, 9-26.

Feldheim, D.L., Foss, C.A., 2002. Metal nanoparticles: synthesis, characterization, and applications. Boca Raton, FL: CRC Press.

Gade, A.K., Bonde, P., Ingle, A.P., Marcato, P.D., Duran, N., Rai, M.K., 2008. Exploitation of *Aspergillus niger* for synthesis of silver nanoparticles. *Journal of Biobased Material and Bioenergy* 2, 243-247.

Ghosh, S., Patil, S., Ahire, M., Kitture, R., Gurav, D.D., Jabgunde, A.M., Kale, S., Pardesi, K., Shinde, V., Bellare, J.J., 2012. *Gnidia glauca* flower extract mediated synthesis of gold nanoparticles and evaluation of its chemocatalytic potential. *J Nanobiotechnology* 10, 17.

Holan, Z.R., Volesky, B., 1995. Accumulation of cadmium, lead and nickel by fungal and wood biosorbents. *Applied Biochemistry and Biotechnology* 53(2), 133-146.

Hutchison, J.E., 2008. Greener nanoscience: a proactive approach to advancing applications and reducing implications of nanotechnology. *ACS Nano* 2, 395–402.

Huang, X., Wu, H., Pu, S., Zhang, W., Liao, X., Shi, B., 2011. One-step room-temperature synthesis of Au@Pd core–shell nanoparticles with tunable structure using plant tannin as reductant and stabilizer. *Green Chem* 13, 950–7.

Herguth, W.R., Nadeau, G., 2004. Applications of scanning electron microscopy and energy dispersive spectroscopy (SEM/EDS) to practical tribology problems, Herguth Lab Inc CA; 94590.

Hussain, I., Brust, M., Papworth, A.J., Cooper, A.I., 2003. Preparation of Acrylate-Stabilized Gold and Silver Hydrosols and Gold-Polymer Composite Films. *Langmuir* 19, 4831–4835.

Jiang, J., Oberdörster, G., Biswas, P., 2009. Characterization of size, surface charge, and agglomeration state of nanoparticle dispersions for toxicological studies. *J Nanopart Res* 11, 77–89.

Kathiresan, K., Manivannan, S., Nabeel, M.A., Dhivya, B., 2009. Studies on silver nanoparticles synthesized by a marine fungus, *Penicillium fellutanum* isolated from coastal mangrove sediment. *Colloids Surf.* B 71, 133.

Khan, M., Adil, S.F., Tahir, M.N., Tremel, W., Alkhathlan, H.Z., Al-Warthan, A., Siddiqui, M.R., 2013. Green synthesis of silver nanoparticles mediated by *Pulicaria glutinosa* extract. *Int. J. Nanomedicine* 8, 1507–1516.

Kesharwani, J., Yoon, K.Y., Hwang, J., Rai, M., 2009. Phytofabrication of silver nanoparticles by leaf extract of Datura metel: hypothetical mechanism involved in synthesis. *J Bionanosci* 3, 39–44.

Klaus-Joerger, T., Joerger, R., Olsson, E., Granqvist, C.G., 2001. Bacteria as workers in the living factory: metal-accumulating bacteria and their potential for materials science. *Trends Biotechnol* 19, 15.

Kowshik, M., Ashtaputre, S., Kharrazi, S., Vogel, W., Urban, J., Kulkarni, S.K., 2003. Extracellular synthesis of silver nanoparticles by a silver-tolerant yeast strain MKY3. *Nanotechnology* 14, 95.

Krishnaraj, C., Jagan, E., Rajasekar, S., Selvakumar, P., Kalaichelvan, P., Mohan, N., 2010. Synthesis of silver nanoparticles using *Acalypha indica* leaf extracts and its antibacterial activityagainst water borne pathogens. *Colloids Surf B Biointerfaces* 76, 50–6.

Kumar, V., Yadav, S.K., 2009. Plantmediated synthesis of silver and gold nanoparticles and their applications. *J Chem Technol Biotechnol* 84, 151–7.

Leela, A., Vivekanandan, M., 2008. Tapping the unexploited plant resources for the synthesis of silver nanoparticles. *Afr. J. Biotechnol* 7, 3162-3165.

Lee, H.J., Lee, G., Jang, N.R., Yun, J.H., Song, J.Y., Kim, B.S., 2011. Biological synthesis of copper nanoparticles using plant extract. *Nanotechnology* 1, 371–4.

Lengke, M.F., Fleet, M.E., Southam, G., 2007. *Langmuir* 23, 2624.

Linga Rao, M., Savithramma, N., 2011. Biological synthesis of silver nanoparticles using *Svensonia hyderabadensis* leaf extract and evaluation of their antimicrobial efficacy, *J Pharm Sci Res* 3, 1117-1121.

Mallikarjuna, K., Narasimha, G., Dillip, G.R., Praveen, B., Shreedhar, B., Shree Lakshmi, C., Reddy, B.V.S., Deva Prasada Raju, B., 2011. Green synthesis of silver nanoparticles using *Ocimum* leaf extract and their characterization, *Digest J Nanomaterials Biostructures* 6, 181-186.

Mohanpuria, P., Rana, K.N., Yadav, S.K., 2008. Biosynthesis of nanoparticles: technological concepts and future applications. *J Nanopart Res* 10, 507–517.

Mukunthan, K., Balaji, S., 2012. Cashew apple juice (*Anacardium occidentale* L.) speeds up the synthesis of silver nanoparticles. *Int J Green Nanotechnol* 4, 71–9.

Mirkin, C.A., Letsinger, R.L., Mucic, R.C., Storhoff, J.J., 1996. *Nature* 382, 607.

Mucalo, M., Bullen, C., Manely-Harris, M., Mc Intire, T., 2002. Arabinogalactan from the Western larch tree: a new, purified and highly water-soluble polysaccharide-based protecting agent for maintaining precious metal nanoparticles in colloidal suspension. *J Material Sci* 37, 493-504.

Nithya, R., Ragunathan, R., 2009. Synthesis of silver nanoparticle using Pleurotus sajor caju and its antimicrobial properties. *Digest Journal of Nanomaterials and Biostructures* 4(4), 623 – 629.

Pazirandeh, M., Baral, S., Campbell, J.R., 1992. Metallized nanotubules derived from bacteria. *Biomimetics* 1:41.

Priya, M.M., Selvi, B.K., John Paul, J.A., 2011. Green synthesis of silver nanoparticles from the leaf extracts of *Euphorbia hirta* and *Nerium indicum*. *Digest J Nanomaterials Biostructures* 6, 869-877.

Prasad, T.N.V.K.V., Elumalai, E., 2011. Biofabrication of Ag nanoparticles using Moringa oleifera leaf extract and their antimicrobial activity. *Asian Pac J Trop Biomed* 1, 439–42.

Rai, M., Yadav, A., Bridge, P., Gade, A., 2009. Myco nanotechnology: A new and emerging science. In: Rai, M Gupta *et al.*, 38 K and Bridge, P D. eds. Applied Mycology. New York, CAB International, 14, 258-267.

Rai, M., Yadav, A., 2013. Plants as potential synthesiser of precious metal nanoparticles: progress and prospects. *IET Nanobiotechnol* 7(3), 117–124.

Raveendran, P., Fu, J., Wallen, S.L., 2003. Completely "green" synthesis and stabilization of metal nanoparticles. *J Am Chem Soc* 125, 13940–1.

Sastry, M., Ahmad, A., Khan, M.I., Kumar, R., 2003. Biosynthesis of metal nanoparticles using fungi and actinomycetes. *Current Science* 85(2), 162-170.

Sanghi, R., Preetiverma, 2009. Biomimetic synthesis and characterization of protein capped silver nanoparticle. *Bio. Res. Technol* 100, 501.

Sastry, M., Ahmad, A., Islam khan, M., Kumar, R., 2003. Biosynthesis of metal nanoparticles using fungi and actinomycete. *Curr. Sci* 85, 162.

Saxena, A., Tripathi, R.M., Singh, R.P., 2010. Biological synthesis of silver nanoparticles by using onion (*Allium cepa*) extract and their antibacterial activity. *Digest J Nanomaterials Biostructures* 5, 427-432.

Sharma, H.S., Ali, S.F., Hussain, S.M., Schlager, J.J., Sharma, A., 2009. *J. Nanosci. Nanotechnol* 9(8), 5055–5072.

Shahverdi, A.R., Minaeian, S., Shahverdi, H.R., Jamalifar, H., Nohi, A.A., 2007. Rapid synthesis of silver nanoparticles using culture supernatants of Enterobacteria: A novel biological approach. *Process Biochemistry* 42, 919-923.

Shaligram, N.S., Bule, M., Bhambure, R., Singhal, R.S., Singh, S.K., Szakacs, G., Pandey, A., 2009. Biosynthesis of silver nanoparticles using aqueous extract from the compactin producing fungal strain. *Proc Biochem* **44**, 939–943.

Shahverdi, A.R., Shakibaie, M., Nazari, P., 2011. Basic and practical procedures for microbial synthesis of nanoparticles. In: Rai M, Duran N, editors. Metal nanoparticles in microbiology. Berlin: Springer, 177–97.

Schmid, G., 1992. *Chem Rev*; 92:1709.

Shankar, S.S., Rai, A., Ahmad, A., Sastry, M., 2004. Rapid synthesis of Au, Ag, and bimetallic Au core–Ag shell nanoparticles using Neem (*Azadirachta indica*) leaf broth. *J Colloid Interface Sci* 275,496–502.

Sepeur, S., 2008. Nanotechnology: technical basics and applications. Hannover: Vincentz.

Shenton, W., Pum, D., Sleytr, U.B., Mann, S., 1997. Synthesis of cadmium sulphide superlattices using self-assembled bacterial S-layers. *Nature* 389, 585–587

Shenton, W., Douglas, T., Young, M., Stubbs, G., Mann, S., 1999. Inorganic–organic nanotube composites from template mineralization of tobacco mosaic virus *Adv. Mater.* 11, 253–6.

Subramanian, V., 2012. Green synthesis of silver nanoparticles using Coleus amboinicus lour, antioxitant activity and invitro cytotoxicity against Ehrlich's Ascite carcinoma. *J Pharm Res,* 5, 1268–72.

Tripathy, A., Raichur, Ashok, M., Chandrasekaran, N., Prathna, T.C., Mukherjee, Amitava, 2010. *Process variables in biomimetic synthesis of silver nanoparticles by aqueous extract of Azadirachta indica (Neem) leaves.* In: Journal *of Nanoparticle Research* 12(1), 237-246.

Vanaja, M., Annadurai, G., 2012. *Coleus aromaticus* leaf extract mediated synthesis of silver nanoparticles and its bactericidal activity. *Appl Nanosci.* http://dx.doi.org/10. 1007/s13204-012-0121-9.

Vasenka, J., Manne, S., Giberson, R., Marsh, T., Henderson, E., 1993. Colloidal gold particles as an incompressible AFM imaging standard for assessing the compressibility of biomolecules, *Biophysical J* 65, 992-997.

Verma, V.C., Kharwar, R.N., Gange, A.C., 2010. Biosynthesis of antimicrobial silver nanoparticles by the endophytic fungus *Aspergillus clavatus. Nanomedicine* (*Lond*) *5,* 33–40.

Vineet Kumar, C., Yadav, S.C., Yadav, S.K., 2010. *J. Chem. Technol. Biotechnol.* 85, 1301.

Vigneshwaran, N., Ashtaputre, N.M., Varadarajan, P.V., Nachane, R.P., Panikar, K.M., Balasubrahmanya, R.H., 2007. *Mater. Lett.* 61, 1413.

Vigneshwaran, A., Kathe, A.A., Varadarajan, P.V., Nachne, R.P., Balasubramanya, R.H., 2006. Biomimetics of silver nanoparticles by white rot fungus, Phaenerochaete chrysosporium.Colloids Surf B Biointerfaces; 53:55–59.

Vijayaraghavan, K., Nalini, S., Prakash, N.U., Madhankumar, D., 2012. One step green synthesis of silver nano/microparticles using extracts of *Trachyspermum ammi* and *Papaver somniferum. Colloids Surf B Biointerfaces* 94, 114–7.

Wong, K.K.W., Douglas, T., Gider, S., Awschalom, D.D., Mann, S., 1998. Biomimetic synthesis and characterization of magnetic proteins (magnetoferritin), *Chem. Mater.,* 10, 279–285, doi:10.1021/cm970421o.

Xu, Z.P., Zeng, Q.H., Lu, G.Q., Yu, A.B., 2006. Inorganic Nanoparticles as carriers for efficient cellular delivery. *Chem. Eng. Sci.* 61, 1027.

Zargar, M., Hamid, A.A., Bakar, F.A., Shamsudin, M.N., Shameli, K., Jahanshiri, F., Farahani, F., 2011. Green synthesis and antibacterial effect of silver nanoparticles using *Vitex negundo* L. *Molecules;* 16, 6667-6676.

2015, Modern Methods in Phytomedicine *Pages* ***389–402***
Editor: **T. Parimelazhagan**
Published by: **DAYA PUBLISHING HOUSE, NEW DELHI**

26

Evaluation of Nutrition and Antioxidant Property of *Canavalia mollis* Wight and Arn.

*Murugaiyan Iniyavan, Sivaraj Dhivya and Thangaraj Parimelazhagan**

Bioprospecting Laboratory, Department of Botany, Bharathiar University, Coimbatore – 641 046, Tamil Nadu

1.0 INTRODUCTION

Medicinal plants are known to contain innumerable biologically active compounds. Several active compounds have been discovered from plants on the basis of ethnobotanical information and used directly as patented drugs. India has an ancient heritage of traditional medicine which is based on various systems including Ayurveda, Siddha, Unani and Homeopathy (Vikas *et al.,* 2010). Many studies have demonstrated that medicinal plants contain various bioactive compounds with antioxidant activity, which are responsible for their beneficial health effects. There is a natural balance between pro-oxidant and antioxidant amount in the body (Nose, 2000). But excess amount of pro-oxidants cannot be scavenged by endogenous antioxidants and thus free radical and antioxidant imbalance transpire leading to several dysfunctions (Dringen, 2000). Oxidative stress plays a major part

* *Corresponding Author*

in the development of chronic and degenerative ailments such as cancer, inflammation, rheumatoid arthritis, cataract, aging, cardiovascular and neurodegenerative diseases (Pham-Huy *et al.*, 2008).

Canavalia mollis Wight and Arn belongs to the family Fabaceae which has high nutritional and medicinal value but it is being meagerly used by the indigenous people only. The prefix Cana refers to white while the suffix valia probably is derived from the term, valid meaning strong and true referring to its medicinal uses (Purseglove, 1981). Canavalia sp. is well known for its protein content and high nutritional value. Lectin derived from Canavalia was widely used in cancer treatment. *C. mollis* (Tamil: Kaatuthampattai) is widely distributed at Cauvery river basins and Kolli hills of Tamil Nadu. Matured seeds were consumed by malayali tribes of kolli hills for health ailment.

2.0 MATERIALS AND METHODS

2.1 Plant Collection and Extraction

C. mollis was gathered as mature pods (nearly 2 kg) from Cauvery river basin, Trichy, Tamil Nadu at appropriate season (April – July). The seeds were removed from the mature pods, air-dried, washed and oven-dried at 50°C overnight (12 h). Kernel and seed coat were separated, dried and powdered. The Kernel and seed coat flour of *C. mollis* was well packed and extracted in soxhlet apparatus by using different solvent from non-polar to polar. The Extracts were collected in screw cap tubes and stored for further phytochemical and pharmacological analysis.

2.2 Proximate Analysis

Triplicates of all processed samples (2.0 g) were heated in an oven at 105°C in pre-dried, cooled and pre-weighed crucibles, until a constant weight was attained (AOAC, 1984) and the moisture content calculated. Ash and total nitrogen content were determined by the micro- Kjeldahl method (Humphries, 1986). Crude lipid, crude fibre and ash contents (AOAC, 1990) were determined as per the standard procedures. Carbohydrate and amino acid content were determined by Standard methods (Sadasivam and Manickam, 2000).

2.3 Extract Recovery Percentage

The amount of crude extracts recovered after successive extraction were weighed and the percentage of the yield was calculated by using the following formula,

$$\text{Recovery Percent} = \frac{\text{Extract + container (g) – Empty container (g)}}{\text{Weight of plant sample (g)}} \times 100$$

2.4 Quantification Assays

2.4.1 Quantification of Total Phenolic and Tannin Contents

The total phenolic content of the extracts was determined according to the method described by Siddhuraju and Becker (2003). Known concentration of each extract was taken in test tubes and made up to the volume of 1 mL with distilled water. Then 0.5 mL of Folin-Ciocalteu phenol reagent (1:1 with water) and 2.5 mL of sodium carbonate solution (20 per cent) were added sequentially in each tube. Soon after vortexing the reaction mixture, the test tubes were placed in dark for 40 min and the absorbance was recorded at 725 nm against the reagent blank. The analysis was performed in triplicate and the results were expressed as the gallic acid equivalents (GAE).

The tannins were estimated in the extracts with polyvinyl polypyrrolidone (PVPP) treatment. One hundred milligrams of PVPP was weighed into a 100 × 12 mm test tube and to this 1.0 mL of distilled water and then 1.0 mL of tannin containing phenolic extract were added. The content was vortexed and incubated at 4°C for 4 h. Then the sample was centrifuged (3000 x g for 10 min at room temperature) and the supernatant was collected. This supernatant has only simple phenolics other than tannins (the tannins would have been precipitated along with the PVPP). The phenolic content of the supernatant was measured as mentioned above and expressed as the content of non-tannin phenolics. From the above results, the tannin content of the sample was calculated as the difference between total phenolics and non-tannin phenolics.

Tannin (per cent) = Total phenolics (per cent) – Non-tannin phenolics (per cent).

2.4.2 Quantification of Flavonoid Content

The flavonoid content of the sample extracts was determined by the use of a slightly modified colorimetric method described previously (Zhishen *et al.,* 1999). A 0.5 mL extract was mixed with 2 mL of distilled water and subsequently with 0.15 mL of 5 per cent $NaNO_2$ solution. After 6 min, 0.15 mL of 10 per cent $AlCl_3$ was added and allowed to stand for 6 min, then 2 mL of 4 per cent NaOH solution was added to the mixture. Immediately distilled water was added to bring the final volume to 5 mL, and then the mixture was thoroughly mixed and allowed to stand for another 15 min. Absorbance of the mixture was determined at 510 nm versus prepared water blank. Rutin was used as the standard compound for the quantification of total flavonoids. All the values were expressed as milligram of rutin equivalents (RE) per gram of extract.

2.5 *In vitro* Antioxidant Studies

2.5.1 Antiradical Activity Using DPPH$^\bullet$ Method

The antioxidant activity of the extract was determined in terms of hydrogen donating or radical scavenging ability using the stable radical DPPH, according to the method of Blois (1958). Sample extract at various concentrations was taken and the volume was adjusted to 100 µL with methanol. 5 mL of 0.1 mM methanolic solution of DPPH was added and shaken vigorously. The tubes were allowed to stand for 20 min at 27°C. The absorbance of the sample was measured at 517 nm. Radical scavenging activity of the samples was expressed as IC_{50}. Concentration of the sample necessary to decrease initial concentration of DPPH$^\bullet$ by 50 per cent (IC_{50}) under the experimental condition was determined. Therefore, lower value of IC_{50} indicates a higher antioxidant activity. BHT and rutin were used as standard antioxidants in DPPH$^\bullet$ assay.

2.5.2 Antioxidant Activity by ABTS•+ Assay

The total antioxidant activity of the samples was measured by ABTS radical cation decolorization assay according to the method of Re *et al.* (1999). ABTS•+ was produced by reacting 7 mM ABTS aqueous solution with 2.4 mM potassium persulfate in the dark for 12-16 h at room temperature. Prior to assay, this solution was diluted in ethanol (about 1:89, v/v) and equilibrated at 30°C to give an absorbance of 0.700±0.02 at 734 nm. The stock solution of the sample extracts were diluted such that after introduction of 10 µL aliquots into the assay, they produced between 20 per cent and 80 per cent inhibition of the blank absorbance. After the addition of 1 mL of diluted ABTS solution to 10 µL of sample or Trolox standards (final concentration 0-15 µM) in ethanol, absorbance was measured at 30° C exactly 30 min after the initial mixing. Appropriate solvent blanks were also run in each assay. Triplicate determinations were made at each dilution of the standard, and the percentage inhibition was calculated with the blank absorbance at 734 nm and then was plotted as a function of Trolox concentration. The unit of Trolox equivalent antioxidant capacity (TEAC) is defined as the concentration of Trolox having equivalent antioxidant activity expressed as µM/g sample extract on dry matter.

2.5.3 Ferric Reducing Antioxidant Power (FRAP) Assay

The antioxidant potential of various extracts of *C. mollis* was estimated from their ability to reduce TPTZ–Fe (III) complex to TPTZ–Fe (II) complex and the results are expressed as the concentration of substance having ferric-TPTZ reducing ability equivalent to that of 1 mM concentration of Fe (II). The antioxidant capacity of the extracts was estimated according to the procedure described by Pulido *et al.* (2000). FRAP reagent (900 µL), prepared freshly and incubated at 37 °C, was mixed with 90 µL of distilled water and 30 µL of test sample or methanol (for the reagent blank). The test samples and reagent blank were incubated at 37 °C for 30 min in a water bath. The final dilution of the test sample in the reaction mixture was 1/34. The FRAP reagent contained 2.5 mL of 20 mM TPTZ solution in 40 mM HCl, 2.5 mL of 20 mM $FeCl_3.6H_2O$ and 25 mL of 0.3 M acetate buffer (pH 3.6) as described by Siddhuraju and Becker (2003). At the end of incubation the absorbance readings were taken immediately at 593 nm, using a spectrophotometer. Methanolic solutions of known Fe (II) concentration, ranging from 100 to 2000 µM ($FeSO_4.7H_2O$) were used for the preparation of the calibration curve. The FRAP value is expressed as µmol Fe (II) equivalent/g extract.

2.5.4 Metal Chelating Assay

The binding of ferrous ions by the sample extracts was estimated according to the method of Dinis *et al.* (1994). Briefly, 100 µL of sample extract was incubated with 20 µM Fe^{2+} in 5 per cent ammonium acetate, pH 6.9. The reaction was initiated by the addition of 100 µM ferrozine and after an incubation period for 10 min, the absorbance was read at 562 nm. The chelating activity of the samples was evaluated using EDTA as standard. The results were expressed as mg EDTA equivalent/g extract.

2.5.5 Lipid Peroxidation

TBARS (thiobarbituric acid reactive species) assay was employed to quantify lipid peroxidation (Esterbauer *et al.*, 1990) and an adapted TBARS method was used to measure the antioxidant capacity of extracts using egg yolk homogenate as lipid rich substrate. Briefly, egg yolk was homogenized (10 per cent v/v) in 20 mM phosphate buffer (pH 7.4), 500 µL of homogenate was blend well and then homogenized with 500 µL of extracts. Lipid peroxidation was induced by addition of 50 µL of $FeSO_4$ solution (0.07M) and 20 µL of L- Ascorbic acid (0.1M). Then incubate at 37°C for 1 hour. Add 0.2 mL of 0.1M EDTA and 1.5 mL of TBA and again incubate at 100°C for 15 min. Quercetin was used as reference antioxidant molecule (positive control); negative control was vehicle alone (water). Reactions were carried out for 30 min at 37°C. After cooling, the samples were centrifuged at 3000 rpm for 10 min. Samples absorbance was measure using a spectrophotometer at 532 nm. The results were expressed in percentage of inhibition.

2.6 Statistical Analyses

All the experiments were done in triplicates and the results were expressed as mean±standard deviation (SD).

3.0 RESULTS AND DISCUSSION

3.1 Extract Yield Percentage

The extraction of any crude drug with a particular solvent yields a solution containing different phytoconstituents. Extractive value is also useful for evaluation of crude drug, which gives an idea about the nature of the chemical constituents present in a crude drug and is useful for the estimation of specific constituents, soluble in that particular solvent used for extraction. The yield percentage of *C. mollis* extracts are shown in Table 26.1. The hot water extracts of *C. mollis* seed kernel (16.43 g/100 g) and *C. mollis* seed coat (11.07 g/100 g) showed higher percentage recovery over other solvent extracts. High polarity nature of hot water leads to higher amount of yield in extracts. On the other hand, chloroform extract of *C. mollis* seed kernel (0.46 g/100 g) showed lower extract yield, compared to the other solvents. The trend observed was in the following order of hot water > methanol > acetone > petroleum ether > chloroform. In earlier reports, extracts obtained with relatively polar solvents (methanol or aqueous methanol mixtures) had the highest content of polyphenolic compounds (Dufour *et al.*, 2007). The higher recovery percent in hot water and methanol reveals that the *C. mollis* may contain more polar compounds compared to non polar compounds.

Table 26.1: Recovery Percentage of *C. mollis* (g/100 g of dried powder)

Solvents	*C. mollis Seed Kernel*	*C. mollis Seed Coat*
Petroleum ether	1.40	0.62
Chloroform	0.46	0.52
Acetone	0.73	0.66
Methanol	7.34	7.62
Hot water	16.43	11.07

3.2 Nutritional Evaluation

3.2.1 Proximate Composition Analysis

The results of proximate composition analysis are shown in Table 26.2. The moisture (8.9 per cent) and ash (0.3063 mg/g) content was respectively higher in *C. mollis* seed kernel. The ash content is an indication of the presence of carbon compounds and inorganic components in the form of salts and oxides. This range is similar to that found in literature for Indian food legumes that serve as good sources of certain minerals. High crude fibre could effectively trap and protect a greater proportion of nutrients (protein and carbohydrates) from hydrolytic breakdown, resulting in low digestibility (Balogun and Fetuga, 1986). Crude fiber content was higher in *C. mollis* seed coat (8.12 per cent). Total Protein (15.98 mg BSAE/g) and free amino acid (6.84 mg LE/g) contents was higher in *C. mollis* seed kernel. Amino acids in the form of proteins make up the greatest portion of our body weight. Total carbohydrate (16.54 mg GE/g) and Starch (10.08 mg GE/g) values was reliable in *C. mollis* seed kernel. Carbohydrates are keen energy source for everyday life. The plants have adequate amount of carbohydrates, protein, amino acids constituents, which are the major class of nutrients essential for the maintenance of human and plant life Nitrogen content (16.88 µg/mg) was higher in *C. mollis* seed coat than seed kernel. Higher amount of carbohydrates, proteins and free amino acid present in the *C. mollis* seed kernel and seed coat were promising its nutritional value.

Table 26.2: Proximate Composition Analysis of *C. mollis*

Component	*C. mollis Seed Kernel*	*C. mollis Seed Coat*
Moisture (per cent)	8.9	ND
Ash (mg/kg)	0.3063	0.1076
Total Carbohydrate(g GE/100g)	*16.54*	*12.83*
Total Protein (mg BSAE/g)	15.98	12.68
Starch (g GE/100g)	10.08	9.28
Total free amino acids(mg LE/g)	6.84	4.90
Crude Fiber (per cent)	7.6	8.12
Nitrogen content (µg/mg)	16.88	6.84

ND: Not determined; GE: Glucose equivalent; BSAE: Bovine Serum Albumin equivalent; LE: Lucien equivalent.

3.3 Quantification of Total Phenolics, Tannins and Flavanoids

Solvents such as water, ethanol, methanol, acetone or their mixture are commonly used to extract phytochemicals from plants, and they are another important factor affecting both extraction yield and antioxidant activity of extracts. Phenols and flavanoids are known to be responsible for free radical-scavenging activity. The results of quantitative estimation were showed in Table 26.3. The methanol extract of seed coat showed high level of total phenolics (416.51 g GAE/100 g extract). Petroleum ether extract of seed kernel showed low phenolics content 53.75 g GAE/100 g extract. This may be due to the extractability of phenolics efficiently by high polar organic

Table 26.3: Phenolics, Tannins and Flavonoids Content in *C. mollis* Extracts

Solvents	*Total Phenolics (GAE mg/g extract)*		*Tannins (GAE mg/g extract)*		*Flavanoids (RE mg/g extract)*	
	Seed Kernel	*Seed Coat*	*Seed Kernel*	*Seed Coat*	*Seed Kernel*	*Seed Coat*
Petroleum ether	53.75±0.68	96.57±8.10	50.24±3.66	78.61±9.32	136.22±7.78	167.41±6.48
Chloroform	89.34±12.94	159.81±12.15	106.30±14.23	154.75±12.25	86.22±6.72	109.49±8.92
Acetone	165.57±11.87	315.13±7.92	112.18±6.07	165.58±7.21	186.80±5.87	212.90±7.17
Methanol	254.88±13.78	416.51±4.56	173.42±13.32	211.59±4.51	201.11±13.33	276.48±6.30
Hot water	198.30±7.17	280.67±8.35	100.84±10.22	169.84±13.56	188.56±9.93	204.98±4.73

GAE: Gallic Acid Equivalent; RE: Rutin Equivalent.

solvents (Siddhuraju and Becker, 2003). Polyphenolic contents of all the sample extracts appear to function as good electron and hydrogen atom donor and therefore should be able to terminate radical chain reaction by converting free radicals to more stable products. Several studies have found the phenolic contents of plant extracts contribute to the excellent antioxidant/radical scavenging ability (Srinivasa Rao *et al.,* 2010).

Tannin content was higher in methanol extract of seed coat (211.59 g GAE/100 g extract) and lower in petroleum ether extract of seed kernel (50.24 g GAE/100 g extract). Natural tannins act as powerful antioxidant agents due to the presence of higher number of hydroxyl groups, especially many ortho dihydroxy or galloyl groups (Bouchet, 1998). The better amount of tannins in the methanol extract of *C. mollis* seed kernel can be due to higher polymerization of existing polyphenolic compounds. Recently, it has been reported that the high molecular weight phenolics such as tannins have more ability to quench/scavenge free radicals.

Plants containing flavonoids have been reported to possess strong antioxidant activity (Raj and sahalini, 1999). Methanol extract of *C. mollis* seed coat have higher amount of flavonoid (276.48g RE/100 g extract) content. The chloroform extract of seed kernel represents lower amount of flavanoid content 86.22 g RE/100 g extract. The group of flavonoids from bilberry are potent antioxidants and play an important role to protect oxidative damage due to their role in scavenging free radicals, so-called reactive oxygen species (ROS), that can damage biological systems (Dunn *et al.,* 2005). The result of quantitative estimation of *C. mollis* clearly denotes the seed coat was having higher amount of phenolics than the seed kernel. The higher phenolics, tannin and flavonoid content present in methanol extract of *C. mollis* seed coat assures its antioxidant consequence.

3.4 *In vitro* Antioxidant Assays

3.4.1 DPPH Radical Scavenging Activity of *C. mollis*

Relative stable DPPH radical has been widely used to test the ability of compounds to act as free radical scavengers or hydrogen donors and thus to evaluate the antioxidant activity (Jao and Ko, 2002). The results of DPPH assay were expressed in IC50 values. The lower value of IC_{50} indicates a higher antioxidant activity. Among the extracts of *C. mollis* the various concentrations of methanol extract of seed coat showed better DPPH radical scavenging activity compared to other solvent extracts (Figure 26.1). IC_{50} of methanol extract of seed coat was found to be 20.2 µg/mL. Natural antioxidants rutin and quercetin; together with 2 synthetic antioxidants BHA and BHT were used as reference compounds. Earlier study reported that, the ethanol extract of *Canavalia mollis* leaves had notable action on DPPH radical scavenging activity (Prabhu *et al.,* 2011). The polarity of the solvent plays an important role on phytochemical extraction. This radical scavenging ability of extracts could be related to the nature of phenolics, thus contributing to their electron transfer/hydrogen donating system. This result indicates that *C. mollis* seed coat have significant effects on scavenging free radicals.

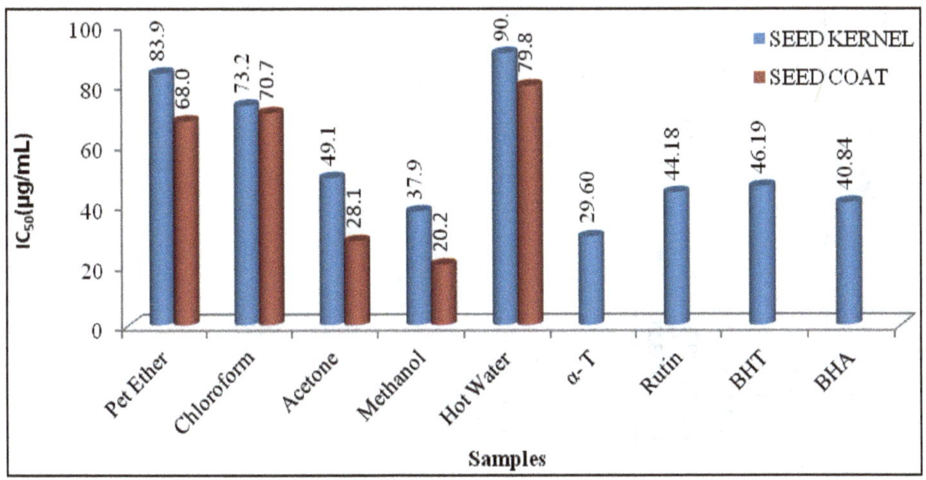

Figure 26.1: DPPH Radical Scavenging Activity of *C. mollis*.

3.4.2 FRAP

The simple and reliable test measures the reducing potential of an antioxidant reacting with a ferric 2, 4, 6- tripyridyl-S-triazine (Fe (III)-TPTZ) complex and producing a colored ferrous 2, 4, 6- tripyridyl-S-triazine (Fe (II))-TPTZ) complex by a reductant at low pH, was adopted. This complex can be monitored at 593 nm. Halvorsen *et al.* (2006) suggested most of the secondary metabolites are redox-active compounds that will be picked up by the FRAP assay. A higher absorbance power indicates a higher ferric reducing power. The results showed (Table 26.4) that the ferric reducing capacity of methanol extract of *C. mollis* seed coat was much higher (3546.88 mM Fe (II)/mg extract) and least in petroleum ether extract of seed kernel (379.87 mM Fe (II)/mg extract). Acetone extract of seed coat can be appreciated for its higher (2454.86 mM Fe (II)/mg extract) reducing activity. Ferric reducing antioxidant power of the plant showed greater variability according to the part used and solvent. The results on reducing powers demonstrate the electron donor properties of *C. mollis* extracts thereby neutralizing free radicals by forming stable products.

3.4.3 ABTS$^{•+}$ Scavenging Activity

The efficiency of ABTS cation radical scavenging activity expressed as Trolox equivalent in different solvent extract of seed kernel and seed coat of *C. mollis* is shown in Table 26.4. The results obtained in the present investigation revealed that the methanol extract of seed coat possess an utmost activity (5406.34 µM TE/g extract). It followed by methanol extract of seed kernel (4223.72 µM TE/g extract). However, the seed coat showed to have more total antioxidant activity in most of its solvent extracts. ABTS$^{•+}$ are more reactive than DPPH radicals and unlike the reactions with DPPH radical which involves the hydrogen atom transfer; the reaction with ABTS radicals involves the electron transfer process (Kaviarasan *et al.*, 2007). Hagerman *et al.* (1998) have reported that the high molecular weight phenolics (tannins) have more ability to quench free radicals (ABTS$^{•+}$) and their effectiveness depends on the molecular weight, the number of aromatic rings and nature of hydroxyl groups

Table 26.4: FRAP, ABTS$^{•+}$, Metal Chelating and Lipid Peroxidation Activity of *C. mollis*

Sample	*Solvents*	*FRAP (mM Fe(II)E/mg extract)*	*ABTS$^{•+}$ (µM TEAC/g extract)*	*Metal chelating (mg EDTA/g extract)*	*Lipid Peroxidation (per cent)*
C. mollis seed kernel	Petroleum ether	379.87 ±22.92	630.04±34.71	50.23±42.17	28.16±14.56
	Chloroform	897.09±28.13	883.16±15.12	79.06±32.63	20.18±13.89
	Acetone	1046.68±13.85	1804.54±21.76	132.17±14.82	52.98±18.90
	Methanol	1895.04±6.34	4223.72±22.49	185.09±28.51	55.39±22.43
	Hot water	904.47±28.41	3393.75±19.60	164.34±33.20	30.63±18.40
C. mollis seed coat	Petroleum ether	564.91±17.73	845.31±28.53	105.37±15.45	16.74±12.56
	Chloroform	1178.44±9.26	1048.75±45.86	166.32±28.04	28.19±10.45
	Acetone	2454.86±23.40	1836.45±18.73	258.65±31.94	57.83±14.29
	Methanol	3546.48±19.65	5406.34±28.34	315.72 ±29.85	61.34± 26.32
	Hot water	1790.64±7.58	3982.84±19.92	290.17±17.48	31.89±17.84

TEAC: Trolox equivalent antioxidant capacity; Fe (II) E: Fe (II) equivalent.

substitution than the specific functional groups. From the result, we conclude that *C. mollis* can act as high $ABTS^{\bullet+}$ radical scavenger.

3.4.4 Metal Chelating

Ferrozine can quantitatively form complexes with Fe^{2+}. In the presence of other chelating agents, the complex formation is disrupted with the result that the red color of the complexes decreases. The Fe^{2+} chelating activity of extracts are shown in Table 26.4. Ferrozine can quantitatively form complex with Fe^{2+}. In the presence of other chelating agents, the complex formation is disrupted which results in the decreased intensity of the red color of the complex. Measurement of the rate of color reduction therefore allows estimation of the chelating activity of the coexisting chelator. In this assay, *C. mollis* extract and EDTA interfered with the formation of ferrous and ferrozine complex suggesting that it has chelating activity and captures ferrous ion before ferrozine. Especially the methanol and acetone extract of seed coat (315.72 and 258.65 mg EDTA/g extract respectively) showing the higher values of ion chelating capacity. At the same time petroleum ether extract of seed kernel was had very poor metal chelating activity (50.23 mg EDTA/g extract) when compared to other parts of the plant. Metal chelating capacity was significant since the extract reduced the concentration of the catalyzing transition metal in lipid peroxidation. It was reported that chelating agents, which form δ-bonds with a metal, are effective as secondary antioxidants because they reduce the redox potential, thereby stabilizing the oxidized form of the metal ion (Gordon, 1990). Results reveal that *C. mollis* extracts has an effective capacity for iron binding, suggesting that its act as antioxidant.

3.4.5 Lipid Peroxidation

Lipid peroxidation is also a good system for assessing antioxidant action of different extracts. One of the degradation products of lipid peroxidation is malondialdehyde which causes cell damage can form a pink colour chromogen with thiobarbituric acid. The antioxidant compounds present in the extract scavenged the hydroxyl radicals generated in the Fenton reaction in the egg yolk. Results of lipid peroxidation activity of *C. mollis* extracts were tabulated in Table 26.4. Among different extracts, methanol and acetone extract of seed coat (61.34 per cent and 57.83 per cent respectively) showed maximum activity. The methanol extract of seed coat showed signicant activity compared to other extracts and positive control (ascorbic acid and BHA). High lipid peroxidation inhibitions showed by acetone extract could be related to the presence of phenolic compound, which has been shown to be correlated to the antioxidant activity of natural plant product (Gulcin *et al.,* 2002). Interestingly, Petroleum ether extract of seed kernel exhibit lowest inhibitory activity. This inhibition of lipid per oxidation may either be due to chelation of Fe ion or by scavenging of the free radicals.

4.0 CONCLUSION

According to the data derived from the present study, *C. mollis* seed kernel has good nutritional value. *C. mollis* extracts were found to be an effective antioxidant in different *in vitro* assays when compared to standard antioxidant compounds. In all the *in vitro* studies, methanol extract of seed coat showed notable radical scavenging activity. Hence further studies should be considered for isolating bioactive compounds from *C. mollis* seed coat which will pave a way in promoting natural drugs for various health diseases.

REFERENCES

AOAC, 1990. In: Helrich, K. (Ed.), Official Methods of Analysis of the Association of Official Analytical Chemists. AOAC, Washington, DC.

Association of Official Analytical Chemists (AOAC), 1984. Official methods of analysis (14th ed.) 14. 009, Virginia, DC USA.

Balogun, A. M., Fetuga, B. L., 1986. Chemical composition of some under-exploited leguminous crop seeds in Nigeria. *Journal of Agriculture and Food Chemistry*, 34: 189–192.

Blois, M.S., 1958. Antioxidant determinations by the use of a stable free radical. *Nature* 181, 1199–1200.

Bouchet, N., Laurence, B., Fauconneau, B., 1998. Radical scavenging activity and antioxidant properties of tannins from *Guiera senegalensis. Phytotherapy Research,* 12(2), 159– 162.

Dinis, T.C.P., Madeira, V.M.C., Almeida, L.M., 1994. Action of phenolic derivatives (acetoaminophen, salycilate and 5-aminosalycilate) as inhibitors of membrane lipid peroxidation and as peroxyl radical scavengers. *Archives Biochemistry and Biophysics*, 315, 161–169.

Dringen, R., 2000. Glutathione metabolism and oxidative stress in neurodegenration. *European Journal of Biochemistry*, 267, 49– 03.

Dufour, D., Pichette, A., Mshvildadze, V., Bradette-Hehert, M.E., Lavoie, S., Longtin, A., Laprise, C., Legault, J., 2007. Antioxidant, anti-inflammatory and anticancer activities of methanolic extracts from *Ledum groenlandicum* Retzius. *Journal of Ethnopharmacology*, 111 (1), 22–28.

Dunn, W. B., Bailey, N. J. C., Johnson, H. E., 2005. Measuring the metabolome: current analytical technologies. *Analyst*, 130, 606-625.

Esterbauer, H., Eckl, P., Ortner, A., 1990. Possible mutagens derived from lipids and lipids precursors. *Mutation Research*, 238, 223–233.

Gordon, M.H., 1990. The mechanism of antioxidant action *in vitro*. pp. 1- 18. In: Food Antioxidants. Hudson BJF (ed). Elsevier Applied Science, London, UK.

Gulcin, I., Buyukokuroglu, M.E., Oktay, M., Kufrevioglu, I.O., 2002. On the in vitro antioxidant properties of melatonin. *Journal of Pineal Research,* 33, 167-171.

Hagerman, A.E., Riedl, K.M., Jones, G.A., Sovik, K.N., Ritchard, N.T., Hartzfeld, P.W., Riechel, T.L., 1998. High molecular weight plant polyphenolics (tannins) as biological antioxidants. *Journal of Agriculture and Food Chemistry,* 46, 1887-1892.

Halvorsen, B.L., Holte, K., Myhrstad, M.C.W., Barikmo, I., Hvattum, E., Remberg, S.F., Wold, A.B., Haffner, K., Baugerod, H., Andersen, L.F., Moskaug, J.O., Jacobs, D.R., and Havsteen, B., 1983. Flavonoids a class of natural products of high pharmacological potency. Biochemical Pharmacology. 32, 1141- 1330Blomhoff, R., 2002. A systematic screening of total antioxidants in dietary plants. *Journal of Nutrition*, 132, 461-471.

Humphries, E.C., 1986. Mineral composition and ash analysis. In K. Peach and M. V. Tracey (Eds.). Modern Methods of Plant Analysis (Vol. 1, pp. 468–502). Berlin: Springer- Verlag.

Jao, C.H., Ko, W.C., 2002. 1, 1 Diphenyl-2-picrylhydrazyl (DPPH) radical scavenging by protein hydrolysaes from tuna cooking juice. *Fisheries Science*, 68, 430-435.

Kaviarasan, S., Naik, G.H., Gangabhagirathi, R., Anuradha, C.V., Priyadarsini K.I., 2007. *In vitro* studies on antiradical and antioxidant activities of fenugreek (*Trigonella foenum* graecum) seeds. *Food Chemistry*, 103: 31–37.

Nose, K., 2000. Role of reactive oxygen species in the regulation of physiological functions. *Biological and Pharmaceutical Bulletin*, 23, 897–903.

Pham-Huy, L.A., He, H., Pham-Huyc, C., 2008. Free Radicals, Antioxidants in Disease and Health. *International Journal of Biomedical Science*, 4(2): 89- 96

Prabhu, S., John Britto, S., Thangavel, P., Joelri Michael Raj, L., Senthilkumar S. R., 2010. Antibacterial and antioxidant activity of leaves of *Canavalia mollis* Wight and Arn. (Horse Bean). *International Journal of Pharmaceutical Sciences and Research*, 2(1), 95-101.

Pulido, R., Bravo, L., Sauro-Calixo, F., 2000. Antioxidant activity of dietary polyphenols as determined by modified ferric reducing antioxidant power assay. *Journal of Agriculture and Food Chemistry*, 48, 3396-3404.

Purseglove, J.W., 1981. Leguminosae. In: Purseglove J.W. (ed.), Tropical Crops, Dicotyledons. Longman Group LTD, Essex, UK.

Raj, K.J., Shalini, K., 1999. Flavonoids - A review of biological activities. *Indian Drugs*, 36, 668-676.

Re, R., Pellegrini, N., Proteggente, A., Pannala, A., Yang, M., Rice, E.C., 1999. Antioxidant activity applying an improved ABTS radical cation decolourization assay. *Free radical biology and Medicine*, 26, 1231-1237

Sadasivam, S., Manickam, A., 2008. Biochemical Methods. New Age International publishers.1- 263.

Siddhuraju, P., Becker, K., 2003. Antioxidant properties of various solvent extracts of total phenolic constituents from three different agroclimatic origins of drumstick tree (*Moringa oleifera* Lam.) leaves. *Journal of Agricultural and Food Chemistry*, 51, 2144-2155.

Srinivasa Rao, K., Chaudhury, P.K, Pradhan, A., 2010. Evaluation of anti-oxidant activities and total phenolic content of *Chromolaena odorata. Food and Chemical Toxicology*, 48, 729–732.

Vikas, V.P., Vijay, R.P., 2010. *Ficus bengalensis* Linn.-an overview. *International Journal of Pharma and Bio Sciences*, 1(2), 1-11.

Zhishen, J., Mengecheng, T., Jianming, W., 1999. The determination of flavonoid contents on mulberry and their scavenging effects on superoxide radical. *Food Chemistry*, 64, 555-559.

Index

N

O

P

T

U

V

W

X

Z

www.ingramcontent.com/pod-product-compliance
Lightning Source LLC
Chambersburg PA
CBHW050506190326
41458CB00005B/1454

* 9 7 8 9 3 5 1 3 0 6 8 4 9 *